Lecture Notes on Data Engineering and Communications Technologies

Volume 29

Series editor

Fatos Xhafa, Technical University of Catalonia, Barcelona, Spain
e-mail: fatos@cs.upc.edu

The aim of the book series is to present cutting edge engineering approaches to data technologies and communications. It will publish latest advances on the engineering task of building and deploying distributed, scalable and reliable data infrastructures and communication systems.

The series will have a prominent applied focus on data technologies and communications with aim to promote the bridging from fundamental research on data science and networking to data engineering and communications that lead to industry products, business knowledge and standardisation.

More information about this series at http://www.springer.com/series/15362

Leonard Barolli · Fatos Xhafa ·
Zahoor Ali Khan · Hamad Odhabi
Editors

Advances in Internet, Data and Web Technologies

The 7th International Conference on Emerging Internet, Data and Web Technologies (EIDWT-2019)

Editors
Leonard Barolli
Department of Information
and Communication Engineering,
Faculty of Information Engineering
Fukuoka Institute of Technology
Fukuoka, Japan

Fatos Xhafa
Technical University of Catalonia
Barcelona, Spain

Zahoor Ali Khan
Computer Information Science (CIS)
Higher Colleges of Technology,
Fujairah Campus
Fujairah, United Arab Emirates

Hamad Odhabi
Faculty of Computer Information Science
Higher Colleges of Technology,
Dubai Campus
Dubai, United Arab Emirates

ISSN 2367-4512 ISSN 2367-4520 (electronic)
Lecture Notes on Data Engineering and Communications Technologies
ISBN 978-3-030-12838-8 ISBN 978-3-030-12839-5 (eBook)
https://doi.org/10.1007/978-3-030-12839-5

Library of Congress Control Number: 2019930656

This Springer imprint is published by the registered company Springer Nature Switzerland AG
The registered company address is: Gewerbestrasse 11, 6330 Cham, Switzerland

Welcome Message of EIDWT-2019 International Conference Organizers

Welcome to the 7th International Conference on Emerging Internet, Data and Web Technologies (EIDWT-2019), which will be held from February 26 to February 28, 2019, at Higher Colleges of Technology, Fujairah Campus, United Arab Emirates (UAE).

EIDWT is dedicated to the dissemination of original contributions that are related to the theories, practices and concepts of emerging Internet and data technologies yet most importantly of their applicability in business and academia toward a collective intelligence approach.

In EIDWT-2019 will be discussed the topics related to Information Networking, Data Centers, Data Grids, Clouds, Crowds, Mashups, Social Networks, Security Issues and other Web 2.0 implementations toward a collaborative and collective intelligence approach leading to advancements of virtual organizations and their user communities. This is because current and future Web and Web 2.0 implementations will store and continuously produce a vast amount of data, which if combined and analyzed through a collective intelligence manner will make a difference in the organizational settings and their user communities. Thus, the scope of EIDWT-2019 includes methods and practices which bring various emerging Internet and data technologies together to capture, integrate, analyze, mine, annotate and visualize data in a meaningful and collaborative manner. Finally, EIDWT-2019 aims to provide a forum for original discussion and prompt future directions in the area. For EIDWT-2019 International Conference, we accepted for presentation 52 papers (about 30% acceptance ratio).

An international conference requires the support and help of many people. A lot of people have helped and worked hard for a successful EIDWT-2019 technical program and conference proceedings. First, we would like to thank all authors for submitting their papers. We are indebted to Program Area Chairs, Program Committee Members and Reviewers who carried out the most difficult work of carefully evaluating the submitted papers. We would like to give our special thanks to Honorary Co-Chairs of EIDWT-2019, for their guidance and support. We would like to express our appreciation to our keynote speakers for accepting our invitation and delivering very interesting keynotes at the conference.

We would like as well to thank the Local Arrangements Chairs for making excellent local arrangements for the conference. We hope you will enjoy the conference and have a great time in Fujairah, UAE.

EIDWT-2019 International Conference Organizers

EIDWT-2019 Steering Committee Co-chairs

Leonard Barolli	Fukuoka Institute of Technology (FIT), Japan
Fatos Xhafa	Technical University of Catalonia, Spain

EIDWT-2019 General Co-chairs

Nasser Nassiri	Higher Colleges of Technology, UAE
Tomoya Enokido	Rissho University, Japan
Zahoor Khan	Higher Colleges of Technology, UAE

EIDWT-2019 Program Committee Co-chairs

Saif Alqaydi	Higher Colleges of Technology, UAE
Juggapong Natwichai	Chiang Mai University, Thailand
Nadeem Javaid	COMSATS Institute of IT, Pakistan

EIDWT-2019 Organizing Committee

Honorary Co-chairs

Khaled Alhammadi	Higher Colleges of Technology, UAE
Makoto Takizawa	Hosei University, Japan
Hamad Odhabi	Higher Colleges of Technology, UAE

General Co-chairs

Nasser Nassiri	Higher Colleges of Technology, UAE
Tomoya Enokido	Rissho University, Japan
Zahoor Khan	Higher Colleges of Technology, UAE

Program Co-chairs

Saif Alqaydi	Higher Colleges of Technology, UAE
Juggapong Natwichai	Chiang Mai University, Thailand
Nadeem Javaid	COMSATS Institute of IT, Pakistan

International Advisory Committee

Adel Zairi	Higher Colleges of Technology, UAE
Janusz Kacprzyk	Polish Academy of Sciences, Poland
Vincenzo Loia	University of Salerno, Italy
Arjan Durresi	IUPUI, USA
Hiroaki Nishino	Oita University, Japan

Publicity Co-chairs

Santi Caballé	Open University of Catalonia, Spain
Pruet Boonma	Chiang Mai University, Thailand
Elis Kulla	Okayama University of Science, Japan
Farookh Hussain	University of Technology Sydney, Australia

International Liaison Co-chairs

Ossama Embarak	Higher Colleges of Technology, UAE
Flora Amato	Naples University Frederico II, Italy
Admir Barolli	Alexander Moisiu University of Durres, Albania
Omar Hussain	University of New South Wales, Australia
Akio Koyama	Yamagata University, Japan

Local Organizing Co-chairs

Mona Almteiri	Higher Colleges of Technology, UAE
Ivana Ercegovac	Higher Colleges of Technology, UAE

Web Administrators

Donald Elmazi	Fukuoka Institute of Technology (FIT), Japan
Yi Liu	Fukuoka Institute of Technology (FIT), Japan
Miralda Cuka	Fukuoka Institute of Technology (FIT), Japan

Finance Chair

Makoto Ikeda	Fukuoka Institute of Technology (FIT), Japan

Steering Committee Co-chairs

Leonard Barolli	Fukuoka Institute of Technology (FIT), Japan
Fatos Xhafa	Technical University of Catalonia, Spain

Program Committee Members

Akimitsu Kanzaki	Shimane University, Japan
Akio Koyama	Yamagata University, Japan
Akira Uejima	Okayama University of Science, Japan
Akshay Uttama Nambi S. N.	Microsoft Research, India
Alba Amato	National Research Council (CNR), Italy

Alberto Scionti	ISMB, Italy
Albin Ahmeti	TU Wien, Austria
Alex Pongpech	National Institute of Development Administration, Thailand
Ali Rodan	Higher Colleges of Technology, UAE
Alfred Miller	Higher Colleges of Technology, UAE
Amelie Chi Zhou	Shenzhen University, China
Amin M. Khan	Pentaho, Hitachi Data Systems, Japan
Ana Azevedo	ISCAP, Portugal
Andrea Araldo	Massachusetts Institute of Technology, USA
Animesh Dutta	National Institute of Technology, Durgapur, India
Anirban Mondal	Shiv Nadar University, India
Anis Yazidi	Oslo and Akershus University College of Applied Sciences, Norway
Antonella Di Stefano	University of Catania, Italy
Arcangelo Castiglione	University of Salerno, Italy
Beniamino Di Martino	Università della Campania "Luigi Vanvitelli", Italy
Benson Raj	Higher Colleges of Technology, UAE
Bhed Bista	Iwate Prefectural University, Japan
Bowonsak Srisungsittisunti	University of Phayao, Thailand
Carmen de Maio	University of Salerno, Italy
Chang Yung-Chun	Taipei Medical University, Italy
Chonho Lee	Osaka University, Japan
Chotipat Pornavalai	King Mongkut's Institute of Technology Ladkrabang, Thailand
Christoph Hochreiner	TU Wien, Austria
Congduc Pham	University of Pau, France
Dalvan Griebler	Pontifícia Universidade Católica do Rio Grande do Sul, Brazil
Dana Petcu	West University of Timisoara, Romania
Danda B. Rawat	Howard University, USA
Debashis Nandi	National Institute of Technology, Durgapur, India
Diego Kreutz	Universidade Federal do Pampa, Brazil
Dimitri Pertin	Inria Nantes, France
Dipankar Das	Jadavpur University, Kolkata, India
Douglas D. J. de Macedo	Federal University of Santa Catarina, Brasil
Dumitru Burdescu	University of Craiova, Romania
Dusit Niyato	NTU, Singapore
Elis Kulla	Okayama University of Science, Japan
Eric Pardede	La Trobe University, Australia
Fabrizio Marozzo	University of Calabria, Italy
Fabrizio Messina	University of Catania, Italy
Farookh Hussain	University of Technology Sydney, Australia
Feng Xia	Dalian University of Technology, China

Francesco Orciuoli	University of Salerno, Italy
Francesco Palmieri	University of Salerno, Italy
Fumiaki Sato	Toho University, Japan
Gabriele Mencagli	University of Pisa, Italy
Gen Kitagata	Tohoku University, Japan
Ghazi Ben Ayed	Higher Colleges of Technology, UAE
Georgios Kontonatsios	Edge Hill University, UK
Giovanni Masala	Plymouth University, UK
Giovanni Morana	C3DNA, USA
Giuseppe Caragnano	ISMB, Italy
Giuseppe Fenza	University of Salerno, Italy
Gongjun Yan	University of Southern Indiana, USA
Guangquan Xu	Tianjin University, China
Gustavo Medeiros de Araujo	Universidade Federal de Santa Catarina, Brazil
Harold Castro	Universidad de Los Andes, Colombia
Hiroaki Nishino	Oita University, Japan
Hiroaki Yamamoto	Shinshu University, Japan
Hiroshi Shigeno	Keio University, Japan
Houssem Chihoub	Grenoble Institute of Technology, France
Ilias Savvas	TEI of Larissa, Greece
Inna Skarga-Bandurova	East Ukrainian National University, Ukraine
Isaac Woungang	Ryerson University, Canada
Ja'far Alqatawna	Higher Colleges of Technology, UAE
Jamal Alsakran	Higher Colleges of Technology, UAE
Jamshaid Ashraf	Data Processing Services, Kuwait
Jaydeep Howlader	National Institute of Technology, Durgapur, India
Jeffrey Ray	Edge Hill University, UK
Ji, Zhang	The University of Southern Queensland, Australia
Jiahong Wang	Iwate Prefectural University, Japan
Kazuyoshi Kojima	Saitama University, Japan
Ken Newman	Higher Colleges of Technology, UAE
Kenzi Watanabe	Hiroshima University, Japan
Kiyoshi Ueda	Nihon University, Japan
Klodiana Goga	ISMB, Italy
Leila Sharifi	Urmia University, Iran
Leyou Zhang	Xidian University, China
Lidia Fotia	Università Mediterranea di Reggio Calabria (DIIES), Italy
Ling Chen	Yangtze University, China
Long Cheng	Eindhoven University of Technology, Netherland
Lucian, Prodan	Polytechnic University Timisoara, Romania
Luiz Fernando Bittencourt	UNICAMP - Universidade Estadual de Campinas, Brazil
Makoto Fujimura	Nagasaki University, Japan

Makoto Nakashima	Oita University, Japan
Mansaf Alam	Jamia Millia Islamia, New Delhi, India
Marcello Trovati	Edge Hill University, UK
Mariacristina Gal	University of Salerno, Italy
Marina Ribaudo	University of Genoa, Italy
Markos Kyritsis	Higher Colleges of Technology, UAE
Marwan Hassani	TU Eindhoven, Netherlands
Massimo Torquati	University of Pisa, Italy
Matthias Steinbauer	Johannes Kepler University Linz, Austria
Matthieu Dorier	Argonne National Laboratory, USA
Mauro Marcelo Mattos	FURB Universidade Regional de Blumenau, Brazil
Mazin Abuharaz	Higher Colleges of Technology, UAE
Minghu Wu	Hubei University of Technology, China
Mingwu Zhang	Hubei University of Technology, China
Minoru Uehara	Toyo University, Japan
Mirang Park	Kanagawa Institute of Technology, Japan
Mohsen Farid	University of Derby, UK
Monther Tarawneh	Higher Colleges of Technology, UAE
Morteza Saberi	UNSW Canberra, Australia
Motoi Yamagiwa	University of Yamanashi, Japan
Mouza Alshemaili	Higher Colleges of Technology, UAE
Muawya Aldalaien	Higher Colleges of Technology, UAE
Mukesh Prasad	University of Technology Sydney, Australia
Muhammad Iqbal	Higher Colleges of Technology, UAE
Naeem Janjua	Edith Cowan University, Australia
Naohiro Hayashibara	Kyoto Sangyo University, Japan
Naonobu Okazaki	University of Miyazaki, Japan
Neha Warikoo	Academia Sinica, Taiwan
Nobukazu Iguchi	Kindai University, Japan
Nobuo Funabiki	Okayama University, Japan
Olivier Terzo	ISMB, Italy
Omar Al Amiri	Higher Colleges of Technology, UAE
Omar Hussain	UNSW Canberra, Australia
Osama Alfarraj	King Saud University, Saudi Arabia
Osama Rahmeh	Higher Colleges of Technology, UAE
P. Sakthivel	Anna University, Chennai, India
Panachit Kittipanya-ngam	Electronic Government Agency, Thailand
Paolo Bellavista	University of Bologna, Italy
Pavel Smrž	Brno University of Technology, Czech Republic
Peer Shah	Higher Colleges of Technology, UAE
Pelle Jakovits	University of Tartu, Estonia
Per Ola Kristensson	University of Cambridge, UK
Philip Moore	Lanzhou University, China
Pietro Ruiu	ISMB, Italy

Pornthep Rojanavasu	University of Phayao, Thailand
Pruet Boonma	Chiang Mai University, Thailand
Raffaele Pizzolante	University of Salerno, Italy
Ragib, Hasan	The University of Alabama at Birmingham, USA
Rao Mikkilineni	C3DNA, USA
Richard Conniss	University of Derby, UK
Ruben Mayer	University of Stuttgart, Germany
Sachin Shetty	Old Dominion University, USA
Sajal Mukhopadhyay	National Institute of Technology, Durgapur, India
Salem Alkhalaf	Qassim University, Saudi Arabia
Salvatore Ventiqincue	University of Campania Luigi Vanvitelli, Italy
Samia Kouki	Higher Colleges of Technology, UAE
Saqib Ali	Sultan Qaboos University, Australia
Sayyed Maisikeli	Higher Colleges of Technology, UAE
Sazia Parvin	Deakin University, Australia
Sergio Ricciardi	BarcelonaTech, Spain
Shadi Ibrahim	Inria Rennes, France
Shigetomo Kimura	University of Tsukuba, Japan
Shinji Sugawara	Chiba Institute of Technology, Japan
Sivadon Chaisiri	University of Waikato, New Zealand
Sofian Maabout	Bordeaux University, France
Sotirios Kontogiannis	University of Ioannina, Greece
Stefan Bosse	University of Bremen
Stefania Boffa	Università dell'Insubria, Italy
Stefania Tomasiello	University of Salerno, Italy
Stefano Forti	University of Pisa, Italy
Stefano Secci	LIP6, University Paris 6, France
Suayb Arslan	MEF University, Turkey
Subhrabrata Choudhury	National Institute of Technology, Durgapur, India
Suleiman Al Masri	Higher Colleges of Technology, UAE
Tatiana A. Gavrilova	St. Petersburg University, Russia
Teodor Florin Fortis	West University of Timisoara, Romania
Thamer AlHussain	Saudi Electronic University, Saudi Arabia
Titela Vilceanu	University of Craiova, Romania
Tomoki Yoshihisa	Osaka University, Japan
Tomoya Kawakami	NAIST, Japan
Toshihiro Yamauchi	Okayama University, Japan
Toshiya Takami	Oita University, Japan
Ugo Fiore	Parthenope University, Italy
Venkatesha Prasad	Delft University of Technology, Netherland
Victor Kardeby	RISE Acreo, Sweden
Vlado Stankovski	University of Ljubljana, Slovenia
Walayat Hussain	University of Technology Sydney, Australia
Xiaoou Song	Engineering University of CAPF, China
Xingzhi Sun	IBM Research, Australia

Xu An Wang	Engineering University of CAPF, China
Xuan Guo	Officer's University of CAPF, China
Xue Li	The University of Queensland, Australia
Yang Lei	Engineering University of CAPF, China
Yongqiang Li	Xi'an High Technology Research Institute, China
Yoshihiro Okada	Kyushu University, Japan
Yoshinobu Tamura	Tokyo City University, Japan
Yue Zhao	National Institutes of Health, USA
Yuechuan Wei	Engineering University of CAPF, China
Yunbo Li	IMT Atlantique, France
Zasheen Hameed	Higher Colleges of Technology, UAE
Zhenhua Chen	Xi'an University of Technology, China
Zhiqiang Gao	Engineering University of CAPF, China
Zia Rehman	COMSATS Institute of Information Technology (CIIT), Pakistan

EIDWT-2019 Reviewers

Ali Khan Zahoor
Alsakran Jamal
Amato Flora
Amato Alba
Barolli Admir
Barolli Leonard
Bista Bhed
Caballé Santi
Chellappan Sriram
Chen Hsing-Chung
Cui Baojiang
Di Martino Beniamino
Embarak Ossama
Enokido Tomoya
Ficco Massimo
Fiore Ugo
Fun Li Kin
Gotoh Yusuke
Hussain Farookh
Hussain Omar
Javaid Nadeem
Ikeda Makoto
Ishida Tomoyuki
Kikuchi Hiroaki
Kolici Vladi
Koyama Akio
Kouki Samia
Kulla Elis
Matsuo Keita
Koyama Akio
Kryvinska Natalia
Nishino Hiroaki
Ogiela Lidia
Ogiela Marek
Palmieri Francesco
Paruchuri Vamsi Krishna
Rahayu Wenny
Sato Fumiaki
Spaho Evjola
Takizawa Makoto
Taniar David
Tarawneh Monther
Terzo Olivier
Uehara Minoru
Venticinque Salvatore
Wang Xu An
Woungang Isaac
Xhafa Fatos

EIDWT-2019 Keynote Talks

How to Build Intelligent Collectives for Prediction Market Tasks

Ngoc-Thanh Nguyen

Wroclaw University of Science and Technology, Wroclaw, Poland

Abstract. This talk presents a general framework for building intelligent collectives for prediction market tasks. It will be shown that two aspects play an important role in this process: diversity and independency. Our findings qualify the positive impact of diversity and independence on collective performance. Particularly, collectives with higher diversity and independence levels will lead to better collective accuracy. Subsequently, expanding the collective cardinality that causes an increase in its diversity will also be positively associated with the collective performance. With some restrictions, the hypothesis "The more diverse the collective, the higher the collective performance" is formally proved.

Cooperative Wireless Relay Networks

Jacek Ilow

Dalhousie University, Halifax, Canada

Abstract. This talk presents principles involved in designing Multiple-Input-Multiple-Output (MIMO) wireless relay networks with different levels of coordination between communicating terminals in ad hoc networks. In particular, cooperative transmission strategies are contrasted with the approaches where terminals transmit independently resulting in high levels of interference. A unifying framework of power control and signal alignment is adopted to demonstrate better power and bandwidth efficiency in cooperative communication systems with compute-and-forward strategies, that exploits interference to obtain significantly higher rates between users in a network. Geometric concepts and abstractions are used to establish the energy-efficient relay choices and to design distributed signal coding to combat fading in networks with randomly distributed nodes such as in Wireless Sensor Networks (WSNs). The benefits of distributed network coding reducing the number of transmissions are examined. While cooperation improves energy efficiency in wireless networks, it introduces a protocol overhead. The balancing of the cooperative gain (in terms of energy and bandwidth efficiency) and the cooperation overhead is also presented in this talk.

Contents

A Fuzzy-Based System for Cloud-Fog-Edge Selection in VANETs

Kevin Bylykbashi[1(✉)], Yi Liu[1], Keita Matsuo[2], Makoto Ikeda[2], Leonard Barolli[2], and Makoto Takizawa[3]

[1] Graduate School of Engineering, Fukuoka Institute of Technology (FIT), 3-30-1 Wajiro-Higashi, Higashi-Ku, Fukuoka 811–0295, Japan
bylykbashi.kevin@gmail.com, ryuui1010@gmail.com

[2] Department of Information and Communication Engineering, Fukuoka Institute of Technology (FIT), 3-30-1 Wajiro-Higashi, Higashi-Ku, Fukuoka 811-0295, Japan
{kt-matsuo,barolli}@fit.ac.jp, makoto.ikd@acm.org

[3] Department of Advanced Sciences, Faculty of Science and Engineering, Hosei University, 3-7-2, Kajino-machi, Koganei-shi, Tokyo 184-8584, Japan
makoto.takizawa@computer.org

Abstract. Vehicular Ad Hoc Networks (VANETs) have gained a great attention due to the rapid development of mobile internet and Internet of Things (IoT) applications. With the evolution of technology, it is expected that VANETs will be massively deployed in upcoming vehicles. However, these kinds of wireless networks face several technical challenges in deployment and management due to variable capacity of wireless links, bandwidth constrains, high latency and dynamic topology. Cloud computing, fog computing and edge computing are considered a way to deal with these communication challenges. In this paper, we propose a Fuzzy-based System for Resource Coordination and Management (FSRCM) in VANETs. The proposed system considers vehicle mobility, data size, time sensitivity and remained storage capacity to select processing layer of the VANETs application data. We evaluated the performance of proposed system by computer simulations. From the simulations results, we conclude that the vehicles choose the appropriate layer to process and keep the data based on their velocity, remained storage and data size.

1 Introduction

Due to the development of computation and wireless communication technologies, the automotive industry is going through a rapid change. Nowadays, every car is likely to be equipped at least with some of the following devices: an on-board computer, a huge capacity storage device, a GPS device, a communication module, an antenna, a short-range rear collision radar device, cameras and various sensors which can gather information about the road and environment conditions to ensure the driving safety. Meanwhile, cloud computing has been

L. Barolli et al. (Eds.): EIDWT 2019, LNDECT 29, pp. 1–12, 2019.
https://doi.org/10.1007/978-3-030-12839-5_1

attracting organizations and individual users to transpose their data and services from local to remote cloud servers. Microsoft, Amazon, IBM, Oracle, Google and many other enterprises have also released their cloud infrastructure services and have enticed millions of users around the world [1]. The cloud is becoming a promising and prevalent service to replace traditional local systems.

Recently, fog computing extends cloud more near to the user. This new architecture analyzes data close to devices for minimizing latency, decision making in real time and offloading massive traffic flow from the core networks. Moreover, to flexibly cover vehicles and avoid frequent handover between vehicles and Roadside Units (RSUs), the usage of fog cells is proposed [2].

With edge computing, resources and services of computing, storage and control capabilities are distributed anywhere along the continuum from the cloud to things [3]. By leveraging the edge computing technology, a significant amount of computing power will be distributed near the vehicles. Therefore, most of data will be processed and stored at the edge, which can minimize latency and ensure better quality of service for connected vehicles [4].

When using only cloud, it becomes intensive processing. In fog computing the data processing device is widely dispersed at a position closer to the vehicles. By utilizing the computing power at the edge, a better content distribution efficiency can be achieved.

It is important for vehicles to choose in which layer to process the data. We are focused on building a cooperative mechanism for this configuration.

In this paper, we propose a fuzzy-based system for service layer selection considering four parameters: Data Size (DS), Vehicle Mobility (VM), Time Sensitivity (TS) and Remained Storage Capacity (RSC). The output parameter is called Service Selection Decision (SSD). We evaluate the performance of the system by computer simulations.

The structure of the paper is as follows. In Sect. 2, we present a short description of Cloud-Fog-Edge Computing. In Sect. 3, we present Vehicular Ad-hoc Networks (VANETs). In Sect. 4, we introduce Fuzzy Logic used for control. In Sect. 5, we present the proposed fuzzy-based system. In Sect. 6, we discuss the simulation results. Finally, conclusions and future work are given in Sect. 7.

2 Cloud-Fog-Edge Computing

The notion of cloud computing started from the realization of the fact that instead of investing in infrastructure, businesses may find it useful to rent the infrastructure and sometimes the needed software to run their applications. One major advantage of cloud computing is its scalable access to computing resources. With cloud computing developers do not need large capital outlays in hardware to deploy their service for Internet applications and services.

Recently, a lot of research is focused on shifting from conventional VANETs to Vehicular Cloud Computing (VCC) by merging VANET with cloud computing [5–7]. In VANET cloud, vehicular nodes leverage and/or share their resources and use/form cloud resources as well. There are three basic architectures of VANET-based clouds in literature: Vehicular Clouds (VC), Vehicles using Clouds (VuC) and Hybrid Vehicular Clouds (HVC). In VANETs, cloud computing can be used as a Network as a Service (NaaS) or Storage as a Service (STaaS). Not all the vehicles on the road have Internet access. In NaaS, the vehicle with Internet access can provide upon request its excess capacity to the other vehicles. For STaaS, the vehicles with ample storage capacity share their storage with other vehicles which need storage capacity for temporary application. For NaaS is believed and expected that many drivers will have persistent connectivity to the Internet through cellular networks and other fixed access points on the road while driving. This network capacity is also expected to be underutilized by many drivers as not all of them will be searching or downloading from the Internet while driving.

However, there still are requirements such as low latency, high throughput, location awareness and mobility support that VCC can barely fulfill. Fog computing is proposed as a solution to overcome these issues between the vehicles and the conventional cloud [8]. Similar to cloud, fog computing provides data, compute, storage and application services at the proximity of the vehicular nodes.

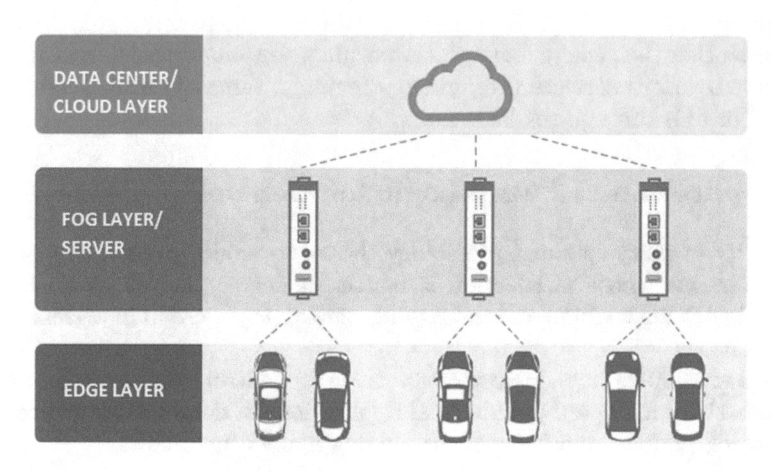

Fig. 1. Architecture of the cloud-fog-edge computing.

The vehicular nodes are the edge of this layered architecture. An illustration of this architecture composed of conventional cloud, fog computing infrastructure, and vehicular nodes is shown in Fig. 1. The fog devices (such as fog servers and RSUs) are located between vehicles and the datacenters of the main cloud environments. The transmission delays in messages, big data analysis and storage management can be satisfied by this paradigm. In addition, with the help of geo-distributed fog servers, the safety messages can be delivered with low latency to certain locations that are in different geographic areas.

3 VANETs

VANETs are considered to have an enormous potential in enhancing road traffic safety and traffic efficiency. Therefore various governments have launched programs dedicated to the development and consolidation of vehicular communications and networking and both industrial and academic researchers are addressing many related challenges, including socio-economic ones, which are among the most important [9].

The VANET technology uses moving vehicles as nodes to form a wireless mobile network. It aims to provide fast and cost-efficient data transfer for the advantage of passenger safety and comfort. To improve road safety and travel comfort of voyagers and drivers, Intelligent Transport Systems (ITS) are developed. The ITS manages the vehicle traffic, support drivers with safety and other information, and provide some services such as automated toll collection and driver assist systems [10].

The VANETs provide new prospects to improve advanced solutions for making reliable communication between vehicles. VANETs can be defined as a part of ITS which aims to make transportation systems faster and smarter, in which vehicles are equipped with some short-range and medium-range wireless communication [11]. In a VANET, wireless vehicles are able to communicate directly with each other (i.e., emergency vehicle warning, stationary vehicle warning) and also served various services (i.e., video streaming, internet) from access points (i.e., 3G or 4G) through roadside units.

4 Application of Fuzzy Logic for Control

The ability of fuzzy sets and possibility theory to model gradual properties or soft constraints whose satisfaction is matter of degree, as well as information pervaded with imprecision and uncertainty, makes them useful in a great variety of applications.

The most popular area of application is Fuzzy Control (FC), since the appearance, especially in Japan, of industrial applications in domestic appliances, process control, and automotive systems, among many other fields.

4.1 FC

In the FC systems, expert knowledge is encoded in the form of fuzzy rules, which describe recommended actions for different classes of situations represented by fuzzy sets.

In fact, any kind of control law can be modeled by the FC methodology, provided that this law is expressible in terms of "if ... then ..." rules, just like in the case of expert systems. However, FL diverges from the standard expert system approach by providing an interpolation mechanism from several rules. In the contents of complex processes, it may turn out to be more practical to get knowledge from an expert operator than to calculate an optimal control, due to modeling costs or because a model is out of reach.

4.2 Linguistic Variables

A concept that plays a central role in the application of FL is that of a linguistic variable. The linguistic variables may be viewed as a form of data compression. One linguistic variable may represent many numerical variables. It is suggestive to refer to this form of data compression as granulation [12].

The same effect can be achieved by conventional quantization, but in the case of quantization, the values are intervals, whereas in the case of granulation the values are overlapping fuzzy sets. The advantages of granulation over quantization are as follows:

- it is more general;
- it mimics the way in which humans interpret linguistic values;
- the transition from one linguistic value to a contiguous linguistic value is gradual rather than abrupt, resulting in continuity and robustness.

4.3 FC Rules

FC describes the algorithm for process control as a fuzzy relation between information about the conditions of the process to be controlled, x and y, and the output for the process z. The control algorithm is given in "if ... then ..." expression, such as:

If x is small and y is big, then z is medium;
If x is big and y is medium, then z is big.

These rules are called *FC rules*. The "if" clause of the rules is called the antecedent and the "then" clause is called consequent. In general, variables x and y are called the input and z the output. The "small" and "big" are fuzzy values for x and y, and they are expressed by fuzzy sets.

Fuzzy controllers are constructed of groups of these FC rules, and when an actual input is given, the output is calculated by means of fuzzy inference.

4.4 Control Knowledge Base

There are two main tasks in designing the control knowledge base. First, a set of linguistic variables must be selected which describe the values of the main control parameters of the process. Both the input and output parameters must be linguistically defined in this stage using proper term sets. The selection of the level of granularity of a term set for an input variable or an output variable plays an important role in the smoothness of control. Second, a control knowledge base must be developed which uses the above linguistic description of the input and output parameters. Four methods [13–16] have been suggested for doing this:

- expert's experience and knowledge;
- modelling the operator's control action;

- modelling a process;
- self organization.

Among the above methods, the first one is the most widely used. In the modeling of the human expert operator's knowledge, fuzzy rules of the form "If Error is small and Change-in-error is small then the Force is small" have been used in several studies [17,18]. This method is effective when expert human operators can express the heuristics or the knowledge that they use in controlling a process in terms of rules of the above form.

4.5 Defuzzification Methods

The defuzzification operation produces a non-FC action that best represent the membership function of an inferred FC action. Several defuzzification methods have been suggested in literature. Among them, four methods which have been applied most often are:

- Tsukamoto's Defuzzification Method;
- The Center of Area (COA) Method;
- The Mean of Maximum (MOM) Method;
- Defuzzification when Output of Rules are Function of Their Inputs.

5 Proposed Fuzzy-Based System

In this work, in order to realise the proposed system, we consider four input parameters: Data Size (DS), Vehicle Mobility (VM), Time Sensitivity (TS) and Remained Storage Capacity (RSC). The output parameter is considered the Service Selection Decision (SSD). The proposed system called Fuzzy System for Resource Coordination and Management (FSRCM) is shown in Fig. 2. These four parameters are fuzzified using fuzzy system. The membership functions for our system are shown in Fig. 3. In Table 1, we show the Fuzzy Rule Base (FRB) of FLC, which consists of 81 rules.

The input parameters for FSRCM are: DS, VM, TS and RSC. The output linguistic parameter is SSD. The term sets of *DS*, *VM*, *TS* and *RSC* are defined respectively as:

$$\begin{aligned} DS &= \{Small,\ Medium,\ Big\} \\ &= \{S,\ M,\ B\}; \\ VM &= \{Slow,\ Middle,\ Fast\} \\ &= \{Sl,\ Mi,\ Fa\}; \\ TS &= \{Small,\ Medium,\ Big\} \\ &= \{Sm,\ Me,\ Bi\}; \\ RSC &= \{Low,\ Middle,\ High\} \\ &= \{Lo,\ Mid,\ Hi\}; \end{aligned}$$

and the term set for the output *SSD* is defined as:

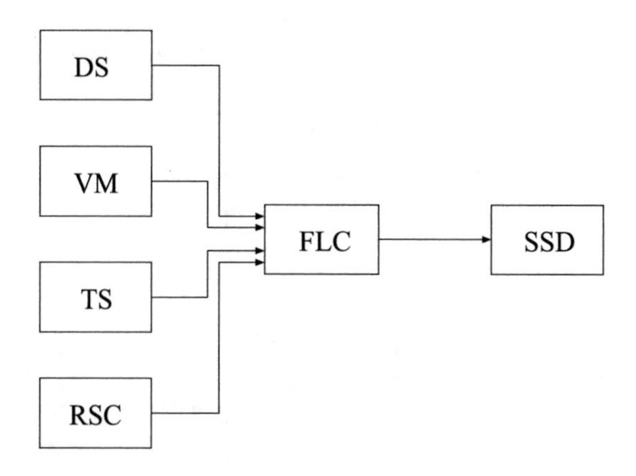

Fig. 2. Proposed FLC.

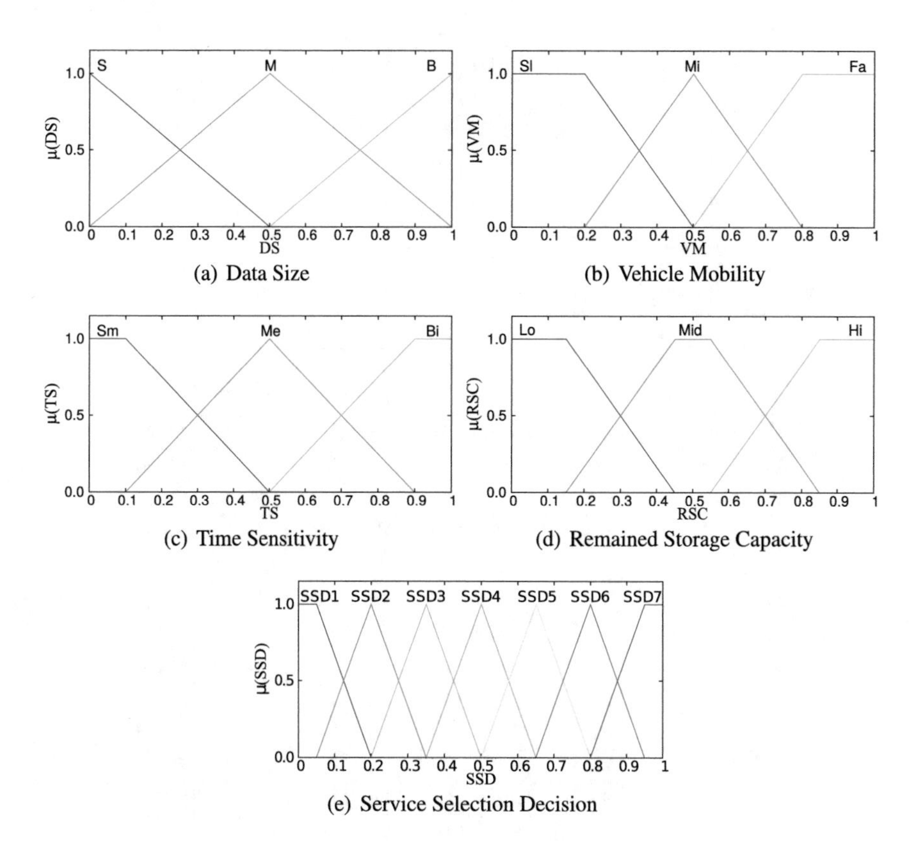

(a) Data Size

(b) Vehicle Mobility

(c) Time Sensitivity

(d) Remained Storage Capacity

(e) Service Selection Decision

Fig. 3. Membership functions.

Table 1. FRB.

Rule	VM	RSC	TS	DS	SSD	Rule	VM	RSC	TS	DS	SSD
1	Sl	Lo	Sm	S	SSD4	41	Mi	Mid	Me	M	SSD4
2	Sl	Lo	Sm	M	SSD5	42	Mi	Mid	Me	B	SSD5
3	Sl	Lo	Sm	B	SSD6	43	Mi	Mid	Bi	S	SSD1
4	Sl	Lo	Me	S	SSD2	44	Mi	Mid	Bi	M	SSD2
5	Sl	Lo	Me	M	SSD4	45	Mi	Mid	Bi	B	SSD4
6	Sl	Lo	Me	B	SSD5	46	Mi	Hi	Sm	S	SSD3
7	Sl	Lo	Bi	S	SSD1	47	Mi	Hi	Sm	M	SSD5
8	Sl	Lo	Bi	M	SSD2	48	Mi	Hi	Sm	B	SSD6
9	Sl	Lo	Bi	B	SSD3	49	Mi	Hi	Me	S	SSD2
10	Sl	Mid	Sm	S	SSD3	50	Mi	Hi	Me	M	SSD3
11	Sl	Mid	Sm	M	SSD4	51	Mi	Hi	Me	B	SSD4
12	Sl	Mid	Sm	B	SSD6	52	Mi	Hi	Bi	S	SSD1
13	Sl	Mid	Me	S	SSD1	53	Mi	Hi	Bi	M	SSD1
14	Sl	Mid	Me	M	SSD2	54	Mi	Hi	Bi	B	SSD2
15	Sl	Mid	Me	B	SSD4	55	Fa	Lo	Sm	S	SSD7
16	Sl	Mid	Bi	S	SSD1	56	Fa	Lo	Sm	M	SSD7
17	Sl	Mid	Bi	M	SSD1	57	Fa	Lo	Sm	B	SSD7
18	Sl	Mid	Bi	B	SSD2	58	Fa	Lo	Me	S	SSD5
19	Sl	Hi	Sm	S	SSD2	59	Fa	Lo	Me	M	SSD6
20	Sl	Hi	Sm	M	SSD3	60	Fa	Lo	Me	B	SSD7
21	Sl	Hi	Sm	B	SSD5	61	Fa	Lo	Bi	S	SSD3
22	Sl	Hi	Me	S	SSD1	62	Fa	Lo	Bi	M	SSD5
23	Sl	Hi	Me	M	SSD2	63	Fa	Lo	Bi	B	SSD6
24	Sl	Hi	Me	B	SSD3	64	Fa	Mid	Sm	S	SSD6
25	Sl	Hi	Bi	S	SSD1	65	Fa	Mid	Sm	M	SSD7
26	Sl	Hi	Bi	M	SSD1	66	Fa	Mid	Sm	B	SSD7
27	Sl	Hi	Bi	B	SSD1	67	Fa	Mid	Me	S	SSD4
28	Mi	Lo	Sm	S	SSD6	68	Fa	Mid	Me	M	SSD6
29	Mi	Lo	Sm	M	SSD7	69	Fa	Mid	Me	B	SSD7
30	Mi	Lo	Sm	B	SSD7	70	Fa	Mid	Bi	S	SSD2
31	Mi	Lo	Me	S	SSD4	71	Fa	Mid	Bi	M	SSD4
32	Mi	Lo	Me	M	SSD5	72	Fa	Mid	Bi	B	SSD5
33	Mi	Lo	Me	B	SSD6	73	Fa	Hi	Sm	S	SSD5
34	Mi	Lo	Bi	S	SSD2	74	Fa	Hi	Sm	M	SSD6
35	Mi	Lo	Bi	M	SSD3	75	Fa	Hi	Sm	B	SSD7
36	Mi	Lo	Bi	B	SSD5	76	Fa	Hi	Me	S	SSD3
37	Mi	Mid	Sm	S	SSD4	77	Fa	Hi	Me	M	SSD4
38	Mi	Mid	Sm	M	SSD6	78	Fa	Hi	Me	B	SSD6
39	Mi	Mid	Sm	B	SSD7	79	Fa	Hi	Bi	S	SSD1
40	Mi	Mid	Me	S	SSD3	80	Fa	Hi	Bi	M	SSD3
						81	Fa	Hi	Bi	B	SSD4

$$SSD = \begin{pmatrix} Service\ Selection\ Decision\ 1 \\ Service\ Selection\ Decision\ 2 \\ Service\ Selection\ Decision\ 3 \\ Service\ Selection\ Decision\ 4 \\ Service\ Selection\ Decision\ 5 \\ Service\ Selection\ Decision\ 6 \\ Service\ Selection\ Decision\ 7 \end{pmatrix} = \begin{pmatrix} SSD1 \\ SSD2 \\ SSD3 \\ SSD4 \\ SSD5 \\ SSD6 \\ SSD7 \end{pmatrix}.$$

6 Simulation Results

In this section, we present the simulation results for our proposed system. In our system, we decided the membership functions and the number of term sets by carrying out many simulations. The simulation results are presented in Figs. 4, 5 and 6. We consider the VM and RSC as constant parameters.

In Fig. 4(a), (b) and (c), we consider the VM value 0.1. We change the RSC value from 0.1 to 0.9. If VM and TS are both 0.1, the processing layer will always be the edge layer. Therefore, high time-sensitive applications will be processed

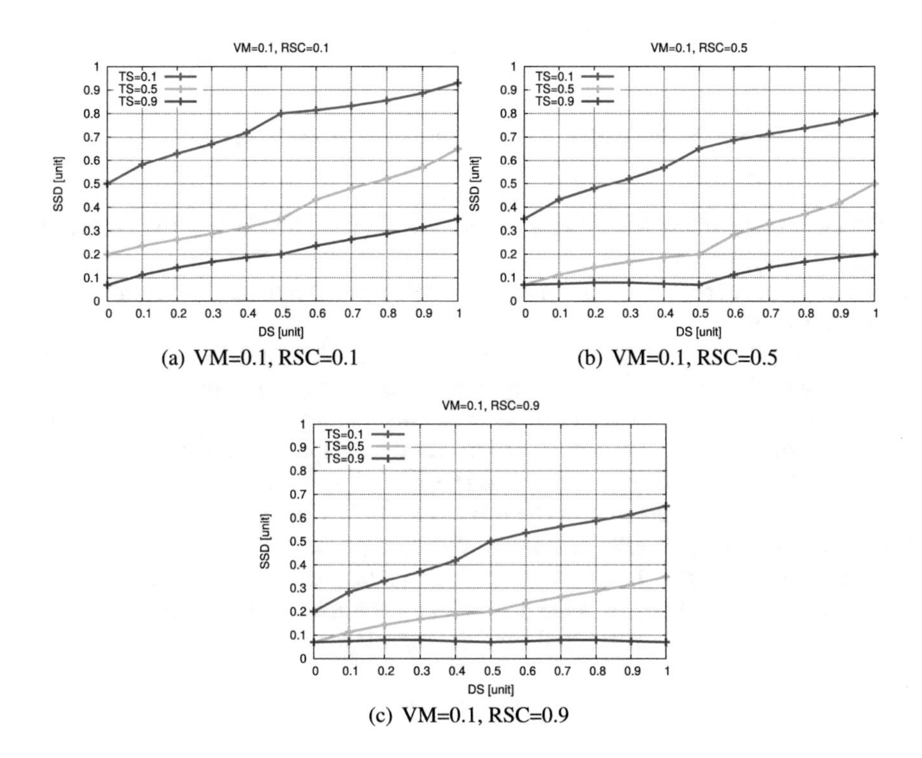

(a) VM=0.1, RSC=0.1

(b) VM=0.1, RSC=0.5

(c) VM=0.1, RSC=0.9

Fig. 4. Simulation results for FSRCM when VM is 0.1.

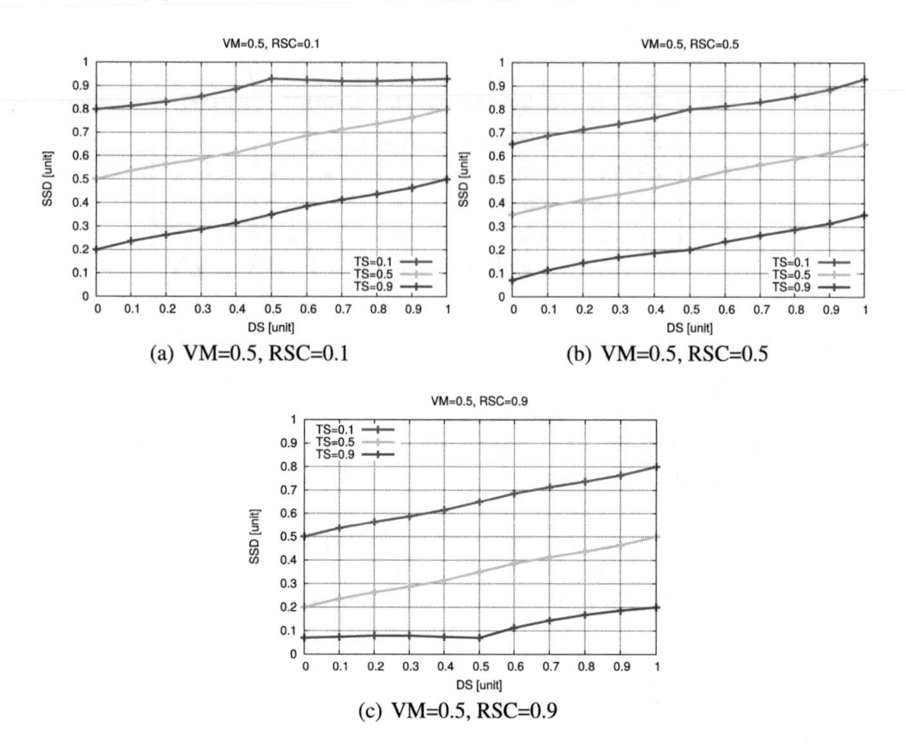

Fig. 5. Simulation results for FSRCM when VM is 0.5.

in the edge layer where the latency is very low. When RSC is low, the processing layer for these applications will be fog or cloud depending on the data size. When RSC is increased, some delay tolerant applications are processed in the edge layer as well.

In Fig. 5(a), (b) and (c), we increase the VM value to 0.5. We can see that by increasing VM value, the edge layer is selected only for time-sensitive applications. Vehicles, at this speed utilize the roadside units to process most of their data. For big data that are not time-sensitive, the cloud layer is selected.

In Fig. 6(a), (b) and (c), we increase the VM value to 0.9. At this speed, it can be seen that vehicles process in the edge layer only time-sensitive data. Also, if the remained storage is low, in this layer will be processed only small-sized data of a real-time application. Other time-sensitive bigger data will be processed in fog layer, where the latency is low as well. If the applications are delay tolerant, vehicles will process these data always in cloud layer.

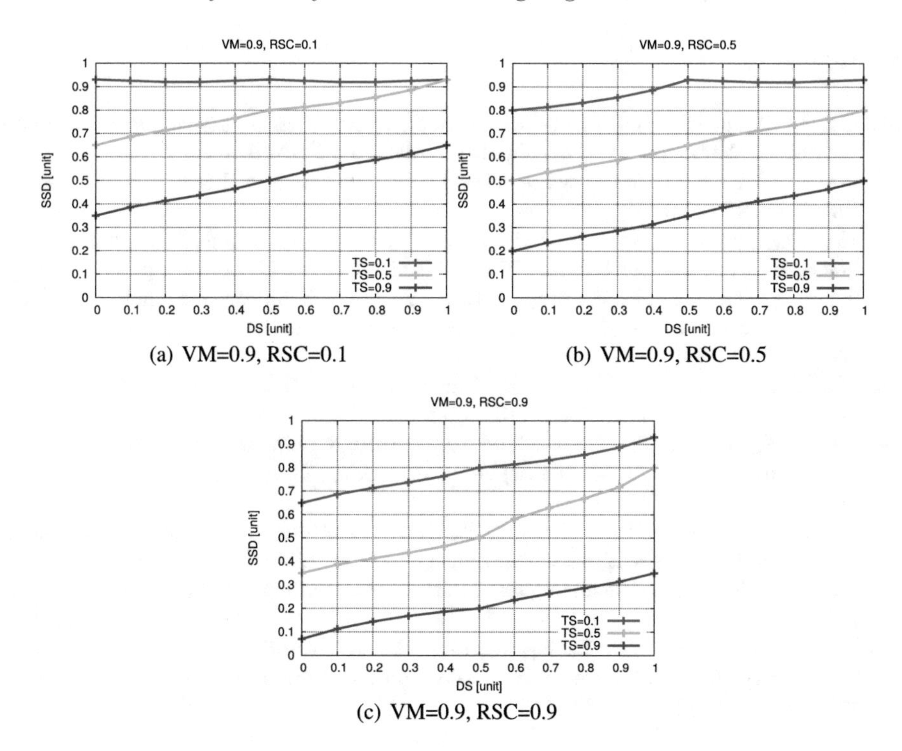

(a) VM=0.9, RSC=0.1

(b) VM=0.9, RSC=0.5

(c) VM=0.9, RSC=0.9

Fig. 6. Simulation results for FSRCM when VM is 0.9.

7 Conclusions

In this paper, we proposed a fuzzy-based system for cloud-fog-edge selection in VANETs. We took into consideration four parameters: DS, VM, TS and RSC. We evaluated the performance of proposed system by computer simulations. From the simulations results, we conclude as follows.

- Highly time-sensitive data are processed in the edge layer.
- Big data are processed in cloud if they are not time-sensitive.
- When the remained storage capacity is low, the possibility of selecting the edge layer for data processing, is decreased.
- At high speed, if the applications are delay tolerant, vehicles will process these data always in cloud layer.

In the future, we would like to make extensive simulations to evaluate the proposed system and compare the performance with other systems.

References

1. Gu, L., Zeng, D., Guo, S.: Vehicular cloud computing: a survey. In: 2013 IEEE Globecom Workshops (GC Wkshps), pp. 403–407, December 2013
2. Ge, X., Li, Z., Li, S.: 5G software defined vehicular networks. IEEE Commun. Mag. **55**(7), 87–93 (2017)
3. Hu, Y.C., Patel, M., Sabella, D., Sprecher, N., Young, V.: Mobile edge computing-a key technology towards 5G. In: ETSI White Paper, vol. 11, no. 11, pp. 1–16 (2015)
4. Yuan, Q., Zhou, H., Li, J., Liu, Z., Yang, F., Shen, X.S.: Toward efficient content delivery for automated driving services: an edge computing solution. IEEE Netw. **32**(1), 80–86 (2018)
5. Olariu, S., Khalil, I., Abuelela, M.: Taking vanet to the clouds. Int. J. Pervasive Comput. Commun. **7**(1), 7–21 (2011)
6. Olariu, S., Hristov, T., Yan, G.: The next paradigm shift: from vehicular networks to vehicular clouds. In: Mobile Ad Hoc Networking: Cutting Edge Directions, vol. 56, no. 6, pp. 645–700 (2013)
7. Hussain, R., Son, J., Eun, H., Kim, S., Oh, H.: Rethinking vehicular communications: merging vanet with cloud computing. In: 4th IEEE International Conference on Cloud Computing Technology and Science Proceedings, pp. 606–609, December 2012
8. Stojmenovic, I., Wen, S., Huang, X., Luan, H.: An overview of fog computing and its security issues. In: Concurrency and Computation: Practice and Experience, vol. 28, no. 10, pp. 2991–3005 (2016)
9. Santi, P.: Mobility Models for Next Generation Wireless Networks: Ad Hoc, Vehicular and Mesh Networks. Wiley, Hoboken (2012)
10. Hartenstein, H., Laberteaux, L.: A tutorial survey on vehicular ad hoc networks. IEEE Commun. Mag. **46**(6), 164–171 (2008)
11. Karagiannis, G., Altintas, O., Ekici, E., Heijenk, G., Jarupan, B., Lin, K., Weil, T.: Vehicular networking: a survey and tutorial on requirements, architectures, challenges, standards and solutions. IEEE Communi. Surv. Tutorials **13**(4), 584–616 (2011)
12. Kandel, A.: Fuzzy Expert Systems. CRC Press Inc., Boca Raton (1992)
13. Zimmermann, H.-J.: Fuzzy control. In: Fuzzy Set Theoryand Its Applications. Springer, pp. 203–240 (1996)
14. McNeill, F.M., Thro, E.: Fuzzy Logic: A Practical Approach. Academic Press Professional Inc., San Diego (1994)
15. Zadeh, L.A., Kacprzyk, J.: Fuzzy Logic for the Management of Uncertainty. Wiley, New York (1992)
16. Procyk, T.J., Mamdani, E.H.: A linguistic self-organizing process controller. Automatica **15**(1), 15–30 (1979)
17. Klir, G.J., Folger, T.A.: Fuzzy Sets, Uncertainty, and Information (1988)
18. Munakata, T., Jani, Y.: Fuzzy systems: an overview. Commun. ACM **37**(3), 69–77 (1994)

Group Speed Parameter Effect for Clustering of Vehicles in VANETs: A Fuzzy-Based Approach

Kosuke Ozera[1(✉)], Kevin Bylykbashi[1], Yi Liu[1], Makoto Ikeda[2], Leonard Barolli[2], and Makoto Takizawa[3]

[1] Graduate School of Engineering, Fukuoka Institute of Technology (FIT), 3-30-1 Wajiro-Higashi, Higashi-Ku, Fukuoka 811–0295, Japan
kosuke.o.fit@gmail.com, kevini_95@hotmail.com, ryuui1010@gmail.com

[2] Department of Information and Communication Engineering, Fukuoka Institute of Technology (FIT), 3-30-1 Wajiro-Higashi, Higashi-Ku, Fukuoka 811–0295, Japan
makoto.ikd@acm.org, barolli@fit.ac.jp

[3] Department of Advanced Sciences, Faculty of Science and Engineering, Hosei University, Kajino-Machi, Koganei-Shi, Tokyo 184-8584, Japan
makoto.takizawa@computer.org

Abstract. In recent years, inter-vehicle communication has attracted attention because it can be applicable not only to alternative networks but also to various communication systems. In this paper, we present the group speed effect in clustering of vehicles in VANETs. We conclude that by selecting vehicles with high SC, high VC and low DCC the possibility that the vehicle will remain in the cluster increases. But in the case of group speed, the GS value should be medium in order that vechile remains in the cluster.

Keywords: VANET · Fuzzy · Cluster

1 Introduction

In recent years, a number of disasters have been occurred around the world. The technologies of disaster management system are improved, however the communication system does not work well in disaster area due to the traffic concentration, device failure, and so on. A key for creating a valuable disaster rescue plan is to prepare alternative communication systems. In disaster area, mobile devices are often disconnected in the network due to network traffic congestion and access point failure. Inter-vehicle communication has attracted attention as an alternative network in disaster situations. In this case, Delay/Disruption/Disconnection Tolerant Networking (DTN) are used as one of a key alternative option to provide the network services [37].

The DTN aims to provide seamless communications with a wide range of network, which have not good performance characteristics [4]. DTN has the

L. Barolli et al. (Eds.): EIDWT 2019, LNDECT 29, pp. 13–24, 2019.
https://doi.org/10.1007/978-3-030-12839-5_2

potential to interconnect vehicles in regions that current networking protocol cannot reach the destination. For inter-vehicle communications, there are different types of communication such as Vehicle-to-Vehicle (V2V), Vehicle-to-Infrastructure (V2I), Vehicle-to-Pedestrian (V2P) and Vehicle-to-X (V2X) communications [2,7,10,14,31]. IEEE 802.11p supports these communications in outdoor environments. It defines enhancements to 802.11 required to support Intelligent Transport System (ITS) applications. The technology operates at 5.9 GHz in various propagation environments to high-speed moving vehicles.

There are different works for Vehicular Ad-Hoc Networks (VANETs). In [16], the authors proposed a Message Suppression Controller (MSC) for V2V and V2I communications. They considered some parameters to control the message suppression dynamically. However, a fixed parameter still is used to calculate the duration of message suppression. To solve this problem, the authors proposed an Enhanced Message Suppression Controller (EMSC) [18] for Vehicular-DTN (V-DTN). The EMSC is an expanded version of MSC [16] and can be used for various network conditions. But, many control packets were delivered in the network.

Security and trust in VANETs is essential in order to prevent malicious agents sabotaging road safety systems built upon the VANET framework, potentially causing serious disruption to traffic flows or safety hazards. Several authors have proposed cluster head metrics which can assist in identifying malicious vehicles and mitigating their impact by denying them access to cluster resources [8].

Security of the safety messages can be achieved by authentication [38]. To make the process of authentication faster [17], vehicles in the communication range of an Road Side Unit (RSU) can be grouped to be in one cluster and a cluster head is elected to authenticate all the vehicles available in the cluster. Formation of clusters in a dynamic VANET and selection of cluster head plays a major role is selected.

In [9] is computed a cluster head selection metric based on vehicle direction, degree of connectivity, an entropy value calculated from the mobility of nodes in the network, and a distrust level based on the reliability of a node's packet relaying. Vehicles are assigned verifiers, which are neighbors with a lower distrust value. Verifiers monitor the network behavior of a vehicle, and confirm whether it is routing packets and advertising mobility and traffic information that is consistent with the verifier's own view of the neighborhood. The distrust value for nodes which behave abnormally is then automatically increased, while it is decreased for nodes which perform reliably. In this way, the trustworthiness of a node is accounted for the cluster head selection process.

Fuzzy Logic (FL) is the logic underlying modes of reasoning which are approximate rather then exact. The importance of FL derives from the fact that most modes of human reasoning and especially common sense reasoning are approximate in nature [19]. FL uses linguistic variables to describe the control parameters. By using relatively simple linguistic expressions it is possible to describe and grasp very complex problems. A very important property of the linguistic variables is the capability of describing imprecise parameters.

In this paper, we present the group speed effect in clustering of vehicles in VANETs. We evaluate the proposed system by simulations. The simulation results shows that the vechiles have high connectivity with other vehicles and are more secure, so they will be remain in the cluster.

The structure of the paper is as follows. In Sect. 2, we present VANETs and DTNs. We give application of Fuzzy Logic for control in Sect. 3. In Sect. 4, we present our proposed systems. In Sect. 5, we show simulation results. Finally, conclusions and future work are given in Sect. 6.

2 VANETs and DTNs

VANETs are considered to have an enormous potential in enhancing road traffic safety and traffic efficiency. Therefore various governments have launched programs dedicated to the development and consolidation of vehicular communications and networking and both industrial and academic researchers are addressing many related challenges, including socio-economic ones, which are among the most important [15,33].

VANET technology uses moving vehicle as nodes to form a wireless mobile network. It aims to provide fast and cost-efficient data transfer for the advantage of passenger safety and comfort. To improve road safety and travel comfort of voyagers and drivers, Intelligent Transport Systems (ITS) are developed. ITS proposes to manage vehicle traffic, support drivers with safety and other information, and provide some services such as automated toll collection and driver assist systems [22].

In essence, VANETs provide new prospects to improve advanced solutions for making reliable communication between vehicles. VANETs can be defined as a part of ITS which aims to make transportation systems faster and smarter in which vehicles are equipped with some short-range and medium-range wireless communication [3]. In a VANET, wireless vehicles are able to communicate directly with each other (i.e., emergency vehicle warning, stationary vehicle warning) and also served various services (i.e., video streaming, internet) from access points (i.e., 3G or 4G) through roadside units [5,22].

The DTN are occasionally connected networks, characterized by the absence of a continuous path between the source and destination [1,13]. The data can be transmitted by storing them at nodes and forwarding them later when there is a working link. This technique is called message switching. Eventually the data will be relayed to the destination. The inspiration for DTNs came from an unlikely source: efforts to send packets in space. Space networks must deal with intermittent communication and very long delays [36]. In [13], the author observed the possibility to apply these ideas for other applications.

The main assumption in the Internet that DTNs seek to relax is that an End-to-End (E2E) path between a source and a destination exists for the entire duration of a communication session. When this is not the case, the normal Internet protocols fail. The DTN architecture is based on message switching. It is also intended to tolerate links with low reliability and large delays. The architecture is specified in RFC 4838 [6].

Bundle protocol has been designed as an implementation of the DTN architecture. A bundle is a basic data unit of the DTN bundle protocol. Each bundle comprises a sequence of two or more blocks of protocol data, which serve for various purposes. In poor conditions, bundle protocol works on the application layer of some number of constituent Internet, forming a store-and-forward overlay network to provide its services. The bundle protocol is specified in RFC 5050 [34]. It is responsible for accepting messages from the application and sending them as one or more bundles via store-carry-forward operations to the destination DTN node. The bundle protocol provides a transport service for many different applications.

3 Application of Fuzzy Logic for Control

The ability of fuzzy sets and possibility theory to model gradual properties or soft constraints whose satisfaction is matter of degree, as well as information pervaded with imprecision and uncertainty, makes them useful in a great variety of applications.

The most popular area of application is Fuzzy Control (FC), since the appearance, especially in Japan, of industrial applications in domestic appliances, process control, and automotive systems, among many other fields [11,12,20,24–28,35].

3.1 FC

In the FC systems, expert knowledge is encoded in the form of fuzzy rules, which describe recommended actions for different classes of situations represented by fuzzy sets.

In fact, any kind of control law can be modeled by the FC methodology, provided that this law is expressible in terms of "if ... then ..." rules, just like in the case of expert systems. However, FL diverges from the standard expert system approach by providing an interpolation mechanism from several rules. In the contents of complex processes, it may turn out to be more practical to get knowledge from an expert operator than to calculate an optimal control, due to modeling costs or because a model is out of reach.

3.2 Linguistic Variables

A concept that plays a central role in the application of FL is that of a linguistic variable. The linguistic variables may be viewed as a form of data compression. One linguistic variable may represent many numerical variables. It is suggestive to refer to this form of data compression as granulation [21].

The same effect can be achieved by conventional quantization, but in the case of quantization, the values are intervals, whereas in the case of granulation the values are overlapping fuzzy sets. The advantages of granulation over quantization are as follows:

- it is more general;
- it mimics the way in which humans interpret linguistic values;
- the transition from one linguistic value to a contiguous linguistic value is gradual rather than abrupt, resulting in continuity and robustness.

3.3 FC Rules

FC describes the algorithm for process control as a fuzzy relation between information about the conditions of the process to be controlled, x and y, and the output for the process z. The control algorithm is given in "if ... then ..." expression, such as:

If x is small and y is big, then z is medium;
If x is big and y is medium, then z is big.

These rules are called *FC rules*. The "if" clause of the rules is called the antecedent and the "then" clause is called consequent. In general, variables x and y are called the input and z the output. The "small" and "big" are fuzzy values for x and y, and they are expressed by fuzzy sets.

Fuzzy controllers are constructed of groups of these FC rules, and when an actual input is given, the output is calculated by means of fuzzy inference.

3.4 Control Knowledge Base

There are two main tasks in designing the control knowledge base. First, a set of linguistic variables must be selected which describe the values of the main control parameters of the process. Both the input and output parameters must be linguistically defined in this stage using proper term sets. The selection of the level of granularity of a term set for an input variable or an output variable plays an important role in the smoothness of control. Second, a control knowledge base must be developed which uses the above linguistic description of the input and output parameters. Four methods [29,32,39,40] have been suggested for doing this:

- expert's experience and knowledge;
- modelling the operator's control action;
- modelling a process;
- self organization.

Among the above methods, the first one is the most widely used. In the modeling of the human expert operator's knowledge, fuzzy rules of the form "If Error is small and Change-in-error is small then the Force is small" have been used in several studies [23,30]. This method is effective when expert human operators can express the heuristics or the knowledge that they use in controlling a process in terms of rules of the above form.

3.5 Defuzzification Methods

The defuzzification operation produces a non-FC action that best represent the membership function of an inferred FC action. Several defuzzification methods have been suggested in literature. Among them, four methods which have been applied most often are:

- Tsukamoto's Defuzzification Method;
- The Center of Area (COA) Method;
- The Mean of Maximum (MOM) Method;
- Defuzzification when Output of Rules are Function of Their Inputs.

4 Proposed System

The proposed system model is show in Fig. 1. For implementing our proposed system, we consider 4 linguistic input parameters: Security (SC), Distance from Cluster Center (DCC), Vehicle Centrality (VC) and Group Speed (GS) to decide the Vehicle Remain or Leave in the Cluster (VRLC) possibility output parameter. These four parameters are not correlated with each other, for this reason we use fuzzy system. The membership functions of proposal system are shown in Fig. 2. In Table 1, we show the Fuzzy Rule Base (FRB) of the proposal system, which consists of 81 rules.

The term sets of *SC*, *DCC*, *VC* and *GS* are defined respectively as:

$$\begin{aligned} SC &= \{Weak,\ Medium,\ Strong\} \\ &= \{We,\ Mi,\ St\}; \\ DCC &= \{Near,\ Middle,\ Far\} \\ &= \{Near,\ Mid,\ Far\}; \\ VC &= \{Low,\ Middle,\ High\} \\ &= \{Lo,\ Mi,\ Hi\}. \\ GS &= \{Slow,\ Medium,\ Fast\} \\ &= \{S,\ M,\ F\}; \end{aligned}$$

and the term set for the output *VRLC* is defined as:

$$VRLC = \begin{pmatrix} Leave \\ WeakLeave \\ VeryWeakLeave \\ NotRemailNotLeave \\ VeryWeakRemain \\ WeakRemain \\ Remain \end{pmatrix} = \begin{pmatrix} Le \\ WLe \\ VWL \\ NRNL \\ VWR \\ WRe \\ Re \end{pmatrix}.$$

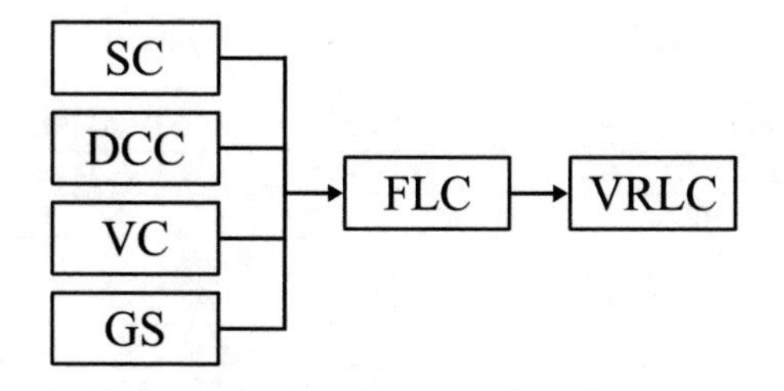

Fig. 1. Proposed system model.

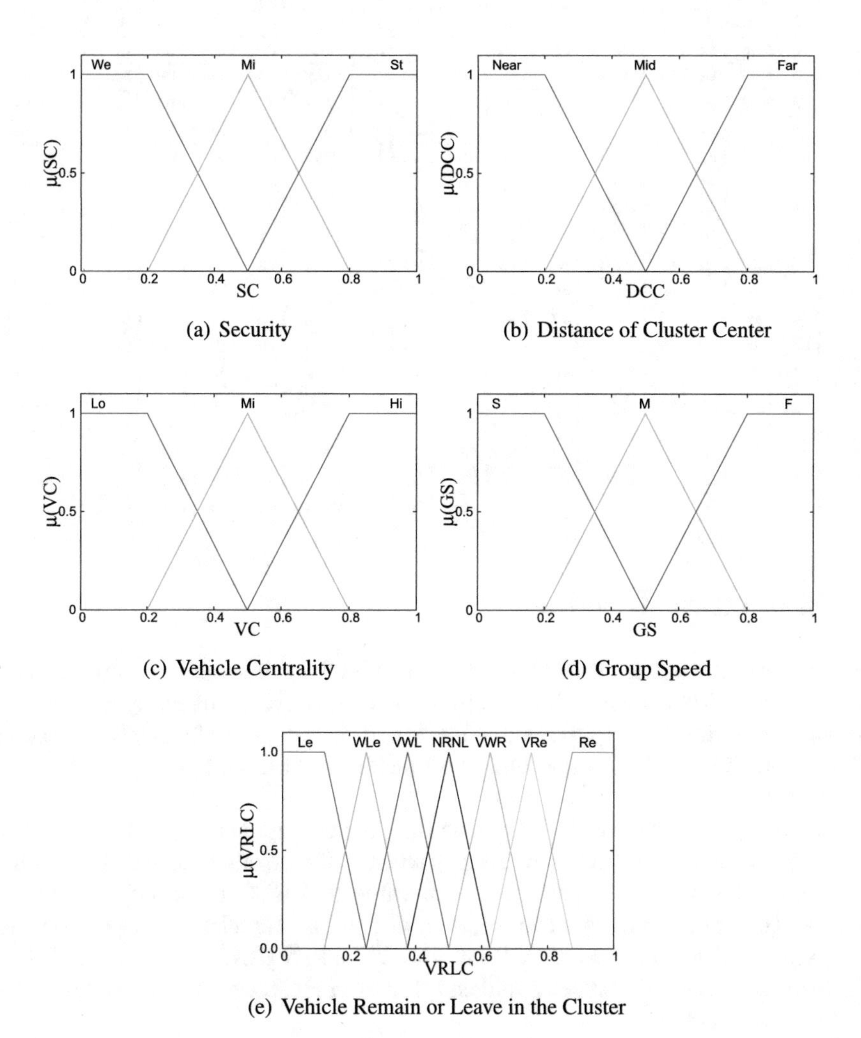

Fig. 2. Membership functions.

Table 1. FRB for proposal system.

Rule	SC	DCC	VC	GS	VRLC	Rule	SC	DCC	VC	GS	VRLC	Rule	SC	DCC	VC	GS	VRLC
1	We	Near	Lo	S	WLe	28	Me	Near	Lo	S	VWL	55	St	Near	Lo	S	VWR
2	We	Near	Lo	M	NRNL	29	Me	Near	Lo	M	VRe	56	St	Near	Lo	M	Re
3	We	Near	Lo	F	WLe	30	Me	Near	Lo	F	VWL	57	St	Near	Lo	F	VWR
4	We	Near	Mi	S	VWL	31	Me	Near	Mi	S	VWR	58	St	Near	Mi	S	VRe
5	We	Near	Mi	M	VWR	32	Me	Near	Mi	M	Re	59	St	Near	Mi	M	Re
6	We	Near	Mi	F	VWL	33	Me	Near	Mi	F	VWR	60	St	Near	Mi	F	VRe
7	We	Near	Hi	S	NRNL	34	Me	Near	Hi	S	VRe	61	St	Near	Hi	S	Re
8	We	Near	Hi	M	VRe	35	Me	Near	Hi	M	Re	62	St	Near	Hi	M	Re
9	We	Near	Hi	F	NRNL	36	Me	Near	Hi	F	VRe	63	St	Near	Hi	F	Re
10	We	Mid	Lo	S	Le	37	Me	Mid	Lo	S	WLe	64	St	Mid	Lo	S	NRNL
11	We	Mid	Lo	M	WLe	38	Me	Mid	Lo	M	NRNL	65	St	Mid	Lo	M	VRe
12	We	Mid	Lo	F	Le	39	Me	Mid	Lo	F	WLe	66	St	Mid	Lo	F	NRNL
13	We	Mid	Mi	S	WLe	40	Me	Mid	Mi	S	VWL	67	St	Mid	Mi	S	VWR
14	We	Mid	Mi	M	VWL	41	Me	Mid	Mi	M	VWR	68	St	Mid	Mi	M	Re
15	We	Mid	Mi	F	WLe	42	Me	Mid	Mi	F	VWL	69	St	Mid	Mi	F	VWR
16	We	Mid	Hi	S	VWL	43	Me	Mid	Hi	S	NRNL	70	St	Mid	Hi	S	VRe
17	We	Mid	Hi	M	VWR	44	Me	Mid	Hi	M	VRe	71	St	Mid	Hi	M	Re
18	We	Mid	Hi	F	VWL	45	Me	Mid	Hi	F	NRNL	72	St	Mid	Hi	F	VRe
19	We	Far	Lo	S	Le	46	Me	Far	Lo	S	Le	73	St	Far	Lo	S	WLe
20	We	Far	Lo	M	Le	47	Me	Far	Lo	M	WLe	74	St	Far	Lo	M	NRNL
21	We	Far	Lo	F	Le	48	Me	Far	Lo	F	Le	75	St	Far	Lo	F	WLe
22	We	Far	Mi	S	Le	49	Me	Far	Mi	S	WLe	76	St	Far	Mi	S	VWL
23	We	Far	Mi	M	WLe	50	Me	Far	Mi	M	NRNL	77	St	Far	Mi	M	VRe
24	We	Far	Mi	F	Le	51	Me	Far	Mi	F	WLe	78	St	Far	Mi	F	VWL
25	We	Far	Hi	S	Le	52	Me	Far	Hi	S	VWL	79	St	Far	Hi	S	VWR
26	We	Far	Hi	M	VWL	53	Me	Far	Hi	M	VWR	80	St	Far	Hi	M	Re
27	We	Far	Hi	F	Le	54	Me	Far	Hi	F	VWL	81	St	Far	Hi	F	VWR

5 Simulation Results

In this section, we present the simulation results for the proposed system. In our system, we decided the membership functions by carrying out many simulations. We show the simulation results in Figs. 3 and 4. We show the relation between VRLC and SC, DCC, VC and GS. We consider the SC and DCC as a constant parameters.

In Fig. 3, we consider the SC value 0.1. We change the GS value from 0 to 1. In Fig. 3(a), the VRLC is increased when GS is from 0 to 0.5 and then is decreased from 0.5 to 1. The maximum value of VRLC is for GS 0.5. This is because the vehicle has the same speed with other vehicles in the cluster. We compare Fig. 3(a) and (b), for GS 0.5 and VC 0.9, VRLC is decreased 17%. In Fig. 3(c), when DCC is 0.9, the vehicle is far from cluster center, thus the vehicle in this case is leaving the cluster.

In Fig. 4(a), (b) and (c), we consider the SC value 0.9. We can see that when the SC value is increased, the VRLC value is also increased. The vehicle remain in the cluster for all scenarios, except the case of Fig. 4(c) when SC is 0.9, DCC is 0.9 and VC is 0.1.

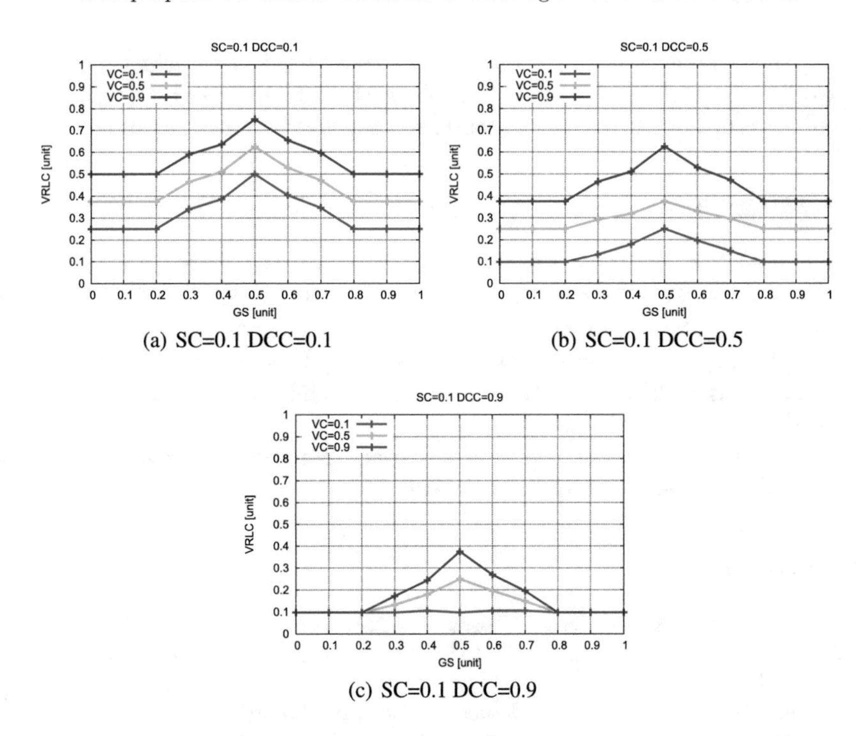

(a) SC=0.1 DCC=0.1

(b) SC=0.1 DCC=0.5

(c) SC=0.1 DCC=0.9

Fig. 3. Simulation results when SC is 0.1.

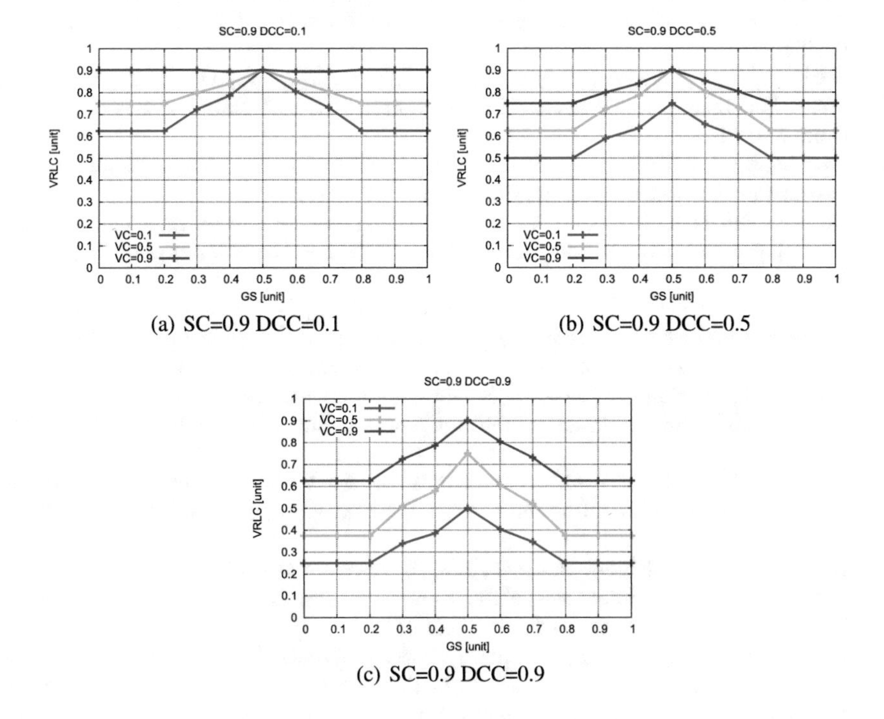

(a) SC=0.9 DCC=0.1

(b) SC=0.9 DCC=0.5

(c) SC=0.9 DCC=0.9

Fig. 4. Simulation results when SC is 0.9.

From the simulation results, we conclude that by selecting vehicles with high SC and VC, low DCC and medium GS values, the vehicles have high connectivity with other vehicles and are more secure, so they will be remained in the cluster.

6 Conclusions

In this paper, we presented the group speed parameter effect for clustering of vehicles in VANETs. We considered four parameters: SC, DCC, VC and GS. We evaluated the systems by simulations. From the simulation results, we conclude that by selecting vehicles with high SC and VC, low DCC and medium GS values, the vechiles have high connectivity with other vechiles and are more secure, so they will be remain in the cluster.

In the future work, we will consider other parameters for the simulation system and carry out extensive simulations.

References

1. Delay- and disruption-tolerant networks (DTNs) tutorial. NASA/JPL's Interplanetary Internet (IPN) Project (2012). http://www.warthman.com/images/DTN_Tutorial_v2.0.pdf
2. Araniti, G., Campolo, C., Condoluci, M., Iera, A., Molinaro, A.: Lte for vehicular networking: a survey. IEEE Commun. Mag. **21**(5), 148–157 (2013)
3. Booysen, M.J., Zeadally, S., van Rooyen, G.J.: Performance comparison of media access control protocols for vehicular ad hoc networks. IET Netw. **1**(1), 10–19 (2012)
4. Burleigh, S., Hooke, A., Torgerson, L., Fall, K., Cerf, V., Durst, B., Scott, K., Weiss, H.: Delay-tolerant networking: an approach to interplanetary internet. IEEE Commun. Mag. **41**(6), 128–136 (2003)
5. Calhan, A.: A fuzzy logic based clustering strategy for improving vehicular ad-hoc network performance. Sadhana **40**(2), 351–367 (2015)
6. Cerf, V., Burleigh, S., Hooke, A., Torgerson, L., Durst, R., Scott, K., Fall, K., Weiss, H.: Delay-tolerant networking architecture. IETF RFC 4838 (Informational), April 2007
7. Cheng, X., Yao, Q., Wen, M., Wang, C.X., Song, L.Y., Jiao, B.L.: Wideband channel modeling and intercarrier interference cancellation for vehicle-to-vehicle communication systems. IEEE J. Sel. Areas Commun. **31**(9), 434–448 (2013)
8. Cooper, C., Franklin, D., Ros, M., Safaei, F., Abolhasan, M.: A comparative survey of vanet clustering techniques. IEEE Commun. Surv. Tutorials **19**(1), 657–681 (2017)
9. Daeinabi, A., Rahbar, A.G.P., Khademzadeh, A.: VWCA: an efficient clustering algorithm in vehicular ad hoc networks. J. Netw. Comput. Appl. **34**(1), 207–222 (2011)
10. Dias, J.A.F.F., Rodrigues, J.J.P.C., Xia, F., Mavromoustakis, C.X.: A cooperative watchdog system to detect misbehavior nodes in vehicular delay-tolerant networks. IEEE Trans. Ind. Electron. **62**(12), 7929–7937 (2015)
11. Elmazi, D., Kulla, E., Oda, T., Spaho, E., Sakamoto, S., Barolli, L.: A comparison study of two fuzzy-based systems for selection of actor node in wireless sensor actor networks. J. Ambient Intell. Humanized Comput. **6**(5), 635–645 (2015)

12. Elmazi, D., Sakamoto, S., Oda, T., Kulla, E., Spaho, E., Barolli, L.: Two fuzzy-based systems for selection of actor nodes in wireless sensor and actor networks: a comparison study considering security parameter effect. Mob. Netw. Appl., 1–12 (2016)
13. Fall, K.: A delay-tolerant network architecture for challenged internets. In: Proceedings of the International Conference on Applications, Technologies, Architectures, and Protocols for Computer Communications, SIGCOMM 2003, pp. 27–34 (2003)
14. Grassi, G., Pesavento, D., Pau, G., Vuyyuru, R., Wakikawa, R., Zhang, L.: VANET via named data networking. In: Proceedings of the IEEE Conference on Computer Communications Workshops (INFOCOM WKSHPS 2014), pp. 410–415 (2014)
15. Hartenstein, H., Laberteaux, L.: A tutorial survey on vehicular ad hoc networks. IEEE Commun. Mag. **46**(6) (2008)
16. Honda, T., Ikeda, M., Ishikawa, S., Barolli, L.: A message suppression controller for vehicular delay tolerant networking. In: Proceedings of the 29th IEEE International Conference on Advanced Information Networking and Applications (IEEE AINA-2015), pp. 754–760 (2015)
17. Huang, J.L., Yeh, L.Y., Chien, H.Y.: ABAKA: an anonymous batch authenticated and key agreement scheme for value-added services in vehicular ad hoc networks. IEEE Trans. Veh. Technol. **60**(1), 248–262 (2011)
18. Ikeda, M., Ishikawa, S., Barolli, L.: An enhanced message suppression controller for vehicular-delay tolerant networks. In: Proceedings of the 30th IEEE International Conference on Advanced Information Networking and Applications (IEEE AINA-2016), pp. 573–579 (2016)
19. Inaba, T., Obukata, R., Sakamoto, S., Oda, T., Ikeda, M., Barolli, L.: Performance evaluation of a QoS-aware fuzzy-based CAC for LAN access. Int. J. Space Based Situated Comput. **6**(4), 228–238 (2016)
20. Inaba, T., Sakamoto, S., Oda, T., Ikeda, M., Barolli, L.: A secure-aware call admission control scheme for wireless cellular networks using fuzzy logic and its performance evaluation. J. Mob. Multimedia **11**(3&4), 213–222 (2015)
21. Kandel, A.: Fuzzy Expert Systems. CRC Press, Boca Raton (1991)
22. Karagiannis, G., Altintas, O., Ekici, E., Heijenk, G., Jarupan, B., Lin, K., Weil, T.: Vehicular networking: a survey and tutorial on requirements, architectures, challenges, standards and solutions. IEEE Commun. Surv. Tutorials **13**(4), 584–616 (2011)
23. Klir, G.J., Folger, T.A.: Fuzzy Sets, Uncertainty, and Information. Prentice Hall, Englewood Cliffs (1988)
24. Kolici, V., Inaba, T., Lala, A., Mino, G., Sakamoto, S., Barolli, L.: A fuzzy-based CAC scheme for cellular networks considering security. In: The 17th International Conference on Network-Based Information Systems (NBiS-2014), pp. 368–373 (2014)
25. Liu, Y., Sakamoto, S., Matsuo, K., Ikeda, M., Barolli, L.: Improving reliability of JXTA-overlay platform: evaluation for E-learning and trustworthiness. J. Mob. Multimedia **11**(2), 34–50 (2015)
26. Liu, Y., Sakamoto, S., Matsuo, K., Ikeda, M., Barolli, L., Xhafa, F.: Improvement of JXTA-overlay P2P platform: evaluation for medical application and reliability. Int. J. Distrib. Syst. Technol. (IJDST) **6**(2), 45–62 (2015)
27. Liu, Y., Sakamoto, S., Matsuo, K., Ikeda, M., Barolli, L., Xhafa, F.: A comparison study for two fuzzy-based systems: improving reliability and security of JXTA-overlay P2P platform. Soft Comput. **20**(7), 2677–2687 (2016)

28. Matsuo, K., Elmazi, D., Liu, Y., Sakamoto, S., Mino, G., Barolli, L.: FACS-MP: a fuzzy admission control system with many priorities for wireless cellular networks and its performance evaluation. J. High Speed Netw. **21**(1), 1–14 (2015)
29. McNeill, F.M., Thro, E.: Fuzzy Logic: A Practical Approach. Academic Press, Boston (1994)
30. Munakata, T., Jani, Y.: Fuzzy systems: an overview. Commun. ACM **37**(3), 68–76 (1994)
31. Ohn-Bar, E., Trivedi, M.M.: Learning to detect vehicles by clustering appearance patterns. IEEE Trans. Intell. Transp. Syst. **16**(5), 2511–2521 (2015)
32. Procyk, T.J., Mamdani, E.H.: A linguistic self-organizing process controller. Automatica **15**(1), 15–30 (1979)
33. Santi, P.: Mobility Models for Next Generation Wireless Networks: Ad Hoc, Vehicular and Mesh Networks. Wiley, Hoboken (2012)
34. Scott, K., Burleigh, S.: Bundle protocol specification. IETF RFC 5050 (Experimental) (2007)
35. Spaho, E., Sakamoto, S., Barolli, L., Xhafa, F., Ikeda, M.: Trustworthiness in P2P: performance behaviour of two fuzzy-based systems for JXTA-overlay platform. Soft Comput. **18**(9), 1783–1793 (2014)
36. Tanenbaum, A.S., Wetherall, D.J.: Computer Networks, 5th edn. Pearson Education Inc., Prentice Hall (2011)
37. Uchida, N., Ishida, T., Shibata, Y.: Delay tolerant networks-based vehicle-to-vehicle wireless networks for road surveillance systems in local areas. Int. J. Space Based Situated Comput. **6**(1), 12–20 (2016)
38. Wen, H., Ho, P.H., Gong, G.: A novel framework for message authentication in vehicular communication networks. In: Global Telecommunications Conference, GLOBECOM 2009, pp. 1–6. IEEE (2009)
39. Zadeh, L.A., Kacprzyk, J.: Fuzzy Logic for the Management of Uncertainty. Wiley, New York (1992)
40. Zimmermann, H.J.: Fuzzy Set Theory and Its Applications. Springer Science & Business Media, New York (1991)

A Fuzzy-Based System for Selection of Actor Nodes in WSANs Considering Actor Reliability and Load Distribution

Donald Elmazi[1(✉)], Miralda Cuka[2], Makoto Ikeda[1], Leonard Barolli[1], and Makoto Takizawa[3]

[1] Department of Information and Communication Engineering, Fukuoka Institute of Technology (FIT), 3-30-1 Wajiro-Higashi, Higashi-Ku, Fukuoka 811-0295, Japan
donald.elmazi@gmail.com, makoto.ikd@acm.org, barolli@fit.ac.jp

[2] Graduate School of Engineering, Fukuoka Institute of Technology (FIT), 3-30-1 Wajiro-Higashi, Higashi-Ku, Fukuoka 811-0295, Japan
mcuka91@gmail.com

[3] Department of Advanced Sciences, Hosei University, 2-17-1 Fujimi, Chiyoda, Tokyo 102-8160, Japan
makoto.takizawa@computer.org

Abstract. Wireless Sensor and Actor Network (WSAN) is formed by the collaboration of micro-sensor and actor nodes. The sensor nodes have responsibility to sense an event and send information towards an actor node. The actor node is responsible to take prompt decision and react accordingly. In order to provide effective sensing and acting, a distributed local coordination mechanism is necessary among sensors and actors. In this work, we consider the actor node selection problem and propose a fuzzy-based system (FBS) that based on data provided by sensors and actors selects an appropriate actor node. We use 4 input parameters: Number of Sensors per Actor (NSA), Distance to Event (DE), Remaining Energy (RE) and Actor Reliability (AR) as new parameter. The output parameter is Actor Selection Decision (ASD). Considering NSA parameter, the ASD has better values when NSA is medium. Thus, when the NSA value is 0.5 the load is distributed better and in this situation the possibility for the actor to be selected is high. Also, for higher values of AR, the actor is selected with high possibility.

1 Introduction

Recent technological advances have lead to the emergence of distributed Wireless Sensor and Actor Networks (WSANs) which are capable of observing the physical world, processing the data, making decisions based on the observations and performing appropriate actions [1].

In WSANs, the devices deployed in the environment are sensors able to sense environmental data, actors able to react by affecting the environment or have both functions integrated. Actor nodes are equipped with two radio transmitters,

L. Barolli et al. (Eds.): EIDWT 2019, LNDECT 29, pp. 25–38, 2019.
https://doi.org/10.1007/978-3-030-12839-5_3

a low data rate transmitter to communicate with the sensor and a high data rate interface for actor-actor communication. For example, in the case of a fire, sensors relay the exact origin and intensity of the fire to actors so that they can extinguish it before spreading in the whole building or in a more complex scenario, to save people who may be trapped by fire [2–4].

To provide effective operation of WSAN, it is very important that sensors and actors coordinate in what are called sensor-actor and actor-actor coordination. Coordination is not only important during task conduction, but also during network's self-improvement operations, i.e. connectivity restoration [5,6], reliable service [7], Quality of Service (QoS) [8,9] and so on.

Sensor-Actor (SA) coordination defines the way sensors communicate with actors, which actor is accessed by each sensor and which route should data packets follow to reach it. Among other challenges, when designing SA coordination, care must be taken in considering energy minimization because sensors, which have limited energy supplies, are the most active nodes in this process. On the other hand, Actor-Actor (AA) coordination helps actors to choose which actor will lead performing the task (actor selection), how many actors should perform and how they will perform. Actor selection is not a trivial task, because it needs to be solved in real time, considering different factors. It becomes more complicated when the actors are moving, due to dynamic topology of the network.

The role of Load balancing in WSAN is to provide a reliable service, in order to have a better performance and effortness. In this paper we consider the Load Balancing issue. Load balancing identifies the optimal load on nodes of the network to increase the network efficiency.

In this paper, different from our previous work [10], we propose and implement a simulation system which considers also the Actor Reliability (AR) parameter. The system is based on fuzzy logic and considers four input parameters for actor selection. We show the simulation results for different values of parameters.

The remainder of the paper is organized as follows. In Sect. 2, we describe the basics of WSANs including research challenges and architecture. In Sect. 3, we describe the system model and its implementation. Simulation results are shown in Sect. 4. Finally, conclusions and future work are given in Sect. 5.

2 WSAN

2.1 WSAN Challenges

Some of the key challenges in WSAN are related to the presence of actors and their functionalities.

- *Deployment and Positioning:* At the moment of node deployment, algorithms must consider to optimize the number of sensors and actors and their initial positions based on applications [11,12].
- *Architecture:* When important data has to be transmitted (an event occurred), sensors may transmit their data back to the sink, which will control the actors' tasks from distance or transmit their data to actors, which can perform actions independently from the sink node [13].

- *Real-Time:* There are a lot of applications that have strict real-time requirements. In order to fulfill them, real-time limitations must be clearly defined for each application and system [14].
- *Coordination:* In order to provide effective sensing and acting, a distributed local coordination mechanism is necessary among sensors and actors [13].
- *Power Management:* WSAN protocols should be designed with minimized energy consumption for both sensors and actors [15].
- *Mobility:* Protocols developed for WSANs should support the mobility of nodes [6,16], where dynamic topology changes, unstable routes and network isolations are present.
- *Scalability:* Smart Cities are emerging fast and WSAN, as a key technology will continue to grow together with cities. In order to keep the functionality of WSAN applicable, scalability should be considered when designing WSAN protocols and algorithms [12,16].

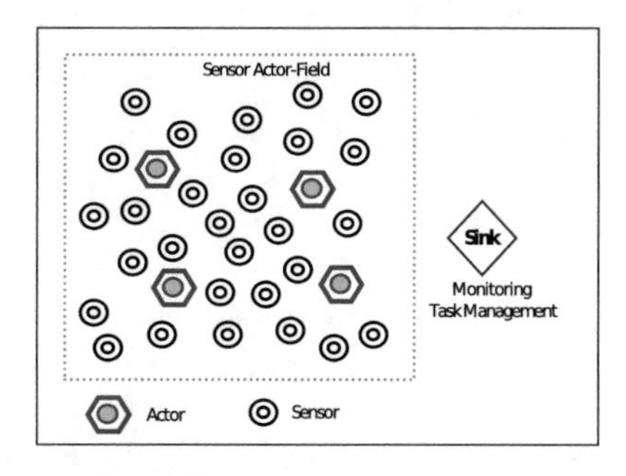

Fig. 1. Wireless Sensor Actor Network (WSAN).

2.2 WSAN Architecture

A WSAN is shown in Fig. 1. The main functionality of WSANs is to make actors perform appropriate actions in the environment, based on the data sensed from sensors and actors. When important data has to be transmitted (an event occurred), sensors may transmit their data back to the sink, which will control the actors' tasks from distance, or transmit their data to actors, which can perform actions independently from the sink node. Here, the former scheme is called Semi-Automated Architecture and the latter one Fully-Automated Architecture (see Fig. 2). Obviously, both architectures can be used in different applications. In the Fully-Automated Architecture are needed new sophisticated algorithms in order to provide appropriate coordination between nodes of WSAN. On the other hand, it has advantages, such as *low latency*, *low energy consumption*, *long network lifetime* [1], *higher local position accuracy*, *higher reliability* and so on.

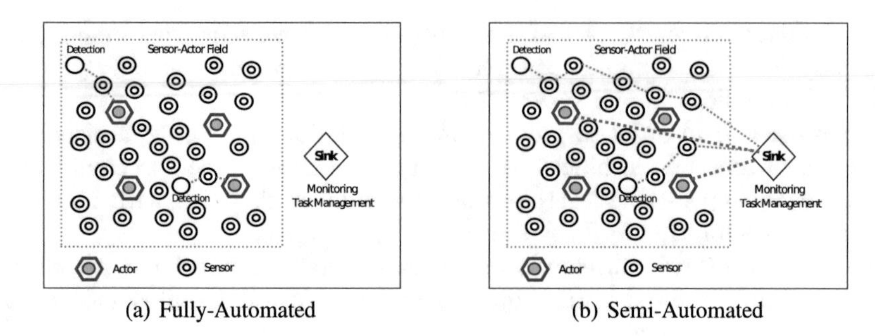

(a) Fully-Automated (b) Semi-Automated

Fig. 2. WSAN architectures.

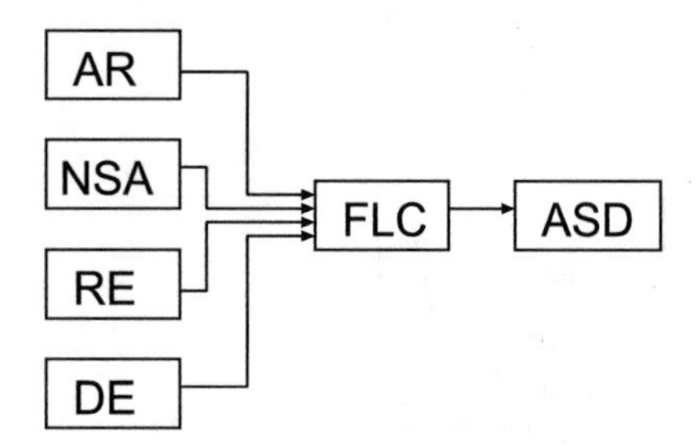

Fig. 3. Proposed system.

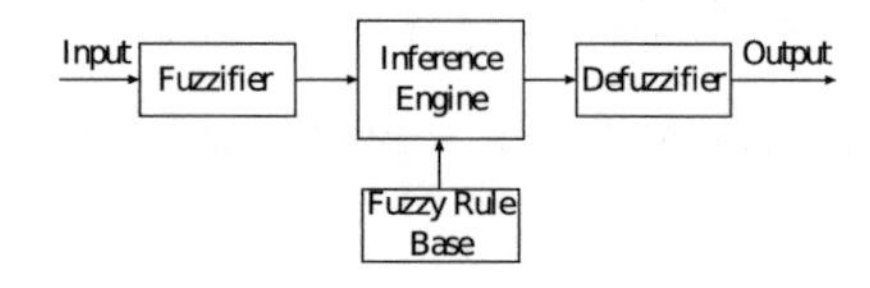

Fig. 4. FLC structure.

3 Proposed Fuzzy-Based System

Based on WSAN characteristics and challenges, we consider the following parameters for implementation of our proposed system.

Actor Reliability (AR): Node reliability can be defined as the level of confidence that one actor node has about another node to assign a task. High reliability is defined by one node about another node based on past interaction and transaction history. The reliability of an actor node increases or decreases over time according to the information provided from trusted neighbour nodes.

Number of Sensors per Actor (NSA): The number of sensors deployed in an area for sensing any event may be in the order of hundreds or thousands. So in order to have a better coverage of these sensors, the number of sensors covered by each actor node should be balanced.

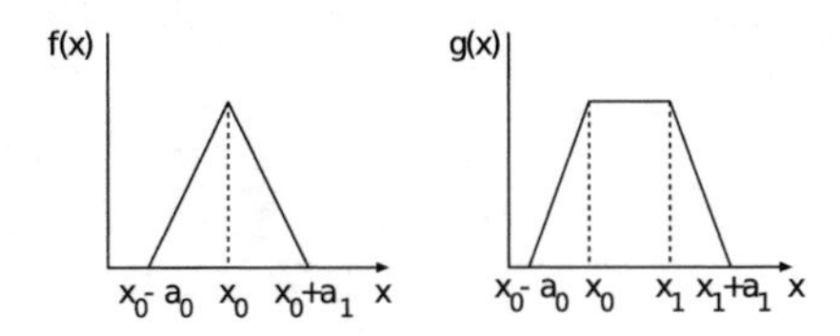

Fig. 5. Triangular and trapezoidal membership functions.

Remaining Energy (RE): As actors are active in the monitored field, they perform tasks and exchange data in different ways. Thus some actors may have a lot of remained power and other may have very little, when an event occurs. It is better that the actors which have more power are selected to carry out a task.

Distance to Event (DE): The number of actors in a WSAN is smaller than the number of sensors. Thus, when an actor is called for action near an event, the distance from the actor to the event is important because when the distance is longer, the actor will spend more energy. Thus, an actor which is close to an event, should be selected.

Actor Selection Decision (ASD): Our system is able to decide the willingness of an actor to be assigned a certain task at a certain time. The actors respond in five different levels, which can be interpreted as:

- Extremely Low Selection Possibility (ELSP) - It is in the minimal condition to be assigned for carrying out a task.
- Very Low Selection Possibility (VLSP) - It is not worth assigning the task to this actor.
- Low Selection Possibility (LSP) - There might be other actors which can do the job better.
- Middle Selection Possibility (MSP) - The Actor is ready to be assigned a task, but is not the "chosen" one.
- High Selection Possibility (HSP) - The actor takes responsibility of completing the task.
- Very High Selection Possibility (VHSP) - Actor has almost all required information and potential and takes full responsibility.
- Extremely High Selection Possibility (EHSP)- Actor is in the perfect condition to carry out the task.

Fuzzy sets and fuzzy logic have been developed to manage vagueness and uncertainty in a reasoning process of an intelligent system such as a knowledge based system, an expert system or a logic control system [17–30].

Table 1. Parameters and their term sets for FLC.

Parameters	Term sets
Actor Reliability (AR)	Unreliable (Ur), Somehow Reliable (ShR), Reliable (Re)
Number of Sensors per Actor (NSA)	Few (Fw), Medium (Me), Many (My)
Remaining Energy (RE)	Low (Lo), Medium (Mdm), High (Hi)
Distance to Event (DE)	Near (Ne), Moderate (Mo), Far (Fa)
Actor Selection Decision (ASD)	Extremely Low Selection Possibility (ELSP), Very Low Selection Possibility (VLSP), Low Selection Possibility (LSP), Middle Selection Possibility (MSP), High Selection Possibility (HSP), Very High Selection Possibility (VHSP), Extremely High Selection Possibility (EHSP)

The structure of the proposed system is shown in Fig. 3. It consists of one Fuzzy Logic Controller (FLC), which is the main part of our system and its basic elements are shown in Fig. 4. They are the fuzzifier, inference engine, Fuzzy Rule Base (FRB) and defuzzifier.

As shown in Fig. 5, we use triangular and trapezoidal membership functions for FLC, because they are suitable for real-time operation [31]. The x_0 in $f(x)$ is the center of triangular function, $x_0(x_1)$ in $g(x)$ is the left (right) edge of trapezoidal function, and $a_0(a_1)$ is the left (right) width of the triangular or trapezoidal function. We explain in details the design of FLC in following.

We use four input parameters for FLC:

- Actor Reliability (AR);
- Number of Sensors per Actor (NSA);
- Remaining Energy (RE);
- Distance to Event (DE).

The term sets for each input linguistic parameter are defined respectively as shown in Table 1.

$$T(AR) = \{Unreliable(Ur), SomehowReliable(ShR), Reliable(Re)\}$$
$$T(NSA) = \{Few(Fw), Medium(Me), Many(My)\}$$
$$T(RE) = \{Low(Lo), Medium(Mdm), High(Hi)\}$$
$$T(DE) = \{Near(Ne), Moderate(Mo), Far(Fa)\}$$

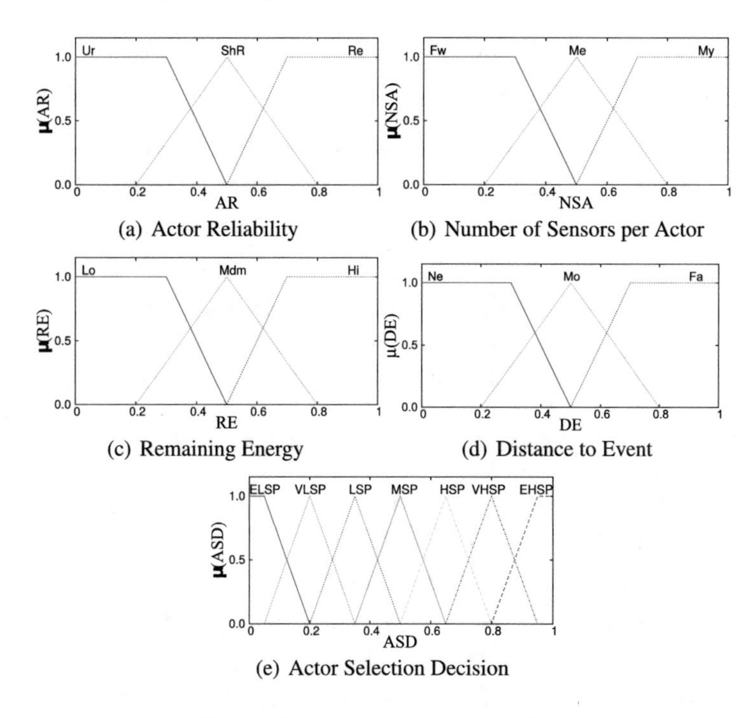

(a) Actor Reliability (b) Number of Sensors per Actor

(c) Remaining Energy (d) Distance to Event

(e) Actor Selection Decision

Fig. 6. Fuzzy membership functions.

The membership functions for input parameters of FLC are defined as:

$$\mu_{Ur}(AR) = g(AR; Ur_0, Ur_1, Ur_{w0}, Ur_{w1})$$
$$\mu_{ShR}(AR) = f(AR; ShR_0, ShR_{w0}, ShR_{w1})$$
$$\mu_{Re}(AR) = g(AR; Re_0, Re_1, Re_{w0}, Re_{w1})$$
$$\mu_{Fw}(NSA) = g(NSA; Fw_0, Fw_1, Fw_{w0}, Fw_{w1})$$
$$\mu_{Me}(NSA) = f(NSA; Me_0, Me_{w0}, Me_{w1})$$
$$\mu_{My}(NSA) = g(NSA; My_0, My_1, My_{w0}, My_{w1})$$
$$\mu_{Lo}(RE) = g(RE; Lo_0, Lo_1, Lo_{w0}, Lo_{w1})$$
$$\mu_{Mdm}(RE) = f(RE; Mdm_0, Mdm_{w0}, Mdm_{w1})$$
$$\mu_{Hi}(RE) = g(RE; Hi_0, Hi_1, Hi_{w0}, Hi_{w1}).$$
$$\mu_{Ne}(DE) = g(DE; Ne_0, Ne_1, Ne_{w0}, Ne_{w1})$$
$$\mu_{Mo}(DE) = f(DE; Mo_0, Mo_{w0}, Mo_{w1})$$
$$\mu_{Fa}(DE) = g(DE; Fa_0, Fa_1, Fa_{w0}, Fa_{w1})$$

The small letters *w0* and *w1* mean left width and right width, respectively.

The output linguistic parameter is the Actor Selection Decision (ASD). We define the term set of ASD as:

$$\begin{aligned}\{&Extremely\ Low\ Selection\ Possibility\ (ELSP),\\ &Very\ Low\ Selection\ Possibility\ (VLSP),\\ &Low\ Selection\ Possibility\ (LSP),\\ &Middle\ Selection\ Possibility\ (MSP),\\ &High\ Selection\ Possibility\ (HSP),\\ &Very\ High\ Selection\ Possibility\ (VHSP),\\ &Extremely\ High\ Selection\ Possibility\ (EHSP)\}.\end{aligned}$$

The membership functions for the output parameter *ASD* are defined as:

$$\begin{aligned}\mu_{ELSP}(ASD) &= g(ASD; ELSP_0, ELSP_1, ELSP_{w0}, ELSP_{w1})\\ \mu_{VLSP}(ASD) &= g(ASD; VLSP_0, VLSP_1, VLSP_{w0}, VLSP_{w1})\\ \mu_{LSP}(ASD) &= g(ASD; LSP_0, LSP_1, LSP_{w0}, LSP_{w1})\\ \mu_{MSP}(ASD) &= g(ASD; MSP_0, MSP_1, MSP_{w0}, MSP_{w1})\\ \mu_{HSP}(ASD) &= g(ASD; HSP_0, HSP_1, HSP_{w0}, HSP_{w1})\\ \mu_{VHSP}(ASD) &= g(ASD; VHSP_0, VHSP_1, VHSP_{w0}, VHSP_{w1})\\ \mu_{EHSP}(ASD) &= g(ASD; EHSP_0, EHSP_1, EHSP_{w0}, EHSP_{w1}).\end{aligned}$$

The membership functions are shown in Fig. 6 and the Fuzzy Rule Base (FRB) is shown in Table 2. The FRB forms a fuzzy set of dimensions $|T(AR)| \times |T(NSA)| \times |T(RE)| \times |T(DE)|$, where $|T(x)|$ is the number of terms on $T(x)$. The FRB has 81 rules. The control rules have the form: IF "conditions" THEN "control action".

4 Simulation Results

The simulation results for our system are shown in Figs. 7, 8 and 9. When an actor moves away from an event, the possibility that this actor will be selected to complete a task, decreases because more energy is needed for longer distances. Also actors with more energy have a higher possibility to be selected. From Fig. 7(a) for DE = 0.1 – RE = 0.1, we see that the ASD is increased 14% from AR = 0.1 to AR = 0.5 and 16% from AR = 0.5 to AR = 0.9, respectively.

In Fig. 7(a), DE is 0.1 and RE is 0.1. In Fig. 7(b), DE is the same and RE increases to 0.5. So, for NSA = 0.5 and AR = 0.1, the ASD is increased 33%. Also, if we compare Fig. 7(b) with Fig. 7(c), for the same values, the ASD is increased 10%.

In WSANs, it is very important that the load should be equally distributed among actors. This provides an uniform numbers of sensors per actor, which causes all the actors to be used for different tasks happening all over the network. For this reason, the possibility of an actor to be selected is increased more for

Table 2. FRB of proposed fuzzy-based system.

No.	DE	RE	AR	NSA	ASD	No.	DE	RE	AR	NSA	ASD
1	Ne	Lo	Ur	Fw	VLSP	41	Mo	Mdm	ShR	Me	VHSP
2	Ne	Lo	Ur	Me	HSP	42	Mo	Mdm	ShR	My	LSP
3	Ne	Lo	Ur	My	VLSP	43	Mo	Mdm	Re	Fw	MSP
4	Ne	Lo	ShR	Fw	LSP	44	Mo	Mdm	Re	Me	VHSP
5	Ne	Lo	ShR	Me	VHSP	45	Mo	Mdm	Re	My	MSP
6	Ne	Lo	ShR	My	LSP	46	Mo	Hi	Ur	Fw	MSP
7	Ne	Lo	Re	Fw	HSP	47	Mo	Hi	Ur	Me	VHSP
8	Ne	Lo	Re	Me	EHSP	48	Mo	Hi	Ur	My	MSP
9	Ne	Lo	Re	My	MSP	49	Mo	Hi	ShR	Fw	HSP
10	Ne	Mdm	Ur	Fw	LSP	50	Mo	Hi	ShR	Me	EHSP
11	Ne	Mdm	Ur	Me	VHSP	51	Mo	Hi	ShR	My	HSP
12	Ne	Mdm	Ur	My	LSP	52	Mo	Hi	Re	Fw	VHSP
13	Ne	Mdm	ShR	Fw	MSP	53	Mo	Hi	Re	Me	EHSP
14	Ne	Mdm	ShR	Me	EHSP	54	Mo	Hi	Re	My	VHSP
15	Ne	Mdm	ShR	My	MSP	55	Fa	Lo	Ur	Fw	EHSP
16	Ne	Mdm	Re	Fw	HSP	56	Fa	Lo	Ur	Me	VLSP
17	Ne	Mdm	Re	Me	EHSP	57	Fa	Lo	Ur	My	EHSP
18	Ne	Mdm	Re	My	HSP	58	Fa	Lo	ShR	Fw	ELSP
19	Ne	Hi	Ur	Fw	HSP	59	Fa	Lo	ShR	Me	LSP
20	Ne	Hi	Ur	Me	EHSP	60	Fa	Lo	ShR	My	ELSP
21	Ne	Hi	Ur	My	HSP	61	Fa	Lo	Re	Fw	VLSP
22	Ne	Hi	ShR	Fw	VHSP	62	Fa	Lo	Re	Me	MSP
23	Ne	Hi	ShR	Me	EHSP	63	Fa	Lo	Re	My	VLSP
24	Ne	Hi	ShR	My	VHSP	64	Fa	Mdm	Ur	Fw	ELSP
25	Ne	Hi	Re	Fw	EHSP	65	Fa	Mdm	Ur	Me	LSP
26	Ne	Hi	Re	Me	EHSP	66	Fa	Mdm	Ur	My	ELSP
27	Ne	Hi	Re	My	EHSP	67	Fa	Mdm	ShR	Fw	ELSP
28	Mo	Lo	Ur	Fw	ELSP	68	Fa	Mdm	ShR	Me	MSP
29	Mo	Lo	Ur	Me	LSP	69	Fa	Mdm	ShR	My	ELSP
30	Mo	Lo	Ur	My	ELSP	70	Fa	Mdm	Re	Fw	VLSP
31	Mo	Lo	ShR	Fw	VLSP	71	Fa	Mdm	Re	Me	HSP
32	Mo	Lo	ShR	Me	HSP	72	Fa	Mdm	Re	My	VLSP
33	Mo	Lo	ShR	My	VLSP	73	Fa	Hi	Ur	Fw	VLSP
34	Mo	Lo	Re	Fw	LSP	74	Fa	Hi	Ur	Me	HSP
35	Mo	Lo	Re	Me	VHSP	75	Fa	Hi	Ur	My	VLSP
36	Mo	Lo	Re	My	LSP	76	Fa	Hi	ShR	Fw	MSP
37	Mo	Mdm	Ur	Fw	VLSP	77	Fa	Hi	ShR	Me	VHSP
38	Mo	Mdm	Ur	Me	MSP	78	Fa	Hi	ShR	My	MSP
39	Mo	Mdm	Ur	My	VLSP	79	Fa	Hi	Re	Fw	HSP
40	Mo	Mdm	ShR	Fw	LSP	80	Fa	Hi	Re	Me	EHSP
						81	Fa	Hi	Re	My	HSP

medium values of NSA. To see this we compare Fig. 8(a) with Fig. 8(b) and Fig. 8(b) with Fig. 8(c). In Fig. 8(a) DE is 0.5 and RE is 0.5. In Fig. 8(b) DE is the same and RE is increased to 0.5. So, for NSA = 0.5 and AR = 0.1, the ASD is increased 10%. Also, comparing Fig. 8(b) with Fig. 8(c), for the same values, the ASD is increased 30%.

In Fig. 9(a), DE is 0.9 and RE is 0.1. In Fig. 9(b), DE is the same and RE is increased to 0.5. So, for NSA = 0.5 and AR = 0.1, we can see that ASD is increased 10%. Also comparing Fig. 9(b) with Fig. 9(c), for the same values, the ASD is increased 24%.

Different actors, based on past history of contacts with other actors, create a level of reliability with each other, which increases or decreases over time. Some actors are more reliable than others to carry out a task. So, highly reliable nodes are more likely to be selected.

In Fig. 7(c), DE is 0.1 and RE is 0.9. In Fig. 8(c), DE is increased to 0.5, while RE is the same. Comparing Fig. 7(c) with Fig. 8(c), for NSA = 0.5 and AR = 0.1, the ASD is decreased 13%. If we compare Fig. 8(c) with Fig. 9(c), DE is increased to 0.9 and RE is the same, so for NSA = 0.5 and AR = 0.1, we can see that the ASD is decreased 10%.

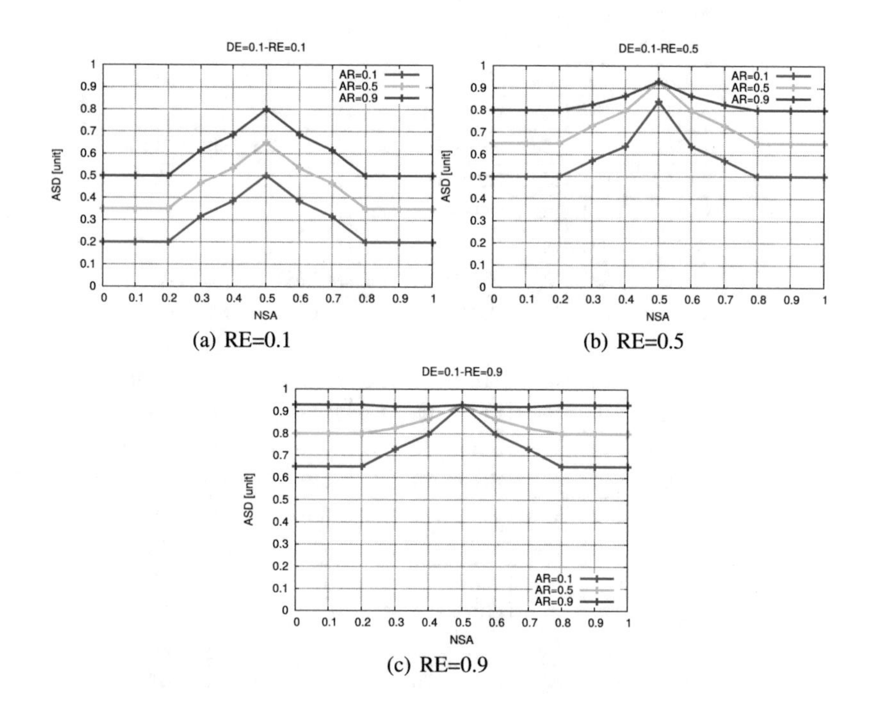

(a) RE=0.1

(b) RE=0.5

(c) RE=0.9

Fig. 7. Results for $DE = 0.1$.

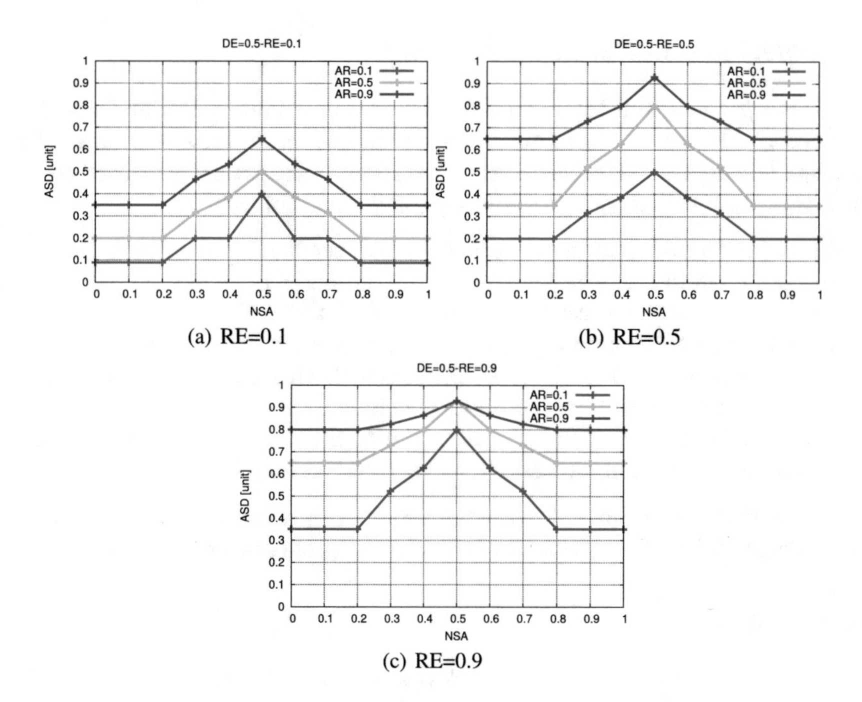

(a) RE=0.1 (b) RE=0.5

(c) RE=0.9

Fig. 8. Results for $DE = 0.5$.

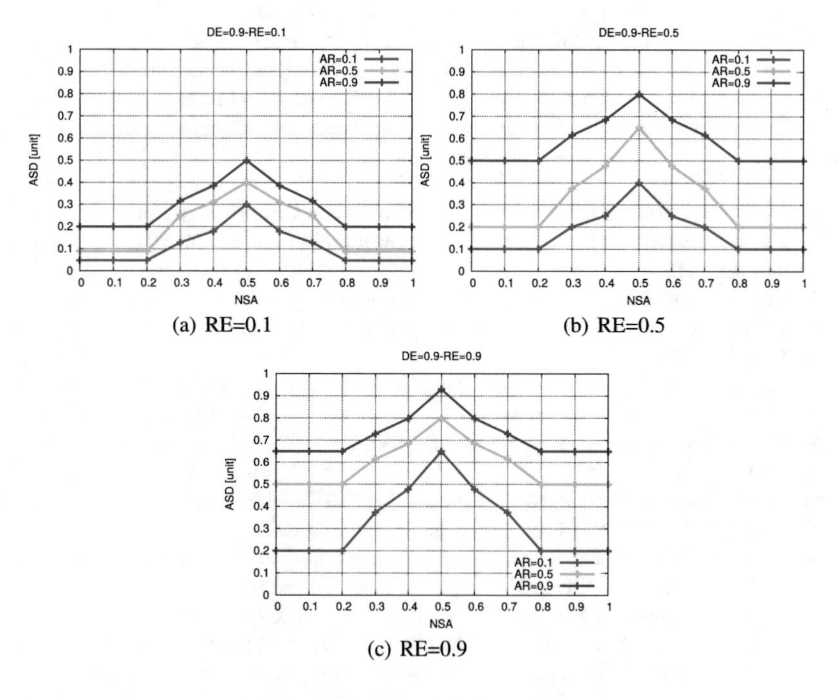

(a) RE=0.1 (b) RE=0.5

(c) RE=0.9

Fig. 9. Results for $DE = 0.9$.

5 Conclusions and Future Work

In this we propose and implement a FBS for actor node selection in WSANs, considering AR as new parameter. Considering that some nodes are more reliable than the others, our system decides which actor nodes are better suited to carry out a task. From simulation results, we conclude as follows.

- When AR and RE parameters are increased, the ASD parameter is increased, so the probability that the system selects an actor node for the job is high.
- When NSA value increases from 0.1 to 0.5, the ASD increases, but when the value of NSA increases from 0.5 to 0.9, the ASD decreases. Thus, when the NSA value is 0.5 the load is distributed better and in this situation the possibility for the actor to be selected is high.
- When the DE parameter is increased, the ASD parameter is decreased, so the probability that an actor node is selected for the required task is low.

In the future work, we will consider also other parameters for actor selection and make extensive simulations to evaluate the proposed system.

References

1. Akyildiz, I.F., Kasimoglu, I.H.: Wireless sensor and actor networks: research challenges. Ad Hoc Netw. J. **2**(4), 351–367 (2004)
2. Akyildiz, I., Su, W., Sankarasubramaniam, Y., Cayirci, E.: Wireless sensor networks: a survey. Computer Netw. **38**(4), 393–422 (2002)
3. Boyinbode, O., Le, H., Takizawa, M.: A survey on clustering algorithms for wireless sensor networks. Int. J. Space Based Situated Comput. **1**(2/3), 130–136 (2011)
4. Bahrepour, M., Meratnia, N., Poel, M., Taghikhaki, Z., Havinga, P.J.: Use of wireless sensor networks for distributed event detection in disaster managment applications. Int. J. Space Based Situated Comput. **2**(1), 58–69 (2012)
5. Haider, N., Imran, M., Saad, N., Zakariya, M.: Performance analysis of reactive connectivity restoration algorithms for wireless sensor and actor networks. In: IEEE Malaysia International Conference on Communications (MICC-2013), pp. 490–495, November 2013
6. Abbasi, A., Younis, M., Akkaya, K.: Movement-assisted connectivity restoration in wireless sensor and actor networks. IEEE Trans. Parallel Distrib. Syst. **20**(9), 1366–1379 (2009)
7. Li, X., Liang, X., Lu, R., He, S., Chen, J., Shen, X.: Toward reliable actor services in wireless sensor and actor networks. In: IEEE 8th International Conference on Mobile Adhoc and Sensor Systems (MASS), pp. 351–360, October 2011
8. Akkaya, K., Younis, M.: COLA: a coverage and latency aware actor placement for wireless sensor and actor networks. In: IEEE 64th Conference on Vehicular Technology (VTC-2006) Fall, pp. 1–5, September 2006
9. Kakarla, J., Majhi, B.: A new optimal delay and energy efficient coordination algorithm for WSAN. In: IEEE International Conference on Advanced Networks and Telecommuncations Systems (ANTS), pp. 1–6, December 2013
10. Elmazi, D., Cuka, M., Ikeda, M., Barolli, L.: A fuzzy-based system for actor node selection in WSANs for improving network connectivity and increasing number of covered sensors. In: The 21st International Conference on Network-Based Information Systems (NBiS-2018) (2018)

11. Akbas, M., Turgut, D.: APAWSAN: actor positioning for aerial wireless sensor and actor networks. In: IEEE 36th Conference on Local Computer Networks (LCN), pp. 563–570, October 2011
12. Akbas, M., Brust, M., Turgut, D.: Local positioning for environmental monitoring in wireless sensor and actor networks. In: IEEE 35th Conference on Local Computer Networks (LCN), pp. 806–813, October 2010
13. Melodia, T., Pompili, D., Gungor, V., AkyildizZX, I.: Communication and coordination in wireless sensor and actor networks. IEEE Trans. Mob. Comput. **6**(10), 1126–1129 (2007)
14. Gungor, V., Akan, O., Akyildiz, I.: A real-time and reliable transport (RT) 2 protocol for wireless sensor and actor networks. IEEE/ACM Trans. Networking **16**(2), 359–370 (2008)
15. Selvaradjou, K., Handigol, N., Franklin, A., Murthy, C.: Energy-efficient directional routing between partitioned actors in wireless sensor and actor networks. IET Commun. **4**(1), 102–115 (2010)
16. Nakayama, H., Fadlullah, Z., Ansari, N., Kato, N.: A novel scheme for WSAN sink mobility based on clustering and set packing techniques. IEEE Trans. Autom. Control **56**(10), 2381–2389 (2011)
17. Inaba, T., Sakamoto, S., Kolici, V., Mino, G., Barolli, L.: A CAC scheme based on fuzzy logic for cellular networks considering security and priority parameters. In: The 9th International Conference on Broadband and Wireless Computing, Communication and Applications (BWCCA-2014), pp. 340–346 (2014)
18. Spaho, E., Sakamoto, S., Barolli, L., Xhafa, F., Barolli, V., Iwashige, J.: A fuzzy-based system for peer reliability in JXTA-overlay P2P considering number of interactions. In: The 16th International Conference on Network-Based Information Systems (NBiS-2013), pp. 156–161 (2013)
19. Matsuo, K., Elmazi, D., Liu, Y., Sakamoto, S., Mino, G., Barolli, L.: FACS-MP: a fuzzy admission control system with many priorities for wireless cellular networks and its performance evaluation. J. High Speed Netw. **21**(1), 1–14 (2015)
20. Grabisch, M.: The application of fuzzy integrals in multicriteria decision making. Eur. J. Oper. Res. **89**(3), 445–456 (1996)
21. Inaba, T., Elmazi, D., Liu, Y., Sakamoto, S., Barolli, L., Uchida, K.: Integrating wireless cellular and ad-hoc networks using fuzzy logic considering node mobility and security. In: The 29th IEEE International Conference on Advanced Information Networking and Applications Workshops (WAINA-2015), pp. 54–60 (2015)
22. Kulla, E., Mino, G., Sakamoto, S., Ikeda, M., Caballé, S., Barolli, L.: FBMIS: a fuzzy-based multi-interface system for cellular and ad hoc networks. In: International Conference on Advanced Information Networking and Applications (AINA-2014), pp. 180–185 (2014)
23. Elmazi, D., Kulla, E., Oda, T., Spaho, E., Sakamoto, S., Barolli, L.: A comparison study of two fuzzy-based systems for selection of actor node in wireless sensor actor networks. J. Ambient Intell. Humanized Comput. **6**(5), 635–645 (2015)
24. Zadeh, L.: Fuzzy logic, neural networks, and soft computing. Commun. ACM **37**, 77–84 (1994)
25. Spaho, E., Sakamoto, S., Barolli, L., Xhafa, F., Ikeda, M.: Trustworthiness in P2P: performance behaviour of two fuzzy-based systems for JXTA-overlay platform. Soft Comput. **18**(9), 1783–1793 (2014)
26. Inaba, T., Sakamoto, S., Kulla, E., Caballe, S., Ikeda, M., Barolli, L.: An integrated system for wireless cellular and ad-hoc networks using fuzzy logic. In: International Conference on Intelligent Networking and Collaborative Systems (INCoS-2014), pp. 157–162 (2014)

27. Matsuo, K., Elmazi, D., Liu, Y., Sakamoto, S., Barolli, L.: A multi-modal simulation system for wireless sensor networks: a comparison study considering stationary and mobile sink and event. J. Ambient Intell. Humanized Comput. **6**(4), 519–529 (2015)
28. Kolici, V., Inaba, T., Lala, A., Mino, G., Sakamoto, S., Barolli, L.: A fuzzy-based CAC scheme for cellular networks considering security. In: International Conference on Network-Based Information Systems (NBiS-2014), pp. 368–373 (2014)
29. Liu, Y., Sakamoto, S., Matsuo, K., Ikeda, M., Barolli, L., Xhafa, F.: A comparison study for two fuzzy-based systems: improving reliability and security of JXTA-overlay P2P platform. Soft Comput. **20**(7), 2677–2687 (2015)
30. Matsuo, K., Elmazi, D., Liu, Y., Sakamoto, S., Mino, G., Barolli, L.: FACS-MP: a fuzzy admission control system with many priorities for wireless cellular networks and its perforemance evaluation. J. High Speed Netw. **21**(1), 1–14 (2015)
31. Mendel, J.M.: Fuzzy logic systems for engineering: a tutorial. Proc. IEEE **83**(3), 345–377 (1995)

IoT Device Selection in Opportunistic Networks: A Fuzzy Approach Considering IoT Device Failure Rate

Miralda Cuka[1(✉)], Donald Elmazi[2], Keita Matsuo[2], Makoto Ikeda[2], Leonard Barolli[2], and Makoto Takizawa[3]

[1] Graduate School of Engineering, Fukuoka Institute of Technology (FIT), 3-30-1 Wajiro-Higashi, Higashi-Ku, Fukuoka 811-0295, Japan
mcuka91@gmail.com

[2] Department of Information and Communication Engineering, Fukuoka Institute of Technology (FIT), 3-30-1 Wajiro-Higashi, Higashi-Ku, Fukuoka 811-0295, Japan
donald.elmazi@gmail.com, kt-matsuo@fit.ac.j, makoto.ikd@acm.org, barolli@fit.ac.jp

[3] Department of Advanced Sciences, Faculty of Science and Engineering, Hosei University, Kajino-Machi, Koganei-Shi, Tokyo 184-8584, Japan
makoto.takizawa@computer.org

Abstract. In opportunistic networks the communication opportunities (contacts) are intermittent and there is no need to establish an end-to-end link between the communication nodes. The enormous growth of devices having access to the Internet, along the vast evolution of the Internet and the connectivity of objects and devices, has evolved as Internet of Things (IoT). There are different issues for these networks. One of them is the selection of IoT devices in order to carry out a task in opportunistic networks. In this work, we implement a Fuzzy-Based System for IoT device selection in opportunistic networks. For our system, we use four input parameters: IoT Device's Number of Past Encounters (IDNPE), IoT Device Storage (IDST), IoT Device Remaining Energy (IDRE) and IoT Device Failure Rate (IDFR). The output parameter is IoT Device Selection Decision (IDSD). The simulation results show that the proposed system makes a proper selection decision of IoT devices in opportunistic networks. The IoT device selection is increased up to 12% and 26% by increasing IDNPE and IDFR, respectively.

1 Introduction

Future communication systems will be increasingly complex, involving thousands of heterogeneous devices with diverse capabilities and various networking technologies interconnected with the aim to provide users with ubiquitous access to information and advanced services at a high quality level, in a cost efficient manner, any time, any place, and in line with the always best connectivity principle.

L. Barolli et al. (Eds.): EIDWT 2019, LNDECT 29, pp. 39–52, 2019.
https://doi.org/10.1007/978-3-030-12839-5_4

The Opportunistic Networks (OppNets) can provide an alternative way to support the diffusion of information in special locations within a city, particularly in crowded spaces where current wireless technologies can exhibit congestion issues. The efficiency of this diffusion relies mainly on user mobility. In fact, mobility creates the opportunities for contacts and, therefore, for data forwarding [1]. OppNets have appeared as an evolution of the MANETs. They are also a wireless based network and hence, they face various issues similar to MANETs such as frequent disconnections, highly variable links, limited bandwidth etc. In OppNets, nodes are always moving which makes the network easy to deploy and decreases the dependence on infrastructure for communication [2].

The Internet of Things (IoT) can seamlessly connect the real world and cyberspace via physical objects embeded with various types of intelligent sensors. A large number of Internet-connected machines will generate and exchange an enormous amount of data that make daily life more convenient, help to make a tough decision and provide beneficial services. The IoT probably becomes one of the most popular networking concepts that has the potential to bring out many benefits [3,4].

OppNets are the variants of Delay Tolerant Networks (DTNs). It is a class of networks that has emerged as an active research subject in the recent times. Owing to the transient and un-connected nature of the nodes, routing becomes a challenging task in these networks. Sparse connectivity, no infrastructure and limited resources further complicate the situation [5,6]. Routing methods for such sparse mobile networks use a different paradigm for message delivery. These schemes utilize node mobility by having nodes carry messages, waiting for an opportunity to transfer messages to the destination or the next relay rather than transmitting them over a path [7]. Hence, the challenges for routing in OppNet are very different from the traditional wireless networks and their utility and potential for scalability makes them a huge success.

In mobile OppNet, connectivity varies significantly over time and is often disruptive. Examples of such networks include interplanetary communication networks, mobile sensor networks, vehicular adhoc networks (VANETs), terrestrial wireless networks, and under-water sensor networks. While the nodes in such networks are typically delay-tolerant, message delivery latency still remains a crucial metric, and reducing it is highly desirable [8].

However, most of the proposed routing schemes assume long contact durations such that all buffered messages can be transferred within a single contact. For example, when hand-held devices communicate via Bluetooth that has a typical wireless range of about 10 m, the contact duration tends to be as short as several seconds if the users are walking. For high speed vehicles that communicate via WiFi (802.11g), which has a longer range (up to 38 m indoors and 140 m outdoors), the contact duration is still short. In the presence of short contact durations, there are two key issues that must be addressed. First is the relay selection issue. We need to select relay nodes that will contact the message's destination long enough so that the entire message can be successfully transmitted. Second is the message scheduling issue. Since not all messages can be exchanged between nodes within a single contact, it is important to schedule the transmission of messages in such a way that will maximize the network delivery ratio [9].

The Fuzzy Logic (FL) is unique approach that is able to simultaneously handle numerical data and linguistic knowledge. The fuzzy logic works on the levels of possibilities of input to achieve the definite output. Fuzzy set theory and FL establish the specifics of the nonlinear mapping.

In this paper, we propose and implement a fuzy-based simulation system for selection of IoT devices in OppNet considering as a new parameter the failure rate of an IoT device. For our system we consider four parameters for IoT device selection. We show the simulation results for different values of parameters.

The remainder of the paper is organized as follows. In the Sect. 2, we present IoT and OppNet. In Sect. 3, we introduce the proposed system model and its implementation. Simulation results are shown in Sect. 4. Finally, conclusions and future work are given in Sect. 5.

2 IoT and OppNets

2.1 IoT

IoT allows to integrate physical and virtual objects. Virtual reality, which was recently available only on the monitor screens, now integrates with the real world, providing users with completely new opportunities: interact with objects on the other side of the world and receive the necessary services that became real due the wide interaction [10]. The IoT will support substantially higher number of end users and devices. In Fig. 1, we present an example of an IoT network architecture. The IoT network is a combination of IoT devices which are connected with different mediums using IoT Gateway to the Internet. The data transmitted through the gateway is stored, proccessed securely within cloud server. These new connected things will trigger increasing demands for new IoT

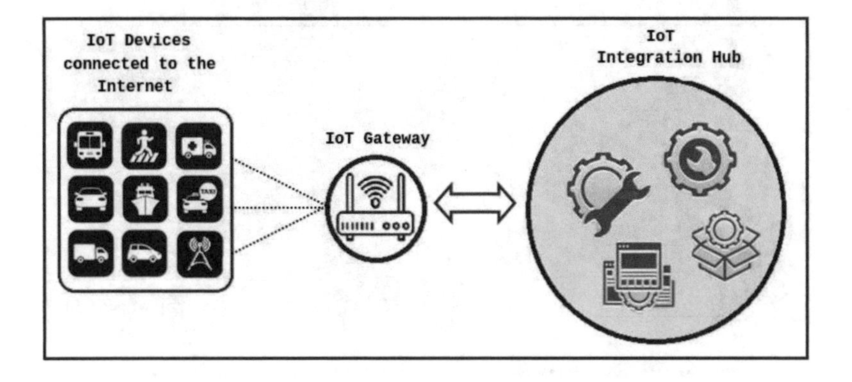

Fig. 1. An Iot network architecture.

applications that are not only for users. The current solutions for IoT application development generally rely on integrated service-oriented programming platforms. In particular, resources (e.g., sensory data, computing resource, and control information) are modeled as services and deployed in the cloud or at the edge. It is difficult to achieve rapid deployment and flexible resource management at network edges, in addition, an IoT system's scalability will be restricted by the capability of the edge devices [11].

2.2 OppNets

In Fig. 2 we show an OppNet scenario. OppNets comprises a network where nodes can be anything from pedestrians, vehicles, fixed devices and so on. The data is sent from the sender to receiver by using communication opportunity that can be Wi-Fi, Bluetooth, cellular technologies or satellite links to transfer the message to the final destination. In such scenario, IoT devices might roam and opportunistically encounter several different statically deployed networks and perform either data collection or dissemination as well as relaying data between these networks, thus introducing further connectivity for disconnected networks. For example, as seen in the figure, a car could opportunistically encounter other IoT devices, collect information from them and relay it until it finds an available

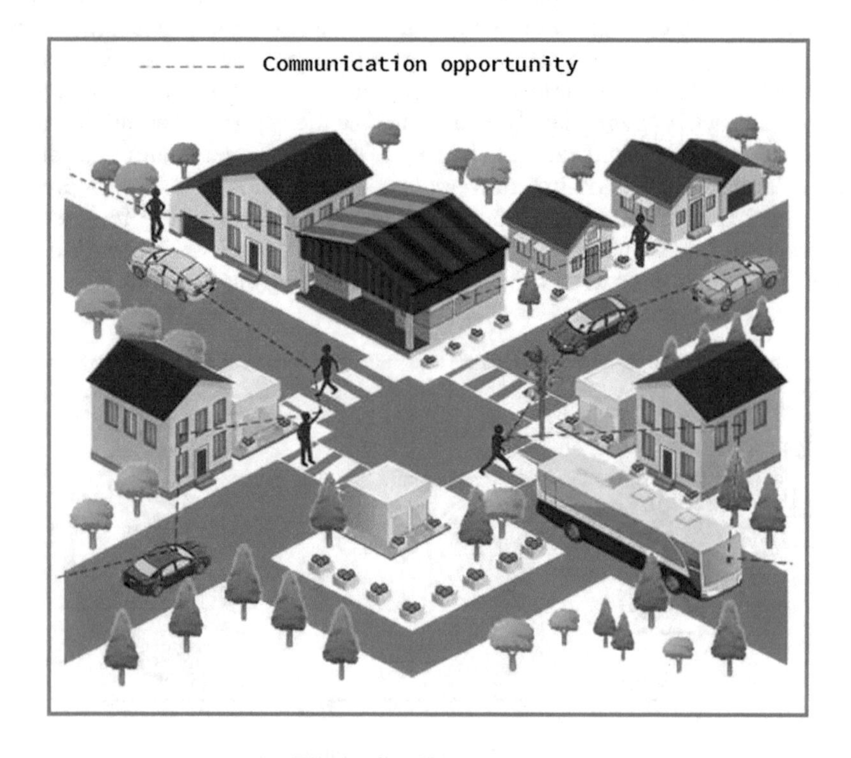

Fig. 2. OppNets scenario.

access point where it can upload the information. Similarly, a person might collect information from home-based weather stations and relay it through several other people, cars and buses until it reaches its intended destination [12].

OppNets are not limited to only such applications, as they can introduce further connectivity and benefits to IoT scenarios. In an OppNet, due to node mobility network partitions occur. These events result in intermittent connectivity. When there is no path existing between the source and the destination, the network partition occurs. Therefore, nodes need to communicate with each other via opportunistic contacts through store-carry-forward operation. There are two specific challenges in an OppNet: the contact opportunity and the node storage.

- *Contact Opportunity:* Due to the node mobility or the dynamics of wireless channel, a node can make contact with other nodes at an unpredicted time. Since contacts between nodes are hardly predictable, they must be exploited opportunistically for exchanging messages between some nodes that can move between remote fragments of the network. Mobility increases the chances of communication between nodes. When nodes move randomly around the network, where jamming signals are disrupting the communication, they may pass through unjammed area and hence be able to communicate. In addition, the contact capacity needs to be considered [13,14].
- *Node Storage:* As described above, to avoid dropping packets, the intermediate nodes are required to have enough storage to store all messages for an unpredictable period of time until next contact occurs. In other words, the required storage space increases as a function of the number of messages in the network. Therefore, the routing and replication strategies must take the storage constraint into consideration [15].

3 Proposed Fuzzy-Based System

3.1 System Parameters

We consider the following parameters for implementation of our proposed system.

IoT Device's Number of Past Encounters (IDNPE): Mobility of the IoT devices creates uncertainty about their location. IoT device's history of past encounters with different devices plays a significant role for making a decision on IoT Device selection. This is because if an IoT device has encountered other devices in the past, then it is more likely to meet them again in the future. Past encounters are probably a good estimate to determine the probability of a future encounter.

IoT Device Storage (IDST): In delay tolerant networks data is carried by the IoT device until a communication opportunity is available. Considering that different IoT devices have different storage capabilities, the selection decision should consider the storage capacity.

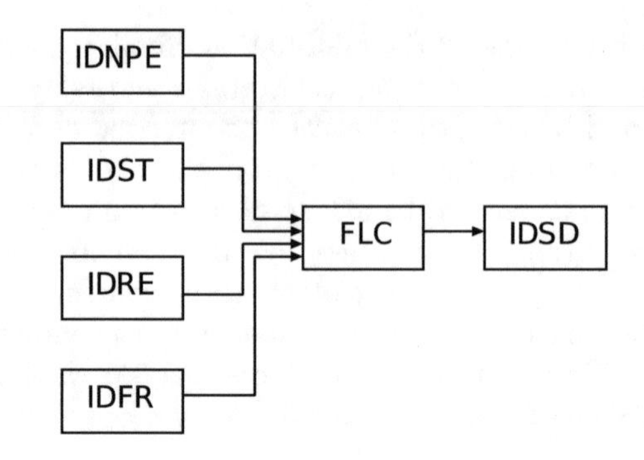

Fig. 3. Proposed system model.

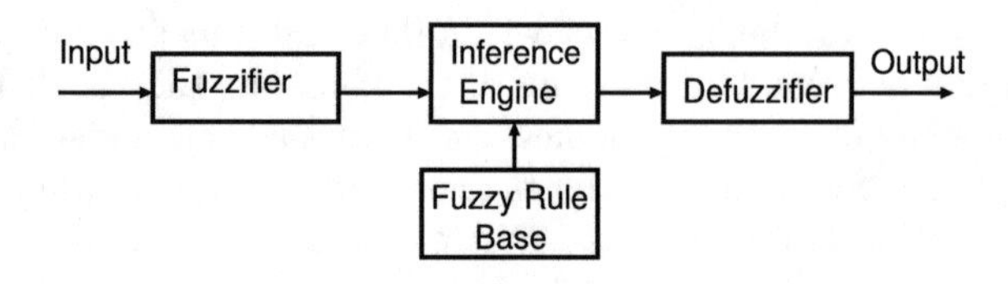

Fig. 4. FLC structure.

IoT Device Remaining Energy (IDRE): The IoT devices in opportunistic networks are active and can perform tasks and exchange data in different ways from each other. Consequently, some IoT devices may have a lot of remaining power and other may have very little, when an event occurs.

IoT Device Failure Rate (IDFR): IoT devices in OppNets are vulnerable to failures due to battery drainage, hardware defects or harsh environments. The capacity of the IoT Devices in the network is limited, a node failure leads to redistribution of the network load and load redistribution makes some nodes load over the load capacity and failure, the node failure may lead to other nodes "cascading failure"

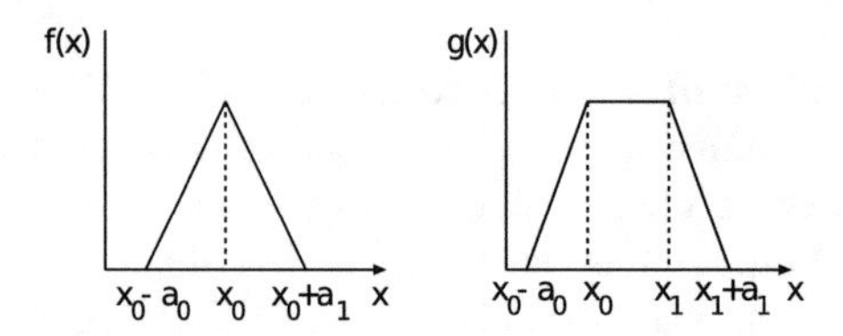

Fig. 5. Triangular and trapezoidal membership functions.

IoT Device Selection Decision (IDSD): The proposed system considers the following levels for IoT device selection:

- Very Low Selection Possibility (VLSP) - The IoT device will have very low probability to be selected.
- Low Selection Possibility (LSP) - There might be other IoT devices which can do the job better.
- Middle Selection Possibility (MSP) - The IoT device is ready to be assigned a task, but is not the "chosen" one.
- High Selection Possibility (HSP) - The IoT device takes responsibility of completing the task.
- Very High Selection Possibility (VHSP) - The IoT device has almost all the required information and potential to be selected and then allocated in an appropriate position to carry out a job.

Table 1. Parameters and their term sets for FLC.

Parameters	Term sets
IoT Device's Number of Encounters (IDNPE)	Few (Fw), Moderate (Mo), Many (Mn)
IoT Device Storage (IDST)	Small (Sm), Medium (Me), High (Hi)
IoT Device Remaining Energy (IDRE)	Low (Lo), Medium (Mdm), High (Hgh)
IoT Device Failure Rate (IDFR)	Low (Lw), Medium (Med), High (Hg)
IoT Device Selection Decision (IDSD)	Very Low Selection Possibility (VLSP), Low Selection Possibility (LSP), Medium Selection Possibility (MSP), High Selection Possibility (HSP), Very High Selection Possibility (VHSP)

3.2 System Implementation

Fuzzy sets and fuzzy logic have been developed to manage vagueness and uncertainty in a reasoning process of an intelligent system such as a knowledge based system, an expert system or a logic control system [16–29]. In this work, we use fuzzy logic to implement the proposed system.

The structure of the proposed system is shown in Fig. 3. It consists of one Fuzzy Logic Controller (FLC), which is the main part of our system and its basic elements are shown in Fig. 4. They are the fuzzifier, inference engine, Fuzzy Rule Base (FRB) and defuzzifier.

As shown in Fig. 5, we use triangular and trapezoidal membership functions for FLC, because they are suitable for real-time operation [30]. The x_0 in $f(x)$ is the center of triangular function, $x_0(x_1)$ in $g(x)$ is the left (right) edge of trapezoidal function, and $a_0(a_1)$ is the left (right) width of the triangular or trapezoidal function. We explain in details the design of FLC in following.

We use four input parameters for FLC: IoT Device's Number of Past Encounters (IDNPE), IoT Device Storage (IDST), IoT Device Remaining Energy (IDRE), Device Failure Rate (IDFR).

The term sets for each input linguistic parameter are defined respectively as shown in Table 1.

Table 2. FRB.

No.	IDNPE	IDST	IDRE	IDFR	IDSD	No.	IDNPE	IDST	IDRE	IDFR	IDSD
1	Fw	Sm	Lo	Lw	VLSP	41	Mo	Me	Mdm	Med	LSP
2	Fw	Sm	Lo	Med	LSP	42	Mo	Me	Mdm	Hg	HSP
3	Fw	Sm	Lo	Hg	HSP	43	Mo	Me	Hgh	Lw	MSP
4	Fw	Sm	Mdm	Lw	LSP	44	Mo	Me	Hgh	Med	MSP
5	Fw	Sm	Mdm	Med	MSP	45	Mo	Me	Hgh	Hg	VHSP
6	Fw	Sm	Mdm	Hg	VHSP	46	Mo	Hi	Lo	Lw	VLSP
7	Fw	Sm	Hgh	Lw	MSP	47	Mo	Hi	Lo	Med	VLSP
8	Fw	Sm	Hgh	Med	HSP	48	Mo	Hi	Lo	Hg	LSP
9	Fw	Sm	Hgh	Hg	VHSP	49	Mo	Hi	Mdm	Lw	VLSP
10	Fw	Me	Lo	Lw	VLSP	50	Mo	Hi	Mdm	Med	LSP
11	Fw	Me	Lo	Med	VLSP	51	Mo	Hi	Mdm	Hg	MSP
12	Fw	Me	Lo	Hg	LSP	52	Mo	Hi	Hgh	Lw	LSP
13	Fw	Me	Mdm	Lw	VLSP	53	Mo	Hi	Hgh	Med	MSP
14	Fw	Me	Mdm	Med	VLSP	54	Mo	Hi	Hgh	Hg	HSP
15	Fw	Me	Mdm	Hg	MSP	55	Mn	Sm	Lo	Lw	HSP
16	Fw	Me	Hgh	Lw	LSP	56	Mn	Sm	Lo	Med	HSP
17	Fw	Me	Hgh	Med	LSP	57	Mn	Sm	Lo	Hg	VHSP
18	Fw	Me	Hgh	Hg	HSP	58	Mn	Sm	Mdm	Lw	HSP
19	Fw	Hi	Lo	Lw	VLSP	59	Mn	Sm	Mdm	Med	VHSP
20	Fw	Hi	Lo	Med	VLSP	60	Mn	Sm	Mdm	Hg	VHSP
21	Fw	Hi	Lo	Hg	VLSP	61	Mn	Sm	Hgh	Lw	VHSP
22	Fw	Hi	Mdm	Lw	VLSP	62	Mn	Sm	Hgh	Med	VHSP
23	Fw	Hi	Mdm	Med	VLSP	63	Mn	Sm	Hgh	Hg	VHSP
24	Fw	Hi	Mdm	Hg	LSP	64	Mn	Me	Lo	Lw	LSP
25	Fw	Hi	Hgh	Lw	VLSP	65	Mn	Me	Lo	Med	MSP
26	Fw	Hi	Hgh	Med	LSP	66	Mn	Me	Lo	Hg	HSP
27	Fw	Hi	Hgh	Hg	MSP	67	Mn	Me	Mdm	Lw	MSP
28	Mo	Sm	Lo	Lw	LSP	68	Mn	Me	Mdm	Med	LSP
29	Mo	Sm	Lo	Med	MSP	69	Mn	Me	Mdm	Hg	HSP
30	Mo	Sm	Lo	Hg	VHSP	70	Mn	Me	Hgh	Lw	VHSP
31	Mo	Sm	Mdm	Lw	MSP	71	Mn	Me	Hgh	Med	VHSP
32	Mo	Sm	Mdm	Med	HSP	72	Mn	Me	Hgh	Hg	VHSP
33	Mo	Sm	Mdm	Hg	VHSP	73	Mn	Hi	Lo	Lw	VLSP
34	Mo	Sm	Hgh	Lw	HSP	74	Mn	Hi	Lo	Me d	LSP
35	Mo	Sm	Hgh	Med	VHSP	75	Mn	Hi	Lo	Hg	MSP
36	Mo	Sm	Hgh	Hg	VHSP	76	Mn	Hi	Mdm	Lw	LSP
37	Mo	Me	Lo	Lw	VLSP	77	Mn	Hi	Mdm	Med	MSP
38	Mo	Me	Lo	Med	VLSP	78	Mn	Hi	Mdm	Hg	HSP
39	Mo	Me	Lo	Hg	MSP	79	Mn	Hi	Hgh	Lw	MSP
40	Mo	Me	Mdm	Lw	LSP	80	Mn	Hi	Hgh	Med	HSP
						81	Mn	Hi	Hgh	Hg	VHSP

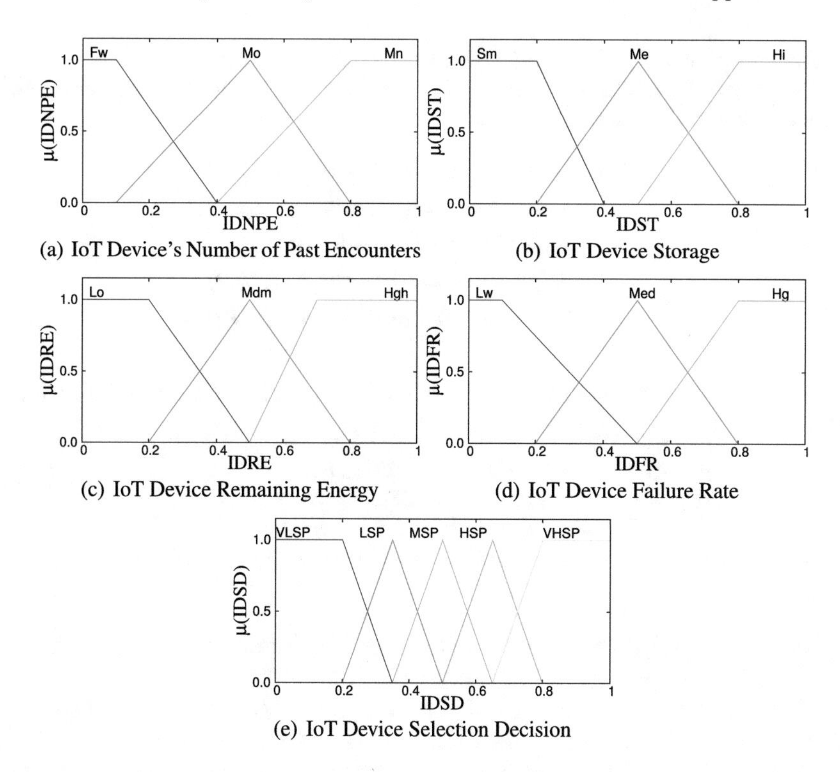

(a) IoT Device's Number of Past Encounters

(b) IoT Device Storage

(c) IoT Device Remaining Energy

(d) IoT Device Failure Rate

(e) IoT Device Selection Decision

Fig. 6. Fuzzy membership functions.

$$
\begin{aligned}
T(IDNPE) &= \{Few(Fw), Moderate(Mo), Many(Mn)\}\\
T(IDST) &= \{Small(Sm), Medium(Me), High(Hi)\}\\
T(IDRE) &= \{Low(Lo), Medium(Mdm), High(Hgh)\}\\
T(IDFR) &= \{Low(Lw), Medium(Med), High(Hg)\}
\end{aligned}
$$

The membership functions for input parameters of FLC are defined as:

$$
\begin{aligned}
\mu_{Fw}(IDNPE) &= g(IDNPE; Fw_0, Fw_1, Fw_{w0}, Fw_{w1})\\
\mu_{Mo}(IDNPE) &= f(IDNPE; Mo_0, Mo_{w0}, Mo_{w1})\\
\mu_{Mn}(IDNPE) &= g(IDNPE; Mn_0, Mn_1, Mn_{w0}, Mn_{w1})\\
\mu_{Sm}(IDST) &= g(IDFR; Sm_0, Sm_1, Sm_{w0}, Sm_{w1})\\
\mu_{Me}(IDST) &= f(IDFR; Me_0, Me_{w0}, Me_{w1})\\
\mu_{Hi}(IDST) &= g(IDFR; Hi_0, Hi_1, Hi_{w0}, Hi_{w1})\\
\mu_{Lo}(IDRE) &= g(IDRE; Lo_0, Lo_1, Lo_{w0}, Lo_{w1})\\
\mu_{Mdm}(IDRE) &= f(IDRE; Mdm_0, Mdm_{w0}, Mdm_{w1})\\
\mu_{Hgh}(IDRE) &= g(IDRE; Hgh_0, Hgh_1, Hgh_{w0}, Hgh_{w1})\\
\mu_{Lw}(IDFR) &= g(IDFR; Lw_0, Lw_1, Lw_{w0}, Lw_{w1})\\
\mu_{Med}(IDFR) &= f(IDFR; Med_0, Med_{w0}, Med_{w1})\\
\mu_{Hg}(IDFR) &= g(IDFR; Hg_0, Hg_1, Hg_{w0}, Hg_{w1})
\end{aligned}
$$

The small letters *w0* and *w1* mean left width and right width, respectively.

The output linguistic parameter is the IoT device Selection Decision (IDSD). We define the term set of IDSD as:

$$\{Very\ Low\ Selection\ Possibility\ (VLSP),\ Low\ Selection\ Possibility\ (LSP),\ Middle\ Selection\ Possibility\ (MSP),\ High\ Selection\ Possibility\ (HSP),\ Very\ High\ Selection\ Possibility\ (VHSP)\}.$$

The membership functions for the output parameter *IDSD* are defined as:

$$\begin{aligned}
\mu_{VLSP}(IDSD) &= g(IDSD; VLSP_0, VLSP_1, VLSP_{w0}, VLSP_{w1})\\
\mu_{LSP}(IDSD) &= f(IDSD; LSP_0, LSP_{w0}, LSP_{w1})\\
\mu_{MSP}(IDSD) &= f(IDSD; MSP_0, MSP_{w0}, MSP_{w1})\\
\mu_{HSP}(IDSD) &= f(IDSD; HSP_0, HSP_{w0}, HSP_{w1})\\
\mu_{VHSP}(IDSD) &= g(IDSD; VHSP_0, VHSP_1, VHSP_{w0}, VHSP_{w1}).
\end{aligned}$$

The membership functions are shown in Fig. 6 and the Fuzzy Rule Base (FRB) for our system are shown in Table 2.

The FRB forms a fuzzy set of dimensions $|T(IDNPE)| \times |T(IDST)| \times |T(IDRE)| \times |T(IDFR)|$, where $|T(x)|$ is the number of terms on $T(x)$. We have four input paramteres, so our system has 81 rules. The control rules have the form: IF "conditions" THEN "control action".

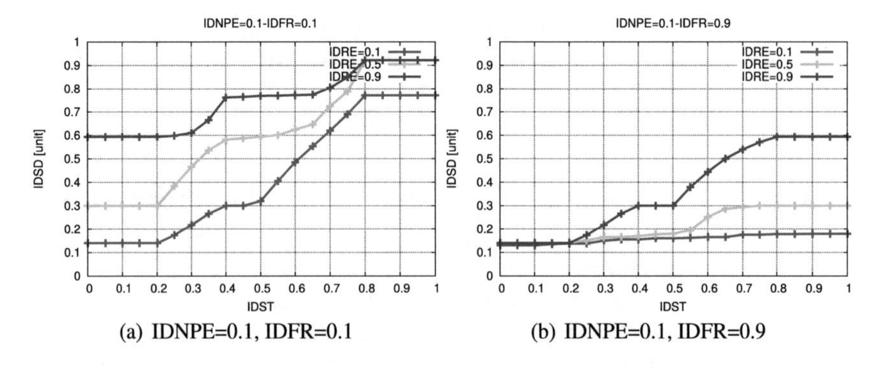

(a) IDNPE=0.1, IDFR=0.1

(b) IDNPE=0.1, IDFR=0.9

Fig. 7. Results for different values of $IDST$, $IDRE$ and $IDFR$ when $IDNPE = 0.1$.

4 Simulation Results

We present the simulation results in Figs. 7, 8 and 9. In these figures, we show the relation between the probability of an IoT device to be selected (IDSD) to

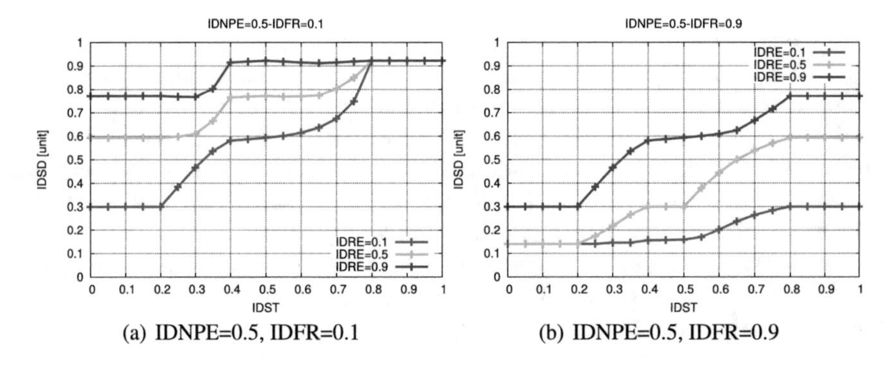

(a) IDNPE=0.5, IDFR=0.1 (b) IDNPE=0.5, IDFR=0.9

Fig. 8. Results for different values of $IDST$, $IDRE$ and $IDFR$ when $IDNPE = 0.5$.

carry out a task, versus IDNPE, IDST, IDRE and IDFR. We consider IDNPE and IDFR constant and change the values of IDRE and IDST. We see that IoT devices with more remaining energy, have a higher possibility to be selected for carrying out a job. In Fig. 7(a), when IDRE is 0.1 and IDST is 0.7, the IDSD is 0.62. For IDRE 0.5, the IDSD is 0.73 and for IDRE 0.9, IDSD is 0.8, thus the IDSD is increased about 13% and 18%, for IDRE 0.5 and IDRE 0.9, respectively.

In Figs. 7(a) and (b), we increase the IDFR value to 0.1 and 0.9, respectively and keep IDNPE constant. From the figures we can see that for IDST 0.7 and IDRE 0.9, the IDSD is decreased 26%. We see that we have a significant decrease because the IoT device with a higher failure rate is less likely to be selected since it has a higher possibility to fail thus disconnecting the network.

Another important parameter that affects the increase of IDSD is IDST as shown in Fig. 7(b), because devices with more storage capacity are more likely to carry the message until there is a contact opportunity. This parameter directly influences the capacity of OppNets because lower storage limits the amount of data that can be transferred between nodes.

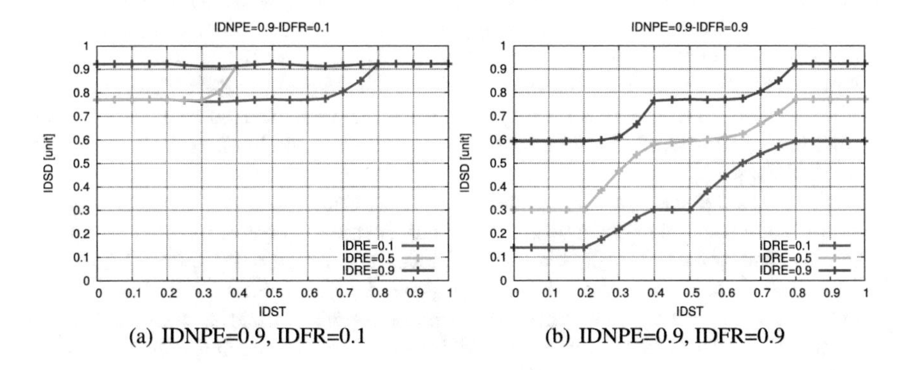

(a) IDNPE=0.9, IDFR=0.1 (b) IDNPE=0.9, IDFR=0.9

Fig. 9. Results for different values of $IDST$, $IDRE$ and $IDFR$ when $IDNPE = 0.9$.

To see the effect that number of past encounters has on IDSD, we compare Figs. 7(a), 8(a) and 9(a) for IDST 0.7, where IDNPE increases from 0.1 to 0.5 and 0.9, respectively. For IDRE 0.1, comparing Fig. 7(a) with Figs. 8(a) and 8(a) with Fig. 9(a), we see that the IDSD is increased 8% and 12% respectively. When it comes to IDNPE, it is preferred that an IoT device has previous encounters with another device, as this increases the chances of predicting a contact opportunity, and a future connection. The larger the number of past encounters, the highest is the possibility of IoT device selection. This because a more efficient management of resources is made by already knowing the presence of another IoT device waiting to make a connection.

5 Conclusions and Future Work

In this paper, we proposed and implemented a fuzzy-based IoT device selection system for OppNets considering the failure rate of an IoT device.

We evaluated the proposed system by computer simulations. The simulation results show that the IoT devices with a lower failure rate, are more likely to be selected for carrying out a job, so with the decrease of IDFR the possibility of an IoT device to be selected increases. We can see that by increasing IDNPE, IDST and IDRE, the IDSD is also increased.

In the future work, we will also consider other parameters for IoT device selection such as Node Computational Time, Interaction Probability and make extensive simulations to evaluate the proposed system.

References

1. Mantas, N., Louta, M., Karapistoli, E., Karetsos, G.T., Kraounakis, S., Obaidat, M.S.: Towards an incentive-compatible, reputation-based framework for stimulating cooperation in opportunistic networks: a survey. IET Networks **6**(6), 169–178 (2017)
2. Sharma, D.K., Sharma, A., Kumar, J., et al.: KNNR: K-nearest neighbour classification based routing protocol for opportunistic networks. In: 10-th International Conference on Contemporary Computing (IC3), pp. 1–6. IEEE (2017)
3. Kraijak, S., Tuwanut, P.: A survey on internet of things architecture, protocols, possible applications, security, privacy, real-world implementation and future trends. In: 16th International Conference on Communication Technology (ICCT), pp. 26–31. IEEE (2015)
4. Arridha, R., Sukaridhoto, S., Pramadihanto, D., Funabiki, N.: Classification extension based on iot-big data analytic for smart environment monitoring and analytic in real-time system. Int. J. Space-Based Situated Comput. **7**(2), 82–93 (2017)
5. Dhurandher, S.K., Sharma, D.K., Woungang, I., Bhati, S.: HBPR: history based prediction for routing in infrastructure-less opportunistic networks. In: 27th International Conference on Advanced Information Networking and Applications (AINA), pp. 931–936. IEEE (2013)
6. Spaho, E., Mino, G., Barolli, L., Xhafa, F.: Goodput and PDR analysis of AODV, OLSR and DYMO protocols for vehicular networks using cavenet. Int. J. Grid Utility Comput. **2**(2), 130–138 (2011)

7. Abdulla, M., Simon, R.: The impact of intercontact time within opportunistic networks: protocol implications and mobility models. TechRepublic White Paper (2009)
8. Patra, T.K., Sunny, A.: Forwarding in heterogeneous mobile opportunistic networks. IEEE Commun. Lett. **22**(3), 626–629 (2018)
9. Le, T., Gerla, M.: Contact duration-aware routing in delay tolerant networks. In: International Conference on Networking, Architecture and Storage (NAS), pp. 1–8. IEEE (2017)
10. Popereshnyak, S., Suprun, O., Suprun, O., Wieckowski, T.: IoI application testing features based on the modelling network. In: The 14-th International Conference on Perspective Technologies and Methods in MEMS Design (MEMSTECH), pp. 127–131. IEEE (2018)
11. Chen, N., Yang, Y., Li, J., Zhang, T.: A fog-based service enablement architecture for cross-domain IoT applications. In: Fog World Congress (FWC), 2017 IEEE, pp. 1–6. IEEE (2017)
12. Pozza, R., Nati, M., Georgoulas, S., Moessner, K., Gluhak, A.: Neighbor discovery for opportunistic networking in internet of things scenarios: a survey. IEEE Access **3**, 1101–1131 (2015)
13. Akbas, M., Turgut, D.: Apawsan: actor positioning for aerial wireless sensor and actor networks. In: IEEE 36th Conference on Local Computer Networks (LCN-2011), October 2011, pp. 563–570 (2011)
14. Akbas, M., Brust, M., Turgut, D.: Local positioning for environmental monitoring in wireless sensor and actor networks. In: IEEE 35th Conference on Local Computer Networks (LCN-2010), October 2010, pp. 806–813 (2010)
15. Melodia, T., Pompili, D., Gungor, V., Akyildiz, I.: Communication and Coordination in Wireless Sensor and Actor Networks. IEEE Trans. Mob. Comput. **6**(10), 1126–1129 (2007)
16. Inaba, T., Sakamoto, S., Kolici, V., Mino, G., Barolli, L.: A CAC scheme based on fuzzy logic for cellular networks considering security and priority parameters. In: The 9-th International Conference on Broadband and Wireless Computing, Communication and Applications (BWCCA-2014), pp. 340–346 (2014)
17. Spaho, E., Sakamoto, S., Barolli, L., Xhafa, F., Barolli, V., Iwashige, J.: A fuzzy-based system for peer reliability in JXTA-overlay P2P considering number of interactions. In: The 16th International Conference on Network-Based Information Systems (NBiS-2013), pp. 156–161 (2013)
18. Matsuo, K., Elmazi, D., Liu, Y., Sakamoto, S., Mino, G., Barolli, L.: FACS-MP: a fuzzy admission control system with many priorities for wireless cellular networks and its performance evaluation. J. High Speed Netwo. **21**(1), 1–14 (2015)
19. Grabisch, M.: The application of fuzzy integrals in multicriteria decision making. Eur. J. Oper. Res. **89**(3), 445–456 (1996)
20. Inaba, T., Elmazi, D., Liu, Y., Sakamoto, S., Barolli, L., Uchida, K.: Integrating wireless cellular and ad-hoc networks using fuzzy logic considering node mobility and security. In: The 29th IEEE International Conference on Advanced Information Networking and Applications Workshops (WAINA-2015), pp. 54–60 (2015)
21. Kulla, E., Mino, G., Sakamoto, S., Ikeda, M., Caballé, S., Barolli, L.: FBMIS: a fuzzy-based multi-interface system for cellular and ad hoc networks. In: International Conference on Advanced Information Networking and Applications (AINA-2014), pp. 180–185 (2014)
22. Elmazi, D., Kulla, E., Oda, T., Spaho, E., Sakamoto, S., Barolli, L.: A comparison study of two fuzzy-based systems for selection of actor node in wireless sensor actor networks. J. Ambient Intell. Hum. Comput. **6**(5), 635–645 (2015)

23. Zadeh, L.: Fuzzy logic, neural networks, and soft computing. ACM Commun. **37**, 77–84 (1994)
24. Spaho, E., Sakamoto, S., Barolli, L., Xhafa, F., Ikeda, M.: Trustworthiness in P2P: performance behaviour of two fuzzy-based systems for JXTA-overlay platform. Soft Comput. **18**(9), 1783–1793 (2014)
25. Inaba, T., Sakamoto, S., Kulla, E., Caballe, S., Ikeda, M., Barolli, L.: An integrated system for wireless cellular and ad-hoc networks using fuzzy logic. In: International Conference on Intelligent Networking and Collaborative Systems (INCoS-2014), pp. 157–162 (2014)
26. Matsuo, K., Elmazi, D., Liu, Y., Sakamoto, S., Barolli, L.: A multi-modal simulation system for wireless sensor networks: a comparison study considering stationary and mobile sink and event. J. Ambient Intell. Hum. Comput. **6**(4), 519–529 (2015)
27. Kolici, V., Inaba, T., Lala, A., Mino, G., Sakamoto, S., Barolli, L.: A fuzzy-based CAC scheme for cellular networks considering security. In: International Conference on Network-Based Information Systems (NBiS-2014), pp. 368–373 (2014)
28. Liu, Y., Sakamoto, S., Matsuo, K., Ikeda, M., Barolli, L., Xhafa, F.: A comparison study for two fuzzy-based systems: improving reliability and security of JXTA-overlay P2P platform. Soft Comput. **20**(7), 2677–2687 (2015)
29. Matsuo, K., Elmazi, D., Liu, Y., Sakamoto, S., Mino, G., Barolli, L.: FACS-MP: a fuzzy admission control system with many priorities for wireless cellular networks and its perforemance evaluation. J. High Speed Netw. **21**(1), 1–14 (2015)
30. Mendel, J.M.: Fuzzy logic systems for engineering: a tutorial. Proc. IEEE **83**(3), 345–377 (1995)

A Comparison Study for Chi-Square and Uniform Client Distributions by WMN-PSOSA Simulation System for WMNs

Shinji Sakamoto[1(✉)], Leonard Barolli[2], and Shusuke Okamoto[1]

[1] Department of Computer and Information Science, Seikei University, 3-3-1 Kichijoji-Kitamachi, Musashino-shi, Tokyo 180-8633, Japan
shinji.sakamoto@ieee.org, okam@st.seikei.ac.jp

[2] Department of Information and Communication Engineering, Fukuoka Institute of Technology, 3-30-1 Wajiro-Higashi, Higashi-Ku, Fukuoka 811-0295, Japan
barolli@fit.ac.jp

Abstract. Wireless Mesh Networks (WMNs) have many advantages such as low cost and increased high-speed wireless Internet connectivity, therefore WMNs are becoming an important networking infrastructure. In our previous work, we implemented a Particle Swarm Optimization (PSO) based simulation system for node placement in WMNs, called WMN-PSO. Also, we implemented a simulation system based on Simulated Annealing (SA) for solving node placement problem in WMNs, called WMN-SA. Then, we implemented a hybrid simulation system based on PSO and SA, called WMN-PSOSA. In this paper, we analyse the performance of WMNs by using WMN-PSOSA considering two types of mesh clients distributions. Simulation results show that a good performance is achieved for Chi-square distribution compared with the case of Uniform distribution.

1 Introduction

The wireless networks and devices are becoming increasingly popular and they provide users access to information and communication anytime and anywhere [3,8,10–12,15,21,27,28,30,34]. Wireless Mesh Networks (WMNs) are gaining a lot of attention because of their low cost nature that makes them attractive for providing wireless Internet connectivity. A WMN is dynamically self-organized and self-configured, with the nodes in the network automatically establishing and maintaining mesh connectivity among them-selves (creating, in effect, an ad hoc network). This feature brings many advantages to WMNs such as low up-front cost, easy network maintenance, robustness and reliable service coverage [1]. Moreover, such infrastructure can be used to deploy community networks, metropolitan area networks, municipal and corporative networks, and to support applications for urban areas, medical, transport and surveillance systems.

L. Barolli et al. (Eds.): EIDWT 2019, LNDECT 29, pp. 53–65, 2019.
https://doi.org/10.1007/978-3-030-12839-5_5

Mesh node placement in WMN can be seen as a family of problems, which are shown (through graph theoretic approaches or placement problems, e.g. [6,16]) to be computationally hard to solve for most of the formulations [38]. In fact, the node placement problem considered here is even more challenging due to two additional characteristics:

(a) locations of mesh router nodes are not pre-determined, in other wards, any available position in the considered area can be used for deploying the mesh routers.
(b) routers are assumed to have their own radio coverage area.

Here, we consider the version of the mesh router nodes placement problem in which we are given a grid area where to deploy a number of mesh router nodes and a number of mesh client nodes of fixed positions (of an arbitrary distribution) in the grid area. The objective is to find a location assignment for the mesh routers to the cells of the grid area that maximizes the network connectivity and client coverage. Node placement problems are known to be computationally hard to solve [13,14,39]. In some previous works, intelligent algorithms have been recently investigated [4,7,17,19,22–24,32,33].

In [25], we implemented a Particle Swarm Optimization (PSO) based simulation system, called WMN-PSO. Also, we implemented a simulation system based on Simulated Annealing (SA) for solving node placement problem in WMNs, called WMN-SA [20,21].

In our previous work, we implemented a hybrid simulation system based on PSO and SA. We call this system WMN-PSOSA. In this paper, we analyse the performance of hybrid WMN-PSOSA system considering Chi-square and Uniform clients distributions.

The rest of the paper is organized as follows. The mesh router nodes placement problem is defined in Sect. 2. We present our designed and implemented hybrid simulation system in Sect. 3. The simulation results are given in Sect. 4. Finally, we give conclusions and future work in Sect. 5.

2 Node Placement Problem in WMNs

For this problem, we have a grid area arranged in cells we want to find where to distribute a number of mesh router nodes and a number of mesh client nodes of fixed positions (of an arbitrary distribution) in the considered area. The objective is to find a location assignment for the mesh routers to the area that maximizes the network connectivity and client coverage. Network connectivity is measured by Size of Giant Component (SGC) of the resulting WMN graph, while the user coverage is simply the number of mesh client nodes that fall within the radio coverage of at least one mesh router node and is measured by Number of Covered Mesh Clients (NCMC).

An instance of the problem consists as follows.

- N mesh router nodes, each having its own radio coverage, defining thus a vector of routers.

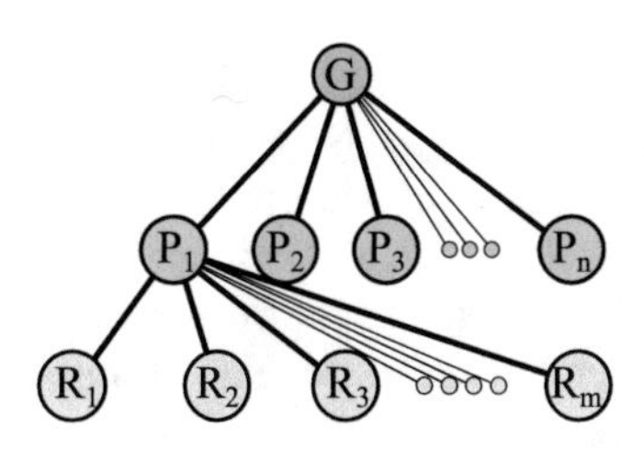

Fig. 1. Relationship among global solution, particle-patterns and mesh routers.

- An area $W \times H$ where to distribute N mesh routers. Positions of mesh routers are not pre-determined and are to be computed.
- M client mesh nodes located in arbitrary points of the considered area, defining a matrix of clients.

It should be noted that network connectivity and user coverage are among most important metrics in WMNs and directly affect the network performance.

In this work, we have considered a bi-objective optimization in which we first maximize the network connectivity of the WMN (through the maximization of the SGC) and then, the maximization of the NCMC.

In fact, we can formalize an instance of the problem by constructing an adjacency matrix of the WMN graph, whose nodes are router nodes and client nodes and whose edges are links between nodes in the mesh network. Each mesh node in the graph is a triple $\boldsymbol{v} = < x, y, r >$ representing the 2D location point and r is the radius of the transmission range. There is an arc between two nodes $\boldsymbol{u}$ and $\boldsymbol{v}$, if $\boldsymbol{v}$ is within the transmission circular area of $\boldsymbol{u}$.

3 Proposed and Implemented Simulation System

3.1 PSO Algorithm

In PSO a number of simple entities (the particles) are placed in the search space of some problem or function and each evaluates the objective function at its current location. The objective function is often minimized and the exploration of the search space is not through evolution [18]. However, following a widespread practice of borrowing from the evolutionary computation field, in this work, we consider the bi-objective function and fitness function interchangeably. Each particle then determines its movement through the search space by combining some aspect of the history of its own current and best (best-fitness) locations with those of one or more members of the swarm, with some random perturbations. The next iteration takes place after all particles have been moved. Eventually the swarm as a whole, like a flock of birds collectively foraging for food, is likely to move close to an optimum of the fitness function.

Each individual in the particle swarm is composed of three $\mathcal{D}$-dimensional vectors, where $\mathcal{D}$ is the dimensionality of the search space. These are the current position $\vec{x}_i$, the previous best position $\vec{p}_i$ and the velocity $\vec{v}_i$.

The particle swarm is more than just a collection of particles. A particle by itself has almost no power to solve any problem; progress occurs only when the particles interact. Problem solving is a population-wide phenomenon, emerging from the individual behaviors of the particles through their interactions. In any case, populations are organized according to some sort of communication structure or topology, often thought of as a social network. The topology typically consists of bidirectional edges connecting pairs of particles, so that if j is in i's neighborhood, i is also in j's. Each particle communicates with some other particles and is affected by the best point found by any member of its topological neighborhood. This is just the vector $\vec{p_i}$ for that best neighbor, which we will denote with $\vec{p}_g$. The potential kinds of population "social networks" are hugely varied, but in practice certain types have been used more frequently.

In the PSO process, the velocity of each particle is iteratively adjusted so that the particle stochastically oscillates around $\vec{p_i}$ and $\vec{p}_g$ locations.

3.2 Simulated Annealing

3.2.1 Description of Simulated Annealing

SA algorithm [9] is a generalization of the metropolis heuristic. Indeed, SA consists of a sequence of executions of metropolis with a progressive decrement of the temperature starting from a rather high temperature, where almost any move is accepted, to a low temperature, where the search resembles Hill Climbing. In fact, it can be seen as a hill-climber with an internal mechanism to escape local optima. In SA, the solution s' is accepted as the new current solution if $\delta \leq 0$ holds, where $\delta = f(s') - f(s)$. To allow escaping from a local optimum, the movements that increase the energy function are accepted with a decreasing probability $\exp(-\delta/T)$ if $\delta > 0$, where T is a parameter called the "temperature". The decreasing values of T are controlled by a *cooling schedule*, which specifies the temperature values at each stage of the algorithm, what represents an important decision for its application (a typical option is to use a proportional method, like $T_k = \gamma \cdot T_{k-1}$). SA usually gives better results in practice, but uses to be very slow. The most striking difficulty in applying SA is to choose and tune its parameters such as initial and final temperature, decrements of the temperature (cooling schedule), equilibrium and detection.

In our system, cooling schedule (γ) will be calculated as:

$$\gamma = \left(\frac{SA\ Ending\ temperature}{SA\ Starting\ temperature} \right)^{1.0/Total\ iterations}.$$

3.2.2 Acceptability Criteria

The acceptability criteria for newly generated solution is based on the definition of a threshold value (accepting threshold) as follows. We consider a succession t_k such that $t_k > t_{k+1}$, $t_k > 0$ and t_k tends to 0 as k tends to infinity. Then, for any two solutions s_i and s_j, if $fitness(s_j) - fitness(s_i) < t_k$, then accept solution s_j.

For the SA, t_k values are taken as accepting threshold but the criterion for acceptance is probabilistic:

- If $fitness(s_j) - fitness(s_i) \leq 0$ then s_j is accepted.
- If $fitness(s_j) - fitness(s_i) > 0$ then s_j is accepted with probability $\exp[(fitness(s_j) - fitness(s_i))/t_k]$ (at iteration k the algorithm generates a random number $R \in (0,1)$ and s_j is accepted if $R < \exp[(fitness(s_j) - fitness(s_i))/t_k]$).

In this case, each neighbour of a solution has a positive probability of replacing the current solution. The t_k values are chosen in way that solutions with large increase in the cost of the solutions are less likely to be accepted (but there is still a positive probability of accepting them).

3.3 WMN-PSOSA Hybrid Simulation System

3.3.1 WMN-PSOSA System Description

Here, we present the initialization, particle-pattern, fitness function.

Initialization

Our proposed system starts by generating an initial solution randomly, by *ad hoc* methods [40]. We decide the velocity of particles by a random process considering the area size. For instance, when the area size is $W \times H$, the velocity is decided randomly from $-\sqrt{W^2 + H^2}$ to $\sqrt{W^2 + H^2}$. Our system can generate many client distributions. In this paper, we consider Normal and Uniform distribution of mesh clients.

Particle-pattern

A particle is a mesh router. A fitness value of a particle-pattern is computed by combination of mesh routers and mesh clients positions. In other words, each particle-pattern is a solution as shown is Fig. 1. Therefore, the number of particle-patterns is a number of solutions.

Fitness Function

One of most important thing in PSO algorithm is to decide the determination of an appropriate objective function and its encoding. In our case, each particle-pattern has an own fitness value and compares other particle-pattern's fitness value in order to share information of global solution. The fitness function follows a hierarchical approach in which the main objective is to maximize the SGC in WMN. Thus, the fitness function of this scenario is defined as

$$\text{Fitness} = \alpha \times \text{SGC}(\boldsymbol{x}_{ij}, \boldsymbol{y}_{ij}) + \beta \times \text{NCMC}(\boldsymbol{x}_{ij}, \boldsymbol{y}_{ij}).$$

3.3.2 WMN-PSOSA Web GUI Tool and Pseudo Code

The Web application follows a standard Client-Server architecture and is implemented using LAMP (Linux + Apache + MySQL + PHP) technology (see Fig. 2). Remote users (clients) submit their requests by completing first the parameter setting. The parameter values to be provided by the user are classified into three groups, as follows.

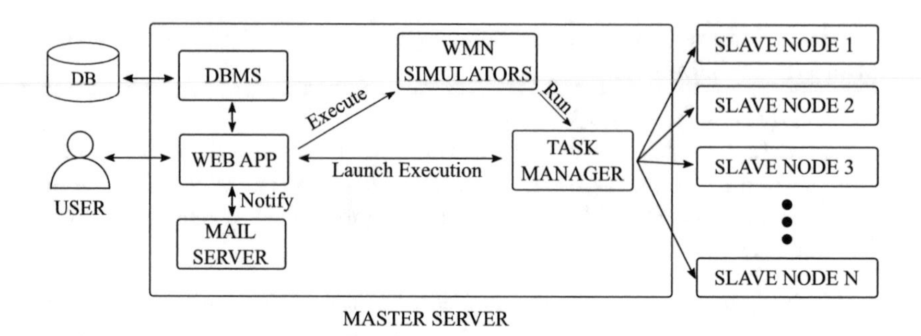

Fig. 2. System structure for web interface.

- Parameters related to the problem instance: These include parameter values that determine a problem instance to be solved and consist of number of router nodes, number of mesh client nodes, client mesh distribution, radio coverage interval and size of the deployment area.
- Parameters of the resolution method: Each method has its own parameters.
- Execution parameters: These parameters are used for stopping condition of the resolution methods and include number of iterations and number of independent runs. The former is provided as a total number of iterations and depending on the method is also divided per phase (e.g., number of iterations in a exploration). The later is used to run the same configuration for the same problem instance and parameter configuration a certain number of times.

We show the WMN-PSOSA Web GUI tool in Fig. 3. The pseudo code of our implemented system is shown in Algorithm 1.

3.3.3 WMN Mesh Routers Replacement Methods

A mesh router has x, y positions and velocity. Mesh routers are moved based on velocities. There are many moving methods in PSO field, such as:

Constriction Method (CM)
: CM is a method which PSO parameters are set to a week stable region ($\omega = 0.729$, $C_1 = C2 = 1.4955$) based on analysis of PSO by Clerc et al. [2,5,36].

Random Inertia Weight Method (RIWM)
: In RIWM, the ω parameter is changing randomly from 0.5 to 1.0. The C_1 and C_2 are kept 2.0. The ω can be estimated by the week stable region. The average of ω is 0.75 [29,36].

Linearly Decreasing Inertia Weight Method (LDIWM)
: In LDIWM, C_1 and C_2 are set to 2.0, constantly. On the other hand, the ω parameter is changed linearly from unstable region ($\omega = 0.9$) to stable region ($\omega = 0.4$) with increasing of iterations of computations [36,37].

Algorithm 1. Pseudo code of PSOSA.

/* Generate the initial solutions and parameters */
Computation maxtime:= T_{max}, $t := 0$;
Number of particle-patterns:= m, $2 \leq m \in \boldsymbol{N}^1$;
Starting SA temperature:= $Temp$;
Decreasing speed of SA temperature:= T_d;
Particle-patterns initial solution:= $\boldsymbol{P}_i^0$;
Global initial solution:= $\boldsymbol{G}^0$;
Particle-patterns initial position:= $\boldsymbol{x}_{ij}^0$;
Particles initial velocity:= $\boldsymbol{v}_{ij}^0$;
PSO parameter:= ω, $0 < \omega \in \boldsymbol{R}^1$;
PSO parameter:= C_1, $0 < C_1 \in \boldsymbol{R}^1$;
PSO parameter:= C_2, $0 < C_2 \in \boldsymbol{R}^1$;
/* Start PSO-SA */
Evaluate($\boldsymbol{G}^0, \boldsymbol{P}^0$);
while $t < T_{max}$ **do**
  /* Update velocities and positions */
  $\boldsymbol{v}_{ij}^{t+1} = \omega \cdot \boldsymbol{v}_{ij}^t$
    $+C_1 \cdot \text{rand}() \cdot (best(P_{ij}^t) - x_{ij}^t)$
    $+C_2 \cdot \text{rand}() \cdot (best(G^t) - x_{ij}^t)$;
  $\boldsymbol{x}_{ij}^{t+1} = \boldsymbol{x}_{ij}^t + \boldsymbol{v}_{ij}^{t+1}$;
  /* if fitness value is increased, a new solution will be accepted. */
  if Evaluate($\boldsymbol{G}^{(t+1)}, \boldsymbol{P}^{(t+1)}$) >= Evaluate($\boldsymbol{G}^{(t)}, \boldsymbol{P}^{(t)}$) **then**
    Update_Solutions($\boldsymbol{G}^t, \boldsymbol{P}^t$);
    Evaluate($\boldsymbol{G}^{(t+1)}, \boldsymbol{P}^{(t+1)}$);
  else
    /* a new solution will be accepted, if condition is true. */
    if Random() $> e^{\left(\frac{Evaluate(G^{(t+1)}, P^{(t+1)}) - Evaluate(G^{(t)}, P^{(t)})}{Temp}\right)}$ **then**
      /* "Reupdate_Solutions" makes particle back to previous position */
      Reupdate_Solutions($\boldsymbol{G}^{t+1}, \boldsymbol{P}^{t+1}$);
    end if
  end if
  $Temp = Temp \times t_d$;
  $t = t + 1$;
end while
Update_Solutions($\boldsymbol{G}^t, \boldsymbol{P}^t$);
return Best found pattern of particles as solution;

Linearly Decreasing Vmax Method (LDVM)

In LDVM, PSO parameters are set to unstable region ($\omega = 0.9$, $C_1 = C_2 = 2.0$). A value of V_{max} which is maximum velocity of particles is considered. With increasing of iteration of computations, the V_{max} is kept decreasing linearly [31,35].

Rational Decrement of Vmax Method (RDVM)

In RDVM, PSO parameters are set to unstable region ($\omega = 0.9$, $C_1 = C_2 = 2.0$). The V_{max} is kept decreasing with the increasing of iterations as

$$V_{max}(x) = \sqrt{W^2 + H^2} \times \frac{T - x}{x}.$$

Where, W and H are the width and the height of the considered area, respectively. Also, T and x are the total number of iterations and a current number of iteration, respectively [26].

3.3.4 Client Distributions

Our proposed system can generate a lot of clients distributions. In this paper, we consider Chi-square and Uniform clients distributions as shown in Fig. 4.

Simulator parameters, Particle Swarm Optimization and Simulated Annealing

Parameter	Value	
Distribution	Uniform	
Number of clients	48 (integer)(min:48 max:128)	
Number of routers	16 (integer) (min:16 max:48)	
Area size (WxH)	32 (positive real number)	32 (positive real number)
Radius (Min & Max)	2 (positive real number)	2 (positive real number)
Independent runs	1 (integer) (min:1 max:100)	
Replacement method	Constriction Method	
Starting SA Temperature value	10 (positive real number)	
Ending SA Temperature value	0.1 (positive real number)	
Number of Particle-patterns	10 (integer) (min:1 max:64)	
Max iterations	800 (integer) (min:1 max:6400)	
Iteration per Phase	4 (integer) (min:1 max:Max iterations)	
Send by mail		

Run

Fig. 3. WMN-PSOSA web GUI tool.

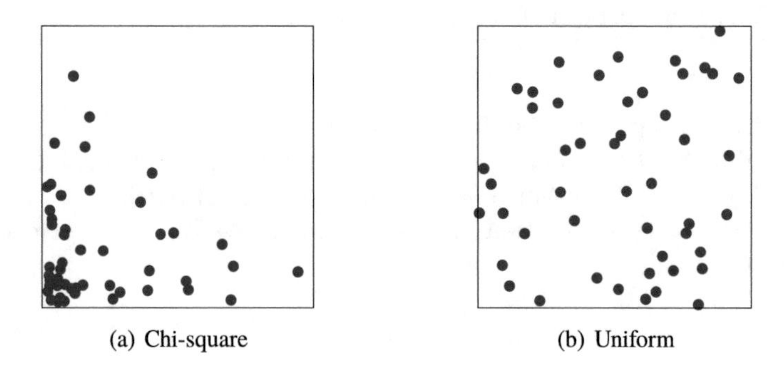

(a) Chi-square (b) Uniform

Fig. 4. Clients distributions.

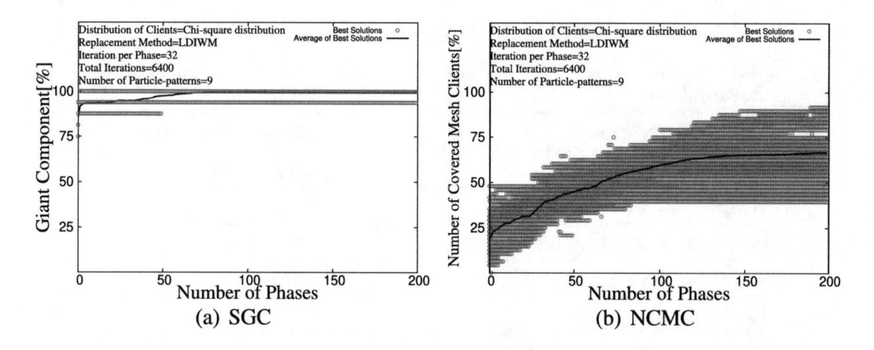

Fig. 5. Simulation results of WMN-PSOSA for Chi-square distribution.

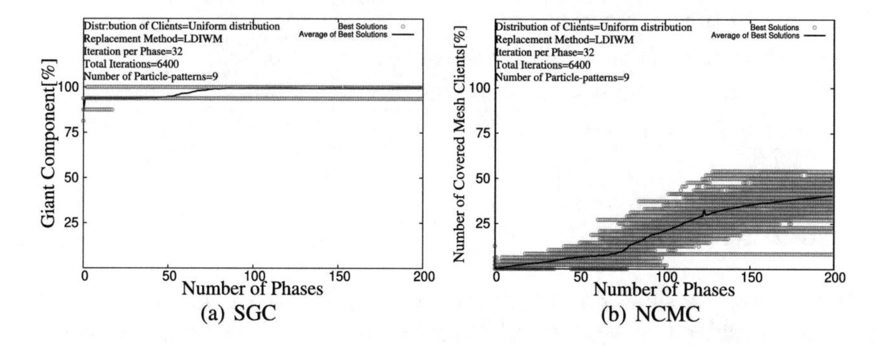

Fig. 6. Simulation results of WMN-PSOSA for Uniform distribution.

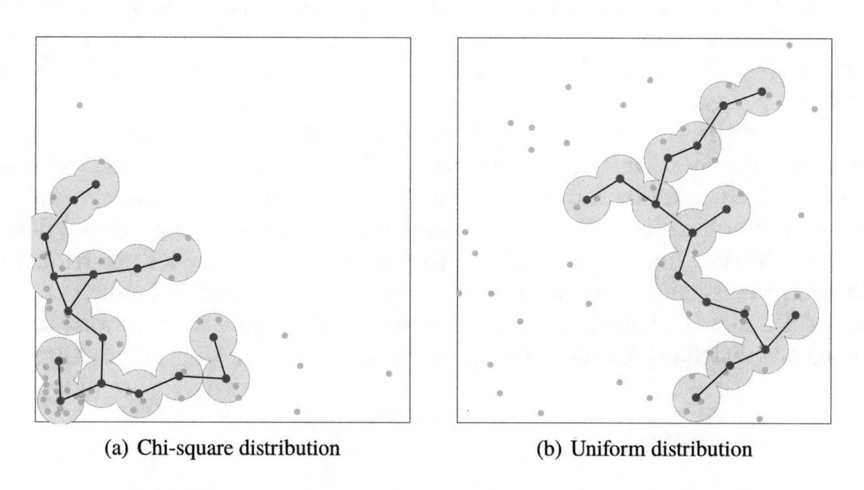

Fig. 7. Visualized image of simulation results for different clients.

Table 1. Parameter settings.

Parameters	Values
Clients distribution	Chi-square, Uniform
Area size	32.0×32.0
Number of mesh routers	16
Number of mesh clients	48
Total iterations	6400
Iteration per phase	32
Number of particle-patterns	9
Radius of a mesh router	2.0
SA starting temperature value	10.0
SA ending temperature value	0.01
Fitness function weight-coefficients (α, β)	0.7, 0.3
Temperature decreasing speed (γ)	0.998921
Replacement method	LDIWM

4 Simulation Results

In this section, we show simulation results using WMN-PSOSA system. In this work, we consider Chi-square and Uniform distributions of mesh clients. The number of mesh routers is considered 16 and the number of mesh clients 48. The total number of iterations is considered 6400 and the iterations per phase is considered 32. We consider the number of particle-patterns 9. We conducted simulations 100 times, in order to avoid the effect of randomness and create a general view of results. We show the parameter setting for WMN-PSOSA in Table 1.

We show the simulation results in Figs. 5, 6 and 7. In Figs. 5 and 6, for both client distributions, the performance for the SGC is almost the same. However, for the NCMC, the performance is better for Chi-square compared with the case of Uniform distribution. We show the visualized results for WMN-PSOSA in Fig. 7. We see that some clients nodes are not covered for both distribution, however, the number of covered mesh clients for Chi-square is better than the case of Uniform distribution. Therefore, we conclude that the performance for Chi-square distribution is better compared with the case of Uniform distribution.

5 Conclusions

In this work, we evaluated the performance of a hybrid simulation system based on PSO and SA (called WMN-PSOSA) considering Chi-square and Uniform distributions of mesh clients. Simulation results show that the performance is better for Chi-square distribution compared with the case of Uniform distribution.

In our future work, we would like to evaluate the performance of the proposed system for different parameters and scenarios.

References

1. Akyildiz, I.F., Wang, X., Wang, W.: Wireless mesh networks: a survey. Comput. Netw. **47**(4), 445–487 (2005)
2. Barolli, A., Sakamoto, S., Ozera, K., Ikeda, M., Barolli, L., Takizawa, M.: Performance evaluation of WMNs by WMN-PSOSA simulation system considering constriction and linearly decreasing Vmax methods. In: International Conference on P2P, Parallel, Grid, Cloud and Internet Computing, pp. 111–121. Springer, Cham (2017)
3. Barolli, A., Sakamoto, S., Barolli, L., Takizawa, M.: Performance analysis of simulation system based on particle swarm optimization and distributed genetic algorithm for WMNs considering different distributions of mesh clients. In: International Conference on Innovative Mobile and Internet Services in Ubiquitous Computing, pp. 32–45. Springer, Cham (2018)
4. Barolli, A., Sakamoto, S., Ozera, K., Barolli, L., Kulla, E., Takizawa, M.: Design and implementation of a hybrid intelligent system based on particle swarm optimization and distributed genetic algorithm. In: International Conference on Emerging Internetworking, Data and Web Technologies, pp. 79–93. Springer, Cham (2018)
5. Clerc, M., Kennedy, J.: The particle swarm-explosion, stability, and convergence in a multidimensional complex space. IEEE Trans. Evol. Comput. **6**(1), 58–73 (2002)
6. Franklin, A.A., Murthy, C.S.R.: Node placement algorithm for deployment of two-tier wireless mesh networks. In: Proceedings of the Global Telecommunications Conference, pp. 4823–4827 (2007)
7. Girgis, M.R., Mahmoud, T.M., Abdullatif, B.A., Rabie, A.M.: Solving the wireless mesh network design problem using genetic algorithm and simulated annealing optimization methods. Int. J. Comput. Appl. **96**(11), 1–10 (2014)
8. Goto, K., Sasaki, Y., Hara, T., Nishio, S.: Data gathering using mobile agents for reducing traffic in dense mobile wireless sensor networks. Mob. Inf. Syst. **9**(4), 295–314 (2013)
9. Hwang, C.R.: Simulated annealing: theory and applications. Acta Appl. Math. **12**(1), 108–111 (1988)
10. Inaba, T., Elmazi, D., Sakamoto, S., Oda, T., Ikeda, M., Barolli, L.: A secure-aware call admission control scheme for wireless cellular networks using fuzzy logic and its performance evaluation. J. Mob. Multimed. **11**(3–4), 213–222 (2015)
11. Inaba, T., Obukata, R., Sakamoto, S., Oda, T., Ikeda, M., Barolli, L.: Performance evaluation of a QoS-aware fuzzy-based CAC for LAN access. Int. J. Space-Based Situated Comput. **6**(4), 228–238 (2016)
12. Inaba, T., Sakamoto, S., Oda, T., Ikeda, M., Barolli, L.: A testbed for admission control in WLAN: a fuzzy approach and its performance evaluation. In: International Conference on Broadband and Wireless Computing, Communication and Applications, pp. 559–571. Springer, Cham (2016)
13. Lim, A., Rodrigues, B., Wang, F., Xu, Z.: k-Center problems with minimum coverage. In: Computing and Combinatorics, pp. 349–359 (2004)
14. Maolin, T., et al.: Gateways placement in backbone wireless mesh networks. Int. J. Commun. Netw. Syst. Sci. **2**(1), 44 (2009)

15. Matsuo, K., Sakamoto, S., Oda, T., Barolli, A., Ikeda, M., Barolli, L.: Performance analysis of WMNs by WMN-GA simulation system for two WMN architectures and different TCP congestion-avoidance algorithms and client distributions. Int. J. Commun. Netw. Distrib. Syst. **20**(3), 335–351 (2018)
16. Muthaiah, S.N., Rosenberg, C.P.: Single gateway placement in wireless mesh networks. In: Proceedings of the 8th International IEEE Symposium on Computer Networks, pp. 4754–4759 (2008)
17. Naka, S., Genji, T., Yura, T., Fukuyama, Y.: A hybrid particle swarm optimization for distribution state estimation. IEEE Trans. Power Syst. **18**(1), 60–68 (2003)
18. Poli, R., Kennedy, J., Blackwell, T.: Particle swarm optimization. Swarm Intell. **1**(1), 33–57 (2007)
19. Sakamoto, S., Kulla, E., Oda, T., Ikeda, M., Barolli, L., Xhafa, F.: A comparison study of simulated annealing and genetic algorithm for node placement problem in wireless mesh networks. J. Mob. Multimed. **9**(1–2), 101–110 (2013)
20. Sakamoto, S., Kulla, E., Oda, T., Ikeda, M., Barolli, L., Xhafa, F.: A comparison study of hill climbing, simulated annealing and genetic algorithm for node placement problem in WMNs. J. High Speed Netw. **20**(1), 55–66 (2014)
21. Sakamoto, S., Kulla, E., Oda, T., Ikeda, M., Barolli, L., Xhafa, F.: A simulation system for WMN based on SA: performance evaluation for different instances and starting temperature values. Int. J. Space-Based Situated Comput. **4**(3–4), 209–216 (2014)
22. Sakamoto, S., Kulla, E., Oda, T., Ikeda, M., Barolli, L., Xhafa, F.: Performance evaluation considering iterations per phase and SA temperature in WMN-SA system. Mob. Inf. Syst. **10**(3), 321–330 (2014)
23. Sakamoto, S., Lala, A., Oda, T., Kolici, V., Barolli, L., Xhafa, F.: Application of WMN-SA simulation system for node placement in wireless mesh networks: a case study for a realistic scenario. Int. J. Mob. Comput. Multimed. Commun. (IJMCMC) **6**(2), 13–21 (2014)
24. Sakamoto, S., Oda, T., Ikeda, M., Barolli, L., Xhafa, F.: An integrated simulation system considering WMN-PSO simulation system and network simulator 3. In: International Conference on Broadband and Wireless Computing, Communication and Applications, pp. 187–198. Springer, Cham (2016)
25. Sakamoto, S., Oda, T., Ikeda, M., Barolli, L., Xhafa, F.: Implementation and evaluation of a simulation system based on particle swarm optimisation for node placement problem in wireless mesh networks. Int. J. Commun. Netw. Distrib. Syst. **17**(1), 1–13 (2016)
26. Sakamoto, S., Oda, T., Ikeda, M., Barolli, L., Xhafa, F.: Implementation of a new replacement method in WMN-PSO simulation system and its performance evaluation. In: The 30th IEEE International Conference on Advanced Information Networking and Applications (AINA 2016), pp. 206–211 (2016). https://doi.org/10.1109/AINA.2016.42
27. Sakamoto, S., Obukata, R., Oda, T., Barolli, L., Ikeda, M., Barolli, A.: Performance analysis of two wireless mesh network architectures by WMN-SA and WMN-TS simulation systems. J. High Speed Netw. **23**(4), 311–322 (2017)
28. Sakamoto, S., Ozera, K., Barolli, A., Ikeda, M., Barolli, L., Takizawa, M.: Implementation of an intelligent hybrid simulation systems for WMNs based on particle swarm optimization and simulated annealing: performance evaluation for different replacement methods. Soft Comput. 1–7 (2017). https://doi.org/10.1007/s00500-017-2948-1

29. Sakamoto, S., Ozera, K., Barolli, A., Ikeda, M., Barolli, L., Takizawa, M.: Performance evaluation of WMNs by WMN-PSOSA simulation system considering random inertia weight method and linearly decreasing Vmax method. In: International Conference on Broadband and Wireless Computing, Communication and Applications, pp. 114–124. Springer, Cham (2017)
30. Sakamoto, S., Ozera, K., Ikeda, M., Barolli, L.: Implementation of intelligent hybrid systems for node placement problem in WMNs considering particle swarm optimization, hill climbing and simulated annealing. Mob. Netw. Appl. **23**(1), 27–33 (2018)
31. Sakamoto, S., Ozera, K., Ikeda, M., Barolli, L.: Performance evaluation of WMNs by WMN-PSOSA simulation system considering constriction and linearly decreasing inertia weight methods. In: International Conference on Network-Based Information Systems, pp. 3–13. Springer, Cham (2017)
32. Sakamoto, S., Ozera, K., Oda, T., Ikeda, M., Barolli, L.: Performance evaluation of intelligent hybrid systems for node placement in wireless mesh networks: a comparison study of WMN-PSOHC and WMN-PSOSA. In: International Conference on Innovative Mobile and Internet Services in Ubiquitous Computing, pp. 16–26. Springer, Cham (2017)
33. Sakamoto, S., Ozera, K., Oda, T., Ikeda, M., Barolli, L.: Performance evaluation of WMN-PSOHC and WMN-PSO simulation systems for node placement in wireless mesh networks: a comparison study. In: International Conference on Emerging Internetworking, Data and Web Technologies, pp. 64–74. Springer, Cham (2017)
34. Sakamoto, S., Ozera, K., Barolli, A., Barolli, L., Kolici, V., Takizawa, M.: Performance evaluation of WMN-PSOSA considering four different replacement methods. In: International Conference on Emerging Internetworking, Data and Web Technologies, pp. 51–64. Springer, Cham (2018)
35. Schutte, J.F., Groenwold, A.A.: A study of global optimization using particle swarms. J. Glob. Optim. **31**(1), 93–108 (2005)
36. Shi, Y.: Particle swarm optimization. IEEE Connect. **2**(1), 8–13 (2004)
37. Shi, Y., Eberhart, R.C.: Parameter selection in particle swarm optimization. In: Evolutionary Programming VII, pp. 591–600 (1998)
38. Vanhatupa, T., Hannikainen, M., Hamalainen, T.: Genetic algorithm to optimize node placement and configuration for WLAN planning. In: Proceedings of the 4th IEEE International Symposium on Wireless Communication Systems, pp. 612–616 (2007)
39. Wang, J., Xie, B., Cai, K., Agrawal, D.P.: Efficient mesh router placement in wireless mesh networks. In: Proceedings of the IEEE International Conference on Mobile Adhoc and Sensor Systems (MASS 2007), pp. 1–9 (2007)
40. Xhafa, F., Sanchez, C., Barolli, L.: Ad hoc and neighborhood search methods for placement of mesh routers in wireless mesh networks. In: Proceedings of the 29th IEEE International Conference on Distributed Computing Systems Workshops (ICDCS 2009), pp. 400–405 (2009)

Performance Analysis of WMNs by WMN-PSODGA Simulation System Considering Exponential and Chi-Square Client Distributions

Admir Barolli[1], Shinji Sakamoto[2(✉)], Leonard Barolli[3], and Makoto Takizawa[4]

[1] Department of Information Technology, Aleksander Moisiu University of Durres, L.1, Rruga e Currilave, Durres, Albania
admir.barolli@gmail.com

[2] Department of Computer and Information Science, Seikei University, 3-3-1 Kichijoji-Kitamachi, Musashino-shi, Tokyo 180-8633, Japan
shinji.sakamoto@ieee.org

[3] Department of Information and Communication Engineering, Fukuoka Institute of Technology, 3-30-1 Wajiro-Higashi, Higashi-Ku, Fukuoka 811-0295, Japan
barolli@fit.ac.jp

[4] Department of Advanced Sciences, Faculty of Science and Engineering, Hosei University, Kajino-Machi, Koganei-Shi, Tokyo 184-8584, Japan
makoto.takizawa@computer.org

Abstract. The Wireless Mesh Networks (WMNs) are becoming an important networking infrastructure because they have many advantages such as low cost and increased high speed wireless Internet connectivity. In our previous work, we implemented a Particle Swarm Optimization (PSO) based simulation system, called WMN-PSO, and a simulation system based on Genetic Algorithm (GA), called WMN-GA, for solving node placement problem in WMNs. Then, we implemented a hybrid simulation system based on PSO and distributed GA (DGA), called WMN-PSODGA. In this paper, we analyze the performance of WMNs using WMN-PSODGA simulation system considering Exponential and Chi-square client distributions. Simulation results show that a good performance is achieved for Chi-square distribution compared with the case of Exponential distribution.

1 Introduction

The wireless networks and devices are becoming increasingly popular and they provide users access to information and communication anytime and anywhere [3,8–11,14,20,26,27,29,33]. Wireless Mesh Networks (WMNs) are gaining a lot of attention because of their low cost nature that makes them attractive for providing wireless Internet connectivity. A WMN is dynamically self-organized and self-configured, with the nodes in the network automatically establishing

L. Barolli et al. (Eds.): EIDWT 2019, LNDECT 29, pp. 66–79, 2019.
https://doi.org/10.1007/978-3-030-12839-5_6

and maintaining mesh connectivity among them-selves (creating, in effect, an ad hoc network). This feature brings many advantages to WMNs such as low up-front cost, easy network maintenance, robustness and reliable service coverage [1]. Moreover, such infrastructure can be used to deploy community networks, metropolitan area networks, municipal and corporative networks, and to support applications for urban areas, medical, transport and surveillance systems.

Mesh node placement in WMN can be seen as a family of problems, which are shown (through graph theoretic approaches or placement problems, e.g. [6,15]) to be computationally hard to solve for most of the formulations [37]. In fact, the node placement problem considered here is even more challenging due to two additional characteristics:

(a) locations of mesh router nodes are not pre-determined, in other wards, any available position in the considered area can be used for deploying the mesh routers.
(b) routers are assumed to have their own radio coverage area.

Here, we consider the version of the mesh router nodes placement problem in which we are given a grid area where to deploy a number of mesh router nodes and a number of mesh client nodes of fixed positions (of an arbitrary distribution) in the grid area. The objective is to find a location assignment for the mesh routers to the cells of the grid area that maximizes the network connectivity and client coverage. Node placement problems are known to be computationally hard to solve [12,13,38]. In some previous works, intelligent algorithms have been recently investigated [4,7,16,18,21–23,31,32].

In [24], we implemented a Particle Swarm Optimization (PSO) based simulation system, called WMN-PSO. Also, we implemented another simulation system based on Genetic Algorithm (GA), called WMN-GA [19], for solving node placement problem in WMNs.

In our previous work, we designed and implemented a hybrid simulation system based on PSO and distributed GA (DGA). We call this system WMN-PSODGA. In this paper, we evaluate the performance of WMNs using WMN-PSODGA simulation system considering Exponential and Chi-square client distributions.

The rest of the paper is organized as follows. The mesh router nodes placement problem is defined in Sect. 2. We present our designed and implemented hybrid simulation system in Sect. 3. The simulation results are given in Sect. 4. Finally, we give conclusions and future work in Sect. 5.

2 Node Placement Problem in WMNs

For this problem, we have a grid area arranged in cells we want to find where to distribute a number of mesh router nodes and a number of mesh client nodes of fixed positions (of an arbitrary distribution) in the considered area. The objective is to find a location assignment for the mesh routers to the area that maximizes the network connectivity and client coverage. Network connectivity is measured by Size of Giant Component (SGC) of the resulting WMN graph, while the user coverage is simply the number of mesh client nodes that fall within the radio coverage of at least one mesh router node and is measured by Number of Covered Mesh Clients (NCMC).

An instance of the problem consists as follows.

- N mesh router nodes, each having its own radio coverage, defining thus a vector of routers.
- An area $W \times H$ where to distribute N mesh routers. Positions of mesh routers are not pre-determined and are to be computed.
- M client mesh nodes located in arbitrary points of the considered area, defining a matrix of clients.

It should be noted that network connectivity and user coverage are among most important metrics in WMNs and directly affect the network performance.

In this work, we have considered a bi-objective optimization in which we first maximize the network connectivity of the WMN (through the maximization of the SGC) and then, the maximization of the NCMC.

In fact, we can formalize an instance of the problem by constructing an adjacency matrix of the WMN graph, whose nodes are router nodes and client nodes and whose edges are links between nodes in the mesh network. Each mesh node in the graph is a triple $\boldsymbol{v} = <x, y, r>$ representing the 2D location point and r is the radius of the transmission range. There is an arc between two nodes $\boldsymbol{u}$ and $\boldsymbol{v}$, if $\boldsymbol{v}$ is within the transmission circular area of $\boldsymbol{u}$.

3 Proposed and Implemented Simulation System

3.1 Particle Swarm Optimization

In PSO a number of simple entities (the particles) are placed in the search space of some problem or function and each evaluates the objective function at its current location. The objective function is often minimized and the exploration of the search space is not through evolution [17]. However, following a widespread practice of borrowing from the evolutionary computation field, in this work, we consider the bi-objective function and fitness function interchangeably. Each particle then determines its movement through the search space by combining some aspect of the history of its own current and best (best-fitness) locations with those of one or more members of the swarm, with some random perturbations. The next iteration takes place after all particles have been moved. Eventually the swarm as a whole, like a flock of birds collectively foraging for food, is likely to move close to an optimum of the fitness function.

Each individual in the particle swarm is composed of three $\mathcal{D}$-dimensional vectors, where $\mathcal{D}$ is the dimensionality of the search space. These are the current position $\vec{x}_i$, the previous best position $\vec{p}_i$ and the velocity $\vec{v}_i$.

The particle swarm is more than just a collection of particles. A particle by itself has almost no power to solve any problem; progress occurs only when the particles interact. Problem solving is a population-wide phenomenon, emerging from the individual behaviors of the particles through their interactions. In any case, populations are organized according to some sort of communication structure or topology, often thought of as a social network. The topology typically consists of bidirectional edges connecting pairs of particles, so that if j is in i's neighborhood, i is also in j's. Each particle communicates with some other particles and is affected by the best point found by any member of its topological neighborhood. This is just the vector $\vec{p}_i$ for that best neighbor, which we will denote with $\vec{p}_g$. The potential kinds of population "social networks" are hugely varied, but in practice certain types have been used more frequently. We show the pseudo code of PSO in Algorithm 1.

Algorithm 1. Pseudo code of PSO.

/* Initialize all parameters for PSO */
Computation maxtime:= Tp_{max}, $t := 0$;
Number of particle-patterns:= m, $2 \leq m \in \boldsymbol{N}^1$;
Particle-patterns initial solution:= $\boldsymbol{P}_i^0$;
Particle-patterns initial position:= $\boldsymbol{x}_{ij}^0$;
Particles initial velocity:= $\boldsymbol{v}_{ij}^0$;
PSO parameter:= ω, $0 < \omega \in \boldsymbol{R}^1$;
PSO parameter:= C_1, $0 < C_1 \in \boldsymbol{R}^1$;
PSO parameter:= C_2, $0 < C_2 \in \boldsymbol{R}^1$;
/* Start PSO */
Evaluate($\boldsymbol{G}^0, \boldsymbol{P}^0$);
while $t < Tp_{max}$ **do**
  /* Update velocities and positions */
  $\boldsymbol{v}_{ij}^{t+1} = \omega \cdot \boldsymbol{v}_{ij}^t$
    $+C_1 \cdot \text{rand}() \cdot (best(P_{ij}^t) - x_{ij}^t)$
    $+C_2 \cdot \text{rand}() \cdot (best(G^t) - x_{ij}^t)$;
  $\boldsymbol{x}_{ij}^{t+1} = \boldsymbol{x}_{ij}^t + \boldsymbol{v}_{ij}^{t+1}$;
  /* if fitness value is increased, a new solution will be accepted. */
  Update_Solutions($\boldsymbol{G}^t, \boldsymbol{P}^t$);
  $t = t + 1$;
end while
Update_Solutions($\boldsymbol{G}^t, \boldsymbol{P}^t$);
return Best found pattern of particles as solution;

In the PSO process, the velocity of each particle is iteratively adjusted so that the particle stochastically oscillates around $\vec{p}_i$ and $\vec{p}_g$ locations.

3.2 Distributed Genetic Algorithm

Distributed Genetic Algorithm (DGA) has been focused from various fields of science. DGA has shown their usefulness for the resolution of many computationally hard combinatorial optimization problems. We show the pseudo code of DGA in Algorithm 2.

Algorithm 2. Pseudo code of DSA.

```
/* Initialize all parameters for DGA */
Computation maxtime:= Tg_max, t := 0;
Number of islands:= n, 1 ≤ n ∈ N^1;
initial solution:= P_i^0;
/* Start DGA */
Evaluate(G^0, P^0);
while t < Tg_max do
   for all islands do
      Selection();
      Crossover();
      Mutation();
   end for
   t = t + 1;
end while
Update_Solutions(G^t, P^t);
return Best found pattern of particles as solution;
```

Population of individuals: Unlike local search techniques that construct a path in the solution space jumping from one solution to another one through local perturbations, DGA use a population of individuals giving thus the search a larger scope and chances to find better solutions. This feature is also known as "exploration" process in difference to "exploitation" process of local search methods.

Fitness: The determination of an appropriate fitness function, together with the chromosome encoding are crucial to the performance of DGA. Ideally we would construct objective functions with "certain regularities", i.e. objective functions that verify that for any two individuals which are close in the search space, their respective values in the objective functions are similar.

Selection: The selection of individuals to be crossed is another important aspect in DGA as it impacts on the convergence of the algorithm. Several selection schemes have been proposed in the literature for selection operators trying to cope with premature convergence of DGA. There are many selection methods in GA. In our system, we implement 2 selection methods: Random method and Roulette wheel method.

Crossover operators: Use of crossover operators is one of the most important characteristics. Crossover operator is the means of DGA to transmit best genetic features of parents to offsprings during generations of the evolution process. Many methods for crossover operators have been proposed such as Blend Crossover (BLX-α), Unimodal Normal Distribution Crossover (UNDX), Simplex Crossover (SPX).

Mutation operators: These operators intend to improve the individuals of a population by small local perturbations. They aim to provide a component of randomness in the neighborhood of the individuals of the population. In our system, we implemented two mutation methods: uniformly random mutation and boundary mutation.

Escaping from local optima: GA itself has the ability to avoid falling prematurely into local optima and can eventually escape from them during the search process. DGA has one more mechanism to escape from local optima by considering some islands. Each island computes GA for optimizing and they migrate its gene to provide the ability to avoid from local optima (See Fig. 1).

Convergence: The convergence of the algorithm is the mechanism of DGA to reach to good solutions. A premature convergence of the algorithm would cause that all individuals of the population be similar in their genetic features and thus the search would result ineffective and the algorithm getting stuck into local optima. Maintaining the diversity of the population is therefore very important to this family of evolutionary algorithms.

3.3 WMN-PSODGA Hybrid Simulation System

In this subsection, we present the initialization, particle-pattern, fitness function and client distributions. Also, our implemented simulation system uses Migration function (see Algorithm 3) as shown in Fig. 2. The Migration function swaps solutions among lands including PSO part.

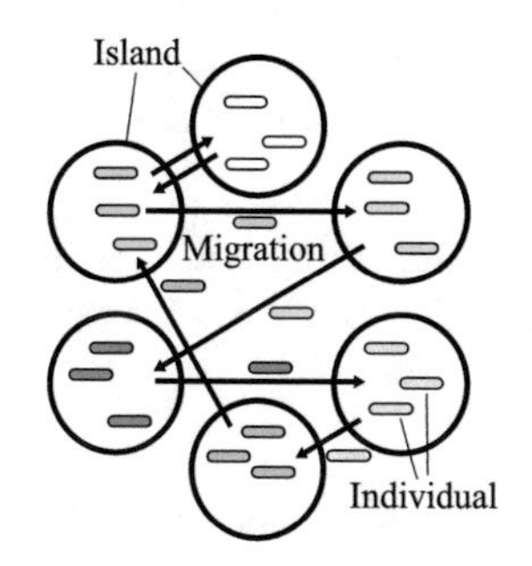

Fig. 1. Model of Migration in DGA.

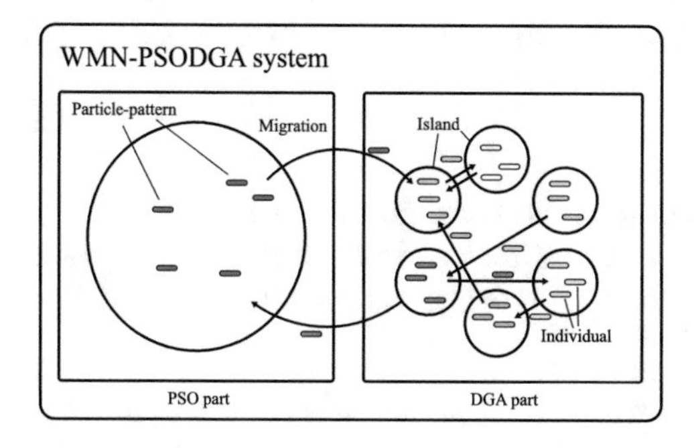

Fig. 2. Model of WMN-PSODGA migration.

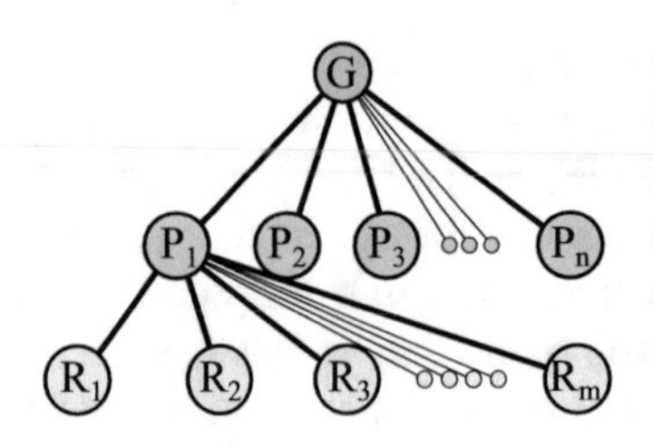

Fig. 3. Relationship among global solution, particle-patterns and mesh routers in PSO part.

Initialization
We decide the velocity of particles by a random process considering the area size. For instance, when the area size is $W \times H$, the velocity is decided randomly from $-\sqrt{W^2 + H^2}$ to $\sqrt{W^2 + H^2}$. In our system, many kinds of client distributions are generated. In this paper, we consider Exponential and Chi-square distributions for mesh clients.

Particle-pattern
A particle is a mesh router. A fitness value of a particle-pattern is computed by combination of mesh routers and mesh clients positions. In other words, each particle-pattern is a solution as shown is Fig. 3.

Gene Coding
A gene describes a WMN. Each individual has its own combination of mesh nodes. In other words, each individual has a fitness value. Therefore, the combination of mesh nodes is a solution.

Fitness Function
One of most important thing in PSO algorithm is to decide the determination of an appropriate objective function and its encoding. In our case, each particle-pattern has an own fitness value and compares it with other particle-pattern's fitness value in order to share information of global solution. The fitness function follows a hierarchical approach in which the main objective is to maximize the SGC in WMN. Thus, the fitness function of this scenario is defined as

$$\text{Fitness} = 0.7 \times \text{SGC}(\boldsymbol{x}_{ij}, \boldsymbol{y}_{ij}) + 0.3 \times \text{NCMC}(\boldsymbol{x}_{ij}, \boldsymbol{y}_{ij}).$$

Routers Replacement Method for PSO Part
A mesh router has x, y positions and velocity. Mesh routers are moved based on velocities. There are many moving methods in PSO field, such as:

Constriction Method (CM)
CM is a method which PSO parameters are set to a week stable region ($\omega = 0.729$, $C_1 = C2 = 1.4955$) based on analysis of PSO by Clerc et al. [2,5,35].

Random Inertia Weight Method (RIWM)
In RIWM, the ω parameter is changing randomly from 0.5 to 1.0. The C_1 and C_2 are kept 2.0. The ω can be estimated by the week stable region. The average of ω is 0.75 [28,35].

Linearly Decreasing Inertia Weight Method (LDIWM)
In LDIWM, C_1 and C_2 are set to 2.0, constantly. On the other hand, the ω parameter is changed linearly from unstable region ($\omega = 0.9$) to stable region ($\omega = 0.4$) with increasing of iterations of computations [35,36].

Linearly Decreasing Vmax Method (LDVM)
In LDVM, PSO parameters are set to unstable region ($\omega = 0.9$, $C_1 = C_2 = 2.0$). A value of V_{max} which is maximum velocity of particles is considered. With increasing of iteration of computations, the V_{max} is kept decreasing linearly [30,34].

Rational Decrement of Vmax Method (RDVM)
In RDVM, PSO parameters are set to unstable region ($\omega = 0.9$, $C_1 = C_2 = 2.0$). The V_{max} is kept decreasing with the increasing of iterations as

$$V_{max}(x) = \sqrt{W^2 + H^2} \times \frac{T - x}{x}.$$

Where, W and H are the width and the height of the considered area, respectively. Also, T and x are the total number of iterations and a current number of iteration, respectively [25].

Client Distributions. Our proposed system can generate many client distributions. In this paper, we consider Exponential and Chi-square distributions as shown in Fig. 4.

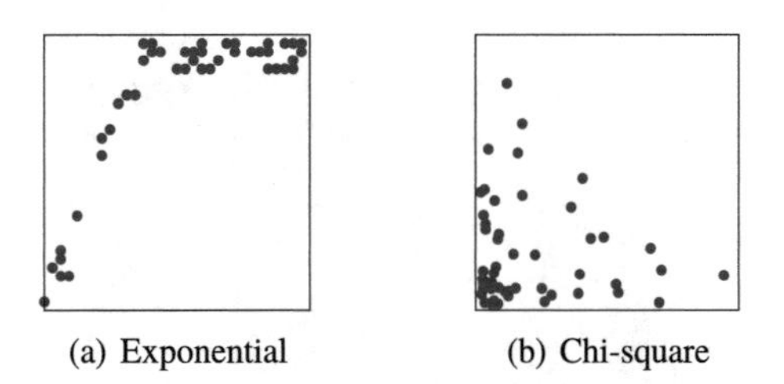
(a) Exponential (b) Chi-square

Fig. 4. Clients distributions.

3.4 WMN-PSODGA Web GUI Tool and Pseudo Code

The Web application follows a standard Client-Server architecture and is implemented using LAMP (Linux + Apache + MySQL + PHP) technology (see Fig. 5). Remote users (clients) submit their requests by completing first the parameter setting. The parameter values to be provided by the user are classified into three groups, as follows.

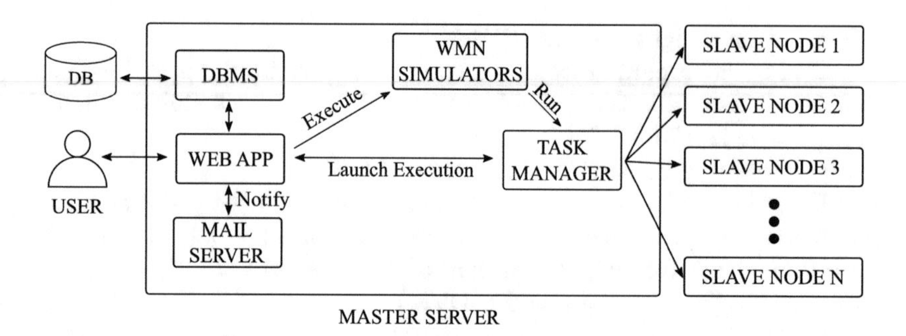

Fig. 5. System structure for web interface.

- Parameters related to the problem instance: These include parameter values that determine a problem instance to be solved and consist of number of router nodes, number of mesh client nodes, client mesh distribution, radio coverage interval and size of the deployment area.
- Parameters of the resolution method: Each method has its own parameters.
- Execution parameters: These parameters are used for stopping condition of the resolution methods and include number of iterations and number of independent runs. The former is provided as a total number of iterations and depending on the method is also divided per phase (e.g., number of iterations in a exploration). The later is used to run the same configuration for the same problem instance and parameter configuration a certain number of times.

We show the WMN-PSODGA Web GUI tool in Fig. 6. The pseudo code of our implemented system is shown in Algorithm 3.

Simulator parameters, Distributed Genetic Algorithm and Particle Swarm Optimization

Parameter	Value	
Distribution	Uniform	
Number of clients	48 (integer)(min:48 max:128)	
Number of routers	16 (integer) (min:16 max:48)	
Area size (WxH)	32 (positive real number)	32 (positive real number)
Radius (Min & Max)	2 (positive real number)	2 (positive real number)
Number of migration	200 (integer)	
Number of islands	200 (integer)	
Populations parameter	1 (integer)	
Independent runs	1 (integer) (min:1 max:100)	
Replacement method	Constriction Method	
Number of evolution steps	10 (integer) (min:1 max:64)	
Crossover rate	0.8 (positive real number)	
Mutation rate	0.2 (positive real number)	
Select method	Random Selection	
Crossover method	BLX-a Method	
Mutation method	Uniform Mutation	
Send by mail	☐	

Run

Fig. 6. WMN-PSODGA Web GUI tool.

Algorithm 3. Pseudo code of WMN-PSODGA system.

Computation maxtime:= T_{max}, $t := 0$;
Initial solutions: $\boldsymbol{P}$.
Initial global solutions: $\boldsymbol{G}$.
/* Start PSODGA */
while $t < T_{max}$ **do**
  Subprocess(PSO);
  Subprocess(DGA);
  WaitSubprocesses();
  Evaluate($\boldsymbol{G}^t, \boldsymbol{P}^t$)
  /* Migration() swaps solutions (see Fig. 2). */
  Migration();
  $t = t + 1$;
end while
Update_Solutions($\boldsymbol{G}^t, \boldsymbol{P}^t$);
return Best found pattern of particles as solution;

Table 1. WMN-PSODGA parameters.

Parameters	Values
Clients distribution	Exponential, Chi-square
Area size	32.0×32.0
Number of mesh routers	16
Number of mesh clients	48
Number of migrations	200
Evolution steps	9
Number of GA islands	16
Radius of a mesh router	2.0–3.5
Replacement method	LDIWM
Selection method	Roulette wheel method
Crossover method	SPX
Mutation method	Boundary mutation
Crossover rate	0.8
Mutation rate	0.2

4 Simulation Results

In this section, we show simulation results using WMN-PSODGA system. In this work, we analyse the performance of WMNs considering Exponential and Chi-square distributions of mesh clients. The number of mesh routers is considered 16 and the number of mesh clients 48. We conducted simulations 100 times, in order to avoid the effect of randomness and create a general view of results. We show the parameter setting for WMN-PSODGA in Table 1.

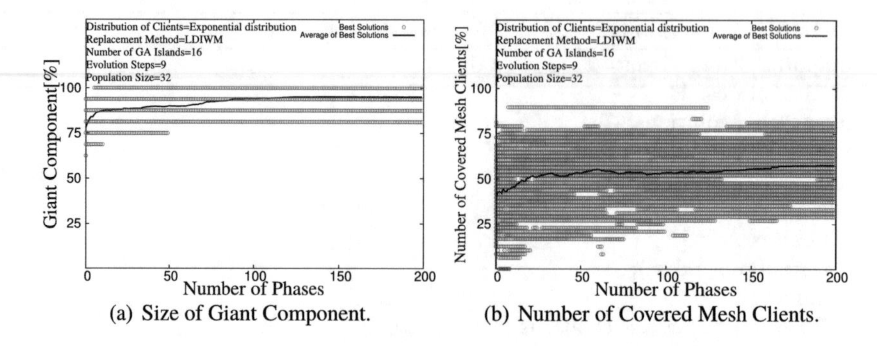

(a) Size of Giant Component.

(b) Number of Covered Mesh Clients.

Fig. 7. Simulation results of WMN-PSODGA for Exponential distribution of mesh clients.

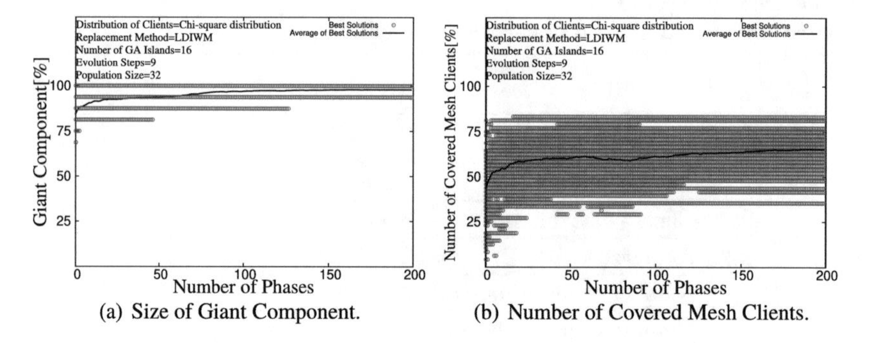

(a) Size of Giant Component.

(b) Number of Covered Mesh Clients.

Fig. 8. Simulation results of WMN-PSODGA for Chi-square distribution of mesh clients.

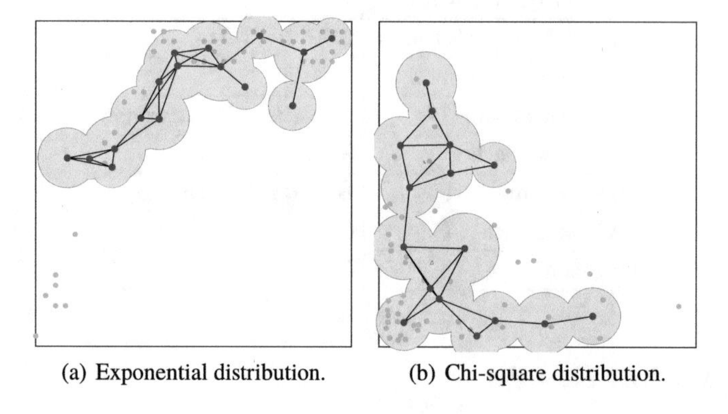

(a) Exponential distribution.

(b) Chi-square distribution.

Fig. 9. Visualized simulation results of WMN-PSODGA for different client distributions.

We show simulation results in Figs. 7, 8 and 9. In Figs. 7 and 8, we see that for both SGC and NCMC, the performance of Chi-square distribution is better than Exponential distribution. We show the visualized simulation results in

Fig. 9. For both distributions, all mesh routers are connected, but some clients are not covered. Thus, we couclude that the performance is better when the clients distribution is Chi-square distribution compared with the Exponential distribution.

5 Conclusions

In this work, we evaluated the performance of WMNs using a hybrid simulation system based on PSO and DGA (called WMN-PSODGA). Simulation results show that the performance is better for Chi-square distribution compared with the case of Exponential distribution.

In our future work, we would like to evaluate the performance of the proposed system for different parameters and patterns.

References

1. Akyildiz, I.F., Wang, X., Wang, W.: Wireless mesh networks: a survey. Comput. Netw. **47**(4), 445–487 (2005)
2. Barolli, A., Sakamoto, S., Ozera, K., Ikeda, M., Barolli, L., Takizawa, M.: Performance evaluation of WMNs by WMN-PSOSA simulation system considering constriction and linearly decreasing Vmax methods. In: International Conference on P2P, Parallel, Grid, Cloud and Internet Computing, pp. 111–121. Springer, Cham (2017)
3. Barolli, A., Sakamoto, S., Barolli, L., Takizawa, M.: Performance analysis of simulation system based on particle swarm optimization and distributed genetic algorithm for WMNs considering different distributions of mesh clients. In: International Conference on Innovative Mobile and Internet Services in Ubiquitous Computing, pp. 32–45. Springer, Cham (2018)
4. Barolli, A., Sakamoto, S., Ozera, K., Barolli, L., Kulla, E., Takizawa, M.: Design and implementation of a hybrid intelligent system based on particle swarm optimization and distributed genetic algorithm. In: International Conference on Emerging Internetworking, Data & Web Technologies, pp. 79–93. Springer, Cham (2018)
5. Clerc, M., Kennedy, J.: The particle swarm-explosion, stability, and convergence in a multidimensional complex space. IEEE Trans. Evol. Comput. **6**(1), 58–73 (2002)
6. Franklin, A.A., Murthy, C.S.R.: Node placement algorithm for deployment of two-tier wireless mesh networks. In: Proceedings of Global Telecommunications Conference, pp. 4823–4827 (2007)
7. Girgis, M.R., Mahmoud, T.M., Abdullatif, B.A., Rabie, A.M.: Solving the wireless mesh network design problem using genetic algorithm and simulated annealing optimization methods. Int. J. Comput. Appl. **96**(11), 1–10 (2014)
8. Goto, K., Sasaki, Y., Hara, T., Nishio, S.: Data gathering using mobile agents for reducing traffic in dense mobile wireless sensor networks. Mob. Inf. Syst. **9**(4), 295–314 (2013)
9. Inaba, T., Elmazi, D., Sakamoto, S., Oda, T., Ikeda, M., Barolli, L.: A secure-aware call admission control scheme for wireless cellular networks using fuzzy logic and its performance evaluation. J. Mob. Multimed. **11**(3&4), 213–222 (2015)

10. Inaba, T., Obukata, R., Sakamoto, S., Oda, T., Ikeda, M., Barolli, L.: Performance evaluation of a QoS-aware fuzzy-based CAC for LAN access. Int. J. Space-Based Situated Comput. **6**(4), 228–238 (2016)
11. Inaba, T., Sakamoto, S., Oda, T., Ikeda, M., Barolli, L.: A Testbedfor admission control in WLAN: a fuzzy approach and its performance evaluation. In: International Conference on Broadband and Wireless Computing, Communication and Applications, pp. 559–571. Springer, Cham (2016)
12. Lim, A., Rodrigues, B., Wang, F., Xu, Z.: k-center problems with minimum coverage. In: Computing and Combinatorics, pp. 349–359 (2004)
13. Maolin, T., et al.: Gateways placement in backbone wireless mesh networks. Int. J. Commun. Netw. Syst. Sci. **2**(1), 44 (2009)
14. Matsuo, K., Sakamoto, S., Oda, T., Barolli, A., Ikeda, M., Barolli, L.: Performance analysis of WMNs by WMN-GA simulation system for two WMN architectures and different TCP congestion-avoidance algorithms and client distributions. Int. J. Commun. Netw. Distrib. Syst. **20**(3), 335–351 (2018)
15. Muthaiah, S.N., Rosenberg, C.P.: Single gateway placement in wireless mesh networks. In: Proceedings of 8th International IEEE Symposium on Computer Networks, pp. 4754–4759 (2008)
16. Naka, S., Genji, T., Yura, T., Fukuyama, Y.: A hybrid particle swarm optimization for distribution state estimation. IEEE Trans. Power Syst. **18**(1), 60–68 (2003)
17. Poli, R., Kennedy, J., Blackwell, T.: Particle swarm optimization. Swarm Intell. **1**(1), 33–57 (2007)
18. Sakamoto, S., Kulla, E., Oda, T., Ikeda, M., Barolli, L., Xhafa, F.: A comparison study of simulated annealing and genetic algorithm for node placement problem in wireless mesh networks. J. Mob. Multimed. **9**(1–2), 101–110 (2013)
19. Sakamoto, S., Kulla, E., Oda, T., Ikeda, M., Barolli, L., Xhafa, F.: A comparison study of hill climbing, simulated annealing and genetic algorithm for node placement problem in WMNs. J. High Speed Netw. **20**(1), 55–66 (2014)
20. Sakamoto, S., Kulla, E., Oda, T., Ikeda, M., Barolli, L., Xhafa, F.: A simulation system for WMN based on SA: performance evaluation for different instances and starting temperature values. Int. J. Space-Based Situated Comput. **4**(3–4), 209–216 (2014)
21. Sakamoto, S., Kulla, E., Oda, T., Ikeda, M., Barolli, L., Xhafa, F.: Performance evaluation considering iterations per phase and SA temperature in WMN-SA system. Mob. Inf. Syst. **10**(3), 321–330 (2014)
22. Sakamoto, S., Lala, A., Oda, T., Kolici, V., Barolli, L., Xhafa, F.: Application of WMN-SA simulation system for node placement in wireless mesh networks: a case study for a realistic scenario. Int. J. Mob. Comput. Multimed. Commun. (IJMCMC) **6**(2), 13–21 (2014)
23. Sakamoto, S., Oda, T., Ikeda, M., Barolli, L., Xhafa, F.: An integrated simulation system considering WMN-PSO simulation system and network simulator 3. In: International Conference on Broadband and Wireless Computing, Communication and Applications, pp. 187–198. Springer, Cham (2016)
24. Sakamoto, S., Oda, T., Ikeda, M., Barolli, L., Xhafa, F.: Implementation and evaluation of a simulation system based on particle swarm optimisation for node placement problem in wireless mesh networks. Int. J. Commun. Netw. Distrib. Syst. **17**(1), 1–13 (2016)

25. Sakamoto, S., Oda, T., Ikeda, M., Barolli, L., Xhafa, F.: Implementation of a new replacement method in WMN-PSO simulation system and its performance evaluation. In: The 30th IEEE International Conference on Advanced Information Networking and Applications (AINA-2016), pp. 206–211 (2016). https://doi.org/10.1109/AINA.2016.42
26. Sakamoto, S., Obukata, R., Oda, T., Barolli, L., Ikeda, M., Barolli, A.: Performance analysis of two wireless mesh network architectures by WMN-SA and WMN-TS simulation systems. J. High Speed Netw. **23**(4), 311–322 (2017)
27. Sakamoto, S., Ozera, K., Barolli, A., Ikeda, M., Barolli, L., Takizawa, M.: Implementation of an intelligent hybrid simulation systems for WMNs based on particle swarm optimization and simulated annealing: performance evaluation for different replacement methods. In: Soft Computing, pp. 1–7 (2017)
28. Sakamoto, S., Ozera, K., Barolli, A., Ikeda, M., Barolli, L., Takizawa, M.: Performance evaluation of WMNs by WMN-PSOSA simulation system considering random inertia weight method and linearly decreasing vmax method. In: International Conference on Broadband and Wireless Computing, Communication and Applications, pp. 114–124. Springer, Cham (2017)
29. Sakamoto, S., Ozera, K., Ikeda, M., Barolli, L.: Implementation of intelligent hybrid systems for node placement problem in WMNs considering particle swarm optimization, hill climbing and simulated annealing. Mob. Netw. Appl. 1–7 (2017)
30. Sakamoto, S., Ozera, K., Ikeda, M., Barolli, L.: Performance evaluation of WMNs by WMN-PSOSA simulation system considering constriction and linearly decreasing inertia weight methods. In: International Conference on Network-Based Information Systems, pp. 3–13. Springer, Cham (2017)
31. Sakamoto, S., Ozera, K., Oda, T., Ikeda, M., Barolli, L.: Performance evaluation of intelligent hybrid systems for node placement in wireless mesh networks: a comparison study of WMN-PSOHC and WMN-PSOSA. In: International Conference on Innovative Mobile and Internet Services in Ubiquitous Computing, pp. 16–26. Springer, Cham (2017)
32. Sakamoto, S., Ozera, K., Oda, T., Ikeda, M., Barolli, L.: Performance evaluation of WMN-PSOHC and WMN-PSO simulation systems for node placement in wireless mesh networks: a comparison study. International Conference on Emerging Internetworking. Data & Web Technologies, pp. 64–74. Springer, Cham (2017)
33. Sakamoto, S., Ozera, K., Barolli, A., Barolli, L., Kolici, V., Takizawa, M.: Performance evaluation of WMN-PSOSA considering four different replacement methods. International Conference on Emerging Internetworking. Data & Web Technologies, pp. 51–64. Springer, Cham (2018)
34. Schutte, J.F., Groenwold, A.A.: A study of global optimization using particle swarms. J. Glob. Optim. **31**(1), 93–108 (2005)
35. Shi, Y.: Particle swarm optimization. IEEE Connect. **2**(1), 8–13 (2004)
36. Shi, Y., Eberhart, R.C.: Parameter selection in particle swarm optimization. In: Evolutionary programming VII, pp. 591–600 (1998)
37. Vanhatupa, T., Hannikainen, M., Hamalainen, T.: Genetic algorithm to optimize node placement and configuration for WLAN planning. In: Proceedings of The 4th IEEE International Symposium on Wireless Communication Systems, pp. 612–616 (2007)
38. Wang, J., Xie, B., Cai, K., Agrawal, D.P.: Efficient mesh router placement in wireless mesh networks. In: Proceedings of IEEE International Conference on Mobile Adhoc and Sensor Systems (MASS-2007), pp. 1–9 (2007)

Evaluation of a Protocol to Prevent Illegal Information Flow Based on Maximal Roles in the RBAC Model

Shohei Hayashi[1(✉)], Shigenari Nakamura[1], Dilawaer Duolikun[1], Tomoya Enokido[2], and Makoto Takizawa[1]

[1] Hosei University, Tokyo, Japan
shohei.hayashi.9f@stu.hosei.ac.jp, nakamura.shigenari@gmail.com, dilewerdolkun@gmail.com, makoto.takizawa@computer.org
[2] Rissho University, Tokyo, Japan
eno@ris.ac.jp

Abstract. In the access control models to make a system secure, a transaction is allowed to read and write an object like a file only if access rights on the object are granted. Suppose a transaction T_1 reads data d from a file f_1 and then writes the data d to another file f_2. Here, another transaction T_2 can get the data d by reading the file f_2 even if T_2 is not granted a read right on the file f_1. Here, the read operation issued by the transaction T_2 is illegal. In our previous studies, a condition to detect an illegal read operation is defined based on the role-based access control (RBAC) model. Here, once a transaction issues an illegal read operation, the transaction is aborted. However, even if the illegal condition is satisfied for a transaction issuing a read operation, illegal information flow may not occur. In this paper, we newly propose a modified read abortion (MRA) protocol which uses a new condition on maximal roles of role sets. In addition, we consider only maximal roles which include a read right on an object which a transaction can read. In the evaluation, we show the number of transactions aborting can be reduced.

Keywords: Illegal information flow · Role-based access control (RBAC) model · Maximal roles · MRA protocol

1 Introduction

Information systems are required to be secure in presence of multiple transactions manipulating objects. A system is composed of *objects* like databases [2] and *subjects* like transactions and users. A transaction is allowed to read and write an object only if the transaction is granted access rights [11] to read and write data in the object, respectively. Suppose a transaction T_1 reads data d in an object o_1 and writes the data d in another object o_2. We also suppose a transaction T_2

L. Barolli et al. (Eds.): EIDWT 2019, LNDECT 29, pp. 80–91, 2019.
https://doi.org/10.1007/978-3-030-12839-5_7

is granted a read right on the object o_2 while not granted a read right on the object o_1. Here, the transaction T_2 can get the data d from the object o_2. Here, information in the object o_1 illegally flows to the transaction T_2 [3]. The role-based access control (RBAC) model [5] is widely used in information systems like database systems [2]. Here, a role is a subset of access rights. Each transaction is granted roles. A subset of the roles granted to a transaction is the *purpose* of the transaction [4].

In our previous studies [7–10], synchronization protocols to prevent illegal information flow are proposed based on the RBAC model. Here, a pair of variables $o_i.R$ and $T_t.R$ are manipulated for each object o_i and each transaction T_t, respectively. Initially, for each transaction T_t, the variable $T_t.R$ is the purpose $T_t.P$ of roles granted to the transaction T_t. The variable $o_i.R$ is initially empty for each object o_i. If the transaction T_t writes data into an object o_i, the roles in the variable $T_t.R$ are added to the variable $o_i.R$. Next, suppose a transaction T_t issues a read operation to an object o_i. If no illegal information flow occurs from every role in the variable $o_i.R$ to every role in the variable $T_t.R$, the transaction T_t is allowed to read data from the object o_i. Roles in the variable $o_i.R$ are added to the variable $T_t.R$. Otherwise, the transaction T_t aborts in the RBS (role-based synchronization) protocol [6]. In order to make the illegal information flow check simpler, maximal roles of $o_i.R$ and minimal roles of $T_t.R$ are checked in the previous synchronization protocols [6].

Even if the illegal condition is satisfied for a transaction, illegal information flow may not occur by performing the transaction while every transaction which causes illegal information flow aborts in the RBS protocol [6]. That is, the transaction may unnecessarily aborts. In this paper, we try to reduce the number of transactions unnecessarily aborting. We newly propose a modified read-abortion (MRA) protocol where only maximal roles $max(o_i.R)$ and $max(T_t.R)$ of role subsets $o_i.R$ and $T_t.R$ are checked. In addition, if a transaction T_t reads an object o_i, only roles in $o_i.R$ which include read rights on objects which T_1 is allowed to be read. A transaction T_t aborts if the transaction T_t issues a read operation to an object o_i where $o_i.R$ illegally precedes $T_t.R$, i.e. maximal roles of role sets $o_i.R$ and $T_t.R$ are checked if $max(o_i.R)$ illegally precede $max(T_t.R)$,

In the evaluation, we show the number of transactions unnecessarily aborting can be reduced in the MRA protocol while every transaction issuing illegal read operations aborts. We also show the computation overhead of the MRA protocol is smaller than other protocol.

In Sect. 2, we present a system model. In Sect. 3, we discuss the illegally precedent relation among subsets of roles. In Sect. 4, we propose the MRA protocol. In Sect. 5, we evaluate the MRA protocol.

2 System Model

An information system is composed of entities. There are two types of *roles*, *subject* and *object*, which each entity plays in a system. A subject s issues an operation to an object o. A transaction is an example of a subject, which issues operations to objects. On receipt of an operation *op*, the operation *op* is performed on an object *o*. A database is an example of an object.

Let S and O be sets of subjects and objects, respectively, in a system. In this paper, each object o_i supports a pair of basic read (rd) and write (wr) operations. Let OP be a set of read (rd) and write ($write$) operations supported by objects in O. An *access rule* is specified in a tuple $\langle s, o, op\rangle \in S \times O \times OP$, which means that a subject s is allowed to manipulate an object o in an operation op [3]. A pair $\langle o, op\rangle$ ($\in O \times OP$) of an object o and an operation op is referred to as *access right*. A subject s is allowed to manipulate an object o in an operation op only if an access right $\langle o, op\rangle$ is granted to the subject s. An access request to manipulate an object o in an operation op is written in a pair $\langle o, op\rangle$ ($\in O \times OP$). If an access right $\langle o, op\rangle$ is not granted to a subject s, an access request $\langle o, op\rangle$ issued by the subject s is rejected.

A role r is a collection of access rights, i.e. $r \in 2^{o \times op}$. A role r shows a job function in real communities, which a person who plays the role r can do in a system. In the role-based access control (RBAC) model [5], a subject s is granted a role r. Here, the subject s is allowed to issue an operation op to an object o only if an access right $\langle o, op\rangle$ is in the role r. Let R ($\subseteq O \times OP$) be a set of roles in a system. In this paper, we assume the mandatory policy [3] is taken, i.e. an authorizer defines roles on objects in a system and grant roles to subjects.

A subject s issues a transaction T_t to manipulate objects. A transaction is a unit of work of a subject and is a sequence of operations on objects. Let $s.R$ ($\subseteq R$) be a subset of roles granted to a subject s. Here, the subject s grants the transaction T_t a subset of roles which are granted to the subject s. The subset $T_t.P$ ($\subseteq s.R$) of roles is referred to as *purpose* of the transaction T_t. A transaction T_t is allowed to issue an access request $\langle o, op\rangle$, i.e. an operation op to an object o only if an access right $\langle o, op\rangle$ is included in the purpose $T_t.P$.

3 Information Flow Relations

We discuss types of information flow relations among roles in the RBAC model. Let R be a set of roles in a system, i.e. $R \in 2^{o \times op}$. Let O be a set of objects in a system. Let $In(r_i)$ be a set of objects which are allowed to be read by a subject granted a role r_i in the role set R, i.e. $In(r_i) = \{o \mid \langle o, r\rangle \in r_i\} \subseteq O$. Let $Out(r_i)$ be a set of objects which are allowed to be written by a subject granted a role r_i ($\in R$), i.e. $Out(r_i) = \{o \mid \langle o, w\rangle \in r_i\} \subseteq O$. Objects in $In(r_i)$ and $Out(r_i)$ are referred to as *input* and *output* objects of a role r_i, respectively. A pair of different roles r_i and r_j are *equivalent* ($r_i \equiv r_j$) iff (if and only if) $In(r_i) = In(r_j)$ and $Out(r_i) = Out(r_j)$. That is, a pair of equivalent roles r_i and r_j ($r_i \equiv r_j$) are composed of the same access rights.

Definition. A role r_i *precedes* a role r_j with respect to information flow ($r_i \rightarrow r_j$) iff (if and only if) $Out(r_i) \cap In(r_j) \neq \phi$.

Suppose a transaction T_t is granted a role r_i and another transaction T_u is granted a role r_j as the purposes $T_t.P$ and $T_u.P$, i.e. $T_t.P = \{r_i\}$ and $T_u.P = \{r_j\}$. We also suppose the role r_i precedes the role r_j ($r_i \rightarrow r_j$). Suppose the transaction T_u is performed after the transaction T_t. Here, the transaction T_t reads some data x in an object o_1 in $In(r_i)$ and writes the data x in an object o_2 in $Out(r_i)$. If $o_2 \in Out(r_i) \cap In(r_j)$, the transaction T_u may read the data x in the object o_2. Thus, the transaction T_u may read data in an object which the transaction T_t writes.

A role r_i is *independent* of a role r_j ($r_i \mid r_j$) iff $r_i \not\rightarrow r_j$, i.e. $Out(r_i) \cap In(r_j) = \phi$. It is noted the precedent relation $\rightarrow$ and the independent relation $\mid$ on roles are neither symmetry nor transitive.

We define the legally and illegally precedent relations on roles.

Definition

1. A role r_i *primarily legally precedes* a role r_j with respect to information flow ($r_i \rightharpoonup r_j$) iff $r_i \rightarrow r_j$, i.e. $Out(r_i) \cap In(r_j) \neq \phi$ and $In(r_i) \subseteq In(r_j)$.
2. A role r_i *primarily illegally precedes* a role r_j with respect to information flow ($r_i \rightharpoondown r_j$) iff $\sim(r_i \mid r_j$ or $r_i \rightharpoonup r_j)$, i.e. $r_i \rightarrow r_j$ and $In(r_i) \not\subseteq In(r_j)$.

Suppose a pair of transactions T_t and T_u are granted roles r_i and r_j, respectively, and the role r_i primarily legally precedes the role $r_j (r_i \rightharpoonup r_j)$. Suppose a pair of transactions T_t and T_u are granted roles r_i and r_j, respectively, and the role r_i primarily legally precedes the role $r_j (r_i \rightharpoonup r_j)$. Suppose the transaction T_u is performed after the transaction T_t. The condition "$In(r_i) \subseteq In(r_j)$" means every object read by the transaction T_t is allowed to be read by the transaction T_u. Next, suppose the role r_i primarily illegally precedes the role r_j ($r_i \rightharpoondown r_j$). If the transaction T_t is performed before the transaction T_u, some object read by the transaction T_t may not be allowed to be read by the transaction T_u.

Definition

1. A role r_i *legally precedes* a role r_j ($r_i \Rightarrow r_j$) iff one of the following conditions holds:
   (1) $r_i \rightharpoonup r_j$ and
   (2) for some role r_k, $r_i \rightharpoonup r_k$ and $r_k \Rightarrow r_j$.
2. A role r_i *illegally precedes* a role r_j ($r_i \mapsto r_j$) iff $\sim(r_i \mid r_j$ or $r_i \Rightarrow r_j)$, i.e. $r_i \rightarrow r_j$ and $r_i \nRightarrow r_j$.

The legally precedent relation $\Rightarrow$ is transitive. It is trivial that $In(r_i) \subseteq In(r_j)$ if $r_i \Rightarrow r_j$. It is also trivial that $r_i | r_j$ or $r_i \Rightarrow r_j$ holds iff $r_i \not\mapsto r_j$.

Example 1. Let us consider a system which includes four file objects f, g, h, and e. Suppose a role r_1 is composed of access rights $\langle f, rd \rangle$ and $\langle g, wr \rangle$, another role r_2 is $\{\langle g, rd \rangle, \langle h, wr \rangle\}$, and the other role r_3 is $\{\langle f, rd \rangle, \langle g, rd \rangle, \langle h, rd \rangle, \langle e, wr \rangle\}$. Here, $In(r_1) = \{f\}$ and $Out(r_1) = \{g\}$ for the role r_1, $In(r_2) = \{g\}$ and $Out(r_2) = \{h\}$ for the role r_2, and $In(r_3) = \{f, g, h\}$ and $Out(r_3) = \{e\}$ for the role r_3 as shown in Fig. 1. The role r_1 precedes the roles r_2 and r_3 ($r_1 \rightarrow r_2, r_1 \rightarrow r_3$) since $Out(r_1) \cap In(r_2) \neq \phi$ and $Out(r_1) \cap In(r_3) \neq \phi$ and the role r_2 precedes the role r_3 ($r_2 \rightarrow r_3$). The role r_1 illegally precedes the role r_2 ($r_1 \mapsto r_2$) since the file object f is in $In(r_1)$ but not in $In(r_2)$. On the other hand, the roles r_1 and r_2 legally precede the role r_3 ($r_1 \Rightarrow r_2, r_1 \Rightarrow r_3$) and $r_2 \Rightarrow r_3$. Here, suppose three subjects s_1, s_2, and s_3 are granted roles r_1, r_2, and r_3, respectively. A pair of the subjects s_2 and s_3 are allowed to read the file object g written by the subject s_1. The subject s_3 is allowed to read the file object h written by the subject s_2. First, the subject s_1 reads the file object f and then writes the file object g. Next, suppose the subject s_2 reads the file object g and then writes the file object h. We also suppose the subject s_3 reads the file objects f, g, and h and then writes the file object e. If the subject s_2 manipulates a pair of the file objects f and g after the subject s_1, the subject s_2 can obtain some data in the file object f by reading the file object g which the subject s_1 writes. However, the subject s_2 is not allowed to read the file object f. Suppose the subject s_3 manipulates a pair of the file objects g and h in the same way as the subject s_2. Here, the role r_2 legally precedes the role r_3 ($r_2 \Rightarrow r_3$). Since the subject s_3 is allowed to read the file object f, the subject s_3 can read data in the file object f without illegal information flow from the file object f to the file object g.

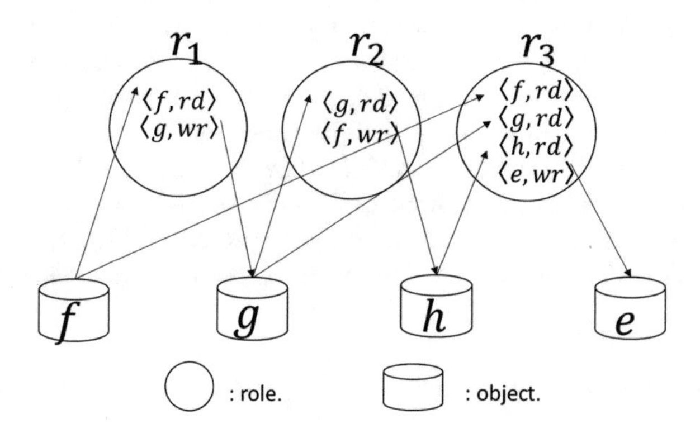

Fig. 1. Illegal information flow.

Let R be a set of roles in a system. Let R_i ($\subseteq R$) be a subset of the role set R. A role r_i is *maximal* in the role subset R_i iff there is no role r_j such that $r_i \Rightarrow r_j$ in the role subset R_i. $max(R_i)$ ($\subseteq R_i$) shows a subset of maximal roles in the role subset R_i. A role r_i is *minimal* in the role subset R_i iff there is no role r_j such that $r_j \Rightarrow r_i$ in the role subset R_i. $min(R_i)$ ($\subseteq R_i$) is a subset of minimal roles in the role subset R_i. A *least upper bound* (*lub*) of roles r_i and r_j is a role r_k such that $r_i \Rightarrow r_k$ and $r_j \Rightarrow r_k$ and there is no role r_h such that $r_i \Rightarrow r_h \Rightarrow r_k$ and $r_j \Rightarrow r_h \Rightarrow r_k$. A *greatest lower bound* (*glb*) of roles r_i and r_j is a role r_k such that $r_k \Rightarrow r_i$, $r_k \Rightarrow r_j$, and there is no role r_h such that $r_k \Rightarrow r_h \Rightarrow r_i$ and $r_k \Rightarrow r_h \Rightarrow r_j$. $lub(R_i)$ and $glb(R_i)$ show *lub* and *glb* roles of every role in a role subset R_i, respectively.

$In(R_i)$ and $Out(R_i)$ are $\cup_{r_i \in R_i} In(r_i)$ and $\cup_{r_i \in R_i} Out(r_i)$, respectively. Let r_l and r_g be a pair of $lub(R_i)$ and $glb(R_i)$ of a role subset R_i, respectively. $Out(r_l)$ is $Out(R_i)$ and $In(r_l)$ is $In(R_i)$. On the other hand, $In(r_g)$ is $\cap_{r_i \in R_i} In(r_i)$ and $Out(r_g)$ is $\cap_{r_i \in R_i} Out(r_i)$.

Definition. Let R_i and R_j be a pair of subsets of roles in the role set R (R_i, $R_j \subseteq R$).

1. R_i is *compatible* with R_j ($R_i | R_j$) iff $Out(R_i) \cap In(R_j) = \phi$.
2. R_i *precedes* R_j ($R_i \rightarrow R_j$) iff $R_i \not| R_j$, i.e. $Out(R_i) \cap In(R_j) \neq \phi$.
3. R_i *legally precedes* R_j ($R_i \Rightarrow R_j$) iff $R_i \rightarrow R_j$ and $In(R_i) \subseteq In(R_j)$.
4. R_i *illegally precedes* R_j ($R_i \mapsto R_j$) iff $R_i \rightarrow R_j$ and $In(R_i) \not\subseteq In(R_j)$, i.e. neither $R_i \Rightarrow R_j$ nor $R_i | R_j$.

As shown in paper [1], a role subset R_i legally precedes a role subset R_j ($R_i \Rightarrow R_j$) iff $lub(R_i) \Rightarrow glb(R_j)$. Here, only maximal roles $max(R_i)$ of a role subset R_i and minimal roles $min(R_j)$ of a role subset R_j are used to obtain lub and glb of the role subsets R_i and R_j, respectively. This means, it is enough for each role subset R_i to just hold maximal roles $max(R_i)$ and minimal roles $min(R_i)$.

In order to simply check if $R_i \mapsto R_j$ for a pair of role subsets R_i and R_j, we propose a new condition in this paper.

Property. For every pair of subsets R_i and R_j of the role set R ($R_i, R_j \subseteq R$), if R_i illegally precedes R_j ($R_i \mapsto R_j$), $max(R_i) \mapsto max(R_j)$ or $max(R_i) | max(R_j)$.

It is trivial $max(R_i) \Rightarrow max(R_j)$ iff $R_i \Rightarrow R_j$ for every pair of role subsets R_i and R_j. Here, $max(R_i) \mapsto max(R_j)$ means $R_i \mapsto R_j$ or $R_i | R_j$. Thus, even if $max(R_i) \mapsto max(R_j)$, R_i may not illegally precede R_j.

Example 2. Let R be a set $\{r_1, r_2, r_3, r_4, r_5, r_6, r_7\}$ of seven roles in system as shown in Fig. 2. Let R_i and R_j be a pair of role subsets $\{r_1, r_2, r_3, r_4\}$ and $\{r_5, r_6, r_7\}$ of the role set R, respectively. Suppose $r_1 \Rightarrow r_3$ and $r_2 \Rightarrow r_3$ in the role subset R_i. In addition, suppose the role r_4 and each of the roles r_1, r_2 and r_3 are compatible with one another ($r_4 \mid r_1, r_2, r_3$) in the role subset R_i. Suppose $r_5 \Rightarrow r_6$ and $r_5 \Rightarrow r_7$ in the role subset R_j. Here, a pair of the roles r_3 and r_4 are maximal and the role r_5 is minimal in the role subset R_j, i.e. $max(R_i) = \{r_3, r_4\}$ and $min(R_j) = \{r_5\}$. Here, $lub(R_i) = max(R_i)$. In addition, $max(R_j)$ is $\{r_7, r_6\}$. In the RBS protocol, it is checked if $lub(R_i) \mapsto glb(R_j)$. Suppose $r_3 \Rightarrow r_5$ and the role r_4 is compatible with the roles r_5 ($r_4 \mid r_5$). Hence, $lub(R_i) \Rightarrow glb(R_j)$.

Next, suppose $r_2 \mapsto r_7$. Here, if $r_3 \Rightarrow r_5$, $r_2 \Rightarrow r_7$ has to hold according to the transitivity of the legally precedent relation $\Rightarrow$. Hence, $r_3 \mapsto r_5$ or $r_3 | r_5$.

In this paper, we consider the maximal roles $max(R_j)$ of the role subset R_j. Here, $max(R_j) = \{r_7, r_6\}$. Here, the $max(R_i)$ ($= \{r_3, r_4\}$) $\mapsto max(R_j)$ ($= \{r_7, r_6\}$). Hence, the role subset R_i illegally precedes the role subset R_j ($R_i \mapsto R_j$) in this paper (Fig. 2).

4 A Modified Read-Abortion (MRA) Protocol

We would like to present the MRA protocol. Let R is a set of roles in a system. There is a variable $o_i.R$ for each object o_i, whose domain is a set R of roles. Initially, $o_i.R = \phi$ for each role o_i. There is a variable $T_t.R$ for each transaction

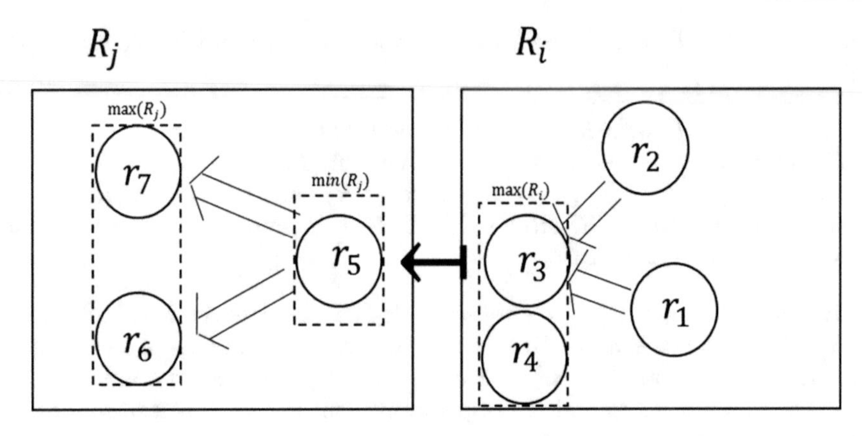

Fig. 2. Illegally precedent relation $\mapsto$ among role subsets.

T_t, whose domain is R. $T_t.P \subseteq R$ is a purpose of a transaction. If a transaction T_t is initiated, $T_t.R = \phi$.

A transaction T_t issues an operation *op* to an object o_i by manipulating the variables $T_t.R$ and $o_i.R$ as follows:

[MRA Protocol]
[Initial]
  $o_i.R = \phi$ for every object o_i;
[Transaction T_t start]
  $T_t.R = \phi$;
[Read operation (rd)] *op* is a read (rd) operation.
  if $o_i.R \mapsto T_t.R$, i.e. T_t illegally reads o_i, **abort** T_t;
  else {
    read o_i;
    $T_t.R = T_t.R \cup \{r_i \in max(T_t.P) | (\{o_i\} \cup_{r \in o_i.R} In(r)) \cap In(r_i) \neq \phi\}$;

[Write operation (wr)] *op* is a write (wr) operation.
  write o_i;
  if $T_t.R = \phi$, $\{o_i.R = \phi;\}$
  else $o_i.R = max(o_i.R \cup T_t.R)$;

Each time a transaction T_t issues a read operation rd to an object o_i, the variables $o_i.R$ and $T_t.R$ are checked. If $o_i.R$ illegally precedes $T_t.R$ ($o_i.R \mapsto T_t.R$), the transaction T_t aborts. Otherwise, the transaction T_t reads data in the object o_i. In the RBS protocol, every role in the variable $o_i.R$ is added to the variable $T_t.R$, i.e. $T_t.R = T_t.R \cup o_i.R$. Differently from the RBS protocol [6], not all the roles in the variable $o_i.R$ are added to the variable $T_t.R$ in the MRA protocol. Only roles in the variable $o_i.R$ which include read rights on objects which the transaction T_t is allowed to read are added to the variable $T_t.R$, i.e. $T_t.R = T_t.R \cup \{r_i \in max(T_t.P) | (\{o_i\} \cup_{r \in o_i.R} In(r)) \cap In(r_i) \neq \phi\}$.

If a transaction T_t issues a write operation wr to an object o_i, data is written to the object o_i. Then, roles in the variable $T_t.R$ are added to the variable $o_i.R$, i.e. $o_i.R = max(o_i.R \cup T_t.R)$ in the same way as the RBS protocol.

Example 3. Let R_1 and R_2 be a pair of role subsets $\{r_1, r_2, r_3, r_4\}$ and $\{r_5, r_6, r_7\}$ of a role set R, respectively. In addition, we consider a system which includes ten file objects $o_1, \ldots, o_{10}$. Suppose a role r_1 is composed of access rights $\langle o_1, rd\rangle$ and $\langle o_2, wr\rangle$, another role r_2 is $\{\langle o_3, rd\rangle, \langle o_2, wr\rangle\}$, another role r_3 is $\{\langle o_1, rd\rangle, \langle o_2, rd\rangle, \langle o_3, rd\rangle, \langle o_4, wr\rangle,\}$ and the other role r_4 is $\{\langle r_{10}, rd\rangle, \langle r_{10}, wr\rangle\}$ in the role subset R_1. Suppose a role r_5 is composed of access rights $\{\langle o_1, rd\rangle, \langle o_2, rd\rangle, \langle o_3, rd\rangle, \langle o_4, rd\rangle, \langle o_5, wr\rangle\}$, another role r_6 is $\{\langle o_1, rd\rangle, \langle o_2, rd\rangle, \langle o_3, rd\rangle, \langle o_4, rd\rangle, \langle o_5, rd\rangle, \langle o_6, wr\rangle\}$, and the other role r_7 is $\{\langle o_1, rd\rangle, \langle o_2, rd\rangle, \langle o_3, rd\rangle, \langle o_4, rd\rangle, \langle o_5, rd\rangle, \langle o_7, wr\rangle\}$ in the role subset R_2. Suppose $r_1 \Rightarrow r_3$, $r_2 \Rightarrow r_3$, and $r_3 \Rightarrow r_5$ in the role subset R_1. In addition, suppose the role r_4 and each of the roles r_1, r_2, r_3, r_5, r_6, and r_7 are compatible with one another ($r_4 \mid r_1, r_2, r_3, r_5, r_6, r_7$) in the role set R. Suppose $r_5 \Rightarrow r_6$ and $r_5 \Rightarrow r_7$ in the role subset R_2. Here, a pair of the roles r_3 and r_4 are maximal and the roles r_1, r_2, and r_4 are minimal in the role subset R_1, i.e. $max(R_1) = \{r_3, r_4\}$ and $min(R_1) = \{r_1, r_2, r_4\}$. In addition, a pair of the roles r_6 and r_7 are maximal and the role r_5 is minimal in the role subset R_2, i.e. $max(R_2) = \{r_6, r_7\}$ and $max(R_2) = \{r_5\}$. $lub(R_2) = min(R_2)$ and $glb(R_1) = max(R_1)$. Suppose a pair of transactions T_1 and T_2 are granted purposes $T_1.P = R_1$ and $T_2.P = R_2$, respectively, as shown in Figs. 3 and 4. First, the transaction T_1 is performed where the object o_2 is read and then the object o_4 is written. Secondly, the transaction T_2 is performed where the object o_4 is read and the object o_7 is written. In the RBS protocol, once issuing a read operation rd to the object o_4, the illegal precedent relation $o_4.R \mapsto T_2.R$ is checked. Here, since $glb(R_1) \mapsto lub(R_2)$, the transaction T_2 aborts.

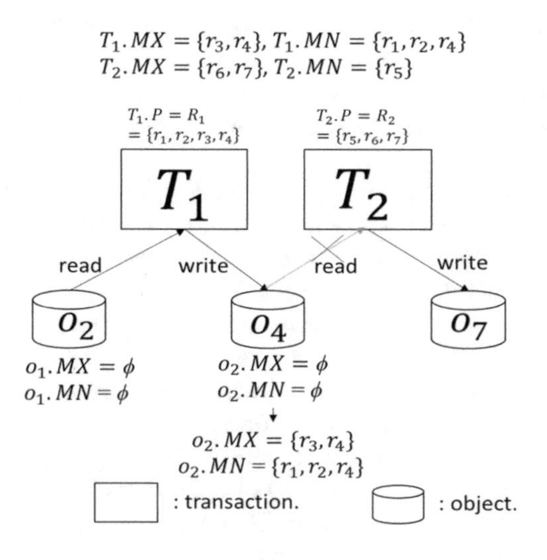

Fig. 3. RBS protocol.

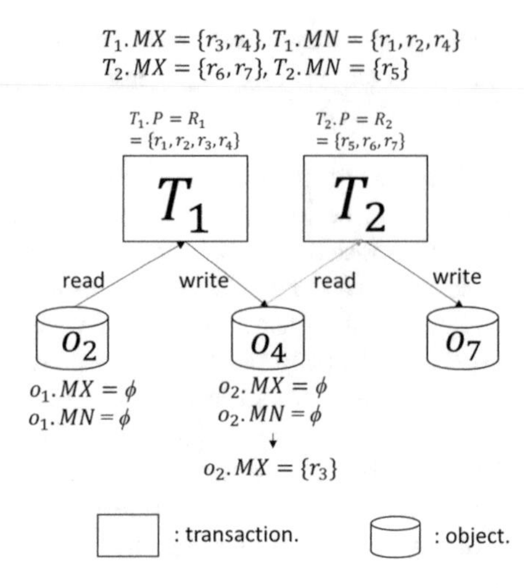

Fig. 4. MRA protocol.

On the other hand, in the MRA protocol, once issuing a read operation rd to the object o_4, the condition $o_4.R \Rightarrow T_2.R$ is checked. Here, since $max(R_1) \Rightarrow max(R_2)$, the read rd is performed and the roles in the variable $o_4.R$ are added to the variable $T_2.R$.

We implement the MRA protocol on a relational database system Sybase [1].

5 Evaluation

We evaluate the MRA protocol in terms of number of transactions aborting to prevent illegal information flow. A system is composed of objects $o_1 \ldots o_n (n \geq 1)$. Let O be a set $\{o_1, \ldots, o_n\}$ of objects in the system. In the evaluation, we consider two hundred objects, i.e. $n = 200$. Each object o_i supports a pair of operations supported by objects, rd (read) and wr (write) operations. Let OP be a set of operations, i.e. $OP = \{rd, wr\}$.

Let R be a set $\{r_1, \ldots, r_m\}$ in the system, $R \subseteq 2^{O \times OP}$. Each role r_i is defined on the objects in the object set O and operations in the operation set OP, i.e. $r_i \in 2^{O \times OP}$. Each role r_i includes rn_i (≥ 1) access rights. For each role r_i in the role set R, the number rn_i of access rights is randomly decided from 1 to 8. Then, rn_i access rights are randomly taken in the direct product $O \times OP$ so that no duplicate role is included in a role r_i.

Totally, t_n (≤ 1) transactions are serially performed on the object set O. In the evaluation, $t_n = 64$. That is, 64 transactions $\{T_1, \ldots, T_{64}\}$ are randomly created. Each transaction T_t is a sequence of tn_t (≤ 1) operations on objects. For each transaction T_t, the number tn_t of operations is randomly decided from 1 to 10. In addition, each transaction T_t is randomly granted tr_n roles in the role set R as the purpose $T_t.P$. The number tr_t is randomly taken from 1 to m. Then, tr_t roles are randomly taken in R as $T_t.P$, so that no duplicate role is included. Then, a sequence of tn_t operations are randomly taken in $O \times OP$ where every operation op on an object o, i.e. $\langle o, op \rangle$ satisfies the purpose $T_t.P$. Here, duplicate operations can be included in each transaction T_t. Then, a sequence of the transactions $T_1, \ldots, T_{64}$ are also randomly decided. Let T be a sequence of the transactions which shows an execution sequence of the transactions.

The role set R and transaction sequence T are thus randomly generated on the object set O. In the evaluation, the role set R is generated and the transaction sequence T is generated for each role set R. For a tuple $\langle O, R, T \rangle$ of the object set O, the role set R, and the transaction sequence T, the OBS [6], RBS [6], and MRA protocols are performed. Here, the transactions in T are serially performed on the objects according to the sequence of T. Each time a transaction T_t issues a read operation rd to an object o_i, it is checked if the read operation rd is illegal in each of the protocols. If the read operation rd is illegal, the transaction T_t aborts. In the OBS protocol, objects manipulated by each transaction are kept in record. We can find data in which object is written to each object by a transaction. Here, only and every transaction which issues illegal read operation can abort in the OBS protocol.

Figure 4 shows the numbers of transactions aborting in the OBS, MRA, and RBS protocols. The horizontal axis shows the ratio $ra = n/m$ where there are m roles on n objects, i.e. $ra = 200/m$. For example, $ra = 4$ means there are 50 roles on 200 objects. The vertical axis indicates the ratio of the number of transactions aborting to the total number of transactions, i.e. 64. OBS shows the number of transactions which issue illegal read operations on objects. In the OBS protocol, every and only transaction issuing an illegal read aborts. In the MRA and RBS protocols, every transaction aborts if a read operation satisfies the illegal read condition. However, even a transaction T_t which does not issue

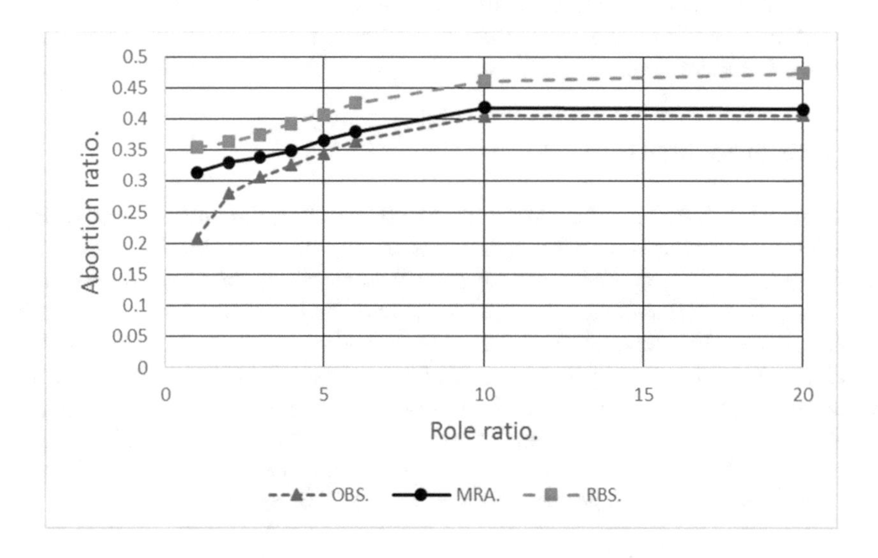

Fig. 5. Abortion ratio.

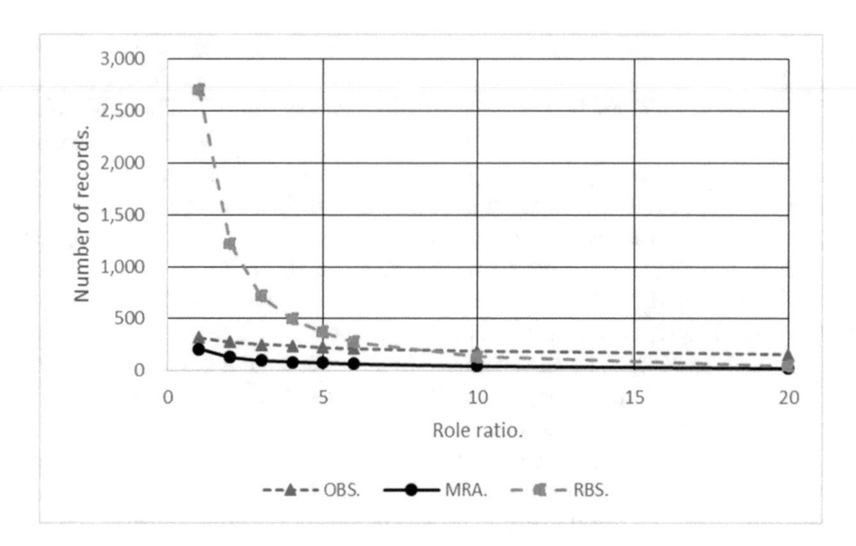

Fig. 6. Number of records.

an illegal read operation aborts, i.e. T_t unnecessarily aborts. The number of transactions unnecessarily aborting in the MRA protocol is fewer than the RBS protocol.

Each time a transaction T_t reads and writes on object o_i, role are added to the variables $T_t.R$ and $o_i.R$ in the protocols. Next, we measure the number of roles added to the variables $T_t.R$ of each transaction T_t and $o_i.R$ of each object o_i. Figure 6 shows number of roles added to the transactions and objects. As shown in Fig. 6, the number of roles added in the MRA protocol is fewer than the OBS and RBS protocol. This means, the computation overhead of the MRA protocol is smaller than the OBS and RBS protocols.

6 Concluding Remarks

In information systems, information in objects may illegally flow to other objects even if every transaction issues operations to objects according to access rules. In this paper, we proposed the new condition to check the illegally precedent relation ($R_i \mapsto R_2$) among role subsets R_1 and R_2 based on the maximal roles of R_1 and R_2. We proposed the MRA protocol where a transaction aborts if the condition is satisfied once issuing a read operation. We also implemented the MRA protocol by using SQL on the relational database. In the evaluation, we showed the number of transactions aborting in the MRA protocol can be reduced compared with the RBS protocol. We also showed the computation overhead of the MRA protocol is smaller than the RBS and OBS protocols.

Acknowledgements. The work was supported by JSPS KAKENHI grant number 15H0295.

References

1. Database management system sybase. http://infocenter.sybase.com/help/index.jsp
2. Date, C.J.: An Introduction to Database Systems, 8th edn. Addison-Wesley, Reading (2013)
3. Denning, D.E.R.: Cryptography and Data Security. Addison-Wesley, Reading (1982)
4. Enokido, T., Takizawa, M.: Purpose-based information flow control for cyber engineering. IEEE Trans. Ind. Electron. **58**(6), 2216–2225 (2011)
5. Ferraiolo, D.F., Kuhn, D.R., Chandramouli, R.: Role-Based Access Controls, 2nd edn. Artech House, Norwood (2007)
6. Nakamura, S., Duolikun, D., Aikebaier, A., Enokido, T., Takizawa, M.: Role-based information flow control models. In: Proceedings of the IEEE the 28th International Conference on Advanced Information Networking and Applications (AINA 2014), pp. 1140–1147 (2014)
7. Nakamura, S., Duolikun, D., Enokido, T., Takizawa, M.: A flexible read-write abortion protocol to prevent illegal information flow among objects. J. Mob. Multimed. **11**(3–4), 263–280 (2015)
8. Nakamura, S., Duolikun, D., Enokido, T., Takizawa, M.: A write abortion-based protocol in role-based access control systems. Int. J. Adapt. Innov. Syst. **2**(2), 142–160 (2015)
9. Nakamura, S., Duolikun, D., Enokido, T., Takizawa, M.: A read-write abortion (RWA) protocol to prevent illegal information flow in role-based access control systems. Int. J. Space-Based Situated Comput. **6**(1), 43–53 (2016)
10. Nakamura, S., Duolikun, D., Takizawa, M.: Read-abortion (RA) based synchronization protocols to prevent illegal information flow. J. Comput. Syst. Sci. **81**(8), 1441–1451 (2015)
11. Wang, M., Wang, J., Guo, K.: Extensible markup language keywords search based on security access control. Int. J. Grid Util. Comput. **9**(1), 43–50 (2018)

Implementation of Fog Nodes in the Tree-Based Fog Computing (TBFC) Model of the IoT

Ryusei Chida[1(✉)], Yinzhe Guo[1], Ryuji Oma[1], Shigenari Nakamura[1], Dilawaer Duolikun[1], Tomoya Enokido[2], and Makoto Takizawa[1]

[1] Hosei University, Tokyo, Japan
{ryusei.chida.7n,yinzhe.guo.8m,ryuji.oma.6r}@stu.hosei.ac.jp, nakamura.shigenari@gmail.com, dilewerdolkun@gmail.com, makoto.takizawa@computer.org

[2] Rissho University, Tokyo, Japan
eno@ris.ac.jp

Abstract. The IoT (Internet of Things) is so scalable that not only computers like servers but also sensors and actuators installed in various things are interconnected in networks. In the cloud computing model, application processes to process sensor data are performed on servers, this means networks are congested and servers are overloaded to handle a huge volume of sensor data. The fog computing model is proposed to efficiently realize the IoT. Here, subprocesses of an application process are performed on not only servers but also fog nodes. Servers finally receive data processed by fog nodes. Thus, traffic to process sensor data in severs and to transmit sensor data in networks can be reduced in the fog computing model. In this paper, we take the tree-based fog computing (TBFC) model where fog nodes are hierarchically structured in a height-balanced tree. We implement types of subprocesses of fog nodes in Raspbery PI. In experiment of the implemented TBFC model, we show the total execution time of nodes in the TBFC model is shorter than the cloud computing model.

Keywords: IoT (Internet of Things) · Fog computing model · Tree-based fog computing (TBFC) model · Raspberry Pi

1 Introduction

The IoT (Internet of Things) is composed of not only computers like servers and clients but also devices like sensors and actuators which are interconnected in networks [10]. In the cloud computing model [6], every sensor data is transmitted from sensors to servers of clouds in networks. Sensor data is processed by application processes on servers and then servers send actions to actuators. The IoT is more scalable than traditional information systems since a huge number

L. Barolli et al. (Eds.): EIDWT 2019, LNDECT 29, pp. 92–102, 2019.
https://doi.org/10.1007/978-3-030-12839-5_8

of sensors are interconnected and huge amount of sensor data are transmitted in networks. The network is congested due to heavy network traffic of sensor data and servers are also overloaded to process sensor data.

In order to efficiently realize the IoT, the fog computing model [13] is proposed. Here, subprocesses of an application process to process sensor data are performed on not only servers in clouds but also fog nodes while performed only on servers in the cloud computing model [6]. Data obtained by sensors are first transmitted to edge fog nodes. On receipt of sensor data, a fog node processes the sensor data and outputs processed data to another fog node. For example, a fog node obtains an average value of a collection of humidity data collected by sensors and sends only the average value of the humidity data to another fog node. Thus, fog nodes receive and process input data from other fog nodes and send output data obtained by processing the input data to other fog nodes. Servers finally receive data processed by fog nodes and can be relived from the computations done by fog nodes. In addition to the routing function of a router, a subprocess of an application process to process sensor data is installed in a fog node.

The TBFC (Tree-Based Fog Computing) model is proposed to reduce electric energy consumed by fog nodes and servers in the IoT [11,14]. Here, fog nodes are hierarchically structured in a height-balanced tree. A root fog node indicates a cluster of servers. Each non-root fog node has one parent fog node. Each non-leaf node has one or more than one child fog node. An edge fog node is a leaf node of the tree and communicates with sensors and actuators. An application process to be performed on servers to handle sensor data in the cloud computing model is assumed to be a sequence of subprocesses in this paper. Each subprocess receives input data from a preceding subprocess and sends output data to a succeeding subprocess. In the TBFC model, a same subprocess is installed in fog nodes at each level. Thus, each fog node at the same level performs the same subprocess on input data sent by child fog nodes and sends processed output data to a parent fog node. Sensor data from a huge number of sensors are processed by multiple fog nodes in a distributed and parallel manner. The fault-tolerant tree-based fog computing (FTTBFC) model is also proposed to make the TBFC model tolerant of faults of fog nodes [11,12].

In this paper, we implement each fog node of the TBFC model by using a Raspberry Pi [3] computer in this paper. Each subprocess of each fog node is characterized by the computation complexity $O(x)$ or $O(x^2)$ for size x of input data. We show the experiments of the implemented fog nodes of the TBFC model and show that the total execution time fog nodes and a server of the TBFC model is shorter than the cloud computing model.

In Sect. 2, we present the tree-based fog computing (TBFC) model. In Sect. 3, we discuss the implementation of the TBFC model. In Sect. 4, we show experiments of the implemented TBFC model.

2 Tree-Based Fog Computing (TBFC) Model

2.1 Tree of Fog Nodes

The fog computing model of the IoT is composed of devices, fog nodes, and clouds of servers [10]. Each server in a cloud supports applications with computation and storage services like the cloud computing model [6]. There are networks of fog nodes to interconnect devices and clouds. Devices like sensors and actuators are installed in various types of things. In the tree-based fog computing (TBFC) model [14], fog nodes are hierarchically structured in a height-balanced tree. A root node shows a cloud of servers. Each fog node is interconnected with a parent fog node and child fog nodes in networks. Fog nodes at the bottom layer are *edge* fog nodes. Edge fog nodes communicate with child sensors and actuators. Each device, i.e. sensor and actuator, has a parent edge fog node. A sensor collects sensor data and sends the sensor data to an edge fog node. Each edge fog node first collects sensor data from sensors. Each edge fog node processes sensor data and sends processed output data to a parent fog node. A fog node receives input data from child fog nodes and processes the data. Then, the fog node sends the processed output data to the parent fog node. Thus, servers in clouds finally receive processed data from fog nodes. Servers just process data processed by fog nodes and decide on actions to be done by actions. Servers send actions to fog nodes. Fog nodes forward the actions to their child fog nodes and each edge fog node finally sends actions to child actuators.

Figure 1 shows a tree of fog nodes in the TBFC model. Here, f_0 is a root node which denotes a cloud of servers. The root node f_0 has l_0 (≥ 0) child fog nodes f_{00}, f_{01}, ..., f_{0,l_0-1}. Each child fog node f_{0i} has l_{0i} (≥ 0) child fog nodes f_{0i0}, f_{0i1}, ..., $f_{0i,l_{0i}-1}$. Thus, a fog node f_R is f_0 if f_R is a root node, i.e. $R = 0$. If f_R is an i th child fog node of a fog node $f_{R'}$, $f_R = f_{R'i}$, i.e. $R = R'i$ ($i < l_R$). Thus, the label R of a fog node f_R shows a path from a root fog node f_0 to the fog node f_R. $|R|$ shows the length of label R of a fog node f_R. Here, a fog node f_R is at level $|R| - 1$. For example, a root fog node f_0 is at level 0 and a fog node f_{010} is at level 2. On a fog node f_R, a subprocess $p(f_R)$ of an application process is performed. At each layer, a same subprocess is performed on every fog node.

A subprocess $p(f_R)$ of a fog node f_R receives input data d_{R0}, d_{R1}, ..., d_{R,l_R-1} from child fog nodes f_{R0}, f_{R1} ..., f_{R,l_R-1}, respectively. Let D_R be a set of input data d_{R0}, ..., d_{R,l_R-1} of the fog node f_R. Then the input data is processed and output data d_R is generated. The output data d_R is sent to a parent fog node pt_{f_R}.

2.2 Subprocesses on Fog Nodes

An application process p is assumed to be a sequence $\langle p_0, p_1, ..., p_{h-1}\rangle$ of subprocesses. In the TBFC model, fog nodes are structured in a height-balanced tree of fog nodes with height h. All the subprocesses p_0, p_1, ..., p_{h-1} are performed on servers in the cloud computing model. Only the subprocess p_0 is performed on a root node, i.e. server cloud f_0. The subprocess p_0 is a *root* subprocess. The other subprocesses p_1, ..., p_{h-1} are performed on different fog nodes. The subprocess p_{h-1} is first performed on edge fog nodes of level $h-1$ by receiving data from sensors. The subprocess p_{h-1} is an *edge* subprocess. The output data of the edge subprocess p_{h-1} is sent to the succeeding subprocess p_{h-2}, which is performed on each of fog nodes of level $h-2$. Thus, a subprocess p_i is performed on fog nodes of level i and sends output data to a succeeding subprocess p_{i-1} on fog nodes of level $i-1$. The lower layer, the more amount of input data are sent from the underlying layer but the more number of fog nodes since the TBFC model is tree-structured and the output data of each fog node is generally smaller than the input data. Hence, the processing load of each fog node is equalized.

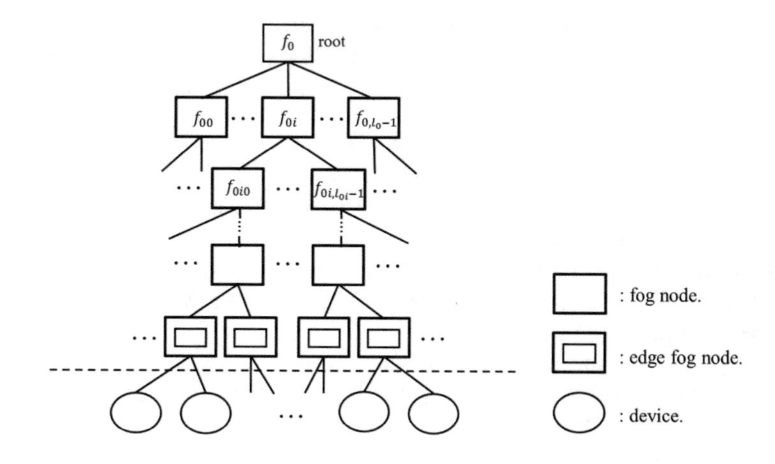

Fig. 1. TBFC model.

In the cloud computing model [6], a fog node is just a router which does only the routing function. In each fog node, not only the routing function but also a subprocess of an application process are performed in the fog computing model. Sensors send every sensor data to servers in a cloud.

There are types of subprocesses characterized in terms of computation complexity. In this paper, we consider the following types of subprocesses to be performed on a fog node f_R to handle input data D_R of size x $(= |D_R|)$ [12]:

1. $O(x)$, e.g. subprocess to calculate an average value of input data D_R.
2. $O(x^2)$, e.g. subprocess to join multiple input data d_{R0}, ..., d_{R,l_R-1} in input data D_R.

For example, a fog node f_R just selects some data in input data D_R of size x $(= |D_R|)$, e.g. a maximum value is selected in a collection D_R of input data from child fog nodes. Here the computation complexity of the subprocess is $O(x)$. The computation complexity of a fog node f_R to merge sorted input data is $O(x)$. The computation complexity of a fog node f_R which joins multiple input data is $O(x^2)$ where multiple input data D_R which have the same attribute values are concatenated.

3 Implementation of the TBFC Model

We discuss the implementation of fog nodes in the TBFC model. Each fog node f_R is implemented in a Raspberry Pi 3 Model B [3] computer with a CPU ARM Cortex-A53, one [GB] memory, and 32 [GB] SD storage. The Raspbian 8.0 [3] is used as an operating system of a fog node. A subprocess of an application process to be performed on each fog node is implemented in C language. Fog nodes are interconnected in a Gbit local area network (LAN). Fog nodes communicate with one another by using the protocol UDP [5]. A *record* is a unit of communication among fog nodes where data is stored. Each record is composed of attributes. Each fog node f_R has one UDP socket, US. A fog node f_R receives records from child fog nodes f_{R0}, ..., f_{R,l_R-1} and sends records at the UDP socket US.

We consider the following types of subprocesses to be performed on each fog node f_R:

1. In a fog node f_R, an aggregate value, e.g. average value of input data D_R is obtained and the aggregate data d_R is output by an *aggregate* subprocess [Fig. 2(1)]. The computation complexity of the fog node f_R is $O(x)$ for size x $(= |D_R|)$ of input data D_R.
2. In a fog node f_R, sorted input data d_{R0}, d_{R1}, ..., f_{R,l_R-1} are merged into a sorted data d_R by a *sort* subprocess [Fig. 2(2)]. The computation complexity of the fog node f_R is $O(x)$.
3. In a fog node f_R, multiple input data d_{R0}, d_{R1}, ..., $d_{R,l_R-1}$are joined to one output data d_R by a *join* subprocess [Fig. 2(3)]. The computation complexity of the fog node is $O(x^2)$.

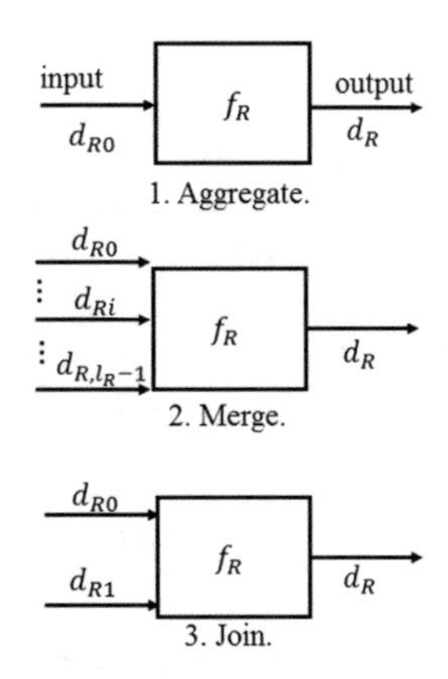

Fig. 2. Types of subprocesses on fog nodes.eps

A fog node f_R receives a record of input data d_{Ri} from each child fog node f_{Ri} and stores the record d_{Ri} in a receipt queue RQ_i. Each record d_{Ri} is composed of attribute values, i.e. tuple $\langle\ v_1, ..., v_l\ \rangle$ $(l > 1)$. On receipt of each record d_{Ri}, a cell c is dynamically allocated in a fog node f_R. The record d_{Ri} is stored in the cell c and the cell c is enqueued into the receipt queue RQ_i. While dequeueing a top record d_{Ri} from each receipt queue RQ_i, records of input data d_{R0}, d_{R1}, ..., d_{R,l_R-1} are processed and a record of output data d_R is generated by performing a subprocess on the input data D_R. The output record d_R is enqueued into an output queue SQ. A top record d_R is dequeued from the output queue SQ and sent to the parent fog node of f_R by using UDP.

Each queue Q is implemented in data structure as shown in Fig. 3. Each queue Q is composed of a control block CBC and doubly linked cells. The variable *no* of the data structure CBC shows the number of cells in the queue Q. The *top* and *tail* fields of the CBC block denote pointers to the top and tail cells of the queue Q. On receipt of a record d_{Ri} from a child fog node f_{Ri}, one cell c is created by a *malloc* system call [7] and the record d_{Ri} is stored in the cell c. Then, the cell c is enqueued in the receipt queue RQ_i, i.e. stored as the tail cell of the receipt queue RQ_i. Cells are linked in bidirectional pointers, *next* and *prior*. The next and prior pointers of a cell c denote cells following and preceding the cell c, respectively.

Each queue Q is manipulated through the following functions:

1. struct CBC *iniqueue( );
2. enqueue (struct CBC **cbc*, struct CELL **c*);
3. struct CELL *dequeue (struct CBC **cbc*);
4. struct CELL *topqueue (struct CBC **cbc*);

First, a control block *cbc* is created by the function *iniqueue*(), i.e. *cbc* = *iniqueue*(). A cell c is dequeued from the queue *cbc* by the function c = *dequeue* (*cbc*). A cell c is enqueued to the queue *cbc* by the function *enqueue* (*cbc*, c). A top cell c in the queue *cbc* is found by c = *topqueue* (*cbc*).

In this paper, every subprocess to be performed on each fog node is implemented on the Raspbian operating system in C language. A subprocess on each fog node f_R communicates with each child fog node and a parent fog node by using the UDP protocol [5].

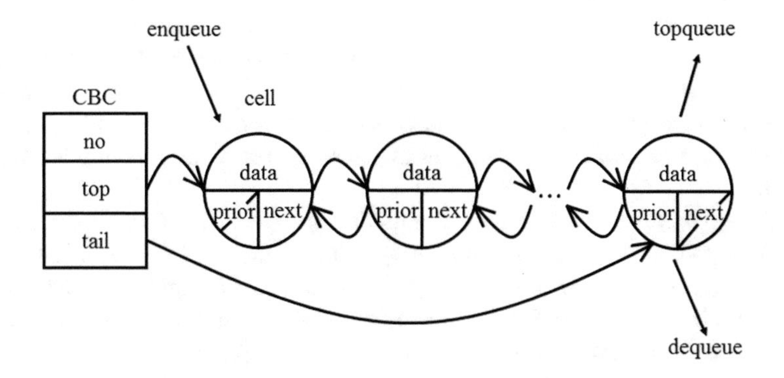

Fig. 3. Queue.

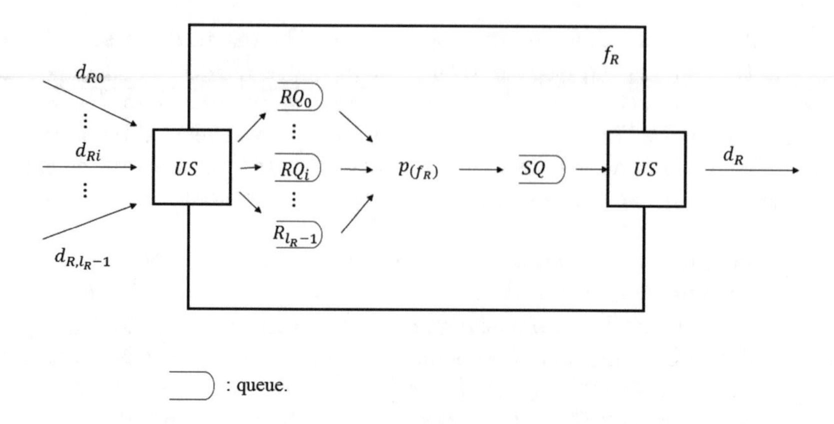

Fig. 4. Fog node.

4 Experiment

4.1 Implemented TBFC Model

We present the experiment of the implemented TBFC model. There are four sensors s_0, s_1, s_2, and s_3, and seven fog nodes f_{00}, f_{000}, f_{001}, f_{0000}, f_{0001}, f_{0010}, and f_{0011} in a tree. The sensors, fog nodes, and sever are interconnected in a Gbit LAN. One root node f_0 is in a cluster. As shown in Fig. 5, the root fog node f_0 is a server in the cloud. The root node f_0 has a single child fog node f_{00}. The fog node f_{00} has a pair of child fog nodes f_{000} and f_{001}. The fog nodes f_{000} and f_{001} of level 2 have pairs of child fog nodes f_{0000} and f_{0001}, and f_{0010} and f_{0011}, respectively. The four fog nodes f_{0000}, f_{0001}, f_{0010}, and f_{0011} are edge fog nodes at level 3, which communicate with sensors s_0, s_1, s_2, and s_3, respectively.

Each of the fog nodes and sensors is implemented in a Raspberry PI BM model computer [3]. A pair of the sensors s_0 and s_1 collect temperature data. A pair of the sensors s_2 and s_3 collect humidity data. Each sensor gets sensor data every one second and sends the data to a parent edge fog node.

The sever f_0 is a PC with a CPU Intel Xeon X3430, 16 [GB] memory, and 2 [TB] HDD, whose operating system is CentOS 7 [1]. In the server f_0, a Sybase [4] database is supported to store data obtained from the fog node f_{00}. In the TBFC model, the *store* subprocess is implemented in C language with transact SQL [4].

In the experiment, each edge fog node is equipped with one sensor as shown in Fig. 4. A pair of the edge fog nodes f_{0000} and f_{0001} collect temperature data from the sensors s_0 and s_1, respectively. Another pair of edge fog nodes f_{0010} and f_{0011} collect humidity data from the sensors s_2 and s_3, respectively. A process of a sensor to collect sensor data is realized in Python [2]. Sensor data is collected by the polling mechanism every one second. A record of collected sensor data is sent by each sensor to an edge fog node, which is composed of *time* and *value*. This means, the *value* is sensor data, i.e. temperature and humidity, which is obtained at the *time*. That is, temperature data is collected by a pair of the sensors s_0 and s_1 and humidity data is collected by another pair of the sensors s_2 and s_3 at *time*. The sensors s_0, s_1, s_2, and s_3 send records of sensor data to the edge fog nodes f_{0000}, f_{0001}, f_{0010}, and f_{001}, respectively, every one second.

Each edge fog node f_{00ij} receives sensor data from a sensor every one second. Then, the edge fog node f_{00ij} calculates an average value of sensor data, temperature or humidity data collected from the sensors for every one minute (i, $j = 0, 1$). A subprocess *aggregate* is performed by each edge fog node f_{00ij} to obtain an average value from input data. Then, the edge fog node f_{00ij} sends the output data d_{00ij} to the parent fog node f_{00i}.

A parent fog node f_{00i} receives a pair of input data d_{00i0} and d_{00i1} from child fog nodes f_{00i0} and f_{00i1}, respectively. The parent fog nodes f_{000} and f_{001} collect temperature data and humidity data from child fog nodes, respectively. A subprocess *merge* is performed on each fog node f_{00i}. Then, a pair of the input data d_{00i0} and d_{00i1} are merged into the output data d_{00i}. If values of input data d_{00i0} and d_{00i1}, whose *time* is the same, are received, the output data d_{00i} is the average value of the input data d_{00i0} and d_{00i1}. Here, the output data d_{000} of temperature is sorted in *time*. The output data d_{001} of humidity is also sorted in *time*. The fog node f_{00i} sends the output data d_{00i} to the fog node f_{00} ($i = 0, 1$).

The fog node f_{00} receives input data from the child fog nodes f_{000} and f_{001}. A pair of input data $d_{000} = \langle v_{000}, t_{000} \rangle$ and $d_{001} = \langle v_{001}, t_{001} \rangle$ are joined, i.e. concatenated by the fog node f_{00} into one output data d_{00}. In the output data d_{00}, temperature data v_{000} and humidity data v_{001} whose time is the same, i.e. $t_{000} = t_{001} = t$ are concatenated to a record $\langle t, v_{000}, v_{001} \rangle$. Thus, a subprocess *join* is performed on the fog node f_{00}. The fog node f_{00} sends the output data d_{00} to the root node f_0.

The root fog node f_0 is a server which receives input data $d_{00} = \langle t, v_{001}, v_{001} \rangle$ from the fog node f_{00}. The server f_0 stores the data d_{00} to a table Data (time, temperature, humidity) in the database DB_0 by SQL insert [7] once the server f_0 receives the data. The database DB_0 is implemented in Sybase [4]. A subprocess *store* is performed on the root node f_0.

In the cloud computing model, the sensors s_0, s_1, s_2, and s_3 are directly interconnect with a sever f_0 as shown in Fig. 5. Sensor data from the sensors s_0, s_1, s_2, and s_3 are sent to the server f_0 by using UDP in a Gbps LAN. All the *aggregate*, *merge*, *join*, and *store* subprocesses are performed on the server f_0. Every sensor data sent by the sensors is processed by a sequence of the *aggregate*, *merge*, *join*, and *store* subprocesses on the server f_0.

4.2 Experiment

We measure total execution time TET [sec] of all the fog nodes and the server since the sensors s_0, s_1, s_2, and s_3 send temperature and humidity data to the edge fog nodes until the server f_0 stores the data to the database DB_0 in the cloud computing model and in the TBFC model. In order to measure the total execution time TET, one fog node f_c is used as shown in Figs. 5 and 6. The fog

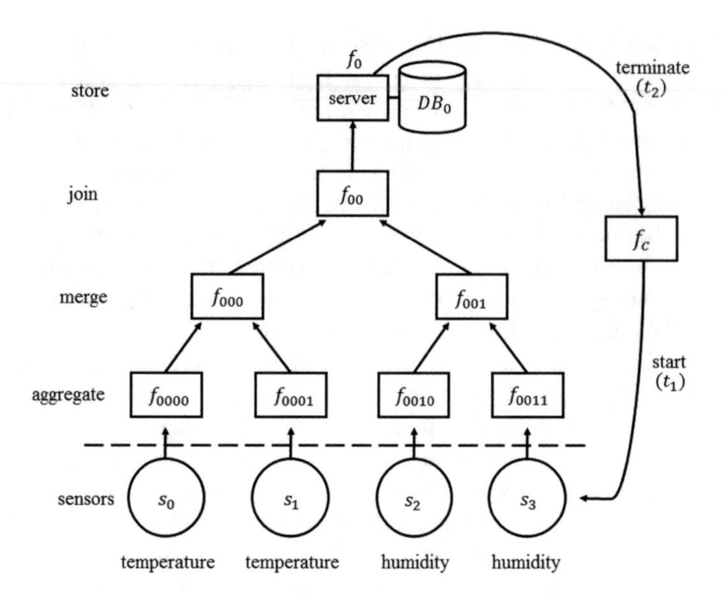

Fig. 5. Experiment of TBFC model.

node f_c sends a *start* message to every sensor at time t_1 and then the sensors start sending sensor data. If every sensor data is stored in the database DB_0, the server f_0 sends a *termination* message to the fog node f_c. The fog node f_c receives the *termination* massage at time t_2. Here, the total execution time TET is $t_2 - t_1$ [sec].

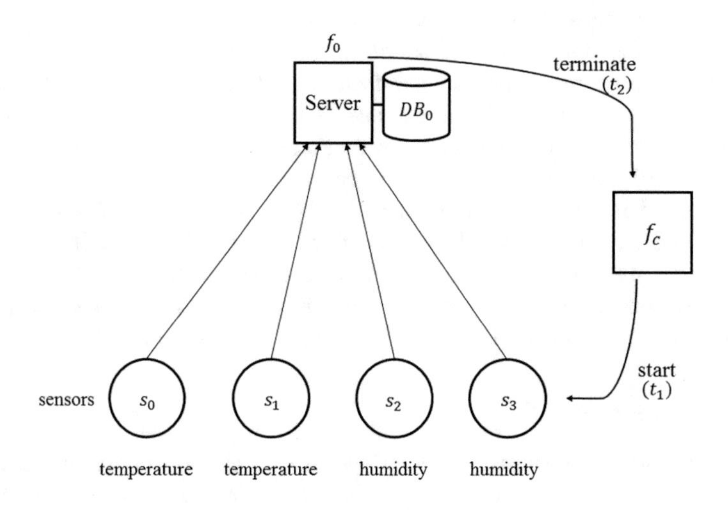

Fig. 6. Cloud computing model.

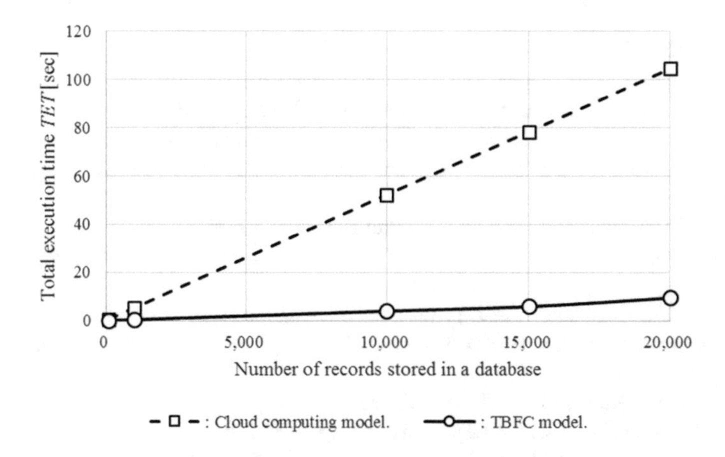

Fig. 7. Total execution time.

Figure 7 shows the total execution time TET of the TBFC model and the cloud computing model for number rn of records stored in the database DB_0. The total execution time TET of the TBFC model and the cloud computing model linearly increases as the number rn of records increases. The total execution time TET of the TBFC model is shorter than the cloud computing model. For example, the total execution time TET of the TBFC model is only 7% and 9% of the cloud computing model, respectively, for $rn = 10,000$ and $rn = 20,000$. This experiment shows the IoT can be efficiently realized by the TBFC model.

5 Concluding Remarks

The fog computing model is useful to efficiently realize the IoT since processes and data are distributed to not only servers but also fog nodes. Then, traffic of servers and networks can be reduced in the fog computing model. In this paper, we discussed the implementation of each fog node in the tree-based fog computing (TBFC) model by a Raspberry Pi 3 Model B computer. We implemented the subprocesses of computation complexity $O(x)$ and $O(x^2)$ for size x of input data, to be performed by each fog node. We showed the experiment of the implemented TBFC model in Raspberry PI.

Here, four sensors, seven fog nodes, and a server are hierarchically structured in a height-balanced tree. In the evaluation, the total execution time of the TBFC model is about 90% to 95% shorter than the cloud computing model. We showed the IoT can be efficiently realized in the TBFC model.

In the IoT, it is critical to reduce the total electric energy consumption [11,12]. We are now evaluating the TBFC model in terms of electric energy [8,9] consumed by fog nodes and servers.

Acknowledgements. This work was supported by JSPS KAKENHI grant number 15H0295.

References

1. The centos linux distribution (centos linux). https://www.centos.org/
2. Python. https://www.python.org/downloads/release/python-2713/
3. Raspberry Pi 3 model B. https://www.raspberrypi.org/products/raspberry-pi-3-model-b/
4. Sybase. https://www.sap.com/products/sybase-ase.html
5. Comer, D.E.: Internetworking with TCP/IP, vol. 1. Prentice Hall, Englewood Cliffs (1991)
6. Creeger, M.: Cloud computing: an overview. Queue **7**(5), 3–4 (2009)
7. Date, C.J.: An Introduction to Database System, 8th edn. Addison Wesley, Reading (2003)
8. Enokido, T., Ailexier, A., Takizawa, M.: An extended simple power consumption model for selecting a server to perform computation type processes in digital ecosystems. IEEE Trans. Industr. Inf. **10**(2), 1627–1636 (2014)
9. Enokido, T., Ailexier, A., Takizawa, M.: A model for reducing power consumption in peer-to-peer systems. IEEE Syst. J. **4**(2), 221–229 (2010)
10. Hanes, D., Salgueiro, G., Grossetete, P., Barton, R., Henry, J.: IoT Fundamentals: Networking Technologies, Protocols, and Use Cases for the Internet of Things. Cisco Press, Indianapolis (2018)
11. Oma, R., Nakamura, S., Duolikun, D., Enokido, T., Takizawa, M.: Evaluation of an energy-efficient tree-based model of fog computing. In: Proceedings of the 21st International Conference on Network-Based Information Systems (NBiS-2018), pp. 99–109 (2018)
12. Oma, R., Nakamura, S., Duolikun, D., Enokido, T., Takizawa, M.: Fault-tolerant fog computing models in the IoT. In: Proceedings of the 13th International Conference on P2P, Parallel, Grid, Cloud and Internet Computing (3PGCIC-2018) (Accepted, 2018)
13. Oma, R., Nakamura, S., Enokido, T., Takizawa, M.: An energy-efficient model of fog and device nodes in IoT. In: Proceedings of IEEE the 32nd International Conference on Advanced Information Networking and Application (AINA-2018), pp. 301–306 (2018)
14. Oma, R., Nakamura, S., Enokido, T., Takizawa, M.: A tree-based model of energy-efficient fog computing systems in IoT. In: Proceedings of the 12th International Conference on Complex, Intelligent, and Software Intensive Systems (CISIS-2018), pp. 991–1001 (2018)

A Framework for Automatically Generating IoT Security Quizzes in a Virtual 3D Environment Based on Linked Data

Wei Shi[1(✉)], Chenguang Ma[2], Srishti Kulshrestha[3], Ranjan Bose[4], and Yoshihiro Okada[1]

[1] The Innovation Center for Educational Resources, Kyushu University Library, Kyushu University, Fukuoka, Japan
shi.wei.243@m.kyushu-u.ac.jp, okada@inf.kyushu-u.ac.jp

[2] NTT Data Global Solutions, Tokyo, Japan
machenguanglala@yahoo.co.jp

[3] School of Information Technology, Indian Institute of Technology Delhi, New Delhi, India
srish.kul@gmail.com

[4] Indraprastha Institute of Information Technology Delhi, New Delhi, India
bose@iiitd.ac.in

Abstract. With the development of information and communication technology, e-learning becomes more and more popular and important. Especially, because the mobile computing devices such as smartphones become popular, we can use e-learning contents at any time and anywhere. One kind of e-learning contents is the quiz games. It is a kind of basic but widely-used e-learning contents. It can be used both to test students and to help them remember knowledge points. Because of the development of 3D technology, some quiz makers have started to provide quizzes in a 3D environment. In this paper, we propose an extended framework which can help quiz makers to create quizzes in a Virtual 3D Environment based on the Linked Data. Such kind of quizzes can train users more effectively. We also discuss how to create IoT (Internet of Things) security quizzes. The generated quiz can also collect users' activity data for the further analyzation.

1 Introduction

In current years, because computers and mobile computing devices have become popular, e-learning plays an important role in education. To develop effective e-learning contents, advanced information and communication technology (ICT) such as 3D Animation and VR/AR technologies are wildly used. As one type of e-learning contents, serious games are helpful for testing and training users. The most widely used serious games are the quiz games. The traditional method of creating quizzes is to store each question, its choices, and maybe the hints as records into a relational database or a plain text file. We suppose that quiz games created by this method have two demerits: (1) Normally, the questions and corresponding choices of quiz games will be stored in

L. Barolli et al. (Eds.): EIDWT 2019, LNDECT 29, pp. 103–113, 2019.
https://doi.org/10.1007/978-3-030-12839-5_9

a database. To build such a database, quiz makers must spend a lot of time. However, these resources are difficult to share and reuse. In fact, many similar kind quiz games are created using such relational databases. It means that to create these quizzes, a lot of storage and time are wasted. (2) The questions' choices are lack of changes. This reduces the training and testing effects of the quiz games. Because many users may answer questions by remembering the position of the right choice or several words in the right choice to obtain the good scores. Although many quiz makers add some changes, it is still not enough. Currently, most quizzes are created using text and 2D media resources. In the future, various of the multi-media resources, such as 3D animation contents, VR/AR contents, and 360-degree photos will be widely used to create the quizzes.

In this paper, we introduce an extended framework extended from our previous research [1]. This framework supports the creation and the customization of quiz games based on the Linked Data in a virtual 3D environment. We believe the virtual 3D environment is closer to reality and can train users more effectively. The quiz games created by our framework can avoid the problems mentioned above. To demonstrate our framework, we introduce an IoT security quiz which is created by using our framework.

The remainder of this paper is organized as follow: Sect. 2 introduces the related work of our research; Sect. 3 explains the detail of our framework; Sect. 4 explains how to extend our framework for supporting the creation of quizzes in a virtual 3D environment. The last section concludes this paper and discusses our future work. In the remaining sections, we use the term "quiz makers" or the "system users" to indicate the people who use our framework to make their quizzes and use the "quiz users" to indicate the people who use the created quiz games.

2 Related Work

In this section, we first introduce the technologies which will be used in our framework, the related researches, and compare our research with them.

2.1 Related Technology

In this research, we use the Linked Data for storing the quiz resources. Linked Data refers to "a set of best practices for publishing structured data on the Web." [2] A data item of Linked Data is defined as a Resource Description Framework (RDF) triple. [3] Each RDF triple is composed by a subject, a predicate, and an object. Every data element has a URI as its identifier. One triple may link to other triples. The features of Linked Data allow users to easily share, extend and reuse the data. Many organizations publish their Linked Data sets such as DBpedia (https://wiki.dbpedia.org/) for open accessing. Comparing to the relational database, the RDF datasets have more flexible structures, and users can query data and relationships among data items by a special language, called SPARQL [4]. In this paper, our framework supports to use the linkages of RDF data triples to realize the automatic generation of quiz games.

In this paper, the generated quiz is composed as a web application which is programmed by the HTML5 and JavaScript. This makes all generated quizzes to be cross-platform and cross-device. The virtual 3D environment in our framework is constructed using Three.js (https://threejs.org/) library. Three.js use WebGL (https://www.khronos.org/webgl/), which "is a cross-platform, royalty-free web standard for a low-level 3D graphics API based on OpenGL ES, exposed to ECMAScript via the HTML5 Canvas element".

2.2 Related Work

In recent years, because of the development of e-learning, some researchers have paid their attention to discuss on how to provide high-quality quizzes. Foulonneau [5] discussed a streamline which can generate assessment items (which has the same means as the "questions" in this paper) using the Linked Open Data from DBpedia. However, in this paper, the RDF data used to generate one question are limited to one triple. The linkages among the things are not used effectively.

Similarly, Liu and Lin [6] proposed the "Sherlock", which can semi-automatically generate quizzes using the Linked Data from DBpedia and BBC. In this paper, they mainly focus on how to "control the difficulty level of the generated quizzes". The templates provided for creating the quizzes are still too basic and may be difficult to improve. The linkages among triples of RDF data are not effectively used as well.

A problem in these researches is that most quiz makers are considered to have the knowledge of the RDF data schema. This is also the reason why the above researches are limited to the special datasets such as the DBpedia. However, there are many different RDF datasets with different schemata. Even some RDF datasets do not have fixed schemata. Such kind of RDF datasets are difficult to be used to generate quizzes, and these two papers have not discussed how to solve this problem. In our paper, we will introduce a tool which can extract and visualize the RDF schema to solve this problem.

Some researchers have also started to use virtual 3D learning environments. Comparing to the 2D e-textbook, the 3D learning environment can greatly improve users' understanding of knowledge, especially in the special fields. For example, in medical education, 3D virtual viscus is much easier to understand than pictures. Students can observe viscus from different angles and get more details. Langi [7] and Livingstone [8] introduce their 3D virtual learning environment. Both systems include quiz functions. But in these two papers, the authors mainly discuss how to construct the learning environments. Rasim also discusses how to collect data and analyze it. Quiz functions are only the additional functions. They did not discuss how to effectively generate quizzes in the 3D environment, and the difference of the normal quizzes and the quizzes in the 3D environments. In this paper, we will discuss how to create quizzes in virtual 3D environments.

3 A Framework for Automatically Generating the Quizzes Based on Linked Data

The framework proposed in this paper is an extension from our previous research [1]. This framework can support the automatic generation of the quizzes through the linkages among the RDF triples in the Linked Data sets. The generated quizzes work as web applications and can collect users' log data for further analyzation. In this section, we will introduce this framework using an example of IoT security quiz.

3.1 An Overview of This Framework

This framework includes two tools. One is an RDF data schema extraction and representation tool. This tool can extract the schema of the RDF data and represent the extracted schema as a graph. By using this tool, instead of knowing the schema of the RDF data which is used to create the quiz, the schema will be visualized and then users can visually manipulate it to define quizzes. The other tool is an authoring tool. This authoring tool provides a set of HTML components for composing quiz page templates. By manually assigning the connections between HTML components and nodes of the visualized RDF schema, the system will automatically generate a set of SPARQL queries. Using these queries, our system can dynamically retrieve the necessary data from our RDF dataset and provide the values to the corresponding HTML components. This will lead to the dynamic generation of the quiz pages. Figure 1 shows an overview of using our framework to create an IoT security quiz. In each question page, there are two questions. For answering the first question, quiz users need to select the name of the IoT device shown in the pictures. The second question is an extension of the first question. Quiz users need to answer what attack may be performed on the selected device in the first question.

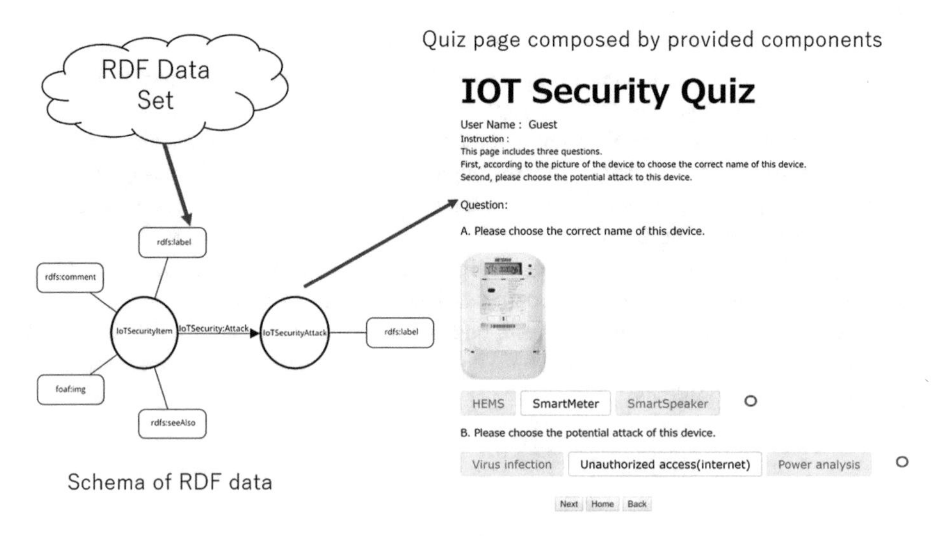

Fig. 1. An IoT security quiz generated by our framework

In the following sections, we will show more details of our framework by explaining how to use our framework to create the quiz in Fig. 1. To simplify the description of the URI in our database, we use the "IoTSecurity:" to replace "http://……./ IoTSecurity/", which is the prefix of the URIs of the RDF data in our database. We also use another two prefixes of the URIs in this paper, the "rdfs:" to replace "http://www.w3.org/2000/01/rdf-schema" and use the "rdf:" to replace "http://www.w3.org/1999/02/22-rdf-syntax-ns", which are defined by W3C.

3.2 RDF Schema Extraction and Representation

To extract the RDF data schema, we reuse the idea of the ESISW framework proposed by Piao [9]. To make the schema easy to read, we modified the graph which used in ESISW framework to represent the schema. The left lower part of Fig. 1 shows the schema of the RDF data for creating our quiz application.

In this framework, the schema is simplified to be composed of edges and two kinds of nodes. One kind of nodes is the Class node, and the other kind is the Property node. The edges in such a graph may be labeled or not. We suppose there is an RDF dataset *R*. As we introduced, the data item *t* in *R* is identified by a unique URI. We suppose *t* is an instance of class *T*. To generate the schema of *R*, we first extract all the classes in this dataset. A class T_c will be represented as a circular Class node (e.g. the node "IoTSecurityItem" in Fig. 1). Next, we extract all the instances of T_c and all the tuples whose subject is these instances. Then, we extract the objects of these tuples. If the value of an object is a fixed value, such as number or text, we add a rectangular Property node to the schema. The label of this node is the predict (e.g., the node "rdfs:comment" in Fig. 1). The edge from the Class node T_c to this property node will be unlabeled. If the value of an object is an instance of another Class T_a, an edge from the Class nodes T_a to T_c will be drawn. The label of this edge is the triple's predict.

3.3 Authoring Tool for Composing Quiz Games

Our framework includes an authoring tool for supporting quiz makers to create quiz games as web applications. This authoring tool is an extension of Ma's research [10]. Ma proposes an authoring tool for creating e-textbooks based on the Linked Data. In our framework, we use such an authoring tool to define the templates of the quiz pages. To simplify users' manipulations, we provide a set of template components. These components are text components, image components, video components sound components, navigation button components, checkbox components, and choice components. Besides these provided components, users can add the necessary HTML elements by directly programming. These components are provided in the form of HTML + JavaScript codes.

In the page template definition, users also need to specify the class nodes which are used as the question, the right choice, and the wrong choice. Then users can associate the component with the Property nodes in the RDF schema. When a quiz page is generated, the values of corresponding properties will be provided to the HTML components and be displayed on the web page. In the first question of our example quiz, we defined the Node "IoTSecurityItem" is used as the class of the question and its choices, the question image is related to the Node "ofaf:img" to the question's picture, the text of each choice is related to its Property Node "rdfs.label".

3.4 Automatic SPARQL Generation and Data Retrieval

After composing a quiz page and associating the components with the nodes of the RDF schema, our system will automatically generate SPARQL queries for retrieving necessary data for generating the quiz pages. The SPARQL queries are generated in two steps. First, users need to specify how to generate the quiz questions and their right answers. Supposing we specify a class A as the question and its linked class B as the right answer, the edge between A and B are labeled as L, and the properties "A.P1" and "B.P2" are associated with the HTML components in the quiz page, the system will generate a query as follows:

```
SELECT ?v1 ?v2 ?v2 ?v3
  WHERE {
        ?v1 rdf:type "A".
        ?v2 L ?v1.
        ?v1 P1 ?v3.
        ?v3 P2 ?v4.
  }
  LIMIT 1
  OFFSET randomNumber .
```

In this query, "randomNumber" is a random number generated by the system to retrieve a random item from the database as the question and the right answer. To generate the first question of our example, the SPARQL query should be

```
SELECT   ?v1 ?v2 ?v3
WHERE {
?v1 rdf:type "IoTSecurityItem
?v1 rdfs:label ?v2.
?v1 pt:IMG ?v3.
}
LIMIT 1
OFFSET randomNumber .
```

Next step, users need to specify how to generate the wrong choices. As a question's wrong choices, they should have one or more common features with the right choice. Users need to define the constraints to specify these common features through the RDF schema, and then our system can translate such specifications into the SPARQL query. Normally, the wrong choices and the right choices should belong to the same class, but users can specify another class for the wrong choices. To define a new constraint, we suppose that the URI of the right answer is *Ur*, the wrong choices is *Uw*, the *Ur*'s property *Pcr* and the *Uw*'s property *Pcw* are used to define the constraint between the wrong answers and the right answer, and the constraint between these two properties is *f*. The constraint is defined as:

$$\mathrm{Uw.Pcw} = \mathrm{f(Ur.Pcr)}.$$

Currently, our framework supports nine kinds of relationship between the right answer and wrong answers. They are "equal", "large than", "less than", "unequal", "union", "and", "minus", "contain", and "not contain". In the future, we will add necessary computations according to the requests from the quiz makers.

In the first question of our quiz example, we only specify the default constraints between the right answer and the wrong answers: the right answer and wrong answers belong to the same class "IoTSecurityItem". Then, we suppose the URI of the right answer is *U* and we need *n* wrong answers, then the SPARQL query for generating one wrong choice is shown as follows:

```
SELECT  ?v1 ?v2 ?v3 ?v4 ?v5
WHERE {
    U rdf:type ?v1.
    ?v2 rdf:type ?v1.
    ?v2 rdfs:label ?v3.
}
LIMIT 1
OFFSET randomNumber
```

This query will be performed *n* times to retrieve *n* wrong answers for this quiz.

4 Extending Our Framework for Generating Quizzes in a Virtual 3D Environment

In this section, we introduce how to extend it to generate the quiz in a virtual 3D environment. In our previous research [11], we propose a framework which supports the development of the VR/AR E-learning materials for IoT Security Education. In this paper, we reuse the same framework for providing the virtual 3D environment. We also use a quiz for IoT security in this section to explain our framework.

Figure 2 shows a screenshot of this virtual 3D environment for the IoT security quiz. In this quiz, we focus to train and to test the knowledge about the attacks on smart devices. In this environment, all objects with no actions are defined as .obj files with corresponding .mtl files, and all objects with actions are defined as .dae files. We use the Blender (https://www.blender.org/) to create .obj and .dae files. There are two textboxes in this environment. One is used for providing necessary information, and the other is used for providing quiz questions. The red character is a non-player character (NPC) who will guide quiz users to finish the quiz in some situations.

Figure 3 shows the construction of the extended framework for creating the 3D quiz environment. In this framework, we use the AnimationCharacter.js to define the class of animation characters. The NPC and IoT devices are instances of this class.

Fig. 2. A screenshot of the virtual 3D IoT security quiz environment

Comparing to these objects, some objects in the environment only work as the background, such as the chairs. They are pre-loaded in the IoTSecurity.js. In this file, we also define the positions, scales, and rotations of all the objects, and pre-load all actions of each IoT devices and the NPC. In the main HTML file, we define how to compose the web application, and define the non-3D objects, such as the textboxes. The quiz.js file includes how to retrieve data from our Linked Data set, the retrieved data for generating quiz pages, the association between the retrieved data and the HTML component, and the mapping from the question to the object's abnormal event.

In this quiz game, one event related with such an abnormal state will be triggered to start a quiz question. According to such event, quiz users need to answer which device is under attack and give a solution. For example, a smart speaker will suddenly start to play a music without the owner's order. Then, the NPC will ask quiz users to find out the abnormal device. Quiz users will answer this question by directly clicking the smart speaker. Then users need to answer the name of the attack. In this framework, we define two kinds of events, auto-triggered events and NPC-triggered events. An auto-triggered event means the event can be triggered automatically. An NPC-triggered event means the NPC need to do some action to trigger the event. The example of the smart speaker is an auto-triggered event. In another situation, the NPC said he cannot control the smart speaker to attract users' attention to start a question. The event of the NPC's speaking action is an NPC-triggered event.

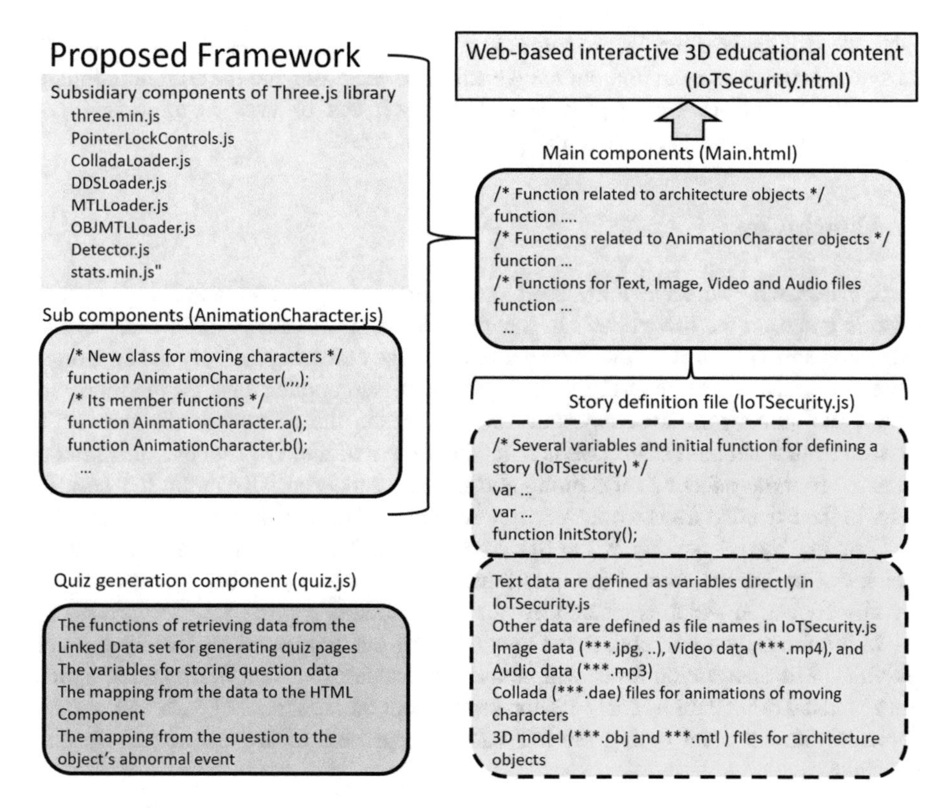

Fig. 3. The construction of our framework for creating the 3D quiz environment

All events are defined as the action of the corresponding objects in the IoTSecurity.js file. In this file, we also need to map the questions to the object (the NPC or devices) actions in the 3D environment. In our database, each attack to a specific device should have a corresponding abnormal event. The abnormal events have a standard format:

$$\#\text{Target} : \text{Action}(\text{Parameters})\,.$$

When users create the 3D quiz, they need to associate the abnormal event with the NPC or the IoT device. Currently, we support three kinds of auto-triggered events on the IoT devices, play sounds, play videos and change positions. We also define three kinds of actions of the NPC. The NPC can speak, walk and do some actions. The functions of handling these events are included in the AnimationCharacter.js file. In the current state, our system only can trigger one event at the same time.

In the future, we will add more functions to this system. One is to realize the complex quiz, we hope to add the support of the multiple abnormal events trigger. Users need to analyze these events to find out the under-attack device. We also want to collect users' activity data and their feedback. These data will be used in two aspects.

First, we will analyze the data to improve users' experiences when they are using the quizzes. Next, we want to find out the weakness of each user and the common difficult points during the answering processes. This result can be used to improve users' learning qualities and to improve the generated quizzes.

5 Conclusion

In this research, we have introduced our framework for automatically generating quizzes based on the linked data. In our framework, it includes an RDF data schema extraction and visualization tool, and a question page authoring tool. By the support of these two tools, users can easily define quizzes as web applications. Furthermore, we extend our framework for supporting users to create the quizzes in the virtual 3D environment. Such kind of quizzes is much closer to real life. We suppose such kind of quizzes are more effective for training and testing users especially in the IoT security field. In the extended framework, we propose a method of mapping the questions to the NPC or IoT device actions in the 3D environment. These actions are called events which can trigger the start of the questions. We also introduce a 3D quiz for the IoT security education which is created by our framework.

In the future, we will collect the data of using our framework and analyze users' activities. The analyzation result will be used to evaluate our framework and to update the generated quiz. Furthermore, we hope to find out the weakness of each user and the common difficult point during the IoT education process to improve users' learning activities.

Acknowledgments. This research was mainly supported by Strategic International Research Cooperative Program, Japan Science and Technology Agency (JST) regarding "Security in the Internet of Things Space", and partially supported by JSPS KAKENHI Grant Number JP16H02923 and JP17H00773.

References

1. Shi, W., Kaneko, K., Ma, C., Okada, Y.: A Framework for Automatically Generating Quiz-Type Serious Games Based on Linked Data (In press)
2. W3C: LinkedData - W3CWiki (2016). https://www.w3.org/wiki/LinkedData
3. Berners-Lee, T.: Linked data – design issue (2009). https://www.w3.org/DesignIssues/LinkedData
4. W3C: SPARQL Query Language for RDF (2008). https://www.w3.org/TR/rdf-sparql-query/
5. Foulonneau, M.: Generating educational assessment items from linked open data: the case of DBpedia. In: ESWC 2011: The semantic Web: ESWC 2011 Workshops, pp. 16–27. Springer, GmbH, Berlin, Heidelberg (2012)
6. Liu, D., Lin, C.: Sherlock: a semi-automatic quiz generation system using linked data. In: Proceedings of the 2014 International Conference on Posters & Demonstrations Track, ISWC-PD 2014, vol. 1272, pp. 9–12. CEUR-WS.org, Aachen, Germany (2014)

7. Langi, A.Z.R., Rosmansyah, Y.: Implementation of 3D virtual learning environment to improve students' cognitive achievement. In: IOP Conference Series: Journal of Physics, vol. 1013, p. 012103 (2018)
8. Livingstone, D., Kemp, J.W.: Integrating web-based and 3D learning environments: second life meets moodle. In: CEPIS (Council of European Professional Informatics Societies)/ Novática, pp. 9–14 (2008)
9. Piao, B., Tanaka, Y.: Interactive framework for exploratory search, integration, and visual analysis of semantic web resources. In: Proceedings of the 16th International Conference on Information Integration and Web-based Applications & Services, pp. 200–206. ACM New York (2014)
10. Ma, C., Srishti, K., Shi, W., Okada, Y., Bose, R.: Educational material development framework based on linked data for IOT security. In: ICERI 2017 Proceedings, pp. 8048–8057. IATED (2017)
11. Ma, C., Kulshrestha, S., Shi, W., Okada, Y., Bose, R: E-learning material development framework supporting VR/AR based on linked data for IoT security education. In: Advances in Internet, Data & Web Technologies, EIDWT 2018. Lecture Notes on Data Engineering and Communications Technologies, vol. 17, pp. 479–491. Springer, Cham (2018)

Authentication Protocols Using Multi-level Cognitive CAPTCHA

Marek R. Ogiela[1(✉)] and Lidia Ogiela[2]

[1] AGH University of Science and Technology,
30 Mickiewicza Avenue, 30-059 Kraków, Poland
mogiela@agh.edu.pl

[2] Pedagogical University of Cracow,
Podchorążych 2 Street, 30-084 Kraków, Poland
lidia.ogiela@gmail.com

Abstract. This paper will present new approaches for creation of authentication protocols, which use multi-level CAPTCHA codes. Proposed codes will create a new class of cognitive CAPTCHA, which require special skills for fast and secure personal authentication, which can be performed is several stages. Authentication protocols can be oriented to allow access for trusted group of persons, based of theirs perception abilities and cognitive skills. For new authentication protocols some possible examples of applications will be also presented.

Keywords: Authentication procedures · Cognitive CAPTCHA · Security procedures · Multi-level authentication

1 Introduction

In advanced security systems we can apply selected personal features or human cognitive abilities for authentication purposes. Such personal features or abilities allow to define user oriented security algorithms, and also extend available security protocols towards creation a new branch called cognitive cryptography. This idea of cognitive cryptography was firstly presented in [1, 2], and in general it is connected with possible application of human oriented protocols and cognitive abilities into security protocols. The broad range of application of such procedures also consist authentication solutions, in which can be used CAPTCHA codes.

CAPTCHA codes are used very often in many application and services. We can describe several approaches for creation of CAPTCHA codes, which depend on the type of signal. Always the main purpose of CATCHA authentication is to guarantee that we receive proper response originating from human user, and not from remote computer systems or intelligent machine. We can define five different categories of CAPTCHA codes like: text, audio, visual, motion, and hybrid solutions combining previously mentioned types [3, 4]. CAPTCHA codes are safe, because distortions or noises, which can be introduced to original patterns, makes the interpretation processes much more difficult than the analysis of original patterns. Possible distortions in visual CAPTCHA may have a form of shadows, outlines, noises, random degradation, stripes, different geometrical transformation etc. [5].

L. Barolli et al. (Eds.): EIDWT 2019, LNDECT 29, pp. 114–119, 2019.
https://doi.org/10.1007/978-3-030-12839-5_10

The aim of this paper is to present more advanced security authentication protocols using CAPTCHA codes, but such codes which depend on the perception capabilities of particular user [6–8]. Cognitive CAPTCHA codes are oriented for specific group of participants, which possess specific skills and knowledge, which can be properly used while authentication processes. Personally oriented verification procedures can play an important role in many applications like services management, IoT, smart technologies etc. [9, 10].

Further in this paper will be presented some new ideas of creation of multi-level cognitive codes proposed for security and authentication purposes. Such technologies seems to be very promising for developing technologies, oriented for personal authentication, which can use specific perception capabilities.

2 Multi-level CAPTCHAs

As was mentioned above, it is possible to create security protocols, which use personal cognitive abilities. Cognitive protocols can join security solutions with perceptual or personal features, what allow to define user oriented security procedures. Below will be described solutions oriented on multi-level cognitive authentication protocols. In general multi-level verification can be realized in several iteration in which different object may be analyzed, different questions can be asked or simply different perception threshold can be used for proper authentication. Further will be described following types of multi-level cognitive CAPTCHAs:

- Attention moving verification codes,
- Multi questions verification codes,
- Fuzzy perception thresholds.

2.1 Attention Moving Approach

The first example of cognitive multi-level authentication codes is connected with attention moving approach. In such type of CAPTCHAs, verification is performed in several consecutive iterations, in which are asked verification questions connected with different parts of visual codes or connected with different objects visible on it.

Figure 1 presents an example with multi-level CAPTCHA answers depending on the division of source visual pattern, or connected with different objects. Depending on the image division for particular parts it is possible to find different possible answers, which may be required sequentially during verification procedure. When we consider only one part of this image, it is possible to find answer for particular question, but for more detailed division it is also possible to find other solutions based on different regions of interest.

2.2 Multi-questions Verification

The second type of multi-level codes can use verification based on several different questions connected with presented pattern, or different objects focused for each question. The main feature of such solution is possibilities to quickly verify if authenticated user has a deep knowledge connected with presented pattern, and also has

Fig. 1. Examples of cognitive CAPTCHA with moving attention verification. In the image presenting Wall Street photo (source: Internet) are visible marked rectangles defining different ROIs (Regions of Interest). Verification question may be in the form: On which part(s) do you see a human figure? Depending on the ROI it is possible to find different answers. The question may also be connected with visible vehicle, or visible color, number of people etc.

skills to quickly answer for several questions with different detail levels. These levels should change for each question, and the whole set of question can be related to different area of expertise. Such protocol allow not only determine if the answering is human, but also determine this user represents a particular area of expertise [11, 12]. Example of such verification is presented in Fig. 2.

2.3 Fuzzy Perception Thresholds

Last example is connected with CAPTCHA codes with different perception thresholds. In this procedures we can use different visual patterns blurred in such way that the main object is not quite easy recognizable. Authenticated user should recognize the content of fuzzy images, what is possible only in situation when he knows something about it, or has a knowledge which allow to guess the proper answer. This type of authentication is also performed in several iteration with different perception thresholds, which can be changed according cognitive and perceptual skills. Below in Fig. 3 is presented an example of image content recognition with different perception thresholds.

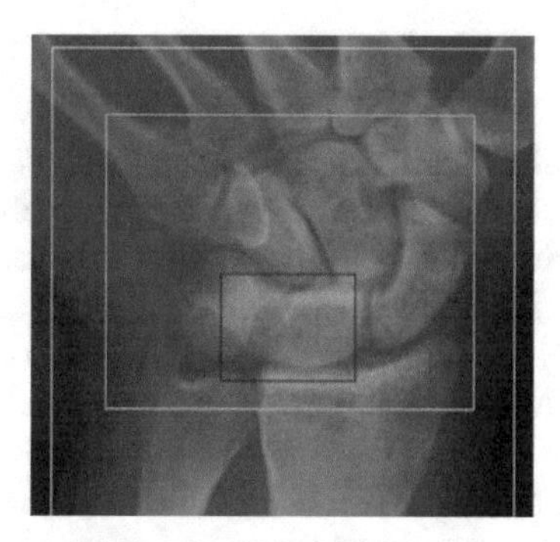

Fig. 2. Examples of cognitive CAPTCHA with multi-questions verification. Image presents wrist bones with some pathologies. In consequent questions, user can be asked about: type of visualization in yellow rectangle (RTG, MRI, CT etc.), presented structure, physiological standards for bones visible in green rectangle, pathological changes marked in red rectangle etc.

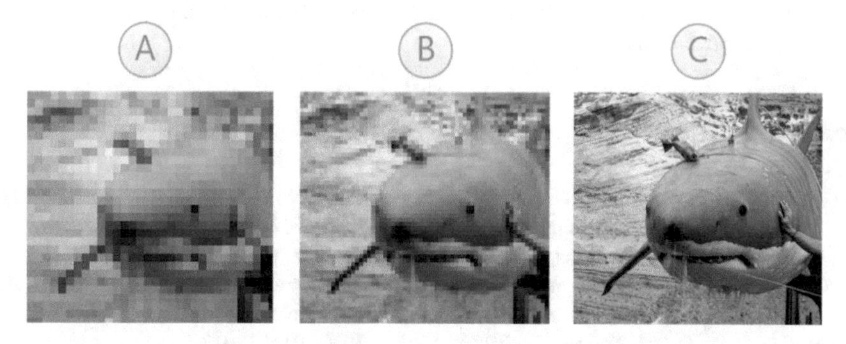

Fig. 3. Examples of cognitive CAPTCHA with fuzzy perception thresholds. On part A user can recognize a fish. On part B it is possible to see a shark. On part C details shows that this object is a submarine.

3 Applications of Multi-level CAPTCHAs

In presented multi-level verification procedures it is possible to provide access for user to systems or services, after answering for questions in proper manner. It means that user should select CAPTCHA parts according questions or content, what finally allow to give him accessing rights. Inappropriate selection of any parts during verification, should prevent the access to data or systems. In all types of presented codes the most important is to correctly select the proper semantic combination of elements, which fulfil requirements or have a specific meaning.

Described multilevel authentication procedures can be applied for verification or authorization instances or users in distributed infrastructures, Big Data repositories, Cloud Computing or many services. Multi-level procedures allow also to obtain access at different stages or levels, depending on the grants and security preferences. For higher security it may be important to perform a larger number of iteration in authorization procedure before providing data or services [13].

4 Conclusions

In this paper have been presented new authentication procedures, which apply cognitive CAPTCHA approaches. Such authentication protocols work in several iteration and use different perception and cognitive skills characteristic for particular users. Security of such techniques lays not only on difficulties in recognition letters or images like in classical CAPTCHA codes, but especially on semantic interpretation of the content, fast meaning evaluation and knowledge on particular area or situation.

It was described three types of multi-level CATCHAs, which allow to verify persons according special skills and knowledge, which is necessary for positive verification. Presented solutions are oriented for groups of people having expert knowledge or high cognitive and perceptual skills.

Acknowledgments. This work has been supported by the AGH University of Science and Technology research Grant.

This work has been supported by the National Science Centre, Poland, under project number DEC-2016/23/B/HS4/00616.

References

1. Ogiela, M.R., Ogiela, L.: On using cognitive models in cryptography. In: IEEE AINA 2016 - The IEEE 30th International Conference on Advanced Information Networking and Applications, Crans-Montana, Switzerland, 23–25 March, pp. 1055–1058 (2016)
2. Ogiela, M.R., Ogiela, L.: Cognitive keys in personalized cryptography. In: IEEE AINA 2017 - The 31st IEEE International Conference on Advanced Information Networking and Applications, Taipei, Taiwan, 27–29 March, pp. 1050–1054 (2017)
3. Alsuhibany, S.: Evaluating the usability of optimizing text-based CAPTCHA generation. Int. J. Adv. Comput. Sci. Appl. **7**(8), 164–169 (2016)
4. Bursztein, E., Bethard, S., Fabry, C., Mitchell, J., Jurafsky, D.: How good are humans at solving CAPTCHAs? a large scale evaluation. In: Proceedings - IEEE Symposium on Security and Privacy, pp. 399–413 (2010)
5. Ogiela, M.R., Krzyworzeka, N., Ogiela, L.: Application of knowledge-based cognitive CAPTCHA in Cloud of Things security. Concurr. Comput. Pract. Exp. **30**(21), e4769, 1–11 (2018). https://doi.org/10.1002/cpe.4769
6. Ogiela, L.: Cognitive computational intelligence in medical pattern semantic understanding. In: Guo, M.Z., Zhao, L., Wang, L.P. (eds.) Proceedings of Fourth International Conference on Natural Computation, ICNC 2008, vol. 6, Jian, Peoples Republic of China, 18–20 October, pp. 245–247 (2008)
7. Ogiela, L.: Cognitive Information Systems in Management Sciences. Elsevier, Academic Press, Amsterdam (2017)
8. Ogiela, L., Ogiela, M.R.: Data mining and semantic inference in cognitive systems. In: Xhafa, F., Barolli, L., Palmieri, F., et al. (eds.) 2014 International Conference on Intelligent Networking and Collaborative Systems (IEEE INCoS 2014) Salerno, Italy, 10–12 September, pp. 257–261 (2014)

9. Ogiela, L., Ogiela, M.R.: Management information systems. In: LNEE, vol. 331, pp. 449–456 (2015)
10. Ogiela, L., Ogiela, M.R.: Insider threats and cryptographic techniques in secure information management. IEEE Syst. J. **11**, 405–414 (2017)
11. Ogiela, M.R., Ogiela, U.: Secure information management in hierarchical structures. In: Kim, T.-h., et al. (eds.) AST 2011, CCIS 195, pp. 31–35 (2011)
12. Ogiela, L., Ogiela, M.R., Ogiela, U.: Efficiency of strategic data sharing and management protocols. In: The 10th International Conference on Innovative Mobile and Internet Services in Ubiquitous Computing (IMIS-2016), 6–8 July, Fukuoka, Japan, pp. 198–201 (2016). https://doi.org/10.1109/imis.2016.119
13. Osadchy, M., Hernandez-Castro, J., Gibson, S., Dunkelman, O., Perez-Cabo, D.: No bot expects the DeepCAPTCHA! introducing immutable adversarial examples, with applications to CAPTCHA generation. IEEE Trans. Inf. Forensics Secur. **12**(11), 2640–2653 (2017)

Linguistic-Based Security in Fog and Cloud Computing

Urszula Ogiela[1], Makoto Takizawa[2], and Lidia Ogiela[3](✉)

[1] Cryptography and Cognitive Informatics Research Group, AGH University of Science and Technology, 30 Mickiewicza Avenue, 30-059 Kraków, Poland
uogiela@gmail.com

[2] Department of Advanced Sciences, Hosei University, 3-7-2, Kajino-cho, Koganei-shi, Tokyo 184-8584, Japan
makoto.takizawa@computer.org

[3] Pedagogical University of Cracow, Podchorążych 2 Street, 30-084 Kraków, Poland
lidia.ogiela@gmail.com

Abstract. This paper will describe new classes of linguistic secret division procedures created with application of different classes of sequential, tree and graph grammars. Description and comparison of such method will be done, as well as possible application for data management in Fog and Cloud Computing will be presented. New classes of linguistic threshold schemes will be proposed for data (service, information, etc.) protection by split secret parts between selected groups of thrusted participants.

Keywords: Secret division protocols · Systems security · Cloud computing · Fog computing · Secure information management

1 Introduction

Data security protocols consist different stages of information securing and privacy preserving. The most important goal is data protection against unauthorized persons. Most cryptographic protocols provide this kind of protection [4, 8, 10, 12]. In the context of data protection, the processes of assessing the significance of protected information sets are also considered.

In this paper we concentrate around one class of the most important secure techniques – data division schemes. This class of cryptographic protocols consist splitting and sharing schemes especially with (m, n)-threshold schemes. This class of cryptographic protocols can be extended by applying new data sharing techniques.

The new idea is extension threshold schemes by linguistic methods used for secret description procedures.

Finally, in this paper we proposed new classes of linguistic-based threshold schemes, dedicated to data division processes.

The most important features of data security is the methodology and algorithms, which we can propose and use for data protection [2, 3, 6]. One of them are linguistic techniques, which analyse meanings of secured data. Data meaning play a very

L. Barolli et al. (Eds.): EIDWT 2019, LNDECT 29, pp. 120–127, 2019.
https://doi.org/10.1007/978-3-030-12839-5_11

important role in all data analysis processes. Allow to understand the meaningful all aspects of the data interpretation processes and their importance in a comprehensive approaches.

Linguistic techniques of data security was presented in [5, 7, 9]. The basic application of linguistic techniques is data analysis process, included description, interpretation and data analysis. The main idea of these methods there is an attempt (possibilities) to understand the analyzed data. Linguistic techniques were used to semantic description and analysis of different data/objects – for example medical images, financial values, etc. [5, 9, 11].

Linguistic technique are very universal, because they allow to describe different types of data and obtain semantic meaning from analysed data. This universality indicates new directions of applications of linguistic methods.

One of the new areas of application of linguistic techniques is the area of data protection. Data protection can be realize by different methods, for example by use cryptographic data encryption techniques or cryptographic data splitting protocols. In both of these groups, we can use threshold schemes for data sharing procedures.

The main feature of linguistic algorithms in data security processes, is application of linguistic formalisms in creation new classes of cryptographic threshold schemes – names linguistic threshold schemes. In this paper we proposed new techniques of data security – linguistic-based security algorithms.

The main features of proposed solution is possibility of application of different types of linguistic grammar to data protection and security processes. New proposed solution based on application linguistic techniques in cryptographic threshold schemes. To our new proposition we used main formal grammar classification.

The main classification of formal grammar was proposed by Noam Chomsky [1]. He defined the following four types of formal grammars:

- unrestricted grammars (type 0)
- context grammars (type 1)
- context-free grammars (type 2)
- regular grammars (type 3).

In our research we extend classical threshold schemes by adding linguistic stage. In this stage we can shared data by used formal grammar. The type of division processes will depend on the type of grammar introduced. In our previous studies and researches was used context-free grammar.

2 New Classes of Linguistic-Based Threshold Schemes

New classes of linguistic-based threshold schemes depend on the type of grammars, which can be used. So, the main classification of linguistic-based threshold schemes are the following:

- context-free threshold schemes,
- regular threshold schemes.

In the first class of threshold schemes – context-free threshold schemes we can select the following subclasses:

- linguistic threshold schemes used sequences – in this class is possible to used bit sequences to describe secret splitting processes.

The new classes of linguistic-based threshold schemes are the following:

- linguistic threshold schemes used trees – in this class is possible to use trees to describe secret splitting processes,
- linguistic threshold schemes used graph – in this class we can use graphs in secret splitting processes.

Both trees and graphs distinguish the way of data division processes, and also the way of secret reconstruction.

In the first class of linguistic threshold schemes we can use trees, and secret can be split by all participant of the protocol. The main feature of this solution is acyclic and consistent structure of undirected graph. The structure of the tree therefore determines the way of distribution parts of the secret between the participants of the protocol. Each of protocol participants belonging to the tree structure, after submitting its parts in accordance with the applicable threshold scheme, it can reproduce the original secret.

In tree structure arbiter distributes secret parts between all his neighbors (directly adjacent to him). All of them can reconstruct original secret by submitting they parts. Also, its possible secret reconstruction by the deputies – deputies means the participants, which can be indicated on the next argument in the tree structure (Fig. 1).

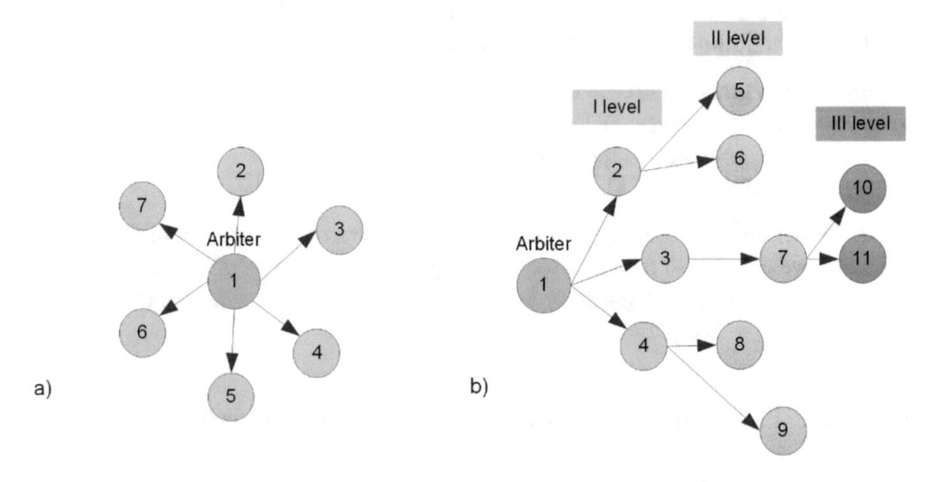

Fig. 1. Examples of tree structures possible to use in linguistic threshold schemes.

Figure 1 presents examples of two (different) possible situations. The first situation is very simple. Arbiter distribute all secret parts (in Fig. 1a is equal seven parts) directly to all participants. Each of them receive one secret part. To reconstruction of original data we have two possibilities:

- in splitting protocols it is required to submit all parts by all participants,
- in sharing protocols it is necessary to assemble selected (determined by used threshold scheme) number of secret parts.

So, the first example presents both situations – splitting and sharing schemes.

The second example presented in Fig. 1b, is dedicated to hierarchical secret sharing procedures. Its means that arbiter distributed secret parts between neighbors directly adjacent to him – the secret was spitted between 2, 3, 4 nodes. Collecting selected number of them (for example any two from three nodes – means application (2, 3)-threshold scheme) can reconstruct original secret at fundamental level I. Also, in this algorithm is possible to reconstruct original secret by combination different nodes from level II. Level II consists nodes derived from previous nodes. So, if the original data is reconstructed by nodes number 3 and 4 (in (2, 3)-threshold scheme), then also submission of all successors can recreate the secret. For example node three can be replaced by node seven, and node four can be replaced by nodes eight and nine. The same situation we see for node seven, which can be replaced by nodes ten and eleven. Many different combinations of possible nodes connections makes this solution very useful and resistant to all external interventions.

In the second class of linguistic threshold schemes we can use graphs. In this type of protocols secret can be split by all participant of the protocol or for selected of them. So, the classifications of possible solutions are the following:

- arbiter distribute secret parts to all participants identified with individual graph nodes – in this protocol original secret can be recreate by nodes (participants) forming a cycle (Fig. 2a),

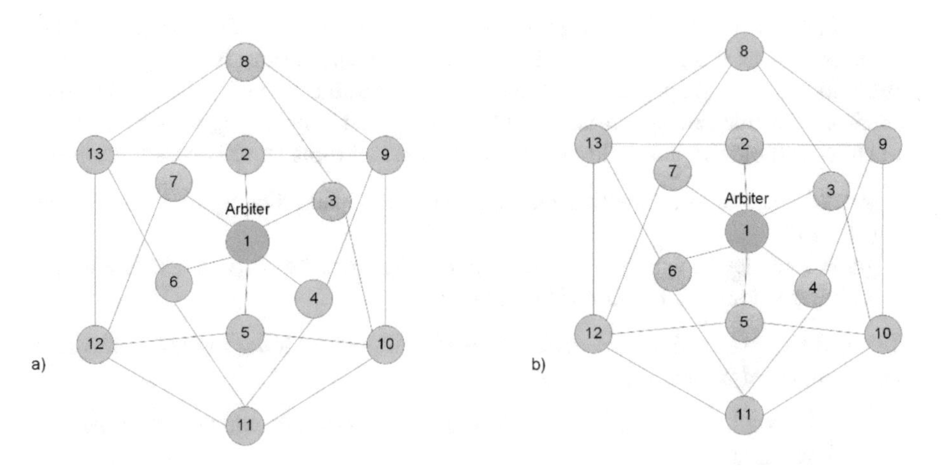

Fig. 2. Examples of graph structures possible to use in linguistic threshold schemes.

- arbiter distribute secret parts to all neighbors directly adjacent to him – in this protocol secret can be recreate by combination of selected secret parts (determined by used (m, n)-threshold scheme) (Fig. 2b).

Figure 2 presents two different protocols. Figure 2a presents the situation, on which it is visible distribution secret parts procedures between all participants (represented by graph nodes). In this protocol secret can be recreated by participants (nodes) forming a cycle – for example by nodes number one, three, seven, and eight. In this protocol it's necessary to find all possible cycles and select one of them to procedure of secret recreation. Another situation presents Fig. 2b. Arbiter distributed secret parts between nodes (neighbors) directly adjacent to him. To secret reconstruction is necessary to combine m from n (all neighbors) nodes. For example (4, 6)-threshold schemes. All of directly adjacent nodes can be replaced by nodes (neighbors) connected in the next argument with the given node. For example node number 2 can be replace by nodes nine and thirteen, etc.

Presented linguistic-based threshold schemes are very universal and they provide great possibilities for their use. One of many possibilities of their application are the areas of cloud computing.

3 Example of Application of Linguistic-Based Threshold Schemes

Proposed algorithms for data protection by using linguistic threshold schemes can be used in different structures dedicated to secret protection at various processing and management levels. The main levels are as follows:

- the level of the structure from which the protection processes take place,
- fog level,
- cloud level.

At the structure level, basic processes of data processing and management take place, as well as the processes of data protection with the discussed linguistic-based threshold schemes may occur. On this level is possible to use linguistic threshold schemes with tree and graph also. Both of those solutions are applied in a specific structure, without the possibility of going outside it. In this level is possible to use:

- tree schemes in which arbiter distribute all secret parts directly to all participants (Fig. 1a)
- graph schemes in which arbiter distribute secret parts to all participants identified with individual graph nodes (Fig. 2a).

The different situation takes place at the second level – at the fog level. The secret can be protected at both algorithms:

- from the fog level – included linguistic threshold schemes which can be realize at this level, and
- from the structure level by dividing the data and distributing the secret between trustees located at the higher level.

The first case illustrates the above described methods of data protection from the structure level. But the second case gives the possibility to manage data by used the following solutions:

- tree schemes in which arbiter distributed secret parts between neighbors directly adjacent to him (Fig. 1b) – with the possibility of chose a given secret holder by his direct neighbors (this means only one tree argument),
- graph schemes in which arbiter distribute secret parts to all neighbors directly adjacent to him (Fig. 2b) – with the possibility of chose a given secret holder by his direct neighbors (this means only one graph argument).

The most interesting solution takes place at the cloud level. At this level secret can be protected by used the following protocols:

- from the cloud level – included linguistic threshold schemes which can be realize at this level, and also at lowers levels,
- from the fog level – included linguistic threshold schemes which can be realize at this level, and at lower level,
- from the structure level by dividing the data and distributing the secret between trustees located at the highest levels.

The those cases are the possibility to manage data by using the following solutions:

- tree schemes in which arbiter distributed secret parts between neighbors directly adjacent to him (Fig. 3a) – and the secret can be recreate by the successors of these neighbors,

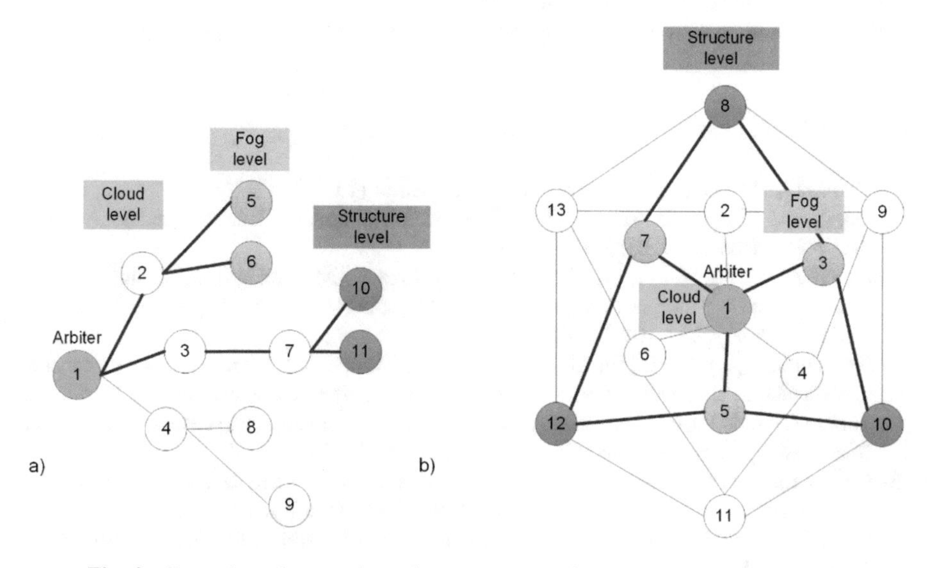

Fig. 3. Examples of tree and graph structures possible to use at the cloud level.

- graph schemes in which arbiter distribute secret parts to all neighbors directly adjacent to him (Fig. 3b) – and the secret can be recreate by the successors of these neighbors.

Figure 3 presents examples of possible secret sharing steps marked with a color convergent with the level to which the trustees belong. Exemplary connections between nodes have been marked with a thick line.

Linguistic-based security techniques can be realized at all of structure levels. Right selection depends on the degree of technological advancement of the structure in which the secret division is realize.

4 Conclusions

In this paper was proposed new idea of using linguistic methods in new areas of application. Linguistic methods can be used as a part of cryptographic schemes, especially in threshold schemes. Development of classic data sharing protocols allows to create protocols in which various data recreation models will be applied.

This paper will describe new classes of linguistic division techniques created with application of different classes of formal grammars, especially tree and graph grammars. Big usefulness of proposed solutions allows their applied in various data encryption tasks, as well as possible application for data protection and management in Fog and Cloud Computing.

Acknowledgments. This work has been supported by the National Science Centre, Poland, under project number DEC-2016/23/B/HS4/00616.

This work was partially supported by JSPS KAKENHI grant number 15H0295.

References

1. Chomsky, N.: Syntactic Structures. Mouton, London (1957)
2. Gregg, M., Schneier, B.: Security Practitioner and Cryptography Handbook and Study Guide Set. Wiley, Hoboken (2014)
3. Duolikun, D., Aikebaier, A., Enokido, T., Takizawa, M.: Design and evaluation of a quorum-based synchronization protocol of multimedia replicas. Int. J. Ad Hoc Ubiquitous Comput. **17**(2/3), 100–109 (2014)
4. Nakamura, S., Ogiela, L., Enokido, T., Takizawa, M.: Flexible synchronization protocol to prevent illegal information flow in peer-to-peer publish/subscribe systems. In: Barolli, L., Terzo, O. (eds.) Complex, Intelligent, and Software Intensive Systems, Advances in Intelligent Systems and Computing, vol. 611, pp. 82–93 (2018)
5. Ogiela, L.: Cognitive computational intelligence in medical pattern semantic understanding. In: Guo, M.Z., Zhao, L., Wang, L.P. (eds.) ICNC 2008: Fourth International Conference on Natural Computation, vol. 6 Proceedings. Jian, Peoples Republic of China, October 18–20, 2008, pp. 245–247 (2008)
6. Ogiela, L.: Cryptographic techniques of strategic data splitting and secure information management. Pervasive Mob. Comput. **29**, 130–141 (2016)

7. Ogiela, L., Ogiela, M.R.: Data mining and semantic inference in cognitive systems. In: Xhafa, F., Barolli, L., Palmieri, F., et al. (eds.) 2014 International Conference on Intelligent Networking and Collaborative Systems (IEEE INCoS 2014) Salerno, Italy, September 10–12, 2014, pp. 257–261 (2014)
8. Ogiela, L., Ogiela, M.R.: Insider threats and cryptographic techniques in secure information management. IEEE Syst. J. **11**(2), 405–414 (2017)
9. Ogiela, M.R., Ogiela, U.: Secure Information splitting using grammar schemes. In: New Challenges in Computational Collective Intelligence, Studies in Computational Intelligence, vol. 244, pp. 327–336 (2009)
10. Ogiela, M.R., Ogiela, U.: Secure information management in hierarchical structures. In: Advanced Computer Science and Information Technology, Communications in Computer and Information Science, vol. 195, pp. 31–35 (2011)
11. Ogiela, U., Takizawa, M., Ogiela, L.: Classification of cognitive service management systems in cloud computing. In: Barolli, L., Xhafa, F., Conesa, J. (eds.) Advances on Broad-Band Wireless Computing, Communication and Applications BWCCA 2017. Lecture Notes on Data Engineering and Communications Technologies, vol. 12, pp. 309–313. Springer International Publishing AG (2018). https://doi.org/10.1007/978-3-319-69811-3_28
12. Yan, S.Y.: Computational Number Theory and Modern Cryptography. Wiley, Hoboken (2013)

Obfuscation Algorithms Based on Congruence Equation and Knapsack Problem

Da Xiao[1,2(✉)], Shouying Bai[1,2(✉)], Qian Wu[3], and Baojiang Cui[1,2]

[1] School of Cyberspace Security,
Beijing University of Posts and Telecommunications, Beijing 100876, China
{xiaoda99, skyxxb}@bupt.edu.cn
[2] National Engineering Lab for Mobile Network, Beijing 100876, China
[3] National Computer Network Emergency Response Technical Team/Coordination Center of China, Beijing 100876, China

Abstract. This paper proposes an obfuscation algorithm based on congruence equation and knapsack problem, for the problem of generating opaque predicate in control flow obfuscation. We binarize the state of the solution of the congruence equation, and then combine the knapsack problem to achieve the binary output of opaque predicate. Compared with the existing chaotic opaque predicate design algorithm, the experimental comparison shows that this paper generates the probability of opaque predicate binary results is nearly 50%, and the time overhead is controlled within 1 s. The experimental results show that the proposed algorithm has better performance in randomness, stability and time overhead, which can effectively resist code reverse analysis.

1 Introduction

Since the beginning of the 21st century, with the development of computer science and the increasing awareness of information security [1, 2], the protection of people's daily use of software has received more attention [3]. With the lingering piracy software, the increasingly serious software copyright issues, and the more frequent data privacy theft [4, 5], how to effectively protect the code has become a hot research topic for security scientists [6].

One of the current effective software protection method is code confusion [7, 8], which is mainly used to resist reverse analysis of software. Collberg proposed the definition of code obfuscation firstly, and Zhang classified code obfuscation in his literature [9]. Among them, the control flow obfuscation has the advantages of high security, simple form and low overhead, which is one of the hotspots in the field of code obfuscation [10]. The degree of the control flow obfuscation scheme mainly depends on opaque predicates. Collberg and his colleagues proposed a renaming construct and a concurrency construct to generate opaque predicates [11], which increased the inverse difficulty of opaque predicates at that time. Arboit used the quadratic residue theory in number theory to construct opaque predicates [12] and applied it to the generation of Java software digital watermarks. Recently, with the development and application of chaos theory, Su et al. [13] constructed opaque predicate clusters based on chaotic mapping Logistic and improved En_Logistic chaotic mapping.

L. Barolli et al. (Eds.): EIDWT 2019, LNDECT 29, pp. 128–136, 2019.
https://doi.org/10.1007/978-3-030-12839-5_12

2 Code Obfuscation

2.1 The Definition of Code Obfuscation

Collberg et al. defined code obfuscation in their literature is $P \xrightarrow{T} P'$ to refer to a transformation of a source program P to a target program P′, where T is a transformation. If P and P′ have the same observable behavior, then P′ is a transformation of P under the condition of T. More precisely, in order for $P \xrightarrow{T} P'$ to be a correct obfuscated transformation, two conditions must be met. If the P is terminated error or not terminated, P′ may or may not terminate; otherwise P′ must be terminated with P terminated, and both of P and P′ generate the same output. The observable behavior can be broadly considered to be the behavior observed by the user, that is, P′ can generate additional behaviors such as creating a file, sending data, etc., as long as P′ is functionally identical to P. At the same time, P′ is not required to be different from P in execution efficiency.

2.2 The Classification of Code Obfuscation

The literature [7] divides code obfuscation into layout obfuscation, control flow obfuscation, data obfuscation and preventive obfuscation from obfuscation technology. Control flow obfuscation adds a function execution branch by inserting conditional jump structures, modifying the control flow graph of the program. For example, inserting a new conditional judgment logic changes the judgment condition of the original program, causing the execution flow to jump to the original code logic or jump to the false code logic, and the false code logic jumps to the conditional judgment again, making the program run complicated. It greatly improves the difficulty of program flow analysis.

2.3 Opaque Predicate

Definition 1. Opaque predicate

A predicate indicates in the computer language the process of seeking a true or false result of a conditional expression. For a predicate OP with the p-point, its output is known before obfuscation, and unknown to the attacker, the OP is said to be an opaque predicate. If the OP is true at the output point, it is recorded as OP^T, the permanent is recorded as OP^F, and if the result of the OP cannot be judged, it is recorded as $OP^?$.

Definition 2. Trapdoor opaque predicate [13]

Let $j \in (1, 2, \ldots, n)$, K_j be the key of predicate P. If K_j is known, it is easy to determine the output of the predicate P at point p in the program before obfuscating; otherwise, if K_j is unknown, it is difficult to determine the output of point p in the program before obfuscating, then the predicate is called an opaque predicate.

Definition 3. Congruence equation [14]

Given a positive integer m, and n times the integer coefficient polynomial (1).

$$f(x) = a_n x^n + a_{n-1} x^{n-1} + \ldots + a_1 x + a_0 \tag{1}$$

Find all integers x such that the congruence formula (2) holds.

$$f(x) \equiv 0 (\text{mod } m) \tag{2}$$

This process is called the solution of the congruence equation, and the Eq. (2) is called the congruence equation of the modulo m. Further, if x = c is the same as the remainder (2), then c is said to be the solution of the congruence equation (2).

3 The Algorithm of Code Obfuscation

3.1 Construct Opaque Predicates

In this paper, Arboit's opaque predicate parameterization concept is proposed, and the opaque predicate at j is defined as P_j. The result of the solution of the random key and several sets of congruence equations is used to parameterize the opaque predicate $\{P_j\}_{j=1}^{j=n} (j = 1, 2, \ldots, n)$.

The Legendre symbol of the prime number p is $(\frac{d}{p})$.

Congruence Equation I.

$$x^2 \equiv -1 (\text{mod } p) \tag{3}$$

When $\left(\frac{-1}{p}\right) = 1$, that is, $p = 4k + 1 (k \in Z)$, there is a solution, and the minimum positive integer solution is x_1.

Congruence Equation II.

$$x^2 \equiv 2 (\text{mod } p) \tag{4}$$

When $\left(\frac{2}{p}\right) = 1$, that is, $p = 8k + 1; p = 8k + 7 (k \in Z)$, there is a solution, and the minimum positive integer solution is x_2.

Congruence Equation III.

$$x^2 \equiv -2 (\text{mod } p) \tag{5}$$

When $\left(\frac{-2}{p}\right) = 1$, that is, $p = 8k + 1; p = 8k + 3 (k \in Z)$, there is a solution, and the minimum positive integer solution is x_3.

Congruence Equation IV.

$$x^2 \equiv 3(\mathrm{mod}\,p) \tag{6}$$

When $\left(\frac{3}{p}\right) = 1$, that is, $p = 12k + 1; p = 12k + 11(k \in Z)$, there is a solution, and the minimum positive integer solution is x_4.

Congruence Equation V.

$$x^2 \equiv -3(\mathrm{mod}\,p) \tag{7}$$

When $\left(\frac{-3}{p}\right) = 1$, that is, $p = 6k + 1(k \in Z)$, there is a solution, and the minimum positive integer solution is x_5.

3.2 Experiment Procedure

The construction of the opaque predicate is as follows:

Step 1:
Let the key of P_j be $K_j, K_j \in Z^*(j = 1, 2, \ldots, n)$, and substitute K_j into (3)–(7) to obtain at least one prime number p, so that the equations of the quadratic congruence equations (3)–(7) have solution.

Step 2:
After determining p, we combine (3)–(7) to solve the corresponding $(x_i)_{i=1}^{i=5}$, classify the solution of x_i and binary $(x_i)_{i=1}^{i=5}$. The binary code with no solution is (0, 0), the binary code with one solution is (0, 1), and the binary code with multiple solutions is (1, 0). $(x_i)_{i=1}^{i=5}$ generate a binary string, which concatenated in the order of the solutions of (3)–(7), encoded as a 10-bit binary sequence $(w_j), w_j \in \{0, 1\}, j = 1, 2, \ldots, 10$ and it is converted into a polynomial with a coefficient of {0, 1}. The polynomial thus produced is the opaque predicate P_j parameterized by the prime number p.

Step 3:
A finite-length super-increment sequence $A = (a_1, a_2, \ldots, a_{10}) \subseteq Z^*$ is randomly generated to ensure that the length of A is consistent with the length of the encoded binary sequence (w_j) and satisfies $\sum_{j<i} a_j < a_i$. Select two random numbers q, r to satisfy $q > \sum_{i=1}^{i=10} a_i, r < q, (r, q) = 1$. For example, $a_j = \sum_1^i a_i + 1, i = (1, 2, \ldots, 10)$; $q = \sum_{i=1}^{i=10} a_i + 1$.

Step 4:
Constructing a backpack sequence $B = (b_1, b_2, \ldots, b_{10})$, satisfy $b_i \equiv (a_i\, r)(\mathrm{mod}\,q)$, then generate $C_j = \sum_{i=1}^{i=10} b_i w_i$.

Step 5:

Pick $k = \sum_{m_i \in M} m_i$, where $M = \{m_1, m_2, m_3, m_4, m_5\} \subset B$, $m_1 \in \{b_1, b_2\}$, $m_2 \in \{b_3, b_4\}$, $m_3 \in \{b_5, b_6\}$, $m_4 \in \{b_7, b_8\}$, $m_5 \in \{b_9, b_{10}\}$.

Step 6:

Compare C_j and k. When $C_j \geq k$, the opaque predicate output is True, otherwise the output is False.

4 Experimental Results and Analysis

This section will analyze the opaque predicate generation algorithm proposed in this paper.

First, the impact of the knapsack sequence on the output is discussed. Here we assume that the encoded sequence $(w_j), w_j \in \{0, 1\}, j = 1, 2, \ldots, 10$ is a fixed value. It can be seen from step 3 and step 4 that if $a_j = \sum_1^i a_i + 1$, $i = (1, 2, \ldots, 10)$; $q = \sum_{i=1}^{i=10} a_i + 1$, the factor affecting the backpack sequence $B = (b_1, b_2, \ldots, b_{10})$ value is the initial elements a_1 of the finite-length super-increment sequence A and r. We take a_1 = 1, 10, 100, 1000, 10000, and for each a_1, randomly generate 3 times r. List all the options for the backpack value, and the resulting opaque predicate results.

As can be seen from the Fig. 1, for the initial elements of different super-incremental sequences A, the output results are related to the selection scheme and are almost independent of the initial elements of A. For the same initial element and selection scheme, the output is related to r. Since r is randomly generated, this will not guarantee the quality of the backpack, but the randomness of the backpack value can be guaranteed to ensure the randomness of the output.

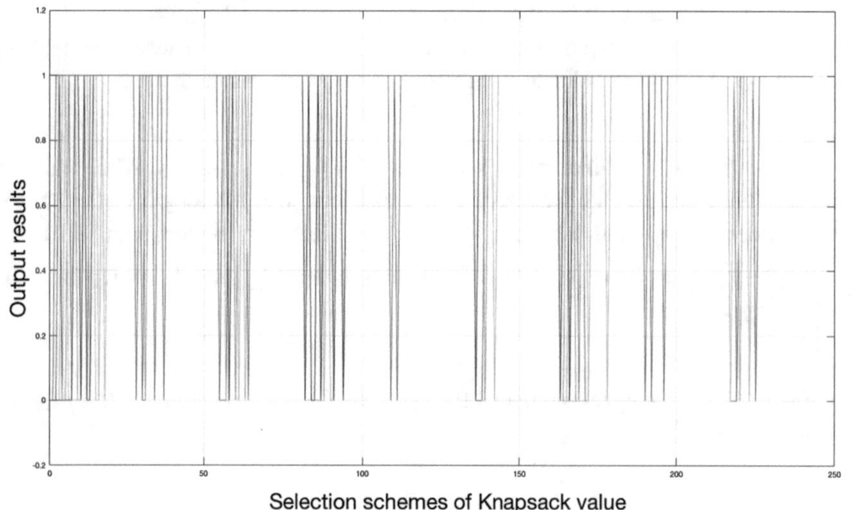

Fig. 1. The impact of the knapsack problem on the outputs.

Then, the opaque predicate generation algorithms designed in this paper and the literature [13] (recorded as algorithm 1) and the literature [15] (recorded as algorithm 2) are compared in the generation of opaque predicate randomness, stability and time overhead.

In the experiment, 10, 20, …, 100, 100, 200, …, 1000 and 10000 opaque predicates are randomly generated, and the proportion of the opaque predicate output as "true" and the time overhead in each of the 21 generations are counted (unit: second) (Fig. 2).

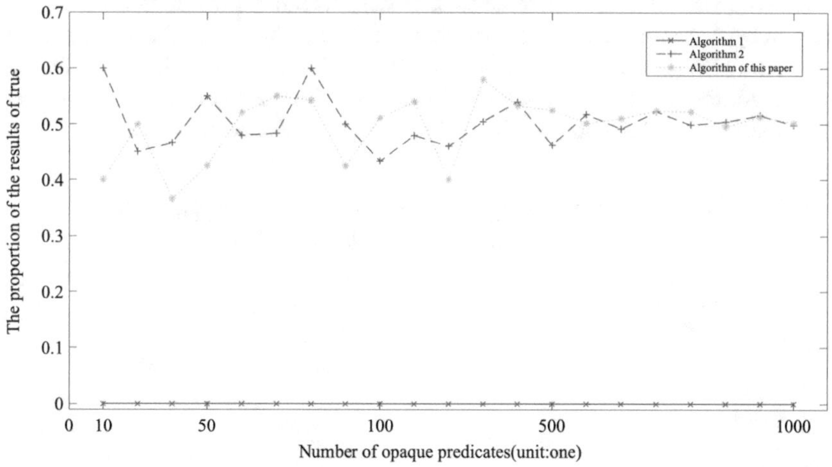

Fig. 2. Three algorithms generate opaque predicates true and false results scale diagram. It shows the proportions in three algorithms of the number of opaque predicates that are "true" to the total number.

Table 1. Evaluation table for the proportion of opaque predicate results generated by three algorithms. It shows the average of the ratio of the true value of each operation to the total number of results, and the mean of the absolute difference between this ratio and 0.5 in three algorithms.

	The ratio of the number of opaque predicates with true to the total number of generations		
	Algorithm 1	Algorithm 2	Algorithm in this paper
Mean	0	0.5154	0.4963
Mean of the difference from the absolute value of 0.5	0.5	0.02726	0.02599

Through comparative analysis by Fig. 1 and Table 1, we can get a few conclusions:

Firstly, when generating the number of opaque predicates is the same, the number of true values generated by algorithm 1 is 0, the number of true or false values generated by algorithm 2 and the algorithm in this paper are almost the same, and the ratio of the output of those algorithms are closer to 50%. Therefore, the randomness of the results of the algorithm of this paper in outputting opaque predicates is better. Simultaneous insertion of such opaque predicates makes the probability of the program executing the original logic and the false branch nearly 50%, increasing the difficulty of the attacker analyzing the program flow, and achieving the purpose of confusing the opaque predicate control flow.

As the number of opaque predicates increases, the ratio of the number of true values generated by Algorithm 2 and the algorithm in this paper to the total number is floating, but the ratio always belongs to the interval [0.4–0.6]. Compared with the difference of 0.5, the algorithm of this paper has less fluctuation with the algorithm 2, so the stability of the algorithm of this paper in the output opaque predicates is better.

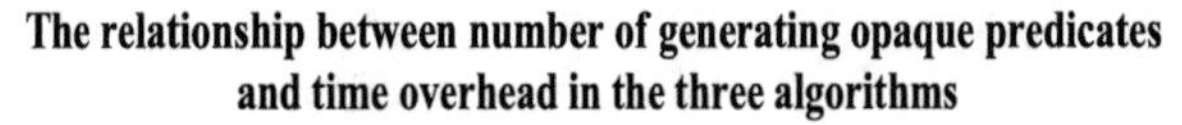

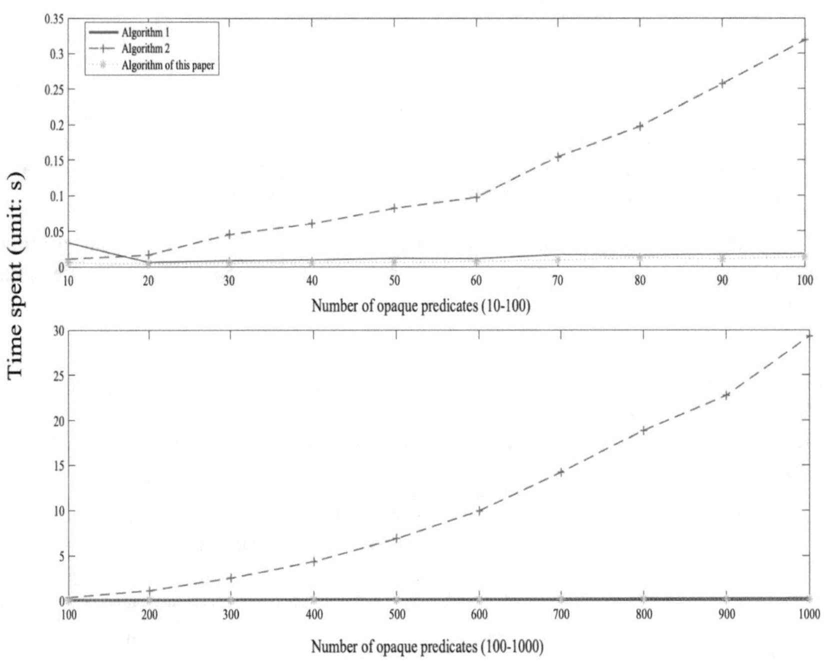

Fig. 3. Three algorithms generate opaque predicate time overhead trend graph.

Table 2. Three algorithms specifically generate opaque predicates and spend timetables.

Generate opaque predicates	Algorithm 1	Algorithm 2	Algorithm in this paper
10	0.0337	0.0109	0.0057
50	0.0115	0.0817	0.0059
100	0.0181	0.3183	0.0135
500	0.0853	6.8611	0.05824
1000	0.1842	29.274	0.1055
10000	1.7631	2822.6	1.1087

Figure 3 and Table 2 show the time cost of generating opaque predicates. As can be seen from the figure:

On the one hand, in addition to generating 10 opaque predicates of time overhead in algorithm 1 is greater than algorithm 2, the time algorithm spends on the algorithm in this paper < algorithm 1 < algorithm 2.

On the other hand, as the number of opaque predicates increases, the time spent by the three algorithms are increasing. The time spent by the algorithm in this paper and algorithm 1 are small, and they take about 0.1 s to generate 1000 opaque predicates, but the growth rate of the time cost in algorithm 2 is larger and much larger than the other two algorithms. For example, it takes 6.86 s to generate 500 opaque predicates, but it takes half a minute to produce 1000 opaque predicates. Even algorithm 1 and the algorithm in this paper take 1 s to generate 10000 opaque predicates, and algorithm 2 reaches nearly 50 min. This is because algorithm 2 generates $N(N \in Z^{*})$ 01 sequences when constructing an opaque predicate. The time we assumed to generate a 01 sequence is T, then the time complexity of algorithm 2 is O(NT), while the time complexity of the other two algorithms is O(T). Since the growth rate of T is much smaller than N, it can be approximated as a constant. As N increases, the time overhead of Algorithm 2 increases linearly.

In summary, the algorithm proposed in this paper produces opaque predicates with a ratio of True to False of almost 1:1, that is, the possibility of executing the original logic and the false branch is the same, increasing the difficulty of control flow analysis, which achieves the purpose of control flow confusion. At the same time, it takes the least amount of time and has good randomness, stability and time overhead.

5 Conclusion

In this paper, based on the control flow obfuscation scheme in code obfuscation technology, an opaque predicate generation algorithm based on the state of the solution of the congruence equation and the knapsack problem is proposed. Compared with the one-way opaque predicate generation method based on hash or number theory conclusion, this paper adopts the chaos theory with higher security and code obfuscation applicability, and combines the congruence equation in number theory with the knapsack problem in the field of computer science algorithms to design opaque predicates. Then compare and analyze the two existing algorithms. Through the above methods, it is proved that the opaque predicates generated by the algorithm proposed in this paper has better randomness, stability and less time overhead, and have good correctness and security to effectively resist reverse analysis.

Acknowledgement. This work was supported by National Natural Science Foundation of China (No. U1536122).

References

1. Yuan, Z., Wen, Q., Mao, M.: Constructing opaque predicates for java programs. In: 2006 International Conference on Computational Intelligence and Security, pp. 895–898 (2006)
2. Schulz, P.: Code Protection in Android. Insititute of Computer Science, Rheinische Friedrich-Wilhelms-Universitgt Bonn, Germany, vol. 110 (2012)
3. Liem, C., Gu, Y.X., Johnson, H.A.: Compiler-based infrastructure for software-protection. In: The Workshop on Programming Languages & Analysis for Security, pp. 33–44 (2008)
4. Shan, L., Emmanuel, S.: Mobile agent protection with self-modifying code. J. Signal Process. Syst. **65**(01), 105–116 (2011)
5. Garg, S., Gentry, C., Halevi, S., et al.: Candidate indistinguishability obfuscation and functional encryption for all circuits. In: Foundations of Computer Science Annual Symposium, vol. 311, no. 02, pp. 40–49 (2013)
6. Larsen, P., Homescu, A., Brunthaler, S., et al.: SoK: automated software diversity. In: IEEE Symposium on Security & Privacy, pp. 276–291 (2014)
7. Collberg, C., Thomborson, C.: Low DA taxonomy of obfuscating transformations. University of Auckland (1997)
8. Collberg, C.S., Thomborson, C.: Watermarking, tamper-proofing, and obfuscation tools for software protection. IEEE Trans. Softw. Eng. **28**(8), 735–746 (2002)
9. Zhang, Y., Zhang, X., Pang, J.: A review of research on code confusion technology. J. Inf. Eng. Univ. (5), 635–640 (2017). https://doi.org/10.3969/j.issn.1671-0673.2017.05.023
10. Zhao, Y.J., Tang, Z.Y., Wang, I.: Evaluation of code obfuscating transformation. J. Softw. **23**(3), 700–711 (2012)
11. Collberg, C., Thomborson, C., Low, D.: Manufacturing cheap, resilient, and stealthy opaque constructs. In: Proceedings of the 25th ACM SIGPLAN-SIGACT Symposium on Principles of Programming Languages, pp. 184–196 (1998)
12. Arboit, G.: A method for watermarking Java programs via opaque predicates. In: Proceedings of International Conference on Electronic Commerce Research, pp. 124–131 (2002)
13. Su, Q., Wu, W., Li, Z., et al.: Research and application of chaos opaque predicate in code obfuscation. Comput. Sci. **40**(6), 155–160 (2013)
14. Chengdong, P., Chengyu, P.: Elementary Algebraic Number Theory, pp. 112–117. Shandong University Press, Jinan (1991)
15. Wang, Y., Huang, Z.-J., Gu, N.-J.: Embedded algorithm based on congruence equation and improved flattened control flow. Comput. Appl. (6), 1803–1807 (2017). https://doi.org/10.11772/j.issn.1001-9081.2017.06.1803

Terminal Access Data Anomaly Detection Based on GBDT for Power User Electric Energy Data Acquisition System

Qian Ma[1,3](✉), Bin Xu[2], Bang Sun[1,3], Feng Zhai[2], and Baojiang Cui[1,3]

[1] School of Cyberspace Security, Beijing University of Posts and Telecommunications, Beijing, China
{maqian95, confuoco, cuibj}@bupt.edu.cn
[2] China Electric Power Research Institute, Beijing 100192, China
{xubinl, zhaifeng}@epri.sgcc.com.cn
[3] National Engineering Laboratory for Mobile Network Security, Beijing, China

Abstract. In recent years, the vulnerability attack on the industrial control system appears more organized and diverse. In this paper, we focus on power user electric energy data acquisition system and its communication protocol, namely 376.1 master station communication protocol. The system is an important infrastructure in national economy and people's livelihood. To efficiently discover abnormal behaviors during its communication, we propose a terminal access data anomaly detection model based on gradient boosting decision tree (GBDT). Firstly, through analyzing the characteristics of the communication protocol and different kinds of terminal access data, we construct a high-quality multidimensional feature set. Then we choose GBDT as the abnormal access data detection model. The experimental result shows that the detection model has a high detection accuracy and outperforms its counterparts.

Keywords: Power user electric energy data acquisition system · 376.1 master station communication protocol · Anomaly detection · Feature extraction · GBDT

1 Introduction

The power user electric energy data acquisition system is a key business system in the residential and enterprise power infrastructure. The system collects and analyzes the electricity consumption data of distribution transformer and the end users through setting the concentrator parameter and collecting data by the master station. Then it accomplishes electricity monitoring, implementation of ladder pricing, load management, line loss analysis, and ultimately achieves automatic meter reading, staggering the peak of electricity using, electricity inspection (anti-stealing), load forecasting and saving electricity costs. Since the master station of the collection system collects a large amount of users' electricity information, and also undertakes controlling the opening/closing of the users' electricity meter, the distribution of power price information and other important tasks. Therefore, it is necessary to adopt strict boundary

L. Barolli et al. (Eds.): EIDWT 2019, LNDECT 29, pp. 137–147, 2019.
https://doi.org/10.1007/978-3-030-12839-5_13

protection measures, such as: setting up a marketing security access zone for collection terminal and configuring firewalls and intrusion prevention system (IPS). Now the system has obtained initial boundary security protection capabilities. As for power user electric energy data acquisition system, it is also necessary to detect abnormal behaviors in the terminal access data from the level of the communication protocol, such as data tampering and implantation of malicious programs (trojans, viruses, malicious code and so on). Once a related security incident occurs, it will seriously affect the normal electricity use of residents and enterprises. Therefore, it is urgent to apply an effective method to detect abnormal behaviors in terminal access data.

Anomaly detection is one of the important research directions in data mining. It aims to find abnormal data in the data set. Abnormal data refers to the data that deviates from the normal behavior mode. In many practical scenarios, it is often easy to obtain a large amount of normal data, but abnormal data is often much lacking due to high acquisition cost and various difficulties in collecting. However, a small amount of abnormal data often contains important behavioral pattern information. Therefore, in addition to traditional anomaly detection methods based on statistical methods and expert systems, artificial intelligence techniques such as machine learning and deep learning have also been introduced into the field of anomaly detection. Such techniques have specific advantages in handling unbalanced datasets. However, there is little research on its use for access data anomaly detection.

There are several inherent drawbacks in rule-based traditional abnormal access data detection methods, such as: they are easy to be bypassed and often have a high false positive rate and false negative rate. To address these problems, we propose a terminal access data anomaly detection model based on gradient boosting decision tree (GBDT) for power user electric energy data acquisition system. We firstly analyze the characteristics of 376.1 master station communication protocol and accomplish feature extraction of terminal access data. Then, we apply hold-out method to select anomaly detection model among several candidate supervised machine learning models. GBDT is selected as the final terminal access data anomaly detection model since it outperforms its counterparts.

2 Related Work

Machine learning-based anomaly detection model can be divided into three types: supervised, semi-supervised and unsupervised model, according to whether the training dataset is pre-labeled. Supervised machine learning models use a labeled dataset to train and test models. k-Nearest Neighbors [1], Bayesian networks [2], support vector machines [3], multi-layer perceptron [4] and decision trees [5] are common supervised machine learning algorithms used in anomaly detection. Moreover, ensemble learning is also widely used in anomaly detection. By constructing and combining multiple single classifiers to complete classification tasks, ensemble learning models often achieve significantly better generalization performance than single ones. Random forest [6], GBDT [7] and other ensemble models are often used in intrusion detection.

In some practical tasks, manually labeling datasets is often very expensive. Therefore, unsupervised models are also used in anomaly detection. Clustering [8] is a common unsupervised learning model. It discovers the latent structures and associations in the dataset by dividing data into different clusters. [9] applies clustering algorithm to anomaly detection. By proposing a more accurate method of finding k clustering center, the anomaly detection model is presented to get better detection performance. In some specific tasks, abnormal samples are scarce or even can't be obtained. So only normal samples can be used for model training and testing. One-class Support Vector Machine (OC-SVM) is an unsupervised learning model commonly used in this case. Rui et al. [10] propose a method for based on OC-SVM for network anomaly detection. They apply OCSVM to learn the data nominal profile. Any data instances deviate from the obtained profile are predicted as abnormal ones. Besides, supervised models are often integrated with unsupervised models to accomplish anomaly detection. Eskin et al. [11] and Honig et al. [12] used an SVM in addition to their clustering methods for unsupervised learning.

Semi-supervised models can be treated as a combination of supervised and unsupervised learning model, which use both labeled data and unlabeled data simultaneously to construct an anomaly detection model. [14] demonstrates that the classification accuracy is significantly improved by semi-supervised detection models, spectral graph transducer and Gaussian fields. Yang et al. [13] proposed an improved semi-supervised algorithm for cooperative training of random forest. The idea of training multiple single classifiers cooperatively not only provides a more accurate method for calculating the confidence of new labeled data, but also makes the labeled results more reliable than the traditional scenarios of two classifiers.

3 A Terminal Access Data Anomaly Detection Model Based on GBDT for Power User Electric Energy Data Acquisition System

3.1 The Framework of Terminal Access Data Anomaly Detection

In this paper, we focus on power user electric energy data acquisition system and its communication protocol, namely 376.1 master station communication protocol. The schematic diagram of the system is shown in Fig. 1. To efficiently detect abnormal behaviors during its communication, we proposed an access data anomaly detection model based on GBDT. The proposed detection method mainly contains two parts: data transformation and machine learning. The core of data transformation module is feature extraction of terminal access data. Machine leaning module accomplishes machine learning model selection and parameter tuning. In this paper, we choose GBDT as the final detection model as it outperforms its counterparts. Figure 2 shows the overall framework of terminal access data anomaly detection. The whole detection process consists of four modules: feature extraction, data deduplication and quantization, machine learning model training, and anomaly detection. In addition, according to the feedback results of the anomaly detection model, the feature set will be updated and optimized.

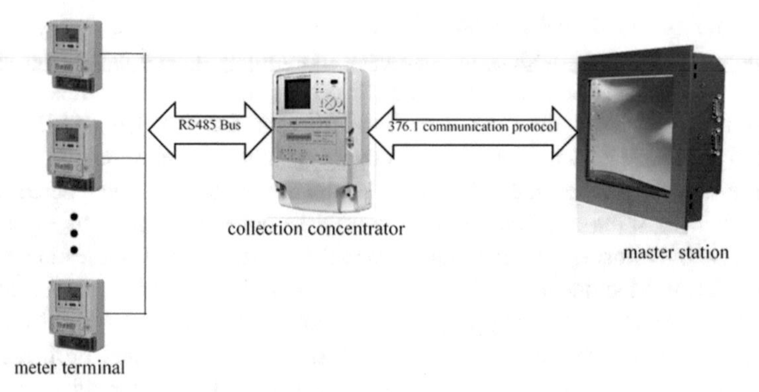

Fig. 1. The schematic diagram of power user electric energy data acquisition system

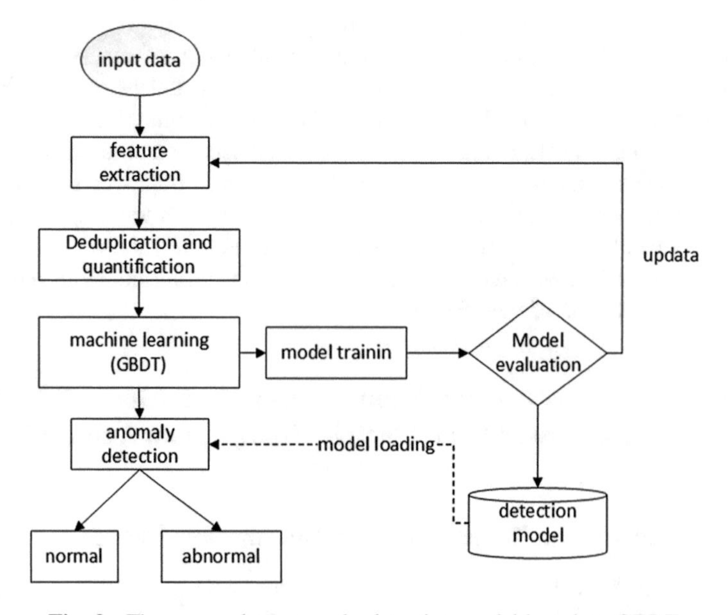

Fig. 2. The access data anomaly detection model based on GBDT

The detail of the four modules in Fig. 2 is described as follows.

(1) Feature extraction module.

Feature extraction is the first key step in terminal access data anomaly detection. We construct a multidimensional feature set through analysis of the attribute fields of 376.1 master station communication protocol. According to the expressive feature set, raw communication message of the protocol is transformed into feature vectors, which can be identified by machine leaning models. Details of aforementioned feature set is presented in Sect. 3.2.

(2) Data deduplication and quantization module.

In order to reduce the size of training set and improve the processing speed of the data, it is necessary to perform deduplication operation on the obtained feature vectors. That is to say, as for data samples which are mapped into a same feature vector, only one feature vector will be used to train and test the subsequent machine learning models. Moreover, for some features with a large range of values, they are quantized using log2 in this module.

(3) Machine learning module.

After feature extraction and optimization of the above modules, a training set with low redundancy is obtained. As for the machine learning model for access data anomaly detection, we choose logistic regression [17], GBDT [16] and random forest [6] as candidate classification models for their significant two-class classification performance. Then we select GBDT as the final terminal access data anomaly detection model through the hold-out method. Detailed information about the selected model are introduced in Sect. 3.3.

(4) Anomaly detection module.

In this module, the model file obtained in the machine learning module is loaded to detect new data instances and then a detection result is generated.

3.2 Feature Extraction

Constructing a high-quality expressive feature set of terminal access data plays an important role in improving the performance of anomaly detection model. Figure 3 shows the frame format of 376.1 master station communication protocol with data acquisition terminal. More details of the communication protocol can be found in [15]. Through analyzing the characteristics of the 376.1 protocol message data and terminal access data, we construct a multi-dimensional feature set. It mainly includes the following five types of features:

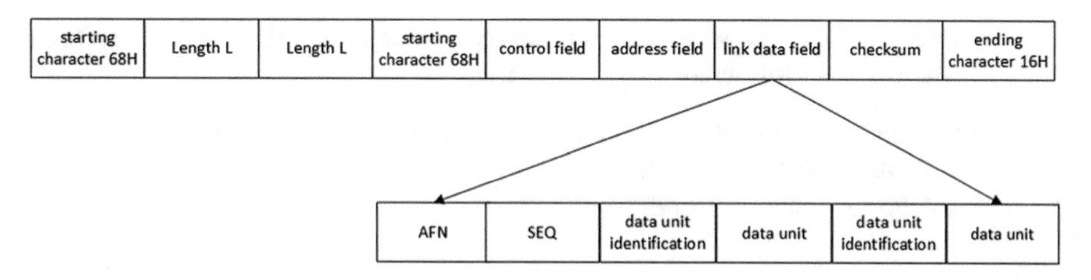

Fig. 3. The frame format of 376.1 master station communication protocol with data acquisition terminal

(1) Abnormal message header format.

Such features are extracted from scenarios where the packet header does not conform to the protocol constraints. This type of features mainly includes the following four features:

a. Whether the first starting character is 68H.
b. Whether the header length of the message is 6 bytes.
c. Whether the second starting character is 68H.
d. Whether the value of protocol identification is 01 and 10.

(2) Abnormal message end format.

This type of feature is obtained from whether the end character of a message is the specified value in the protocol. It mainly includes one feature: whether the ending character is 16H.

(3) Abnormal data unit.

This type of features is extracted from the data unit segment of the packet. It mainly includes the following five features:

a. Whether the length of the data unit identification is in the interval (0,4)
b. Whether the value range of BCD code[1] is beyond $0 \sim 9$.
c. In the uplink packet, whether AFN code[2] is 01H, 04H, or 05H.
d. Whether the value range of Fn[3] is beyond $0 \sim 248$.
e. For a particular Fn, whether the length of the corresponding data unit is less than lower bound.
f. whether the length of the data field of Fn satisfied the communication protocol.

(4) Abnormal event content.

This type of features derives from the event recording segment [15]. It mainly includes one feature that for an event record, whether the difference between the end pointer and the start pointer (indicate the number of uploaded events) is negative.

(5) other features

This type of features is extracted from the reply content of collection concentrator. It mainly includes the following two features:

a. Security authentication failure: whether the concentrator reports the ERC20[4] message authentication failure record.
b. Whether the time stamp is complete.

[1] BCD code is the abbreviation of Binary-Coded Decimal.

[2] AFN code is the application layer function code in the frame format. More detailed information can be found in [15].

[3] Fn is the information class identification code in 376.1 protocol. More detailed information can be found in [15].

[4] ERC20 is the message authentication error record in 376.1 protocol. More detailed information can be found in [15].

3.3 Gradient Boosting Decision Tree Basics

Gradient boosting decision tree is a powerful ensemble model based on decision tree, which is originally proposed by Friedman [16]. Different from random forest [6], which is constructed by bootstrap aggregating, GBDT applies a greedy strategy to ensemble single decision tree, namely gradient boosting. The basic idea of ensemble models is to combine several weak classifiers into a more powerful one. In the training of GBDT, a new decision tree is trained to minimize the error of the whole ensemble model at each iteration. To make the loss function decreases in each iteration, the negative gradient of the current loss function was calculated to train the next tree model. In our experiment, we choose logistic loss function as the loss function. The whole ensemble model is obtained by the weighted summation of the above tree models. A new data point x is classified as:

$$F(x; P) = F\left(x; \{\beta_m, \alpha_m\}_{m=1}^{M}\right) = \sum_{m=1}^{M} \beta_m h(x; \alpha_m) \tag{1}$$

where P denotes the set of all parameters, β_m is the weight of each single tree model and α_m is the parameter set of each single tree model. Details of GBDT can be found in [16].

4 Performance Evaluation

4.1 Introduction to the Dataset

In our experiment, the access data set is provided by China Electric Power Research Institute. After data deduplication, the training dataset consists of 1270 normal samples and 390 malicious samples.

4.2 Model Selection Method

In model selection, we choose GBDT, logistic regression and random forest as candidate models. Then we apply the hold-out method to select the terminal access data anomaly detection model. The hold-out method divides the dataset D into two exclusive sets. Machine learning is trained on one of them and tested on the other. This process is repeated for many times through random dataset partition and the mean value of each index is eventually returned as the evaluation result of the hold-out method.

4.3 Model Evaluation Indexes

To evaluate the performance of candidate models, we use standard measurements, precision, recall rate and f1 score. In our experiment, abnormal data instances are labeled as 0 and normal ones are labeled as 1. The three evaluation indexes are defined as follows in terms of confusion matrix (see Table 1).

(1) precision
   (a) abnormal data: $Pa = \frac{TP}{TP+FN}$. It indicates the proportion of true positive samples in the number of samples predicted to be positive.
   (b) normal data: $Pn = \frac{TN}{TN+FP}$. It indicates the proportion of true negative samples in the number of samples predicted to be negative.
(2) recall rate
   (a) abnormal data: $Ra = \frac{TP}{TP+FP}$. It indicates the proportion of the positive samples which are correctly predicted.
   (b) normal data: $Rn = \frac{TN}{TN+FN}$. It indicates the proportion of the negative samples which are correctly predicted.
(3) f1 score
   (a) abnormal data: $F1a = \frac{2*Pa*Ra}{Pa+Ra}$. It indicates the harmonic mean of the precision and recall rate of the positive samples.
   (b) normal data: $F1n = \frac{2*Pn*Rn}{Pn+Rn}$. It indicates the harmonic mean of the precision and recall rate corresponding to the negative samples.

Table 1. Confusion matrix

		Predicted label	
		0	1
Actual label	0	True Positive (TP)	False Positive (FP)
	1	False Negative (FN)	True Negative (TN)

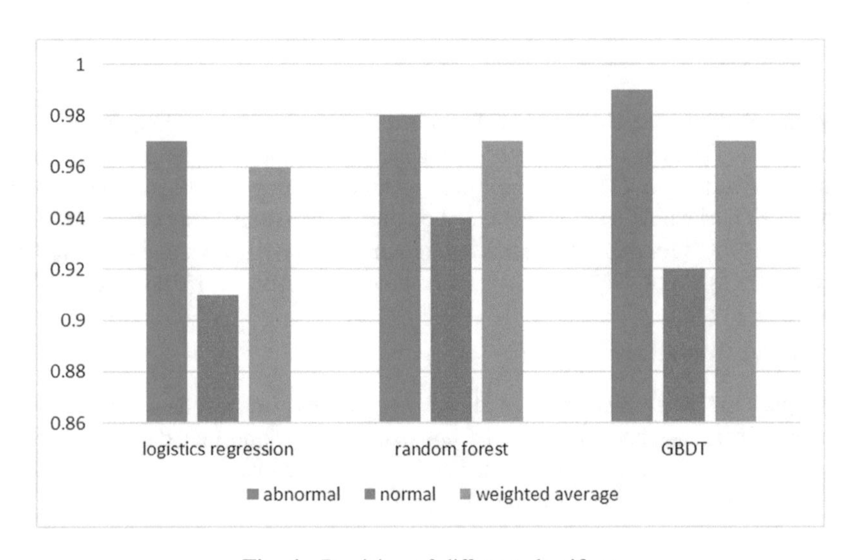

Fig. 4. Precision of different classifiers

4.4 Experimental Results and Analysis

The precision, recall rate and f1 score of the three candidate models are shown in Figs. 4, 5 and 6, respectively. The experimental result shows that both ensemble models, GBDT and random forest outperforms logistic regression model in terms of the aforementioned model evaluation indexes. As for the weighted average of three indexes, GBDT performs as same as random forest, the indexes of which are all above 97%. However, in the actual communication environment of the acquisition system, the

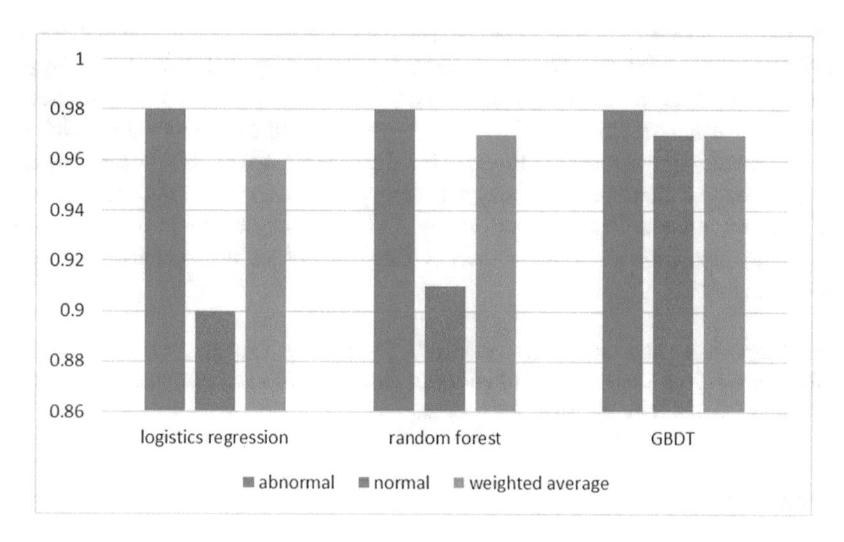

Fig. 5. Recall rate of different classifiers

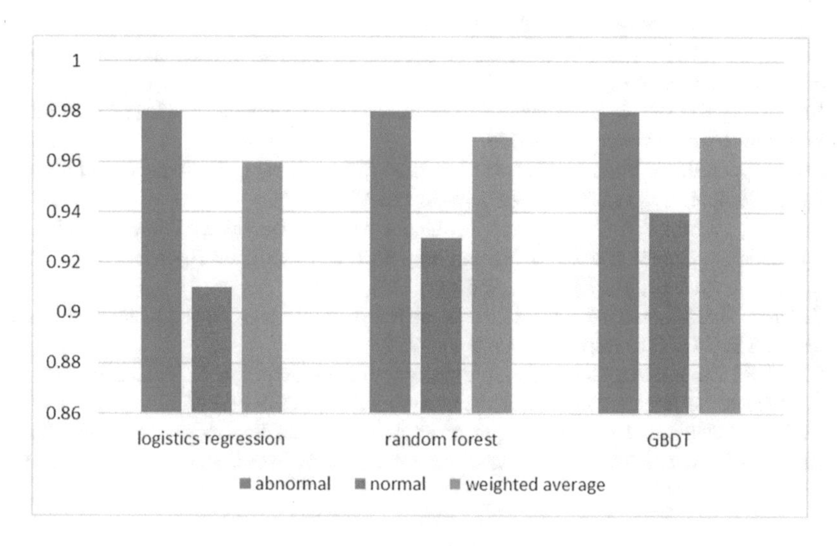

Fig. 6. F1 score of different classifiers

data is highly unbalanced. For example, more than 99% of terminal access data would be not malicious. Considering the fact, we focus on the classification performance over abnormal data in model selection. Both models have same precision. When it comes to the recall rate of abnormal data, GBDT performs better than random forest. Therefore, GBDT is chosen as the terminal access data anomaly detection model.

5 Conclusion

In this paper, we propose a terminal access data anomaly detection model based on GBDT for power user electric energy data acquisition system. We first analyze the attribute fields of the 376.1 master station communication protocol and its packet data. Then we construct a multidimensional feature set of terminal access data, which is the input to the subsequent machine models. Finally, we choose GBDT as the access data anomaly detection model by the hold-out method. The experimental results show that GBDT has significant detection performance. Our work also provides a new idea for the anomaly detection of terminal access data for power user electric energy data acquisition system.

Acknowledgments. This work was supported by Research and Application of Key Technologies for Unified Data Collection of Multi-meter (JL71-17-007) and National Natural Science Foundation of China (No. U1536122).

References

1. Ahmed, M., Naser Mahmood, A., Hu, J.: A survey of network anomaly detection techniques. J. Netw. Comput. Appl. **60**, 19–31 (2016)
2. Manocha, S., Girolami, M.: An empirical analysis of the probabilistic K-nearest Neighbor Classifier. Patt. Recogn. Lett. **28**, 1818–1824 (2007)
3. Moore, D.: Internet traffic classification using Bayesian analysis techniques. In: Proceedings of ACM SIGMETRICS (2005)
4. Kumar, G., Kumar, K., Sachdeva, M.: The use of artificial intelligence based techniques for intrusion detection: a review (2010)
5. Sahar, S., Hashem, M., Taymoor, M.: Intrusion detection using multi-stage neural network. Int. J. Comput. Sci. Inf. Secur. **8**(4), 14–20 (2010)
6. Zhao, Z., Mehrotra, K.G., Mohan, C.K.: Online anomaly detection using random forest. In: International Conference on Industrial, Engineering and Other Applications of Applied Intelligent Systems, pp. 135–147. Springer, Cham (2018)
7. Feng, H., Li, M., Hou, X., et al.: Study of network intrusion detection method based on SMOTE and GBDT. Appl. Res. Comput. (2017)
8. Rawat, S.: Efficient data mining algorithms for intrusion detection. In: Proceedings of the 4th Conference on Engineering of Intelligent Systems (EIS 2004) (2005)
9. Li, H.: Research and implementation of an anomaly detection model based on clustering analysis. In: International Symposium on Intelligent Information Processing and Trusted Computing (2010)

10. Rui, Z., Shaoyan, Z., Yang, L., Jianmin, J.: Network anomaly detection using one class support vector machine. In: Proceedings of the International Multi Conference of Engineers and Computer Scientists (2008)
11. Eskin, E., Arnold, A., Preraua, M., Portnoy, L., Stolfo, S.: A geometric framework for unsupervised anomaly detection: detecting intrusions in unlabeled data. In: Barber, D., Jajodia, S. (eds.) Data Mining for Security Applications. Kluwer Academic Publishers, Boston
12. Honig, A.: Adaptive model generation: an architecture for the deployment of data mining based intrusion detection systems. In: Barbar, D., Jajodia, S. (eds.) Data Mining for Security Applications. Kluwer Academic Publishers, Boston (2002)
13. Yang, J., et al.: Multi-classification for malicious URL based on improved semi-supervised algorithm. In: 2017 IEEE International Conference on Computational Science and Engineering (CSE) and Embedded and Ubiquitous Computing (EUC), vol. 1. IEEE (2017)
14. Chen, C., Gong, Y., Tian, Y.: Semi-supervised learning methods for network intrusion detection. In: 2008 IEEE International Conference on Systems, Man and Cybernetics, SMC 2008, pp. 2603–2608. IEEE (2008)
15. Liu, K., Liao, X.: Design and implementation of Q/GDW 376. 1 protocol and DL/T 645 protocol conversion. Adv. Technol. Electr. Eng. Energy **32**(02), 72–75+81 (2013)
16. Natekin, A., Knoll, A.: Gradient boosting machines, a tutorial. Front. Neurorobot. **7** (2013)
17. Kleinbaum, D.G., Klein, M.: Introduction to logistic regression. Stat. Biol. Health **31**(4), 1–39 (2010)

E-Learning Material Development Framework Supporting 360VR Images/Videos Based on Linked Data for IoT Security Education

Yoshihiro Okada[1(✉)], Akira Haga[1], Shi Wei[1], Chenguang Ma[2], Srishti Kulshrestha[3], and Ranjan Bose[4]

[1] The Innovation Center for Educational Resources, Kyushu University Library, Kyushu University, Fukuoka, Japan
okada@inf.kyushu-u.ac.jp,
{haga.akira.879,shi.wei.243}@m.kyushu-u.ac.jp
[2] NTT Data Global Solutions, Tokyo, Japan
machenguanglala@yahoo.co.jp
[3] Department of Electrical Engineering, Indian Institute of Technology Delhi, New Delhi, India
srish.kul@gmail.com
[4] Indraprastha Institute of Information Technology Delhi, New Delhi, India
bose@iiitd.ac.in

Abstract. This paper proposes an e-learning material development framework using 360VR images/videos for IoT security education. The authors joined a research project about "Security in the Internet of Things space". There are several activities of this project from hardware level to application software level, and IoT security education is also included in this project. The authors of this paper are the members of this IoT security education group, and one of its activities is to provide e-learning materials about IoT security. For preparing the contents of such e-learning materials, the authors have been building a database that contains the IoT security information like various kinds of IoT threats based on Linked Data. The next research activity is to provide e-learning materials themselves using the database that attract students to IoT security. The use of new media like 360VR images/videos may enable e-learning materials to attract students. Therefore, in this paper, the authors propose an e-learning material development framework supporting 360VR images/videos based on Linked Data for IoT security education.

Keywords: e-Learning · Material development · Linked Data · IoT security · 360VR

1 Introduction

This paper proposes an e-learning material development framework using 360VR images/videos for IoT security education. We have been doing a collaborative research project entitled "Security in the Internet of Things Space" between Kyushu University and IIT Delhi, supported by JST (Japan Science & Technology Agency) and DST

L. Barolli et al. (Eds.): EIDWT 2019, LNDECT 29, pp. 148–160, 2019.
https://doi.org/10.1007/978-3-030-12839-5_14

(Department of Science & Technology, India) respectively. Recently, Cybersecurity has become very important as one of the research themes because there have been many serious cyberattacks in the world that have attracted a lot of attention from the researchers and government agencies. One of the reasons for this is that the Internet has become common infrastructure and drastically increased the attack surface. Similarly, IoT (Internet of Things) security has become very important because at present, many things (devices and sensors) are connected to the Internet in order to collect various types of data including human activity data for building a sophisticated society highly supported by ICT. This also causes the demand of IoT security education. Consequently, to include IoT security education and awareness is regarded as a critical objective of our project. Although researches of IoT should cover various fields, this education-research focuses specifically on the 'Security of Smart Building/Smart Home' because these are closely related to our lives.

As described above, in this research project, we focus on IoT security education and consider developing such e-learning materials. Before developing e-learning materials about IoT security, we have been building a database about the IoT security information as Linked Data because the database realized as Linked Data becomes easy to be updated and can be easily shared with other researchers/educators. Therefore, we have already proposed the use of Linked Data for building the database of the IoT security information [1, 2]. After completing the database, we should provide e-learning materials themselves about IoT security using the database. For enhancing the educational efficiency, we should provide very attractive e-learning materials for students using recent ICT like 3D CG. Then, we have already proposed an interactive educational material development framework using 3D CG [3]. In these year, VR (Virtual Reality) and AR (Augmented Reality) have become popular again because of the technological advancements of them, and we should use VR/AR in e-learning materials because such e-learning materials can be attractive for students. Then, we have already proposed an interactive educational material development framework that supports VR/AR [4]. Furthermore, we have proposed the combinatorial use of the framework and the database of IoT security information realized as Linked Data, and introduced a prototype of the e-learning material that supports VR/AR for IoT security education [5].

Recently, many industries release 360VR cameras as a commercial product, and then, 360VR images/videos have become popular as one of the new media. 360VR images/videos are not only attractive for students, but also convenient rather than 3D CG for the development of immersive environments that have higher educational efficiency. Hence, we have already introduced functionalities into the above framework proposed in [4] for supporting 360VR images/videos as an individual system [6], i.e., a web-based 360VR image/video viewer. It is available on various platforms such as iOS and Android smartphones and tables besides standard desktop PCs. This availability is significant for mobile learning in BYOD (Bring Your Own Device) classes. For e-learning materials using 360VR images/videos, it is also necessary to support VR/AR devices. Then, we introduced functionalities for that to the proposed viewer [7]. Those are a stereo view, touch interfaces and device orientation interfaces. Unfortunately, the above web-based 360VR image/video viewer proposed in [7] is used as an e-learning material for Japanese history education, but not for IoT security education. Therefore, in this paper, we propose the combinatorial use of the web-based 360VR image/video viewer and the database of IoT security information realized as Linked Data, and introduce a prototype of the e-learning material that support 360VR images/videos for IoT security education.

The remainder of this paper is organized as follows: Sect. 2 describes related work. In Sect. 3, we explain our Linked Data and implementation of RDF (Resource Description Framework) stores for IoT security information, and introduce the overview of an authoring tool for e-learning materials based on Linked Data. Section 4 introduces the prototype of the e-learning material using 360VR images/videos for IoT security education. Finally, we conclude the paper and discusses our future work in Sect. 5.

2 Related Work

The term Linked Data proposed by Berners-Lee is used for describing a set of best practices for publishing and connecting structured data on the Web [8, 9]. These practices mean the foundation of the evolution of the Web of Documents to the Web of Data, a global data space connecting data from a multitude of different domains. For making it the part of a single global data space, Berners-Lee proposed several technical rules for publishing data on the Web that have become known as Linked Data Principles [9] as follows:

1. Use of URIs as names for things.
2. Use of HTTP URIs so that people can look up those names.
3. Provision of useful information using the standard (RDF, SPARGL) when someone looks up a URI.
4. Inclusion of links to other URIs so that they can discover more things.

Also, Linked Data can be used for data integration [10]. In the area of e-learning, such data can be shared as learning objects (LO) or learning entities of any kind, respectively on the Web. Dicheva explains three generations of Web-based educational systems [11]. The first generation systems provide a central entry-point for accessing learning materials and online course, e.g., LMS and educational portals. The second generation systems employ Web and AI technologies to support intelligently personalization and adaption, those are called educational adaptive hypermedia systems. The third generation systems are included in a class of ontology-aware software using and enabling Semantic Web standards and technologies in order to grant scalability, reusability and interoperability of e-learning material distributed over the Web. Therefore, Linked Data technologies raise high expectations as one of the solutions in a field of e-learning.

There are several development systems for e-learning materials, supporting 3D CG and VR/AR. One of them is *IntelligentBox* proposed by Okada and Tanaka [12] as a constructive visual software development system for 3D graphics applications. There have been many applications actually developed using *IntelligentBox* so far. Therefore, this system seems very useful for also developing e-learning materials. Unfortunately, *IntelligentBox* cannot be used for creating web-based contents although there is the web-version of *IntelligentBox* [13] because its performance is not good for web-based contents of many users like students in a class. Also, *IntelligentBox* does not support 360VR images/videos. Using *Webble World* [14], it may be possible to create web-based interactive e-learning materials through simple operations for authoring and of course, possible to render 3D graphics assets. However, it does not have some functionalities that *IntelligentBox* has, and does not support 360VR images/videos.

There are some electronic publication formats like EPUB, EduPub, iBooks and their authoring tools. Of course, these contents are used as e-learning materials. However, basically, these are not available on the Web browser and do not support 3D graphics except iBooks. iBooks supports rendering functionality of a 3D scene and control functionality of its viewpoint. However, 360VR images/videos cannot be supported by it.

From the above situation, for creating web-based interactive 3D educational contents, usually we must use any dedicated toolkit systems. The most popular tool is Unity [15], one of the game development engines. Actually, using Unity, we have developed 3D educational contents [16, 17] for the medical course students of our university because Unity is a very powerful tool that supports many functionalities and possible to develop applications available on iOS and Android smartphones and tablets. However, Unity is not easy for standard end-users like teachers to use because Unity requires any programming knowledge and skills of the operations for it. Therefore, we propose our framework in this paper.

3 Linked Data of IoT Security Information

In this paper, we propose the combinatorial use of Linked Data for a database of IoT security information and of the web-based 360VR image/video viewer for developing e-learning materials of IoT security education. This section explains Linked Data for realizing a database of IoT security information. For Linked Data, we built up the RDF store [2] for storing designed RDF data based on Apache Jena. Thereby, the integration of IoT security information designed based on RDF, and the retrieval of the information by SPARQL, a query language for RDF store, are implemented.

3.1 Design of RDF Store for IoT Security Information

For designing the linked data for IoT security information, there are three steps as follows:

We designed the schema of RDF store in the first step. Figure 1 shows an example of RDF store schema and we designed a new prefix for IoT security as shown in Fig. 2 because there is not any existing prefix for IoT security field according to the RDF schema document. In the second step, we created an excel file of IoT security according to the schema. In the third step, as shown in Fig. 3, we converted the excel file of IoT security into turtle format or RDF format by using OpenRefine, a free open source, power tool for working with messy data. Meanwhile, IoT security information possesses the contents of Table 1 that are important as IoT threats information.

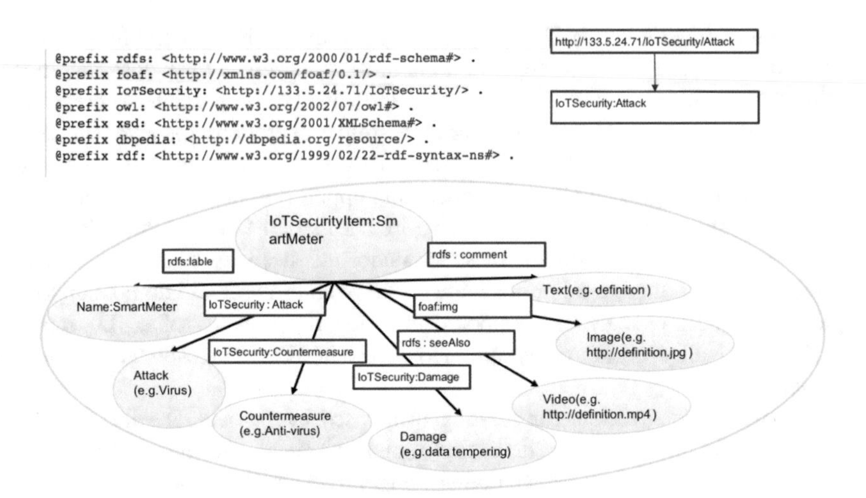

Fig. 1. Example of RDF store schema.

RDF Schema Alignment

The RDF schema alignment skeleton below specifies how the RDF data that will get generated from your grid-shaped data. The cells in each record of your data will get placed into nodes within the skeleton. Configure the skeleton by specifying which column to substitute into which node.

Base URI: http://133.5.24.71/IoTSecurityItem/ edit

RDF Skeleton | RDF Preview

Available Prefixes: rdfs foaf IoTSecurity xsd owl rdf + add prefix manage prefixes

	Property	Object
IoTSecurityItem URI add rdf:type	rdfs:label	**IoTSecurityItem** cell
	rdfs:comment	**Text** cell
	foaf:img	**Image** cell
	rdfs:seeAlso	**Video** cell
	IoTSecurity:Attack	**Attack** cell
	IoTSecurity:Countermeasure	**Countermeasure** cell
	IoTSecurity:Damage	**Damage** cell
	add property	

Fig. 2. Designation of RDF skeleton of IoT security information.

3.2 Implementation of RDF Store for IoT Security Information

As described in the previous subsection, there are three steps for the implementation of RDF store of IoT security information. In the first step, we built up RDF store based on Apache Jena, which is capable to store the integrated data as shown in Fig. 4. In the second step, we stored the exported Turtle File of IoT security information into the RDF store as shown in Fig. 5. In the third step, we can achieve the retrieval of IoT security information through virtuoso, a server works as a RDF store, by using the query language SPARQL as shown in Fig. 6.

```
<rdf:Description rdf:about="http://133.5.24.71/IoTSecurityItem/SmartMeter">
        <rdfs:label>SmartMeter</rdfs:label>
        <rdfs:comment>https://en.wikipedia.org/wiki/Smart_meter</rdfs:comment>
        <foaf:img>https://upload.wikimedia.org/wikipedia/commons/9/9a/Intelligenter_zaehler-_Smart_meter.jpg</foaf:img>
        <rdfs:seeAlso>https://www.youtube.com/watch?v=C4L01fejQJk</rdfs:seeAlso>
        <IoTSecurity:Attack>Unauthorized access(internet)</IoTSecurity:Attack>
        <IoTSecurity:Countermeasure>Vulnerability countermeasure</IoTSecurity:Countermeasure>
        <IoTSecurity:Damage>Measurement data is peeked.</IoTSecurity:Damage>
        <IoTSecurity:Countermeasure>User authentication</IoTSecurity:Countermeasure>
        <IoTSecurity:Damage>Measurement data is tampered with.</IoTSecurity:Damage>
        <IoTSecurity:Countermeasure>Firewall function</IoTSecurity:Countermeasure>
        <IoTSecurity:Damage>Measurement data can not be viewed.</IoTSecurity:Damage>
</rdf:Description>

<rdf:Description rdf:about="http://133.5.24.71/IoTSecurityItem/HERMS">
        <rdfs:label>HERMS</rdfs:label>
        <rdfs:comment>https://en.wikipedia.org/wiki/Energy_management_system</rdfs:comment>
        <foaf:img>https://research.ece.ncsu.edu/adac/wp-content/uploads/2015/10/total_prj.jpg</foaf:img>
        <rdfs:seeAlso>https://www.youtube.com/watch?v=Vi-16b00Pn8</rdfs:seeAlso>
        <IoTSecurity:Attack>Virus infection</IoTSecurity:Attack>
        <IoTSecurity:Countermeasure>Vulnerability countermeasure</IoTSecurity:Countermeasure>
        <IoTSecurity:Damage>The data of device controlled by HEMS is peeked.</IoTSecurity:Damage>
        <IoTSecurity:Countermeasure>Anti- virus</IoTSecurity:Countermeasure>
        <IoTSecurity:Damage>The data of device controlled by HEMS is tampered with.</IoTSecurity:Damage>
        <IoTSecurity:Countermeasure>Software digital signature</IoTSecurity:Countermeasure>
        <IoTSecurity:Damage>The data of device controlled by HEMS can not be viewed.</IoTSecurity:Damage>
        <IoTSecurity:Countermeasure>Secure development</IoTSecurity:Countermeasure>
</rdf:Description>
```


Fig. 3. Converted RDF file of IoT security information.

Table 1. IoT threat information

1. What kinds of IoT (systems, devices, sensors) would be present in a Smart Building/Smart Home.
2. What kinds of attacks can be perpetrated on such IoT spaces.
3. What kinds of preventive measures can be taken to thwart such attacks.
4. What kinds of damages could be caused by such attacks and what would be the associated cost.
5. Textual explanation of IoT threat information.
6. Illustrated explanation of IoT threat information.
7. Video explanation of IoT threat information.

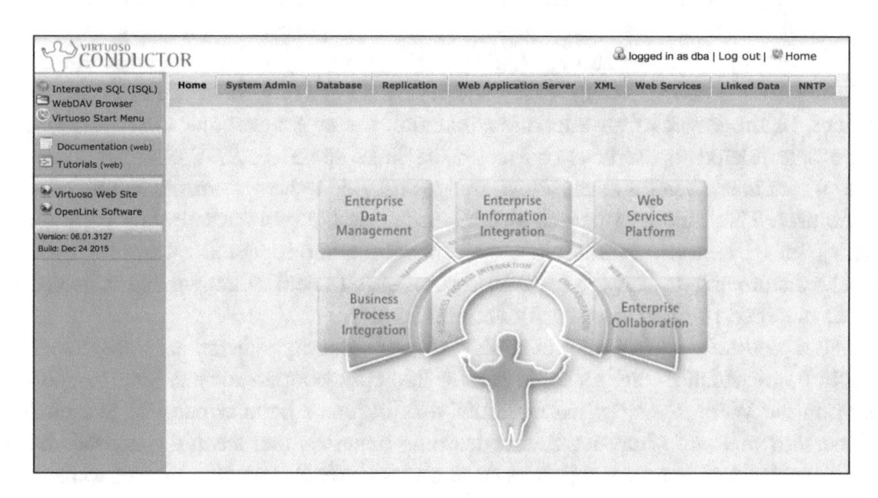

Fig. 4. Management of RDF data of IoT security by virtuoso.

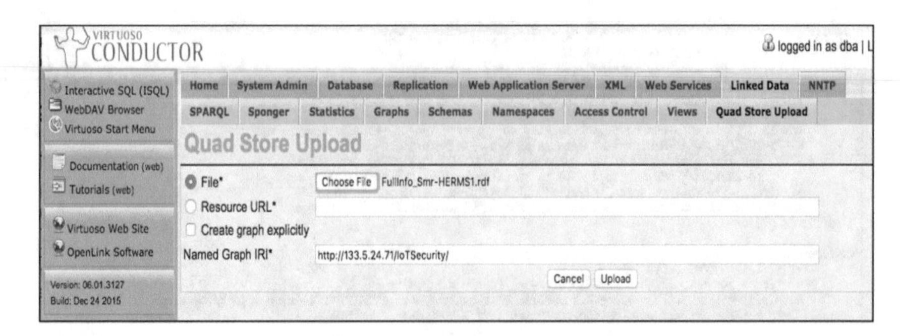

Fig. 5. Storing of RDF file.

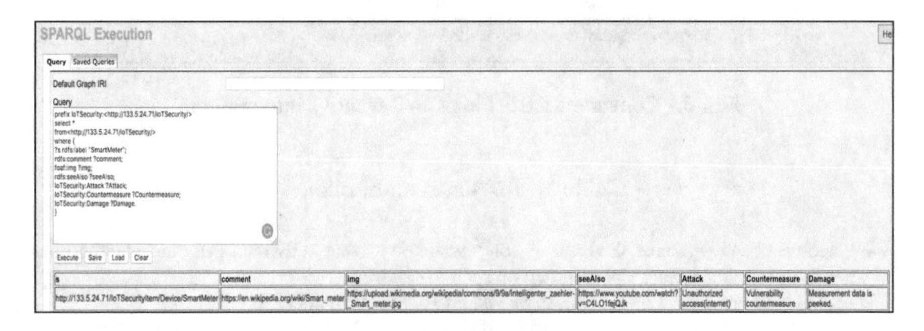

Fig. 6. Retrieval of RDF file.

3.3 Authoring Tool for e-Learning Materials of IoT Security Education

By curating the IoT security information using the links stored in RDF store, it is possible to create e-learning materials. We have proposed our authoring tool in [2] whose overview is shown in Fig. 7. The screen image of our prototype system of the authoring tool is shown in Fig. 8. By specifying a keyword ‘SmartMeter’, one of IoT devices, as the target of an e-learning material, the system retrieves text, image and video data related to the keyword using the links stored in RDF store. If there are several candidate data for each of text, image and video, the system shows one of them to the user. Then, the user can choose his/her desirable one through the operation on [back], [next] buttons. In this way, by using the authoring tool, e-learning material developers are able to edit the material regarding IoT security knowledge managed by the links stored in RDF store easily and efficiently.

After authoring e-learning materials of IoT security education, such materials should be provided in any readable format like epub/edupub for enabling students to learn on the Web. Recently, learning analytics becomes popular because that enables teachers/instructors to improve their e-learning materials and teaching methods based on the analysis of learning activities of students/learners. For that, viewer tools of e-learning materials should have functionality to record such learning activity data and to store them as a database. Our research group has also proposed a prototype system of web-based viewer for e-learning materials provided in epub format that has functionality of learning analytics [18].

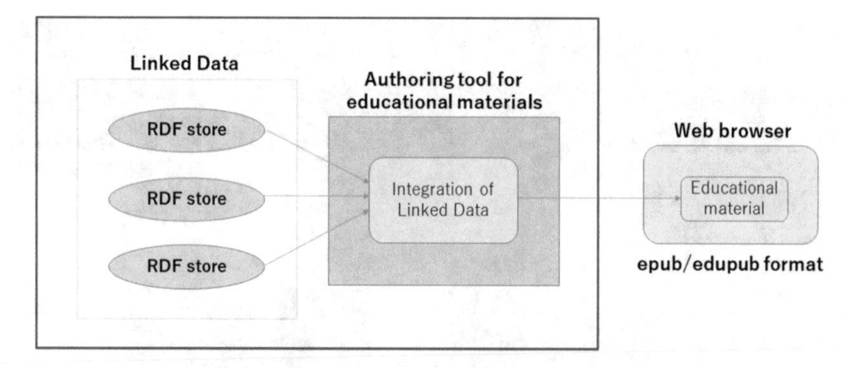

Fig. 7. Overview of authoring tool.

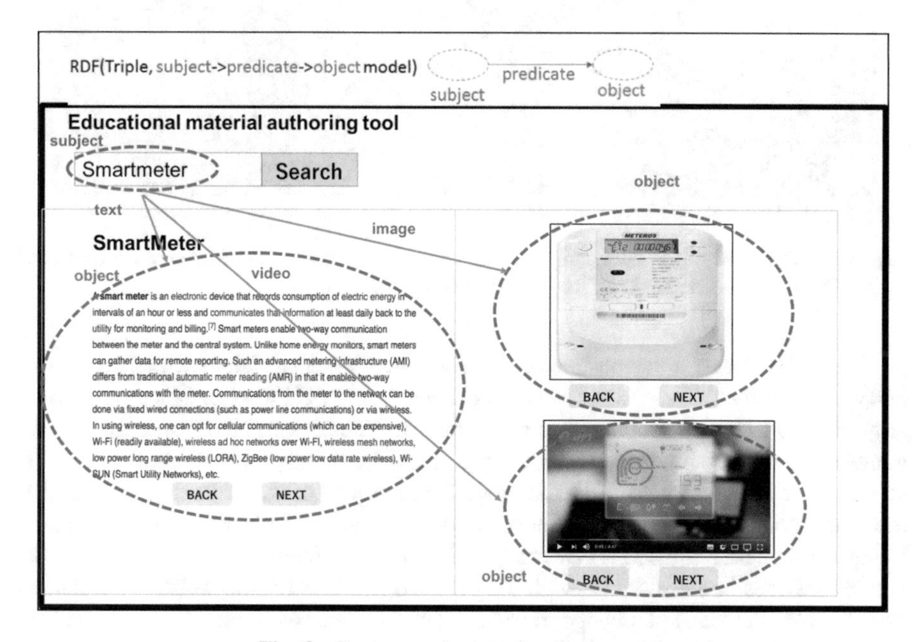

Fig. 8. Prototype system of authoring tool.

4 e-Learning Material Supporting 360VR Images/Videos for IoT Security Education

As one of the previous achievements of our research group, we have developed an e-learning material for IoT security education that employs 3D CG as a model of smart home/building and uses the database of IoT threats information realized as Linked Data as shown in Fig. 9, introduced in [5]. Similarly to this, currently, we have been developing an e-learning material for IoT security education that supports 360VR images/videos using the database of IoT threats information realized as Linked Data and the web-based viewer of 360VR images/videos proposed in [6, 7].

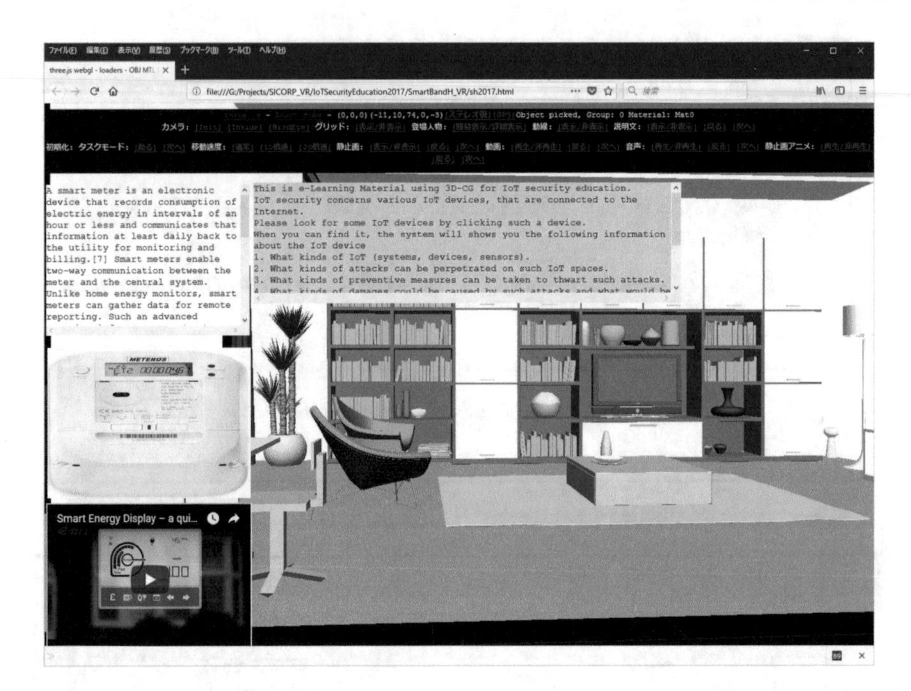

Fig. 9. A screenshot of the e-learning material on a browser for IoT security education realized by 3D CG and based on Linked Data of IoT security information.

Figure 10 shows a screenshot of the e-learning material that is a certain part of 360VR image for IoT security education executed on a PC browser. Since our current target is smart home/building, there are several IoT devices located inside the 360VR image of a room as shown in the figure, e.g., a mobile phone, a smart speaker, PC, and so on, those are connected to the Internet. The purpose of this e-learning material is teaching students the information about IoT devices and their threats shown in Table 1. Using this material, a student can change his/her view-direction freely for finding out any IoT devices in the room through the mouse operations interactively. If the student find an IoT device and change his/her view-direction onto the device, its detail information derived from the contents of Table 1 will appear on the browser window as shown in the left part of the figure similarly to Figs. 8 and 9. In this way, the students will be able to learn IoT threats. For this, we newly added an object picking functionality by changing the view-direction to the web-based 360VR image/video viewer.

At present, unfortunately, the contents of Table 1 about various IoT devices should be entered manually into the e-learning material. We are also developing authoring functionality to the web-based 360VR image/video viewer in order to make easier to enter the contents into it similarly to the authoring tool shown in Fig. 8.

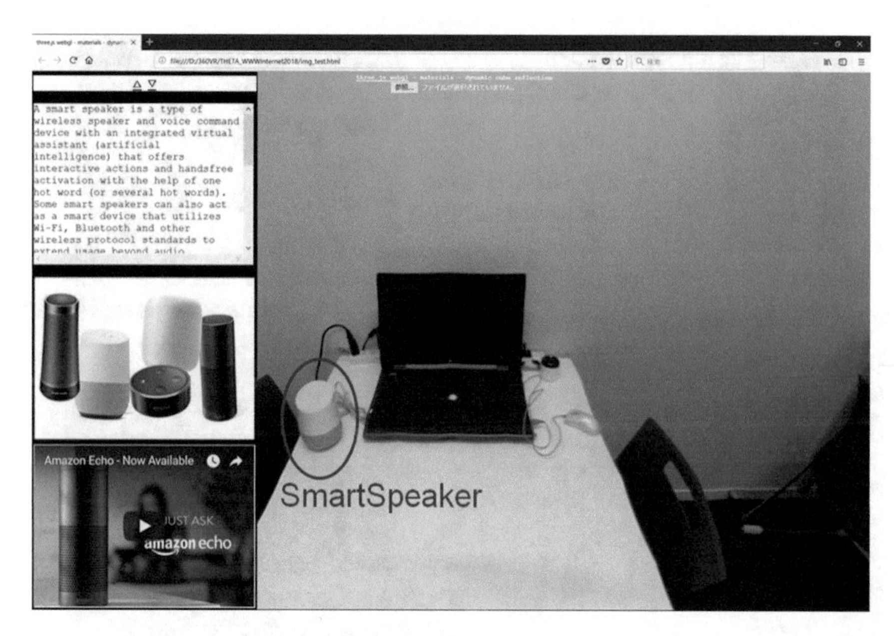

Fig. 10. A screenshot of the e-learning material on a browser for IoT security education based on Linked Data of IoT security information that uses 360VR image/video.

The following is the fundamental mechanism of a web-based 360VR image/video viewer realized using WebGL [19]. Originally, there are example JavaScript programs of Three.js [20] that can display 360VR images/videos so that we extended those programs to make them more convenient. For displaying 360VR images/videos, Equirectangular projection images/videos are prepared as shown in the upper right part of Fig. 11. It is possible to generate such a projection image/video by transforming from the image/video captured with a special camera called 360VR camera of multiple fish-eye lenses, e.g., RICHO Theta of two fish-eye lenses and its fish-eye image shown in the upper left part and lower part of the figure, respectively. Usually, companies of such 360VR cameras provide dedicated software for the transforming. After that, by the texture mapping of such projection image/video frame onto the surface inside of a sphere located in a 3D CG space realized by WebGL and Three.js, we obtain 360VR image or one frame of 360VR video shown in the lower right part of the figure. Furthermore, we integrated this web-based 360VR image/video viewer with the functionality of displaying the text explanation, the image and the video of a target IoT device as shown in Fig. 10.

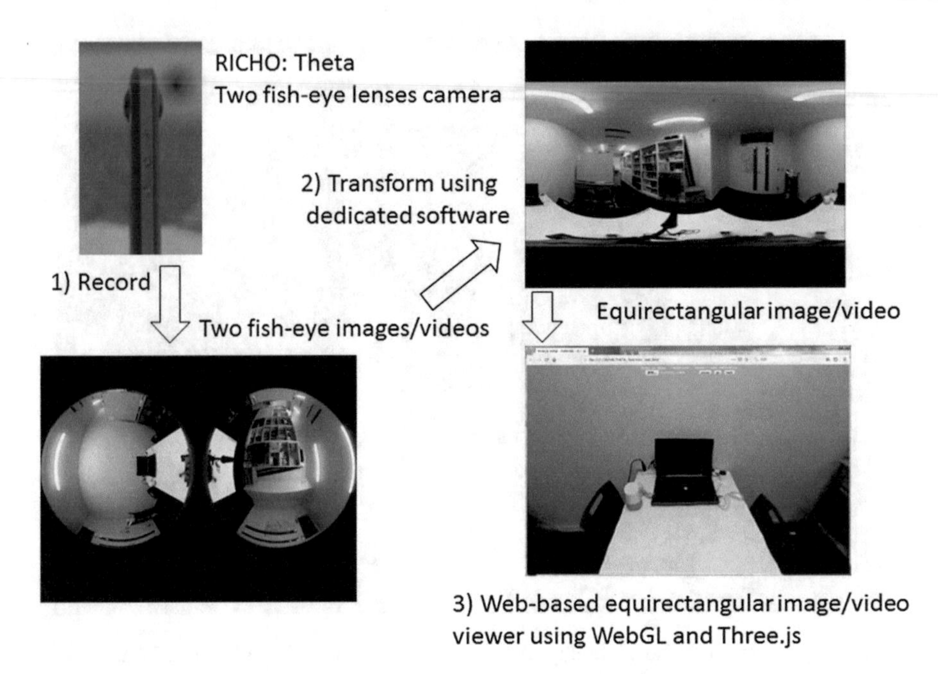

Fig. 11. Fundamental mechanism of web-based equirectangular image/video viewer using WebGL and Three.js.

5 Conclusions

In this paper, we treated a novel framework for the development of educational materials related to IoT security. Especially, we proposed the combinatorial use of the web-based 360VR image/video viewer and the database of IoT threats information realized as Linked Data for developing e-learning materials, and introduced a prototype of the e-learning material that uses 360VR image/video for IoT security education.

There are a lot of future work we have to do. First of all, we will complete the database as Linked Data of IoT threat information as soon as possible. After that, we will finish the development of the e-learning material that uses 3D CG and the e-learning material that uses 360VR image/video. Unfortunately, we have to register threats information of IoT devices appear in 360VR images/videos of a room manually by specifying their location on the surface inside of the sphere whom the 360VR image/video frame is mapped onto. Furthermore, we are supposed to implement image recognition functionality of IoT devices using machine learning such as Deep Neural Network for automatically finding out the IoT devices. Furthermore, we will evaluate the usefulness of the proposed e-learning materials by asking several students to learn IoT security using them.

Acknowledgements. This research was mainly supported by Strategic International Research Cooperative Program, Japan Science and Technology Agency (JST) regarding "Security in the Internet of Things Space", and partially supported by JSPS KAKENHI Grant Number JP16H02923 and JP17H00773.

References

1. Okada, Y., Kaneko, K., Tanizawa, A.: Interactive educational contents development framework based on linked open data technology. In: Proceedings of the 9th Annual International Conference of Education, Research and Innovation (iCERi 2016), pp. 5066–5075, 14–16 November 2016 (2016)
2. Ma, C., Srishti, K., Shi, W., Okada, Y., Bose, R.: Interactive educational material development framework based on linked data for IoT security. In: Proceedings of the 10th Annual International Conference of Education, Research and Innovation (iCERi 2017), pp. 8048–8057, 12–14 November 2017 (2017)
3. Okada, Y., Nakazono, S., Kaneko, K.: Framework for development of web-based interactive 3D educational contents. In: Proceedings of the 10th International Technology, Education and Development Conference (INTED 2016), pp. 2656–2663, 7–9 March 2016 (016)
4. Okada, Y., Kaneko, K., Tanizawa, A.: Interactive educational contents development framework and its extension for web-based VR/AR applications. In: Proceedings of the 18th Annual European GAMEON Conference on Simulation and AI in Computer Games (GameOn 2017), Eurosis, pp. 75–79, 6–8 September 2017 (2017)
5. Ma, C., Kulshrestha, S., Shi, W., Okada, Y., Bose, R.: e-Learning material development framework supporting VR/AR based on linked data for IoT security education. In: Proceedings of 6th International Conference on Emerging Internet, Data & Web Technologies (EIDWT 2018), pp. 479–491, 15–17 March 2018 (2018)
6. Okada, Y., Haga, A., Shi, W.: Web based interactive 3D educational material development framework supporting 360VR images/videos and its examples. In: Proceedings of the 21st International Conference on Network-Based Information Systems (NBiS-2018), pp. 395–406, 5–7 September 2018 (2018)
7. Haga, A., Shi, W., Hayashida, G., Okada, Y.: Web based interactive viewer for 360VR images/videos supporting VR/AR devices with human face detection. In: Proceedings of the 13th International Conference on Broadband and Wireless Computing, Communication and Applications (BWCCA-2018), pp. 260–270, 27–29 October 2018 (2018)
8. Berners-Lee, T.: Design issues for the World Wide Web. "Architectural" and "philosophical points" (1998)
9. Bizer, C., Heath, T., Berners-Lee, T.: Linked Data -the story so far. Int. J. Semant. Web Inf. Syst. (IJSWIS) **5**, 1–22 (2009)
10. Miller, P., Berners-Lee, T.: Semantic web is open for business, The Semantic Web (2008)
11. Dicheva, D.: Ontologies and semantic web for e-Learning. In: Adelsberger, H.H., Kinshuk, Pawlowski, J.M., Sampson, D.G. (eds.) Handbook on Information Technologies for Education and Training, International Handbooks on Information Systems, pp. 47–65. Springer, Heidelberg (2008)
12. Okada, Y., Tanaka, Y.: IntelligentBox: a constructive visual software development system for interactive 3D graphic applications. In: Proceedings of Computer Animation 1995, pp. 114–125. IEEE CS Press (1995)

13. Okada, Y.: Web Version of IntelligentBox (WebIB) and Its Integration with Webble World. Communications in Computer and Information Science, vol. 372, pp. 11–20. Springer (2013)
14. Webble World. https://github.com/truemrwalker/wblwrld3
15. Unity. https://unity3d.com/jp
16. Sugimura, R., Kawazu, S., Tamari, H., Watanabe, K., Nishimura, Y., Oguma, T., Watanabe, K., Kaneko, K., Okada, Y., Yoshida, M., Takano, S., Inoue, H.: Mobile game for learning bacteriology. In: Proceedings of IADIS 10th International Conference on Mobile Learning 2014, pp. 285–288 (2014)
17. Sugimura, R., Tamari, H., Kitaguchi, H., Kawatsu, S., Oya, K., Ono, Y., Kaneko, K., Nakazono, S., Okada, Y., Yoshida, M., Kono, Y.: Serious games for education and their effectiveness for higher education medical students and for junior high school students. In: Proceedings of 4th International Conference on Advanced in Information System, E-Education and Development (ICAISEED 2015), pp. 36–45 (2015)
18. Ma, C., Srishti, K., Shi, W., Okada, Y., Bose, R.: Learning analytics framework for IoT security education. In: Proceedings of 12th Annual International Technology, Education and Development Conference (INTED 2018), pp. 9181–9919, 5–7 March 2018 (2018)
19. WebGL, 20 December 2018. https://unity3d.com/jp
20. Three.js, 20 December 2018. https://threejs.org/

Domain Formalization for Metaphorical Reasoning

Flora Amato[1], Giovanni Cozzolino[1(✉)], and Francesco Moscato[2]

[1] University of Naples Federico II, Naples, Italy
{flora.amato,giovanni.cozzolino}@unina.it
[2] Second University of Naples, Naples, Italy
francesco.moscato@unicampania.it

Abstract. It is commonplace that cultural heritage is always discussed and analysed by using references to figures of speech. In particular, metaphors and allegories are very frequent in ancient and historical documents, painting and sculptures. It is also frequent to have some hints about the assets and their authors by comparing their contents with elements in the domain of the figures of speech to which the asset refer. In order to enable reasoning by figures of speech, we propose here a methodology able to link concepts in the domain of cultural assets with concepts in the domain of the figure of speech. We show here how reasoning in all these domains help in discovering some elements related to the cultural heritage that humans may neglect at first glance. Our approach can be useful in a variety of applications related to cultural heritage such as semantics annotations, linked data, crowd-sensing, etc.

1 Introduction

Cultural heritage assets are full of metaphors and allegories that vary according to what is desired to be highlighted in the allegoric form. For instance, sovereigns were represented by Apollo-Sun in order to emphasise the fact they were the primary source of light for their kingdom and their subjects, or by Mars to exalt their warrior and strategist properties.

Associating the metaphors or allegories to the elements which they are related to makes it easier to discover hidden contents, authors provocations or to clarify unclear elements. This operation is equivalent to reasoning in abstract domain, which is an old problem, but still investigated in literature [1,2]. In [3] authors make a first attempt in coupling metaphors and analysis of assets although they only outlined the need of extending automated reasoning to metaphors. The idea of using semantic technologies (i.e. ontologies) in metaphorical reasoning is really novel and is attracting the attention of researchers in the field. Authors in [4] reported a model where metaphors are associated to attributes in ontologies defined by OWL. While, in [2,5] there is given a first attempt of ontology-based formalization of metaphors.

L. Barolli et al. (Eds.): EIDWT 2019, LNDECT 29, pp. 161–169, 2019.
https://doi.org/10.1007/978-3-030-12839-5_15

In this work we present an approach to address the problem of linking domains represented in cultural heritages with the ones related to metaphors or allegories. It should be noted that this problem has not been fully addressed in previous works in the literature. The methodology presented in this work exploits semantic techniques and technologies [6–8], aiming at enriching the analysis and the abstraction processes of knowledge extraction.

Our approach can be used in a variety of applications related to cultural heritage such as semantics annotations, linked data, crowd-sensing, etc. Our contribution can be useful also in the context of Big Data. Indeed, advanced approaches can be further exploited for extracting information from a huge amount of data, through application-specific design methodologies [9–11] or dedicated solutions [12–15].

The rest of the paper is organised as follows. We briefly present in Sect. 2 the main concepts behind metaphoric abstraction. The methodology is outlined in Sect. 3. We exemplify the approach by a case study in Sect. 4. We end the paper in Sect. 5 with some conclusions and future research work.

2 Metaphoric Abstraction

Metaphors are the basics elements of the methods proposed in our approach. The definition of Metaphor, for instance according to the Oxford Dictionary, is the following:

A figure of speech in which a word or phrase is applied to an object or action to which it is not literally applicable.

or,

A thing regarded as representative or symbolic of something else.

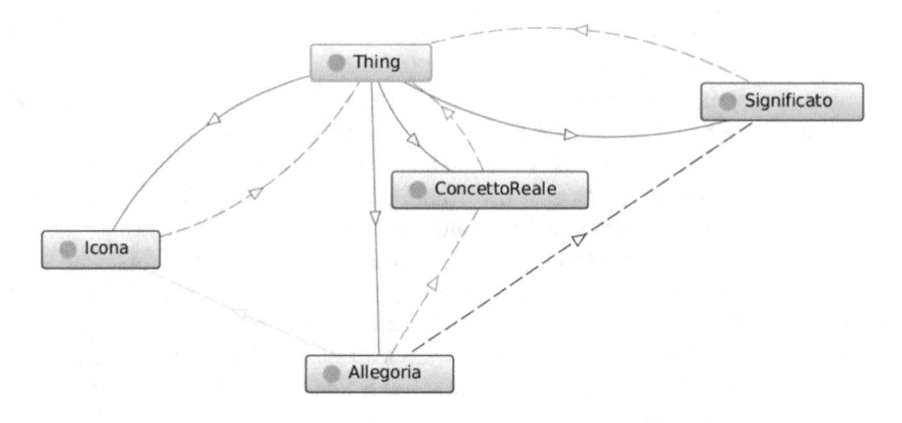

Fig. 1. Allegories Elements

The later definition means that a Metaphor is something having a given semantics in a domain $\mathscr{A}$ that is used to address something else in an another domain $\mathscr{B}$.

Let us start explaining our approach by example. If we use, in the same document, usual concepts and metaphors, undoubtedly the complexity of the document would be increased. Instead, we consider a well-known, well-defined and clear domain $\mathscr{M}$, and a set of (historical) documents, belonging to a knowledge domain $\mathscr{D}$ that we suppose 'metaphor-pruned' (i.e. without content containing metaphoric meanings). Describing semantics of documents in the domain $\mathscr{D}$, in terms of the concepts belonging to the domain $\mathscr{M}$, regardless of the meaning of terms in the documents, we should have the semantics of elements in the documents represented in terms of a clear set of words with defined semantics [16]. Reasoning on the concepts of domain $\mathscr{D}$ is simpler and clearer than reasoning on $\mathscr{D}$ and new concepts can be discovered from source documents.

3 Methodology

In our methodology we apply Abstraction, Pattern Recognition and Generalisation on a set of source documents (S) to be analysed. Let us suppose that the documents belong to different historical periods and that a domain ontology $\mathscr{D}$ exists able to describe the main concepts in the documents. Furthermore, we assume that terms and concepts related to them have changed during the years. In this case domain $\mathscr{D}$ is no longer suitable to describe all the meanings of the documents, since some of them have been lost due to language changes.

The first step of the methodology is retrieving suitable Metaphors for documents contents. We assume that $\mathscr{M}$ is the ontology used for defining metaphoric meanings of concepts. Both $\mathscr{D}$ and $\mathscr{M}$ have to be defined with a formal language and it is extremely important that $\mathscr{M}$ is the more possible detailed. In our example we use OWL in order to define ontologies.

The next steps are the following:

1. we annotate documents by using concepts in $\mathscr{D}$ ontology;
2. we apply the abstraction technique: if we found a metaphor linking concepts (and instances) from $\mathscr{D}$ to $\mathscr{M}$, the link between concepts is formally reported, so it is possible to refer to known concept in $\mathscr{D}$ by using a metaphoric symbol;
3. we adopt a pattern recognition technique in order to retrieve metaphoric connection lost during the years, observing if, through metaphoric linked instances across domains $\mathscr{M}$ and $\mathscr{D}$, (graph) patterns exist in $\mathscr{M}$ with no related pattern in $\mathscr{D}$, even if similar graph structure exists in $\mathscr{D}$;
4. we apply generalisation technique after then lost metaphoric link are re-applied to source documents, recovering lost semantics [17–19] and information.

4 Case Study

This case study reports an example of reasoning about iconography, metaphors and allegories, thus, in this section we illustrate the structure of the ontology and what, at a first analysis, could be the usefulness of a relative semantic annotation [20].

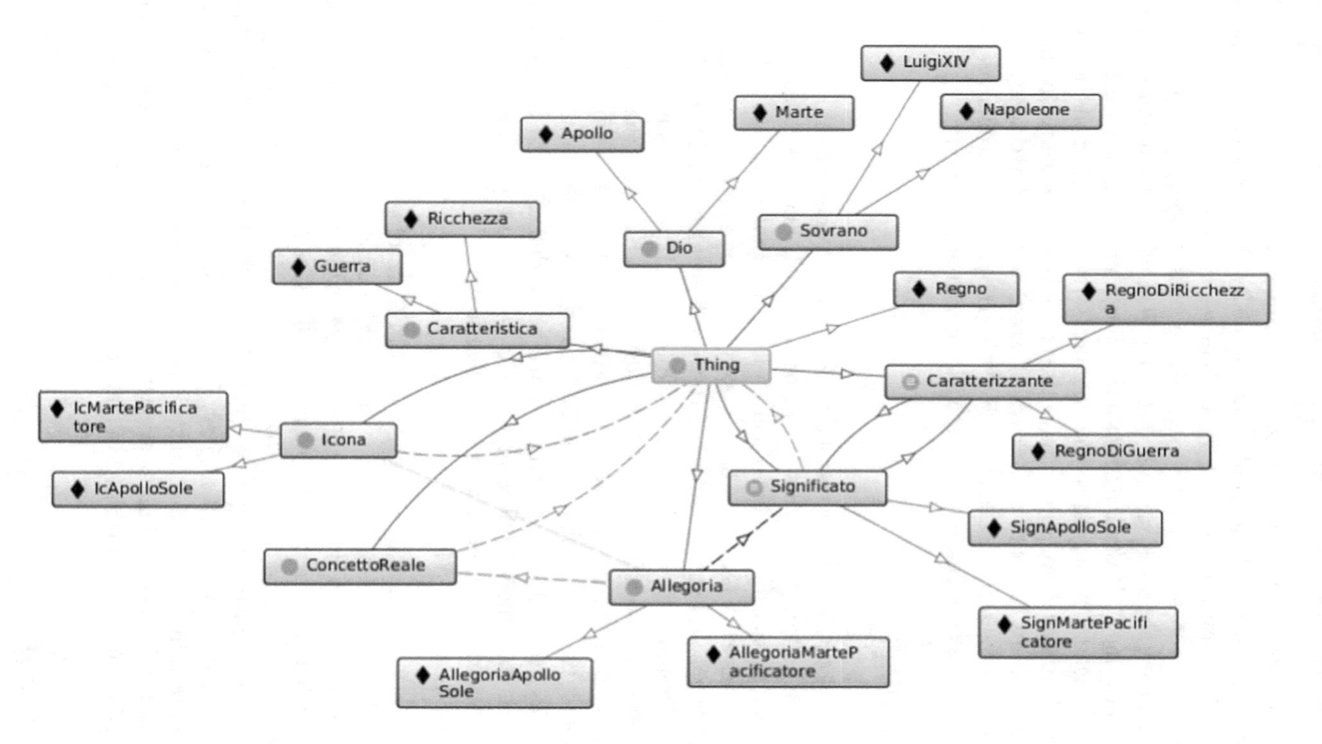

Fig. 2. Apollo Sun and Mars Peacemeaker

Here, we choose the subject of noble iconography, where mythological figures meant to be allegories of sovereignty and of the actions by commanders and sovereigns. Allegories have changed according to what was desired to be highlighted in the allegoric form. Also the association to semi-gods were frequent, like, for instance, Hercules in his various representations, mainly linked to the region where the works of art belonged to. This meant that the every representation was an indication of the region of belonging of the monarch or the noble to whom it was associated.

So, it is clear that allegories associate a symbolic semantic to concepts present in a domain (such as the mythological one): Mars and War, Jupiter father of the Gods, Apollo bringer of Light without which life cannot exist.

We use OWL language to define ontologies, and we choose Protègè[1] as a tool to draw the ontology in an easier way than directly describing it through an XML file. The ontological description of these concepts is interesting because we can distinguish four descriptive levels, belonging to four different domains:

- monarchy, of aristocracy and noble families;
- mythology, describing divinities, their domains of belonging;
- iconography, where allegories are represented; in this domain concepts (such as poses, backgrounds or the colours used to paint objects) are important;
- allegory, which is hidden in the representation and associates a meaning to an iconographic and mythological figure.

4.1 Allegory Description

The fourth ontology allows to colour concepts belonging to the first three domains (associated with a terminological semantic), by associating a meaning to them, a symbolic semantic. Further ontological concepts, associated with this connection (that could be ontologically represented by the graph in Fig. 1), transfer the description of the semiotic-terminological concept to its allegorical meaning.

Continuous lines in Fig. 1 represent a hierarchy relationships ('is-a' kind), while dashed lines represent object properties, i.e. relationships, between concepts. So, we can see that the allegory concept is semantically connected [21,22] (through object properties):

- to the real concepts it represents;
- to the icon used by the allegory to represent the real concept;
- to its semantic-allegoric meaning.

We apply this reasoning to our case study, i.e. the description of the Apollo Sun and Mars Peacemaker allegories. In the Fig. 2 is reported only a piece of the complete ontology.

[1] http://progete.stanford.edu.

Fig. 3. Louis XIV as Apollo Sun (Henri Gissey, Bibliotheque Nationale, Paris)

In the considered scenario, the real concepts are Louis XIV (see Fig. 3) and Napoleon (see Fig. 4) instances, both representing the concept of Sovereign. The analysis of reported pictures' iconography reveals that often Louis XIV was represented by Apollo as a symbol of the splendour of its reign, while Napoleon by Mars Peacemaker to magnify his war campaigns.

So, the two allegoric figure related with two sovereigns appear to be instances of the concept of God. The instances of the allegorical Meaning associated to the two allegories are distinct: the first links the Icon of Mars and the sovereign Napoleon, while the second links Louis XIV as Apollo Sun, both representing the fact of having a reign of war (referring two instances of Characterizing and Meaning: Reign and ReignOfWar).

Definitely, through metaphorical and allegorical connections it is possible to associate different concepts belonging to different domains, simplifying the explanation of meaning hidden below historical representation, and allowing the creation of reasoning methods and techniques based on computational approaches.

Fig. 4. Napoleon as Mars the Peacemaker (Canova, Apsley House, London)

5 Conclusions and Future Works

In this paper we have defined a methodology that aims to analyse the meaning of historical documents through the explanation and clarification of metaphors present in them. A simple cultural heritage scenario has been chosen as case study for applying the defined methodology, to show that it produces significant results, summarized in previous section.

We foresee to introduce, in future works, additional metaphors' ontologies in order to improve obtained results and to implement automatic information extraction techniques in order to annotate the source documents in a semi-automatic way. We will then evaluate our approach in terms of accuracy, usefulness and scalability when applied to large domains of cultural heritage through centuries.

Acknowledgment. This work is supported by CREA European Project: Conflict Resolution with Equitative Algorithms, Grant Agreement number: 766463, CREA, JUST−AG− 2016 /JUST−AG− 2016−05.

References

1. Narayanan, S.: Moving right along: a computational model of metaphoric reasoning about events. In: AAAI/IAAI, pp. 121–127 (1999)
2. Diarra, D., Clouzot, M., Nicolle, C.: Causal reasoning and symbolic relationships in medieval illuminations (2018)
3. Endert, A., North, C., Chang, R., Zhou, M.: Toward usable interactive analytics: coupling cognition and computation. In: Proceedings of the ACM SIGKDD Workshop on Interactive Data Exploration and Analytics (IDEA 2014), pp. 52–56 (2014)
4. Bulat, L., Clark, S., Shutova, E.: Modelling metaphor with attribute-based semantics. In: Proceedings of the 15th Conference of the European Chapter of the Association for Computational Linguistics, Short Papers, vol. 2, pp. 523–528 (2017)
5. Gangemi, A., Alam, M., Presutti, V.: Amnestic forgery: an ontology of conceptual metaphors. arXiv preprint arXiv:1805.12115 (2018)
6. Di Lorenzo, G., Mazzocca, N., Moscato, F., Vittorini, V.: Towards semantics driven generation of executable web services compositions. J. Software **2**(5), 1–15 (2007)
7. Moscato, F.: Exploiting model profiles in requirements verification of cloud systems. Int. J. High Perform. Comput. Networking **8**(3), 259–274 (2015)
8. Di Martino, B., Moscato, F.: An ontology based methodology for automated algorithms recognition in source code, pp. 1111–1116 (2010)
9. Cilardo, A., Durante, P., Lofiego, C., Mazzeo, A.: Early prediction of hardware complexity in hll-to-hdl translation. In: 2010 International Conference on Field Programmable Logic and Applications (FPL), pp. 483–488. IEEE (2010)
10. Fusella, E., Cilardo, A.: Minimizing power loss in optical networks-on-chip through application-specific mapping. Microprocess. Microsyst. **43**, 4–13 (2016)
11. Cilardo, A.: Exploring the potential of threshold logic for cryptography-related operations. IEEE Trans. Comput. **60**(4), 452–462 (2011)
12. Balzano, W., Vitale, F.: Dig-park: a smart parking availability searching method using v2v/v2i and dgp-class problem. In: 2017 31st International Conference on Advanced Information Networking and Applications Workshops (WAINA), pp. 698–703. IEEE (2017)
13. Balzano, W., Murano, A., Vitale, F.: Wifact–wireless fingerprinting automated continuous training. In: 2016 30th International Conference on Advanced Information Networking and Applications Workshops (WAINA), pp. 75–80. IEEE (2016)
14. Balzano, W., Murano, A., Vitale, F.: V2v-en-vehicle-2-vehicle elastic network. Procedia Comput. Sci. **98**, 497–502 (2016)
15. Balzano, W., Murano, A., Vitale, F.: Snot-wifi: Sensor network-optimized training for wireless fingerprinting. J. High Speed Networks **24**(1), 79–87 (2018)
16. Xhafa, F., Barolli, L.: Semantics, intelligent processing and services for big data. Future Gener. Comput. Syst. **37**, 201–202 (2014)
17. Amato, F., Cozzolino, G., Mazzeo, A., Mazzocca, N.: Correlation of digital evidences in forensic investigation through semantic technologies, pp. 668–673 (2017). cited By 1
18. Amato, F., Cozzolino, G., Maisto, A., Mazzeo, A., Moscato, V., Pelosi, S., Picariello, A., Romano, S., Sansone, C.: ABC: a knowledge based collaborative framework for e-health, pp. 258–263 (2015). cited By 1
19. Cozzolino, G.: Using semantic tools to represent data extracted from mobile devices, pp. 530–536 (2018). cited By 0

20. Amato, F., Cozzolino, G., Moscato, V., Picariello, A., Sperlí, G.: Automatic personalization of visiting path based on users behaviour, pp. 692–697 (2017). cited By 1
21. Moore, P., Xhafa, F., Barolli, L.: Semantic valence modeling: Emotion recognition and affective states in context-aware systems. In: 2014 28th International Conference on Advanced Information Networking and Applications Workshops (WAINA), pp. 536–541. IEEE (2014)
22. Xhafa, F., Martinez, A.L., Caballé, S., Kolici, V., Barolli, L.: Mining navigation patterns in a virtual campus. In: 2012 Third International Conference on Emerging Intelligent Data and Web Technologies (EIDWT), pp. 181–189. IEEE (2012)

A Hybrid Approach for Document Analysis in Digital Forensic Domain

Flora Amato[1], Giovanni Cozzolino[1(✉)], Marco Giacalone[2], Francesco Moscato[3], Francesco Romeo[1], and Fatos Xhafa[4]

[1] Università degli Studi di Napoli Federico II, Naples, Italy
{flora.amato,giovanni.cozzolino,francesco.romeo}@unina.it
[2] Vrije Universiteit Brussel (VUB), Brussels, Belgium
marco.giacalone@vub.be
[3] Second University of Naples, Naples, Italy
francesco.moscato@unicampania.it
[4] UPC BarcelonaTech, Barcelona, Spain
fatos@cs.upc.edu

Abstract. Our daily life is characterized by the continuous interaction with many devices and technologies, which build a digital ecosystem around each person. In this context, if, for example, it is necessary to investigate the commission of a crime, the adoption of digital forensic sciences is of fundamental importance. These disciplines are in fact concerned with gathering any information that may be of some interest to the legal system, in full respect of its nature and without its characteristics being altered in any way; in this way the probative value of the information collected and its admissibility in the trial are guaranteed. The complexity of the scenario in which the forensic operators are to operate derives from the widespread diffusion of devices with a large variety of characteristics and, mostly, from the heterogeneous and unstructured representation of relevant information. In this work we present a semantic approach for textual document analysis to enrich the correlation process of a forensics investigation.

1 Introduction

The pervasiveness of new technologies in the everyday life has determined the advent of the so-called "Information Society". A society in which technologies offer new ways of interaction with Institutions and Public Administration, new ways of using services, intervening in political life, new forms of communication (social networks, virtual relations between users, etc.).

All the activities performed in the digital world leave pieces or echoes of themselves, in the form of *digital fingerprint*. The digital fingerprint can be found in a variety of forms such as web browser history, caches and cookies, meta-data, log information, removed fragments, etc.

L. Barolli et al. (Eds.): EIDWT 2019, LNDECT 29, pp. 170–179, 2019.
https://doi.org/10.1007/978-3-030-12839-5_16

Computer forensics and digital forensics are concerned with the research and analysis of "digital evidences" increasingly present in the courtroom, then in civil, administrative, tax and criminal trials. The main concern behind digital forensics is to examine the real data sources, analyse and extract any relevant information for the reconstruction of trace evidence.

In fact, "digital forensics" is used not only in the case of the commission of crimes that are consumed in the digital world, but also in the case of traditional crimes that also leave traces in the digital world. Indeed, we can think of digital data stored in computers, pen-drives, digital media of any kind, as well as digital data contained in mobile phones, email, SMS, MMS, satellite navigators, etc. Such IT tools and data flows, inevitably are subjected to digital analysis by technicians specialized in computer forensics, in order to find digital traces useful for the reconstruction of the facts and the identification of the author of the crime. Since smartphones, computers, and IT devices in general, are more and more pervasive they even more represent the largest growing resource to be exploited for legal purposes in analysing the activities of the parties involved in a criminal case, or by cyber-criminals in collecting information on their victims. So, nowadays Digital Forensics is a fundamental discipline for supporting the law enforcement sector [20, 21].

Digital forensics deals with the collection, acquisition and documentation of data stored on digital devices, but, above all, with the analysis of these information in order to promote them as digital evidences. According to ISO/IEC 27037, a *digital evidence* is any "information or data, stored or transmitted in binary form that may be relied on as evidence". Information correlation is a crucial phase in forensic investigations, because the investigator have take into account many different kinds of information, not necessarily in digital form, and their correlation is the only mean to contextualize them, promoting as clues.

Taking into account the previous considerations, the main challenges that experts have to face during a forensic investigations are:

- the preservation of the admissibility of acquired data;
- the volume and heterogeneity of data requires information extraction;
- the integration and correlation of information extracted with different tools that generate files with different format.

In this work we present a methodology that, through the application of semantic techniques and technologies [11, 12, 18], enrich the analysis process of a forensics investigation. In fact, the annotation of extracted data with additional semantic assertion support and improves the correlation of information by enhancing the searches' mechanism. An advanced approaches can be further exploited to improve performances through application-specific design methodologies [9, 13] or dedicated hardware [4–6, 8].

The paper is structured as follows: Sect. 2 presents basic principles, concepts and definitions of digital evidence. Section 3 describes the proposed methodology covering all phases of the process. In Sect. 4 are outlined the building blocks of the system architecture according to the proposed methodology. Section 5 makes an overview of related work relevant to the studied topic. Finally, in Sect. 6 we present some conclusions and indicate future research work.

2 Digital Evidences

The volatile and easily alterable nature of digital information arise many challenges for digital forensic experts, because they have to guarantee that the collected evidences avoid (even involuntary) alteration or modification that may compromise and contaminate the *scena criminis* status.

In [20] the author has identified three basic properties of a digital evidence, namely fidelity, volatility and latency, that an investigator have to care of:

1. *Fidelity* involves the adoption of techniques that preserve the integrity of evidences;
2. *Volatility* involves the storage support from which evidences are extracted (disk, memory, registers, etc.) and affects the acquisition phase;
3. *Latency* requires the adoption of additional information to contextualize and interpret a digital encoding.

The Budapest Convention on Cybercrime has regulated the principles and procedures to be followed during a forensic investigation, in order to correctly manage evidences through the different phases of collection, analysis and presentation [15,16]. The convention also provides technical measures to adopt in order to preserve the fidelity of acquired data.

Forensic experts identify five main phases of a digital forensic investigation, summarized in the followings:

Identification: involves the choice of the devices or the systems to be examined;
Preservation: requires the adoption of multiple procedures to guarantee the "chain of custody" in order to isolate the examined evidence from external agents and alterations;
Acquisition: collects data from devices selected during Identification phase, though consolidated techniques and tools that avoid alterations;
Analysis and Correlation: involves many different tasks that strictly depend on the actual scenario to be examined and on requirements for searches;
Documentation: reports the results of entire investigation, describing all the operation carried out and the relevant evidence found.

3 Methodology

Evidence analysis can be a very critical process of digital investigations due to the heterogeneity of collected data, from the point of view of media to be represented (text, image, video, audio, and so on) but also due to the different formats of representation.

In this work we deal with the analysis of textual information extracted from a digital device, that introduce many challenges related to Information Extraction and Retrieval and Natural Language Processing. In fact, the analysis of the text saved into a textual document has to face off the complexity of the natural language that make this task very difficult. This section describes the methodology that propose to enrich textual data with other information collected from digital evidences, by applying semantic text processing techniques [2,10].

Semantic processing of documents turns out to be not an easy task to be performed; it depends on many factors such as the personal style to write documents, the explicit or implicit knowledge of the writer, the knowledge of the reader, among others. A text, in fact, is the result of a collaborative process between an author and a reader: the former uses language signs to codify meanings, the latter decodes these signs and interprets their meaning. Much of information encoded and decoded by the actors is implicitly assumed by the context, consisting in relationships at a morphological, syntactic and semantic level [7,22] (infra-textual context) and more generally involving the domain of interest (extra-textual context).

Therefore, the activity of knowledge extraction from textual documents comprehends different text analysis methodologies, aiming at recreating the domain model which the texts belongs to. The model reconstruction also requires information about the properties characterizing each concept within the considered domain as well as the identification of the set of entities the concept refers to.

The term-extraction is a fundamental task in the automatic document processing and derivation of knowledge from texts, because it aims to find a series of peculiar terms in order to detect the set of concepts that allow the resource identification. Terms serve to convey the fundamental concepts of a specific domain, because their relationships constitute the semantic frame both of the documents and of the domain itself. For the term-extraction process we use an hybrid method that combines *linguistic and statistical techniques*. More precisely, we use a *linguistic filter* in order to extract a set of candidate terms and then use a *statistical method* to assign a value to each candidate term. In particular, linguistic filter is based on lexical type operations performed on the words constituting the text, as *part-of-speech tagger* in order to apply a filter on terms, so to extract only the categories of interest, such as nouns and verbs, and *lemmatization*, to restore words to a dictionary form. Statistical methods are based on the analysis of word occurrences within texts, because not all words are equally useful to describe documents.

In order to extract relevant terms from a document collection several steps are required, described next.

3.1 Text Preprocessing: Tokenization and Normalization

Text Preprocessing deals with the extraction of relevant units of lexical elements to be processed in next phases. Text Preprocessing includes Text tokenization and text normalization.

Text tokenization consists in the segmentation of a sentence into *tokens*, defined as minimal units of analysis that correspond to simple or complex lexical items. Text tokenization requires, various sub-steps, as:

- *grapheme analysis*, to verify possible mistake through the definition of the set of alphabetical signs used within the text collection;
- *disambiguation of punctuation marks*, aiming at token separation;
- *separation of continuous strings* to be considered as independent tokens;
- *identification of separated strings* to be considered as complex tokens and, therefore, single units of analysis.

Tokenization can be performed by specialized tool, called *tokenizers*, that adopt *domain glossaries*, that comprehend tokens and well-known expressions of the considered domain, and *mini-grammars* containing heuristic rules regulating token combinations. The combined use of glossaries and mini-grammars ensures higher level of accuracy, even in presence of acronyms or abbreviations.

Text normalization deals with the transformation of multiple variations of the same lexical expression that should be reported in a unique way. For example:

- words that assume different meaning if are written in small or capital letter;
- compounds and prefixed words that can be (or not) separated by a hyphen;
- dates that can be written in different ways;
- acronyms and abbreviations.

3.1.1 Morpho-Syntactic Analysis

Morpho-syntactic analysis enables the extraction of word categories, both in simple and complex forms, in order to obtain a list of candidate terms on which information extraction can be performed.

Part-Of-Speech Tagging. Part of Speech (POS) Tagging consists in the assignment of a grammatical category (noun, verb, adjective, adverb, etc.) to each lexical unit identified within the text collection. Morphological information about the words entail a first semantic distinction among the analysed words, in fact, generally, nouns are indicators of people, things and places; verbs denote actions, states, conditions and processes; adjectives indicate properties or qualities of the noun they refer to; adverbs represent modifiers of other classes (place, time, manner, etc.). Automatic POS tagging assigns the correct category (generally more than one) to each word encountered within a text.

The *word-category disambiguation* involves the finding of all the possible tags for each lexical item and the choosing, among them, the correct one. Hence, the vocabulary of the relevant documents is compared with an external lexical resource, whereas the procedure of disambiguation is carried out through the analysis of the words in their contexts of occurrence. An effective help comes from the *Key-Word In Context (KWIC) Analysis*, a study of the local context where the occurrences of a lexical item appear. For each textual signifier it can be located its occurrences in the text and, so, the textual parts preceding and following each one. This kind of analysis, then, permits to visualize the use of the words in their contexts of occurrence in order to disambiguate their grammar category. In this way, the ambiguity between noun and adjective in the Italian word "pubblico" can be solved by observing the categories of the preceding or following words, ie, the evidence of the presence of an article, a preposition or a noun: in the first two cases the word at issue is a noun, in the last one it is an adjective. Then, the ambiguous form is firstly associated the set of possible POS tags, and then disambiguated by resorting to the KWIC analysis. Further morphological specifications, such as inflectional information[1], are then associated to each word.

Lemmatization. Text Lemmatization is performed in order to reduce all the inflected forms to the respective *lemma*, coinciding with the singular male/female form for nouns, the singular male form for adjectives and the infinitive form for verbs. This operation is performed on the list of tagged terms.

3.1.2 Relevant Terms Recognition

The methodology aims to the identification of the relevant terms, useful to characterize the sections of interest, because not all words are equally useful to describe resources. Some words are semantically more relevant than others, and among these words there are lexical items weighting more than other. In our approach, we compute the semantic relevance by evaluating the TF-IDF index (*Term Frequency - Inverse Document Frequency*) on the corpus vocabulary and by calculating the term frequency and the term distribution within the corpus. TF-IDF index, in fact, takes into account:

- the *term frequency* (*tf*), that is the number of times a given term occurs in a document. Frequent terms are supposed to be more representative of its contents;
- the *inverse document frequency (idf)* that deals with the term distribution within the corpus and relies on the principle that term importance is inversely proportional to the number of documents from the corpus where the given term occurs.

Therefore, we can select relevant concepts, filtering all terms that have a $TF-IDF$ value under an established threshold, that correspond to the terms that are frequent and concentrated on few documents. This statement is summarized in the following equation

$$W_{t,d} = f_{t,d} * log(N/D_t)$$

where:

$W_{t,d}$ is the evaluated weight of term t in resource d;
$f_{t,d}$ is the frequency of term t in the resource d;
N is the total number of occurrences within the examined corpus;
D_t is the number of resources containing the term t.

[1] Inflection is the modification of a word to express different grammatical categories such as tense, case, voice, aspect, person, number, gender, and mood.

4 System Architecture

In this section we show the architecture of a system that implements the proposed methodology, briefly discussing the tools and techniques used by. Our system is based on a custom ontology, build on public and shared domain ontologies and enriched by derived knowledge through ad-hoc reasoning rules, and five modules: Evidences Manager, Semantic Parser, Inference Engine, SWRL Rule Engine and a Query and Visualization module. An overview of the architecture is presented in Fig. 1. Its main components are further discussed below.

Evidence Manager loads all the binary content stored on a digital device and verify its integrity by computing the hash value of the entire bit-stream image. Moreover, the module provides many functions to extract relevant information from the bit-stream image, like user files, browser history, Windows registry, etc. The extracted knowledge is a collection of entities, consisting in files with a set of attributes, often structured and ready to be represented as ontology instances' attributes. Instead, the content of the most number of files is non-structured and then require further processing based on regular expression or NLP techniques.

Semantic Parser generates a semantic representation of entities extracted in the previous step. So, this module is responsible of the creation of ontology instances and of the integration of derived knowledge with public domain ontologies, if available. Ontology population is made by creating an instance for each item of acquired data, and by linking them each other according to object properties defined in the ontology schema. Through integration mechanism, the ad-hoc ontology can be created exploiting the reuse of entities defined in other public ontologies, thus increasing system flexibility, since domain ontologies may not be easily modifiable.

Inference Engine performs automated reasoning, according to domain ontologies definition and OWL specifications. Reasoners can specify the kind of inferences to be made to improve the performance of query execution. Inferences enrich the knowledge base with new facts, increasing the number of explicit information regarding an evidence. For example, it is possible to infer a file type from its extension or to deduce the author of an action from logged account information at a specific time.

SWRL Rule Engine adopts SWRL rules in order to perform inferences that can't be made by the OWL specification. For example, through SWRL rules it can be established relationships among individuals belonging to different ontologies but representing similar concepts.

SPARQL Queries Interface allow the user to run SPARQL queries to be evaluated against a SPARQL query engine, that hosts the set of RDF triples asserted during the semantic parsing, or inferred by the reasoning.

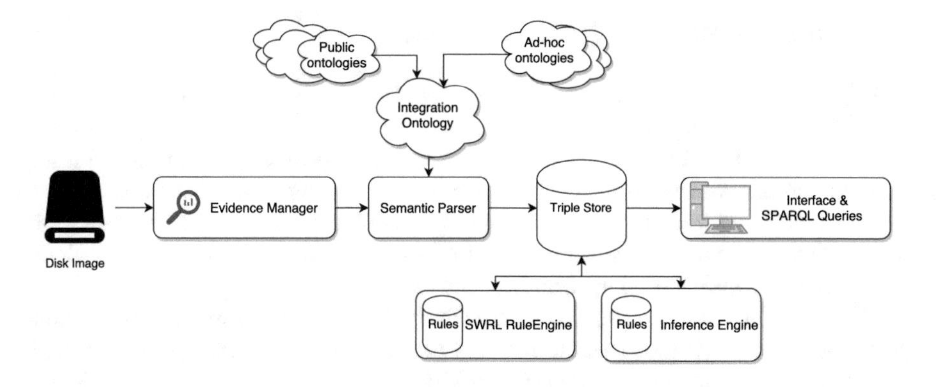

Fig. 1. System architecture for proposed methodology.

5 Related Work

Many approaches adopting Semantic Web technologies have been proposed in literature, but current support for correlation during evidences analysis is poor among forensic tools.

We can distinguish three main kinds of approach:

1. XML-based;
2. RDF-based;
3. Ontology-based.

Among various existing approaches, the two most representative XML-based approach projects are DFXML (Digital Forensics XML) [14] and DEX (Digital Evidence Exchange) [17]. DFXML provides a standardized XML vocabulary for representing evidence's attributes, ie forensics-related XML tags such as `<source>`, `<imagefile>`, `<acquisition_date>`, `<volume>`, etc. DEX also adopt an XML-based representation of the output of various forensics tool, focusing on tracing the sequential order of the tools used and on all the instructions given to them. Although these projects promote interoperability through a standardized representation format, they lack on the semantic content of what they represent.

RDF-based approaches make an important improvement in the creation of attributes with standard or custom types instead of only string types. The Advanced Forensic Format (AFF) project is the most famous adoption of RDF in forensic domain. AFF is a file format able to contain both a copy of the original acquired data as well as arbitrary meta-data, such as system-related information or user-specified ones.

FORE (Forensics for Rich Events) [19] is an ontology-based approach that propose an architecture composed by an ontology, an event log parser and a custom-defined rule language, FR3. Rules expressed in FR3 language are evaluated against the knowledge base [1,3] in order to add new links between instances.

6 Conclusions and Future Work

In this paper, a methodology based on semantic representation, integration and correlation of digital evidence and an architecture that implements it is proposed and analysed. The approach is based on an ad-hoc ontology, defining a forensic domain model, and on a set of modules for collecting and extracting data to be examined, populating the ontology, inferring new facts and analysing the results. The ontology constitutes a unified model that simplifies the knowledge representation and the query building processes.

The approach can be improved, especially for the execution time of the analysis phase and for the collection from a wider set of sources concerns. In fact, in future works, we have planned to add additional devices, such as mobile devices with Android and iOS operating systems, as data sources for the collecting phase.

In conclusion, we have shown that through the proposed methodology is possible to enhance the analysis of a forensic investigation. The enhancement derives from the exploitation of semantic technologies, that alleviate the problem related to the heterogeneity of forensic tools outputs. Additionally, new knowledge can be extracted from analysis of original data sources yielding to asserted facts, which in turn serve as a basis to build the forensic evidences.

References

1. Amato, F., Cozzolino, G., Mazzeo, A., Mazzocca, N.: Correlation of digital evidences in forensic investigation through semantic technologies. In: 2017 31st International Conference on Advanced Information Networking and Applications Workshops (WAINA), pp. 668–673. IEEE (2017)
2. Amato, F., Cozzolino, G., Mazzeo, A., Romano, S.: Detecting anomalies in Twitter stream for public security issues. In: 2016 IEEE 2nd International Forum on Research and Technologies for Society and Industry Leveraging a Better Tomorrow (RTSI), pp. 1–4. IEEE (2016)
3. Amato, F., Moscato, F.: A model driven approach to data privacy verification in e-health systems. Trans. Data Priv. **8**(3), 273–296 (2015)
4. Balzano, W., Murano, A., Vitale, F.: V2V-EN - Vehicle-2-Vehicle elastic network. Procedia Comput. Sci. **98**, 497–502 (2016)
5. Balzano, W., Murano, A., Vitale, F.: Wifact–wireless fingerprinting automated continuous training. In: 2016 30th International Conference on Advanced Information Networking and Applications Workshops (WAINA), pp. 75–80. IEEE (2016)
6. Balzano, W., Vitale, F.: DiG-Park: a smart parking availability searching method using V2V/V2I and DGP-class problem. In: 2017 31st International Conference on Advanced Information Networking and Applications Workshops (WAINA), pp. 698–703. IEEE (2017)
7. Caballé, S., Xhafa, F., Barolli, L.: Using mobile devices to support online collaborative learning. Mob. Inf. Syst. **6**(1), 27–47 (2010)
8. Cilardo, A.: Exploring the potential of threshold logic for cryptography-related operations. IEEE Trans. Comput. **60**(4), 452–462 (2011)
9. Cilardo, A., Durante, P., Lofiego, C., Mazzeo, A.: Early prediction of hardware complexity in HLL-to-HDL translation. In: 2010 International Conference on Field Programmable Logic and Applications (FPL), pp. 483–488. IEEE (2010)

10. Cozzolino, G.: Using semantic tools to represent data extracted from mobile devices. In: 2018 IEEE International Conference on Information Reuse and Integration (IRI), pp. 530–536. IEEE (2018)
11. Di Lorenzo, G., Mazzocca, N., Moscato, F., Vittorini, V.: Towards semantics driven generation of executable web services compositions. J. Softw. **2**(5), 1–15 (2007)
12. Di Martino, B., Moscato, F.: An ontology based methodology for automated algorithms recognition in source code, pp. 1111–1116 (2010)
13. Fusella, E., Cilardo, A.: Minimizing power loss in optical networks-on-chip through application-specific mapping. Microprocess. Microsyst. **43**, 4–13 (2016)
14. Garfinkel, S.L.: Automating disk forensic processing with SleuthKit, XML and Python. In: Fourth International IEEE Workshop on Systematic Approaches to Digital Forensic Engineering, SADFE 2009, pp. 73–84. IEEE (2009)
15. Horng, S.-J., Rosiyadi, D., Fan, P., Wang, X., Khan, M.K.: An adaptive watermarking scheme for e-government document images. Multimed. Tools Appl. **72**(3), 3085–3103 (2014)
16. Horng, S.-J., Rosiyadi, D., Li, T., Takao, T., Guo, M., Khan, M.K.: A blind image copyright protection scheme for e-government. J. Vis. Commun. Image Represent. **24**(7), 1099–1105 (2013)
17. Levine, B.N., Liberatore, M.: DEX: digital evidence provenance supporting reproducibility and comparison. Digit. Investig. **6**, S48–S56 (2009)
18. Moscato, F.: Exploiting model profiles in requirements verification of cloud systems. Int. J. High Perform. Comput. Netw. **8**(3), 259–274 (2015)
19. Schatz, B., Mohay, G.M., Clark, A.: Generalising event forensics across multiple domains. School of Computer Networks Information and Forensics Conference, Edith Cowan University (2004)
20. Schatz, B.L.: Digital evidence: representation and assurance. Ph.D. thesis, Queensland University of Technology (2007)
21. Seo, H., Liu, Z., Choi, J., Kim, H.: Multi-precision squaring for public-key cryptography on embedded microprocessors. In: International Conference on Cryptology in India, pp. 227–243. Springer, Cham (2013)
22. Xhafa, F., Barolli, L.: Semantics, intelligent processing and services for big data. Future Gener. Comput. Syst. **37**, 201–202 (2014)

Evaluating Indoor Location Triangulation Using Wi-Fi Signals

Yasir Javed[1(✉)], Zahoor Khan[2], and Sayed Asif[1]

[1] Higher College of Technology, Al'Ain, UAE
yjaved@hct.ac.ae
[2] Higher College of Technology, Fujairah, UAE
zkhan@hct.ac.ae

Abstract. The advancement in Global Positioning System (GPS), has led to a huge number of location-based applications. Such applications can also be very useful for indoor environment; however, GPS technology struggles with indoor location mapping. Currently, there are various techniques, which are used for indoor localization namely: wireless fidelity-based, Bluetooth, radio frequency identification (RFID), infrared beam, and Sensors. The Wi-Fi access points (APs) are installed at various indoor locations to cover most of the areas, and the smart phones and tablets, are equipped with wireless transceiver modules, which can receive Wi-Fi signals. Therefore, it becomes more practical to use Wi-Fi signal for such application in comparison to infrared beam, Bluetooth and other wireless technologies, as Wi-Fi has significant advantages, including wider range, higher stability, and there are no requirements for additional hardware devices. Literature review confirms that the non-line of sight (NLOS) factors and the multipath effect easily affects most of the existing indoor localization algorithms based on Wi-Fi access points (APs). There also exist many other problems, such as positioning stability and blind spots, which can cause a decline in positioning accuracy at certain positions or even failure of positioning. In this research, we propose to use triangulation of location based on Wi-Fi signals from multiple APs. This method utilizes the received signal strength indications (RSSI) from multiple static APs to determine the location. Based on this, evaluation is done using experiments to measure the accuracy and effectiveness of the new proposed algorithm. The results are promising and can be improved with the use of Artificial intelligence, which is the future work of this project. The proposed method will overcome most of the problems caused by NLOS factors and the multipath effect.

1 Introduction

Indoor location tracking can be very useful for many applications. It can be applied to numerous scenarios e.g. guiding visitors of a large airport, retail industry to locate a specific product or area in large stores, guiding customers or visitors to specific outlets in big shopping malls, location-based communication with the visitors or customers, and providing them with the real time information about the location or products located at that location etc. However, indoor location tracking is a big challenge. A lot of research has been conducted in this area however; they all struggled to provide a

L. Barolli et al. (Eds.): EIDWT 2019, LNDECT 29, pp. 180–186, 2019.
https://doi.org/10.1007/978-3-030-12839-5_17

practical solution. This research proposes to use the Wi-Fi signal strength for indoor location mapping. Previously, research has reported many issues with Wi-Fi based location tracking however; we are proposing to use triangulation from multiple fixed access points to minimize these issues. This will be the first step towards the further research, which will incorporate artificial intelligence techniques to improve the accuracy.

2 Indoor Location Mapping

The location-based services (LBS) have invited wide attention due large number of applications across the industries. The two types of localization technologies are outdoor and the indoor localization. Global positioning system (GPS) [1–3] is commonly used for outdoor localization. However, it is challenging for GPS to provide accurate indoor positioning services [4]. Currently, there are various techniques, which are used for indoor localization namely: wireless fidelity-based, Bluetooth, radio frequency identification (RFID), infrared beam, and Sensors. The Wi-Fi access points (APs) are installed at various indoor locations to cover most of the areas, and the smart phones and tablets, are equipped with wireless transceiver modules, which can receive Wi-Fi signals. Therefore, it is useful to use Wi-Fi signal for such application in comparison to infrared beam, Bluetooth and other wireless technologies, as Wi-Fi has significant advantages, including wider range, higher stability, and no requirements for additional hardware devices [5].

Currently, the algorithms, which are mostly used for Wi-Fi based indoor localization are using angle of arrival localization algorithm (AOA), the time of arrival localization algorithm (TOA), the time difference of arrival localization algorithm (TDOA), the received signal strength indication localization algorithm (RSSI), the fingerprint-based localization algorithm [6–10]. All these algorithms significantly rely on static APs and can have many problems as indicated above. For example, open or closed doors can easily affect the signal propagation, furniture in the room or on the way to the device, human presence, which can cause problems like reflection, refraction. Moreover, signals problems can be due to the non-line of sight (NLOS) factors, the multipath effect. The other indoor wireless signals and electromagnetic interferences can also affect the positioning accuracy [5].

The positioning premise of indoor localization algorithms based on fingerprints is that it must be able to receive no less than 4 specified intensity values of AP signals in the area where the location is going to be measured. This is because the positioning accuracy mainly relies on the number of APs operating during the localization phase, and the positioning results will have smaller errors when the same number of APs are used in both off-line and localization phases. However, this will require large number of static APs [11]. This study will evaluate the use of two or three static APs for triangulating the location.

3 Indoor Location Triangulation Using Wi-Fi

Wi-Fi signal strength or intensity values of wireless signals with the relative positions of static APs were recorded in the indoor environment where the localization is needed. For experimentation purpose, physical locations in the corridors are mapped by recording Wi-Fi signal strength from multiple APs at each door. Therefore, the signal strengths at each door with room identifier e.g. C221 will be say [S_X, S_Y, S_Z] coming from X, Y, and Z, APs respectively where X, Y, and Z are the identifiers for the APs. The signal strengths were recorded according to the identifiers of APs. At each door in the corridor, all this information was recorded and saved in the form of Table 1 shown below; this table will work as a baseline. For the baseline data signal strength can be stored for as many APs in the area as possible. This is useful while determining the location, as system will have more options to compare and moreover, if one or more APs are not working or inaccessible, localization will still be possible.

Table 1. Example baseline data

Location ID	AP 1 ID	AP 1 signal strength	AP 2 ID	AP 2 signal strength	AP 3 ID	AP 3 signal strength
C221	X	S_X	Y	S_Y	Z	S_Z

Table 2. Example of signal strength (dBm) gathered at each location

Location ID	AP mac address (AP ID)	Signal strength (dBm)
C221	6cf37fa296e2	−49
	d8c7c84b23c2	−60
	d8c7c84b23d2	−69
	6cf37fa296f2	−72
	6cf37fa6fc42	−73
C219	6cf37fa296e2	−55
	6cf37fa6fc42	−61
	6cf37fa29602	−71
	6cf37fa29612	−71
	6cf37fa295f2	−73
C220	6cf37fa6fc42	−43
	6cf37fa6fc52	−57
	6cf37fa297a2	−63
	d8c7c84b2d22	−73
	d8c7c84b2542	−76

An application on the mobile device that can read signal strengths from APs in a real time can utilize the baseline data to identify the location with some margin of error. See Table 2 for baseline data recorded at various locations in the corridor using a free mobile application[1].

In Fig. 1, it can be seen that at each location, mobile terminals [MT1, MT2, and MT3] are getting signal strengths from various APs [AP1, AP2, AP3... AP8] depending upon where MT is physically located. Therefore, it will be very rare for any of these mobile terminals to find two or more locations where same signal strengths will exist from the same APs. Even when the APs are placed parallel and signal strength is measured on both sides, the difference in environmental factors will provide us the different signal strength therefore, we will use the same weakness mentioned by other researchers in the previous research as a strength and improved accuracy. Accuracy of the system can have different levels, e.g. when all conditions are met, location is considered to be most accurate however when less conditions are met location can still be guessed to be close to the possible location.

4 Evaluation Methodology

Testing was be done by reading the dynamic signal strength and by comparing it with the recorded data (baseline) to see if it is possible to guess the location correctly using three APs. A simple algorithm depicting the concept is implemented using Python

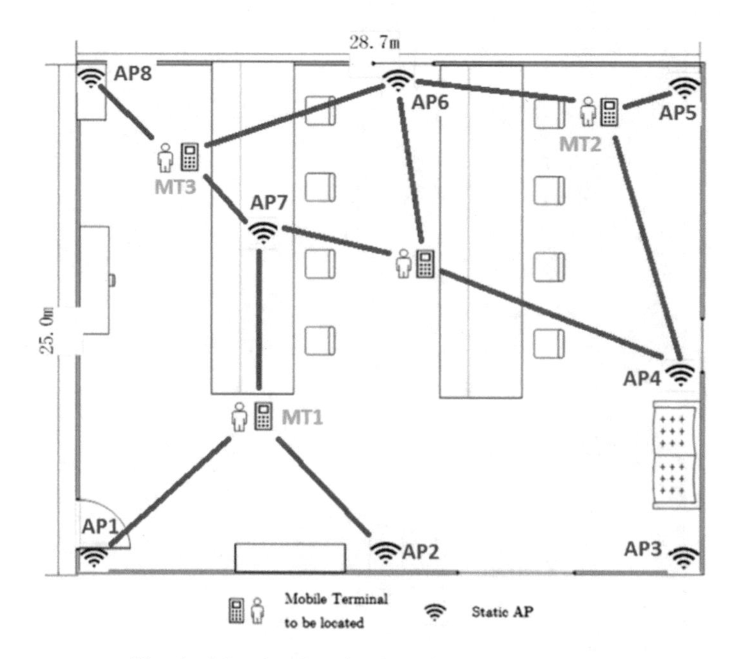

Fig. 1. Mapping location based on signal strength

[1] Wi-Fi Analyzer Android Application.

programing language, see Fig. 2 below. The code is kept simple to express the proposed logic. Same concept can be implemented using arrays, list, or more dominant data structure like dictionaries in Python. The program simply compares the entered identifier for the access point with the ones in the baseline data, if found, it is the starting point for comparison. On the next step signal strength from the same access point is compared with the baseline data for the access point, if it is from the range counter is incremented by 1. Now same data is read for second access point in the area and based on the condition counter is incremented.

```
1  ap1 = input("Enter ID for Acess Point 1> ")
2  ap2 = input("Enter ID for Acess Point 2> ")
3  ap3 = input("Enter ID for Acess Point 3> ")
4
5  count = 0
6
7  SignalStrength1 = int(input("Enter Signal Strength for Acess Point 1> "))
8  SignalStrength2 = int(input("Enter Signal Strength for Acess Point 2> "))
9  SignalStrength3 = int(input("Enter Signal Strength for Acess Point 3> "))
10
11 if ap1 == "6cf37fa296e2":
12     if (SignalStrength1 > 46) and (SignalStrength1 < 52):
13         count=count+1
14 if ap2 == "d8c7c84b23c2":
15     if (SignalStrength2 > 57) and (SignalStrength2 < 63):
16         count=count+1
17 if ap3 == "d8c7c84b23d2":
18     if (SignalStrength3 > 66) and (SignalStrength3 < 72):
19         count=count+1
20
21 if count==3:
22     print ("Location is C221")
23 if count==2:
24     print ("Location is close to C221")
25 count = 0
26
27 if ap1 == "6cf37fa296e2":
28     if SignalStrength1 > 52 and SignalStrength1 < 58:
29         count=count+1
30 if ap2 == "6cf37fa6fc42":
31     if SignalStrength2 > 58 and SignalStrength2 < 64:
32         count=count+1
33 if ap3 == "cf37fa29602":
34     if SignalStrength3 > 68 and SignalStrength3 < 74:
35         count=count+1
36
37 if count==3:
38     print ("Location is C219")
39 if count==2:
40     print ("Location is close to C219")
41 count = 0
42
43
44 if ap1 == "6cf37fa6fc42":
45     if SignalStrength1 > 40 and SignalStrength1 < 46:
46         count=count+1
47 if ap2 == "6cf37fa6fc52":
48     if SignalStrength2 > 54 and SignalStrength2 < 60:
49         count=count+1
50 if ap3 == "6cf37fa297a2":
51     if SignalStrength3 > 60 and SignalStrength3 < 66:
52         count=count+1
53
54 if count==3:
55     print ("Location is C220")
56 if count==2:
57     print ("Location is close to C220")
58
59 if count == 0:
60     print ("Location is not found!")
```

Fig. 2. Python snippet to work with conditions

After reading data from three closest APs, the decision about the location is made. If counter has value three, it means that the identities of all the access points in the baseline data has been matched along with their corresponding signal strength range, therefore location can be accurately guessed as shown in Fig. 3(a) and (b).

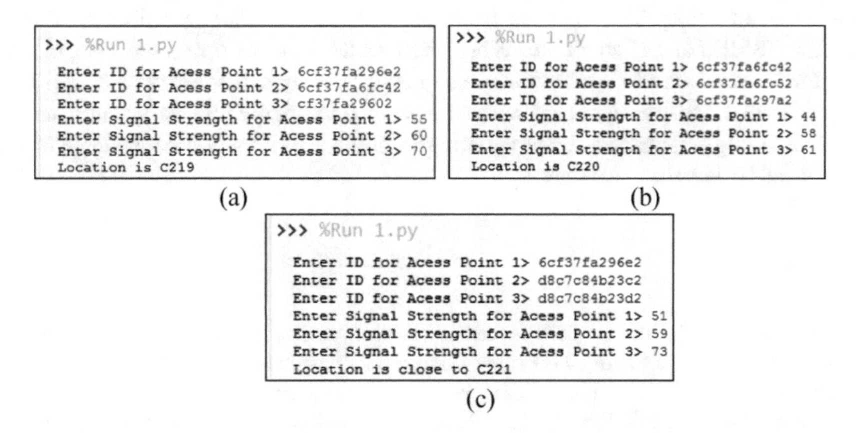

Fig. 3. (a) Finding location C219 (b) Finding location 220 (c) Finding location close ot C221

However, after matching data from two APs, if the third one is not matched the location is mapped as "close to the point of interest", as shown in Fig. 3(c). For the testing purposes, variety of information related to various access points and different levels of signal strengths were dynamically entered to see the results. In almost all the cases it was noticed that the program was able to identify the location based on the baseline data.

Another kind of evaluation was done about consistency of signal strength at different locations. It was done by reading the signal strengths from all the APs at the locations where they were recorded for baseline data and compared. The results for all 18 locations were found within range with margin of error of 3 dBm. Therefore, the method of triangulating the location is possible with real time data being compared to the baseline data.

5 Discussion and Conclusion

This paper tested the triangulation using with normal conditions first by recording the environmental variables like furniture, doors closed/open, number of people in the surroundings etc. However, in real environment we can expect a lot of variability in all these variables. Though, the tests proved to be successful for the simple scenario they have to be varied to see the changes in the outcome. Based on the experiments, it can be concluded that real time incoming data from the access points can be used to identify the location by comparing it with the baseline data. Based on the available data from the APs if identifying the exact location is not possible, we can still identify the close by locations. However, more experiments are needed simulating the real environment with changing environmental variables most importantly number of people in the room which is the most changing variable. Therefore, in future work, variables will be varied to see their impact on the accuracy of location identification. In addition to obtaining RSSIs used in fingerprint-based localization, we can also obtain signal parameters of other APs, such as basic service set identifier (BSSID), service set identifier (SSID) and channel etc. When the mobile terminal opens the hotspot function, the current mobile terminal can also serve as a dynamic AP, which also has a unique BSSID and the signal intensity values of RSSI and SSID. In accordance with the methodology described above variation in number of APs will be evaluated to see the impact on location accuracy.

References

1. Fan, B., Leng, S., Liu, Q.: GPS: a method for data sharing in Mobile Social Networks. In: Proceedings of 2014 IFIP Networking Conference, Trondheim, pp. 1–9 (2014)
2. Huang, J., Tsai, C.: Improve GPS positioning accuracy with context awareness. In: Proceedings of 2008 First IEEE International Conference on Ubi-Media Computing, Lanzhou, pp. 154–159 (2008)
3. Binjammaz, T., Al-Bayatti, A., Al-Hargan, A.: GPS integrity monitoring for an intelligent transport system. In: Proceedings of 2013 10th Workshop on Positioning Navigation and Communication, Dresden, pp. 1–6 (2013)
4. Sheng, L.: Research of indoor wireless positioning system based on GPS, Master thesis, Department Circuits and systems, East China Normal University, China, Shanghai (2012)
5. Liu, H.: Survey of wireless indoor positioning techniques and systems. IEEE Trans. Syst. Man Cybern. **73**(7), 1067–1088 (2007)
6. Zhu, G., Hu, J.: Distributed network localization using angle-of arrival information Part I: continuous-time protocol. In: Proceedings of 2013 American Control Conference, Washington, DC, pp. 1000–1005 (2013)
7. Hwang, S., Kwon, G., Pyun, J.: AOA selection algorithm for multiple GPS signals. In: Proceedings of 2013 Asilomar Conference on Singals, Systems and Computers, Pacific Grove, CA, pp. 481–485 (2013)
8. Huang, B., Xie, L., Yang, Z.: Analysis of TOA localization with heteroscedastic noises. In: Proceedings of 2014 33rd Chinese Control Conference, Nanjing, pp. 327–332 (2014)
9. Sharp, I., Yu, K.: Indoor TOA error measurement, modeling, and analysis. IEEE Trans. Instrum. Meas. **63**(9), 2129–2144 (2014)
10. Shikur, B., Weber, T.: TDOA/AOD/AOA localization in NLOS environments. In: Proceedings of 2014 IEEE International Conference on Acoustics, Speech and Signal Processing, Florence, pp. 6518–6522 (2014)
11. Debnath, S.K., Funabiki, N., Saha, M., Mamun, S.A., Kao, W.: MIMO host location optimisation in active access-point configuration algorithm for elastic WLAN system. Int. J. Space Based Situated Comput. **8**(2), 59–69 (2018)

Big Data Role in Improving Intelligent Transportation Systems Safety: A Survey

Mohammed Arif Amin[1(✉)], Samah Hadouej[1], and Tasneem S. J. Darwish[2]

[1] Higher Colleges of Technology, Abu Dhabi, UAE
{mamin,shadouej}@hct.ac.ae
[2] Universiti Teknologi Malaysia (UTM), Johor Bahru, Malaysia
tasneem83darwish@gmail.com

Abstract. Achieving smart and intelligent transportation requires the use of millions of devices which generate a huge volume of data, termed as Big Data. With the flourishing of Big Data analytics, intelligent transportation management and control is now becoming more data driven. Big data can provide ample information obtained from vehicles, traffic infrastructure, smart phones and weather stations. Such data has promising applications in intelligent transportation systems, especially in the road safety sector. In particular, utilizing Big Data analytics assesses accident prevention and detection, thereby reducing causalities, loses and damage. This survey paper explores the role of big data in shaping the intelligent transportation systems with a focus on the road safety sector. In addition, the limitations of existing studies are discussed and future research directions are suggested.

1 Introduction

In today's age of data abundance, nearly every aspect of social activity is afflicted by the huge amounts of data available. The information is characterized by volume, velocity and variety which is often referred to as "Big Data" [1]. Human life depends greatly on transportation systems, which points to the promises and difficulties caused by the Big data. Intelligent transportation system (ITS), which was introduced few years ago has gathered huge amount of data over a vast geographical range including many countries. However, this data is very rich in information, disorganized and consistently growing. The data could be used by experts to analyze and understand traffic management, its performance and the nature of accidents including transportation patterns. This data could also be used for proactive measures such as road safety, traffic snarls, road blocks and traffic flux in real-time.

Road safety has been considered a high priority issue among highway authorities for many years. One of the most prominent behavioral study application is in the area of proactive road safety diagnosis using surrogate safety methods. This is based on statistical methods using accident data which require log observation periods. Therefore one must wait for enough accidents to occur in order

L. Barolli et al. (Eds.): EIDWT 2019, LNDECT 29, pp. 187–199, 2019.
https://doi.org/10.1007/978-3-030-12839-5_18

to have large data volume for analysis. In 1960s, many attempts were made to predict the number of collisions based on observations without a collision being reached [2].

In recent years, most of safety systems are based on data taken during or after a crash. Crash data analysis has been used to predict accidents and hence many different measures have been taken to avoid accidents. With the involvement of big data, road safety could be achieved or improved at a much higher rate if used properly. In large cities, ITS is being used to check and balance traffic congestion on the roads. However, many real-time monitoring systems have also been installed to either provide security or to monitor irregular activities on the road. These monitoring systems could be used to analyze traffic data and hence analyze accidents occurring on the road.

This paper presents a review of Big Data and its involvement in road safety. The authors have tried to gather as much information as possible on big data and how it can be used to provide road safety to drivers. It also discusses the available methods and type of data available during and after crashes which could be used to avoid or predict accidents. The remainder of this paper is organized as follows: Sect. 2 presents the background, Sect. 3 discusses the challenges of big data. Section 4 shows existing studies limitations. Section 6 introduces a proposed solution. Section 7 concludes the paper and suggests future work directions.

2 Background

Road and traffic accident analysis requires a deep knowledge of the factors effecting the uncertainty and unpredictability of accidents on the road. There are a number of variables and data sets involved in accident data which is discrete in nature. Therefore, the data recorded is heterogeneous and must be treated with variance [3]. Traffic accident analysis is often performed based on statistical or data mining techniques. Multiple studies have employed road accident data analysis using statistical techniques [4,5] in order to identify existing relationships between accident and relevant data. Some other studies used data mining techniques [6,7] in order to identify the main factors associated with a road and traffic accidents. However, most of these techniques can only handle a small subset of traffic accidents. Therefore, certain relationships remain hidden. Now Days, Big data is being increasingly deployed by municipalities and police forces in order to make roads safer. Indeed the use of big data technology allows traffic monitoring systems which can handle a large amount of data. Most road accidents can be mainly categorized into two, statistical and data mining techniques. Multiple studies on road accident data analysis have used traditional techniques and data mining techniques. Therefore, effective strategies should be used to improve traffic operation using Big Data applications. Big Data generated from traffic congestion, car crashes, traffic flow, weather conditions, road design, human behavior and intelligent transport detection systems could be used to perform analysis and avoid accidents, hence improving safety on the road. Therefore, Big Data plays an important role in providing safety drivers.

2.1 Big Data Sources in Transportation

Big Data in the transportation system come from many sources such as traffic surveillance systems on the road and inside the vehicle. In addition sensors installed in the vehicle play an important role in collection of data during normal, uncertain or crash conditions. At the same time technologies like Global Positioning System (GPS), cellular phones, Bluetooth, Ground-based Radio Navigation, Automatic Vehicle Identification, Automatic Vehicle Location and Radio Frequency Detection (RFID) play an important role in addition to the fixed infrastructure [8].

Many other data sources such as mobile devices, social media data, demographic data, weather reporting systems, geometric characteristics, and crash data are extensively used in traffic operation, and safety management. However, efficient data integration and fusion have to be carried out to receive the maximum out put from the data. For example, integration of sensor data with the GPS or road monitoring data. Therefore, a new method is required to filter the most important variables from the available data types, perform analysis and predict road safety. Data integration would play an important role in the future for Big Data applications in the transportation arena.

The benefits of Big Data technologies include direct and indirect applications. Direct applications could be congestion reduction, incident prediction, and travel time estimation. Indirect applications are carried out through enhancement of traffic modeling in the model development, calibration and validation processes. Traffic simulation could also be greatly improved based on the real data collected from field.

2.2 Big Data Applications

Traffic demand could vary from time to time on different roads. Traditionally, the volume to capacity ratio and level of service (LOS) are implemented by the transportation authorities as an indicator of congestion control [9]. However, volume to capacity ratios lack the capability to capture the variability of congestion. There could be an incident and a number of cars could be diverted to other routes depending on the condition on a particular road. Big Data can provide a more comprehensive and accurate data from many different sources in real time about congestion to the authorities. They can zoom into specific location and check the performance of the whole system in order to make decisions.

2.2.1 Real-Time Crash Control

In a normal road transportation system the drivers need to respond to many different complex events and at the same time to maintain high speed. These events include vehicle maneuvering, taking instant decisions regarding routes, reading road signs and maintaining safe distance from other maneuvering vehicles simultaneously. Hence, any additional disruption in the traffic condition may create driving error which can eventually result in a crash. Crashes could be avoided by spotting the disrupted traffic situations as early as possible. This would result in proactive measures such as sending warnings to the driver, applying various traffic smoothing techniques, variable speed limits, maintaining lines, maneuvering of vehicle, such action would bring the traffic back to normal [10–12].

2.2.2 Vehicle Motion Planning

Many different driver braking behaviors are proposed and are based on potential filed model [13–15]. Number of different methods are discussed to avoid other vehicles and keep in lane. A technique to assign a potential field to obstacles in order to prevent crashes when a vehicle enters the potential field area with a certain distance. This could result in automating the deceleration of the vehicle speed or alerting the driver of incidence which might occur in order to take action. This method requires a large amount of data from each vehicle traveling in the same direction on a particular highway. Big Data can play an important role in capturing such data and analyzing it with high speed data processing techniques and sending either alerts to the vehicle or mobile devices to take action before a crash occurs. However, this could be difficult in the case of sudden crashes or if an unusual scene occurs on the road such as blind corner at a intersection. This could effect the speed of the vehicle due to the shape of the intersection or driver deciding to reduce the speed and pass the intersection. Therefore, it is difficult to define the unique value of the speed to predict the action [16].

2.2.3 Vehicle Position Detection

Vehicle position or location can greatly affect the prediction of the potential for crashes if calculated precisely and in advance. However, creating a map of potential vehicles is a challenge because this would also include the velocity of the vehicle in consideration. The detection method of a vehicle based on 3D point clouds is proposed in [17]. The data obtained by the 3D map could then be used to create a road map with multiple driving profiles and compared to predicted a crash. The method based on cloud point has advantages in the accuracy and estimation of the motion. Therefore, every position is mapped on the standard map and sensing data can be compared.

3 Big Data Challenges

Gartner defines big data challenges and opportunities as being three dimensional (3Vs model):

- Volume (increasing amounts of data): There is a huge explosion in the data available. The challenge is no longer the availability, but the management of this data. In transport, the volume of data has increased because of growth in the amount of traffic (all modes) and detectors.

- Velocity (speed of data in and out): The velocity of data has increased in transport due to improved communications technology and media (particularly fiber optic cabling) and increased processing power and speed for monitoring and processing. It is important to keep up with real-time data. This will help build better insights and enhance decision-making capabilities. Currently, there are a few reliable tools, though many still lack the necessary sophistication.
- Variety (range of data types and their sources): Transportation big data can be obtained from many sources. The variety of transport-related data has increased significantly. Modern trains and aircraft report internal system telemetry in real time from anywhere in the world. Along with the rise in unstructured data, there has also been a rise in the number of data formats. Therefore, data integration would play an important role in Big Data applications in future transportation systems.

4 Utilizing Big Data Analysis to Increase Road Safety

Crash occurrences are often regarded as random events affected by human behaviour, roadway design, traffic flow and weather conditions. However, big data generated from the ITS could be leveraged to develop real-time traffic monitoring [18]. One of the key objectives in accident data analysis is to identify the main factors associated with road and traffic accidents. Afterwards, accidents can be prevented by early detection of causal factors or traffic patterns and provide timely alerts to drivers. On the other hand, the early detection of accident occurrence is important to save lives, and it can help to reduce loses and damage.

[19] A system and method for classifying and identifying a driver using driving performance data has been proposed. The system provides an accurate and predictive way to measure and analyze driving behavior. Classification and identification of a driver (e.g., a driving signature which could represent driving patterns in a continuous manner) could be used for insurance purposes and/or driving risk evaluation. The system could also be used to analyze, classify, and/or provide feedback and coaching to drivers and vehicle owners (e.g., for green driving (e.g., fuel efficient and environmentally friendly green driving), personal safety, family safety, fleet safety, etc.). However, such a system does not provide real-time detection for traffic accidents. Moreover, the author did not explain how the data of each individual vehicle is collected and processed.

In [20], the author explores In-Vehicle Data Recorders (IVDRs) information to investigate undesirable driving events (such as hard braking, lane changing, and sharp turning) among 148 individuals. The information was logged over three years. The objective was to gain deeper understanding about the heterogeneity among drivers with respect to behavior change over time, the effect of trip duration and the distribution of events counts. The paper introduced a statistical model that works on each driver's data separately. In some respects drivers are similar, enabling the application of the same methodological approach to most drivers. In other respects, differences among drivers are substantial, and thus personalized examination of the data is advised. Analyzing individuals' data may assist drivers, insurance agents, safety officers, or driving instructors who wish to understand how individuals' behaviors can change over time and what variables explain this change. However, studying driver behavior using precise data with a small number of drivers is not enough to detect dangerous driving patterns. Therefore, more advanced analysis is required where the behavior of thousands of drivers can be studied simultaneously.

In [18], the author introduced a real-time modeling framework to monitor traffic and study the relation between congestion and rear-end crashes. A Microwave Vehicle Detection System (MVDS) deployed on an express-way network was utilized to collect the traffic big data. It was found that congestion on urban express-ways was highly localized and time-specific. Data mining (random forest) and Bayesian inference techniques were implemented in the real-time crash prediction models. The identified effects confirmed the significant impact of congestion on rear-end crash likelihood. In aggregate safety analysis, the issue related to averaging congestion intensity might be the cause of the insignificant effects of congestion found in many crash frequency studies. Real-time congestion measurement based on Big Data is more desirable to identify congestion patterns as it considers both the temporal and spatial dimensions. This work focused on studying the effect of congestion on accident occurrence. However, multiple factors other than congestion could lead to crashes in the real world. Therefore, to fully realize the power of Big Data, more data sources should be utilized especially real-time weather condition as it is an important factor for express- way operation and safety as well.

The heterogeneous nature of road accident data makes the analysis task difficult. In [21] data segmentation has been used to overcome this heterogeneity of accident data. A framework is proposed that used K-modes clustering analysis as a preliminary task for segmentation. In addition, association rule mining is used to identify the various circumstances that are associated with the occurrence of an accident for both the entire data set (EDS) and the clusters identified by the K-modes clustering algorithm. The results reveal that the combination of k mode clustering and association rule mining produced important information that would remain hidden if no segmentation had been performed prior to generate association rules. In ITS every car has its own mobile database of information, which makes the system database distributed. However, the mining technique is centralized in one central server. Subsequently, a real-time system will face the problem of increased overhead on the network communication system to send and receive alerts, new patterns and update patterns.

The author of [22] proposed an analysis method of driving behaviours based on large-scale and long-term vehicle recorder data. The method classifies drivers by their skill, safety, physical/mental fatigue, and aggressiveness. In this study, the ability of a dataset that is sparse but large-scale (over 100 fleet drivers) and long-term (10 months worth) was examined. The focus was on classifying drivers recently involved in accidents through examining the correlation in driving behaviours. The drivers classification was done using long-term records of their driving operations (braking, wheeling, etc.) with several attributes (max speed, acceleration, etc.). Following a machine learning approach, two methods to characterize driver's behaviours were used; entropy-like model, and KL divergence model, where effective features were selected and successfully found some informative outcomes. This work is an example of some existing studies that do not consider real-time collection and processing of data, which makes such studies less efficient in providing timely information that can be used to improve road safety. However, in the future real-time applications will have higher demand in order to serve the intelligent transportation applications.

Early detection of accidents can save lives, provides quicker openings of roads, hence decreases wasted time and resources. In [23] a real-time accident detection model is introduced that utilizes transport big data with computational intelligence techniques. In this model, Istanbul City traffic-flow data for the year 2015 from various sensor locations are populated using big data processing methodologies. The extracted features are then fed into the nearest neighbor model, a regression tree, and a feed-forward neural network model. The acquired raw data is passed through an ETL (Extract-Transform-Load) process through Hadoop distributed file system (HDFS) and Apache Spark. The original data is stored in a SQLServer format and imported to Hadoop environment via Sqoop. The imported data is then processed on a 10-PC cluster using Spark and HDFS. The results revealed that all models are very good in catching accidents, however the number of false positives are considerably high. Indeed, road and traffic accidents are uncertain and it is difficult to predict incidents. Accordingly, some existing prediction models create high false predictions. Such predictions may cause significant disturbances to the ITS. Hence, it is critical to find the features that provide accurate prediction and detection, and design more accurate prediction models.

Self-driving vehicle technology promises to provide many economical and societal benefits and impacts. Safety is on the top of these benefits. Trajectory or path planning is one of the essential and critical tasks in operating an autonomous vehicle. [6] proposed a method for predicting accidents and selecting safe-optimal trajectory in a autonomous cloud based connected vehicle environment. The prediction is done by applying the Distributed Random Forest classification algorithm, the estimated time to arrive is calculated using the Linear Regression (LR) algorithm. All experiments were done using 10-fold cross validation using H2O Big Data analytics software, where selecting the safe trajectory is based on using Big Data mining and analysis of real-life accidents data and real-time connected vehicles data. The decision of selecting a trajectory is done automatically without any human intervention. Human involvement would be only at defining and prioritizing the driving preferences and concerns at the beginning of a planned trip. However, the proposed method still needs to be further tested in a more realistic environment.

Through studying the causes of road accidents using big real-time accidents data, [5] designed an anticipation and alert system of congestion and accidents. It was designed around dividing the roadway into segments, based on the infrastructure availability. The system aims to prevent or at least decrease traffic congestions as well as crashes. It uses DSRC, cellular, wi-fi, and hybrid communication. The data is analyzed and validated by using H2O and R Big Data tools in the cloud infrastructure that combines the historical data and the real-time data received from the vehicles. The system receives online streamed data from vehicles on the road in addition to real-time average speed data from vehicles detectors on the road side to (1) Provide accurate Estimated Time of Arrival (ETA) using a Linear Regression (LR) model (2) Predict accidents and congestions before they happen using Naive Bayes (NB) and Distributed Random Forest (DRF) classifiers (3) Update ETA if an accident or congestion takes place by predicting accurate clearance time. The Lambda Architecture (LA) is considered a good fit for real-time solutions in big Data analytics as it has proved its scalability, robustness, ability for generalization, extensibility, and fault tolerance.

In [24] the author built a prediction model for highway accidents. Data imbalance is one of the major problem encountered in training datasets for data mining. A dataset is imbalanced if the cases of the positive class are outnumbered by cases of the negative class. This can result in high false negatives, mainly harming the minority class, which is the most important class. The primary approach for handling class imbalance is sampling. Sampling transforms the dataset to be more balanced by adding or removing instances until a desired class ratio is reached. To overcome the imbalanced data set, the author employed an oversampling operation to repair the data and prevent the biased result in classification analysis of the imbalanced data. The data used to build the learning model is generated on the Gyeongbu Expressway which connects Seoul and Busan. The data are text files showing traffic data created between Jan 1st, 2011 and June 30th, 2013. Traffic data are created using a vehicle detection sensor (VDS) which measures speeds of cars every 30 s and records the number of cars that run on the road. The Hadoop framework was utilized to process and analyze big traffic data efficiently. The performance of the data mining process was tested using total and target precision.

Studies	Velocity	Variety	Volume
[5]	yes	no	yes
[6]	no	yes	yes
[18]	yes	no	yes
[19]	no	yes	yes
[20]	no	yes	no
[21]	no	yes	yes
[22]	no	yes	yes
[23]	yes	no	yes
[24]	no	yes	yes

Fig. 1. Comparison table based on 3Vs

5 Discussion and Future Research Directions

This section points out the limitations of existing studies. A comparison table of the different studies based on the three Vs is shown in Fig. 1, which shows that the majority of related works do not give a solution to the three big data challenges. Consequently, some critical issues may arise and need to be considered in future research to fully utilize the transportation big data.

- The Big Data generated by the ITS systems is worth further exploration to bring all their full potential for more proactive traffic management. Most of existing studies focus on one or two of the transportation big data sources, whereas there exists many other resources that produce valuable information (e.g. smart phones, traffic lights, weather stations, etc.).
- Transportation big data can be obtained from many sources. However, such data is highly heterogeneous. Therefore, data integration would play an important role in Big Data applications in the future transportation systems.
- Traffic and transportation systems simulation software can be greatly improved based on the real data collected from the intelligent transportation systems. Such simulation software will play an essential role in accelerating the development of ITS while reducing the costs of testing new applications.

6 Proposed Solution

The data collected through the technologies of intelligent transportation systems (ITS) are increasingly complex and are characterized by heterogeneous formats, large volume, nuances in spatial and temporal processes, and frequent real-time

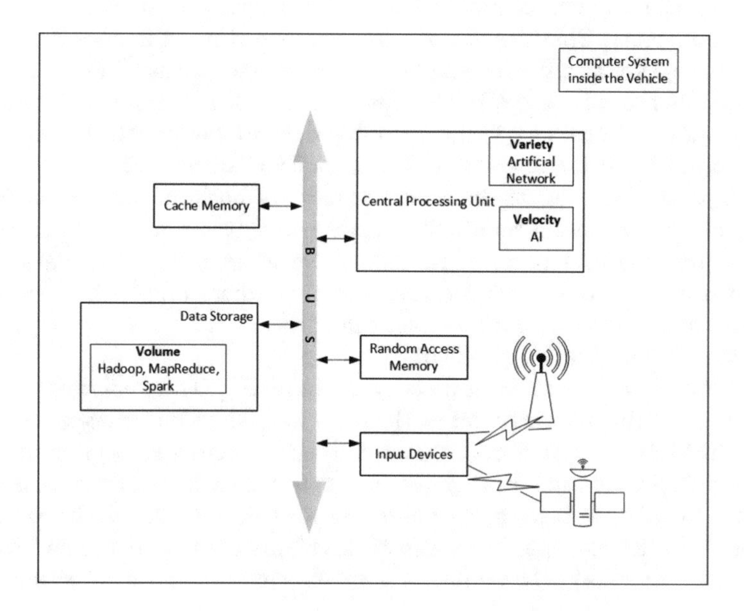

Fig. 2. Proposed solution architecture

processing requirements. Simple data processing, integration, and analytics tools do not meet the needs of complex ITS data processing tasks. Based on our related work research, we can conclude that, an efficient method will need to give a solution to the three main big data challenges (volume, variety and velocity). In the following discussion, we present our approach that combines a number of technologies and tools in order to satisfy the set of big data challenges. The overall architecture of proposed solution is shown in Fig. 2.

- A Solution to the Volume Problem
  A Hadoop system is created with the ability to handle massive amounts of data. It is the most well-known framework for big data processes, it was developed to process large data sets in a distributed manner, data is stored on different nodes. Hadoop was designed for scalable applications and offers its own type of storage. Hadoop is not only a storage system but is a platform for large data storage as well as processing. In fact, Hadoop can be divided into two parts: processing and storage. MapReduce is a programming model which allows the processing huge data stored in Hadoop. However, MapReduce reads and writes from disk, as a result, it slows down the processing speed. Therefore, MapReduce fails when it comes to real-time data processing as it was designed to perform batch processing on voluminous amounts of data.
  Apache Spark is an open-source, distributed processing system used for big data workloads. It utilizes in-memory caching, and optimized query execution for fast analytic queries against data of any size. This is accomplished through in-memory caching, and optimized query execution, Spark can run fast analytic queries against data of any size. It can also process real time data. Spark's strength is its ability to process live streams efficiently. By reducing the number of read/write cycle to disk and storing intermediate data in-memory, Spark makes it possible. However, as it does not have its own storage system, it runs analytics on other storage systems.
  Outside of the differences in the design of Spark and Hadoop MapReduce, many organizations have found these big data frameworks to be complimentary, using them together to solve a broader challenge. Hadoop is used for Batch processing whereas Spark can be used for both. In this regard, Hadoop users can process using MapReduce tasks where batch processing is required. Spark uses the best parts of Hadoop for reading and storing data. Hence, MapReduce and Spark can be used together where MapReduce is used for batch processing and Spark for real-time processing.
- Solution to the Velocity Problem
  Data velocity management is a key component Big Data analytics not due to the speed of the data arriving at the data warehouse but processing of data. Data in ITS may arrive from many different sources such as inbuilt sensors in the vehicle, data monitoring systems, external sensors, live cameras or IoT devices installed either inside or outside the vehicle. Data velocity is not only associated with the speed but also with volume because at the end the data has to be processed. It would be very difficult to send a large amount of

data with very high speed, process it in the data warehouse and then send results to the vehicle for taking decision. There is a big challenge in predicting the road conditions or accidents where the processing is happening. At the moment data is sent to a data warehouse to process and predict or produce an output. This process my take large amount of time and may not be feasible to predict immediate hazards on the roads. A mechanism must be built in which data processing could be executed within the vehicle after obtaining data from internal and external sources. This could involve installing an intelligent artificial neural network system which could analyze the data in real time and predict the situation on the road for the driver or vehicle to take a decision. First of all the data must be filtered within the vehicle to separate good data form a bad data. This will allow the computational systems to process data good data faster while vehicle is in motion. Secondly the data must be processed within the vehicle for immediate action against hazards. This would involve having a large cache size on board the computer system which can drastically reduce the processing time. This data can then be sent to the central location with a very high speed network infrastructure for later analysis. At the same time there must be customizable applications must installed within the vehicle which could help customize the application based on the traffic pattern, data latency, data filtration, and data processing.

- Solution to the Variety Problem
  Data generated by the ITS Systems has different formats, such as numerical data gathered through sensors on both infrastructure and vehicles, multimedia and text data captured from social media, and GIS and image data loaded for digital maps. Therefore, collected data can be structured but also unstructured from a wide variety of sources. One challenge involves finding methods in order to deal with unstructured data. Deep learning networks can figure out how to make sense of the data's various input formats and feed that into other networks to harvest meaning from the data. A second challenge consists in identifying data that pertains to the decision-making process. Techniques related to data searching and filtering can be used.

7 Conclusion and Future Work

Big Data play an important role in the rapid development of intelligent transport systems. Traffic accidents are unpredictable most of the time. However, big data's real time nature, analysis of data at fast speed is vital for the crash prediction in the transportation system. Traditional congestion methods lack the ability to capture variable or dynamics of the congestion. A real-time congestion management based on Big Data is desirable. In this study many techniques such as congestion control, variable mapping, cloud mapping, sensing data, vehicle location, vehicle velocity monitoring are surveyed in order to provide a detail information on how Big Data could be utilized in traffic monitoring systems and allow smooth movement on the high way with very few accidents. Many different methods suggested by authors were discussed and presented to show the importance of Big Data in the intelligent transport system. The authors would like to continue research on the topic and develop a methodology to predict road accidents or safety. The authors aim to collect real-time data from a transportation system running in the country and propose different techniques to predict or avoid accidents on the road. As a conclusion, the application of Big Data for better operation should emphasize real-time monitoring of traffic conditions and a quick response based on the retrieved data. This would be possible only if the used method is able to cope with data variety, velocity and volume at the same time. The authors of this paper proposed a solution combining methods and techniques that address the three big data challenges. A future direction consists of implementing the proposed model and testing it on some ITS applications. Another research direction consists of addressing security and privacy problems. Indeed, although the purpose of accident detection and prevention applications is to improve transportation safety, such applications need to consider the privacy, data protection and security issues of the participating individuals.

References

1. Beyer, M.A., Laney, D.: The Importance of 'Big Data': A Definition, pp. 2014–2018. Gartner, Stamford (2012)
2. Perkins, S.R., Harris, J.I.: Traffic conflict characteristics: accident potential at intersections. Research Laboratories, General Motors Corporation, Warren (1967)
3. Savolainen, P.T., Mannering, F.L., Lord, D., Quddus, M.A.: Accid. Anal. Prev. **43**(5), 1666 (2011)
4. Akagi, Y., Raksincharoensak, P.: IEEE Intelligent Vehicles Symposium (IV), pp. 368–373 (2015)
5. Al Najada, H., Mahgoub, I.: IEEE Symposium Series on Computational Intelligence (SSCI), pp. 1–8. IEEE (2016)
6. Al Najada, H., Mahgoub, I.: IEEE Annual Ubiquitous Computing, Electronics & Mobile Communication Conference (UEMCON), pp. 1–6. IEEE (2016)
7. Amadeo, M., Campolo, C., Molinaro, A.: IEEE Commun. Mag. **54**(2), 98 (2016)
8. Turner, S.: United States. Federal Highway Administration. Office of Highway Information Management, Texas Transportation Institute: Travel Time Data Collection Handbook. Office of Highway Information Management, Federal Highway Administration, U.S. Department of Transportation (1998). https://books.google.ae/books?id=XIDbAQAACAAJ
9. Grant, M., Bowen, B., Day, M., Winick, R., Bauer, J., Chavis, A., Trainor, S.: Congestion management process: a guidebook. Technical report, Transportation Research Board (2011)
10. Abdel-Aty, M., Pande, A.: 83rd Annual Meeting of the Transportation Research Board, Washington, DC (2004)
11. Lee, C., Saccomanno, F., Hellinga, B.: Transp. Res. Rec. J. Transp. Res. Board **1784**(1), 1 (2002)
12. Lee, C., Hellinga, B., Saccomanno, F.: Real-time crash prediction model for application to crash prevention in freeway traffic. Transp. Res. Rec. J. Transp. Res. Board **1840**, 67–77 (2003)

13. Raksincharoensak, P., Akamatsu, Y., Moro, K., Nagai, M.: IFAC Proc. **46**(21), 335 (2013)
14. Wolf, M.T., Burdick, J.W.: IEEE International Conference on Robotics and Automation, pp. 3731–3736 (2008)
15. Bauer, E., Lotz, F., Pfromm, M., Schreier, M., Abendroth, B., Cieler, S., Eckert, A., Hohm, A., Lüke, S., Rieth, P., Willert, V., Adamy, J.: PRORETA 3: an integrated approach to collision avoidance and vehicle automation, vol. 60 (2012). https://www.degruyter.com/view/j/auto.2012.60.issue-12/auto.2012.1046/auto.2012.1046.xml
16. Aoude, G.S., Desaraju, V.R., Stephens, L.H., How, J.P.: IEEE Trans. Intell. Transp. Syst. **13**(2), 724 (2012)
17. Broggi, A., Cattani, S., Patander, M., Sabbatelli, M., Zani, P.: IEEE Conference on Intelligent Transportation Systems, Proceedings, ITSC (2013)
18. Shi, Q., Abdel-Aty, M.: Transp. Res. Part C Emerg. Technol. **58**, 380 (2015)
19. Freiberger, A., Izhaky, D., Shamir, A., Steinberg, O., Tamir, A.: System and method for classifying and identifying a driver using driving performance data. US Patent 14/049,837 (2013)
20. Musicant, O., Bar-Gera, H., Schechtman, E.: Accid. Anal. Prev. **70**, 55 (2014)
21. Kumar, S., Toshniwal, D.: J. Big Data **2**(1), 26 (2015)
22. Yokoyama, D., Toyoda, M.: IEEE International Conference on Big Data (Big Data), pp. 2877–2879. IEEE (2015)
23. Ozbayoglu, M., Kucukayan, G., Dogdu, E.: IEEE International Conference on Big Data (Big Data), pp. 1807–1813. IEEE (2016)
24. Park, S.h., Kim, S.m., Ha, Y.g.: J. Supercomp. **72**(7), 2815 (2016)

Predicting Cancer Survivability: A Comparative Study

Ola Abu Elberak[1], Loai Alnemer[1], Majdi Sawalha[1], and Jamal Alsakran[2](✉)

[1] The University of Jordan, Amman, Jordan
Ola.muneer@hotmail.com, {l.nemer,sawalha.majdi}@ju.edu.jo
[2] Higher Colleges of Technology, Fujariah, UAE
jalsakran@hct.ac.ae

Abstract. The prediction of cancer survivability in patients remains a challenging task due to its complexity and heterogeneity. Nevertheless, studying cancer survivability has been receiving an increasing attention essentially because of the positive impact it has on patients and physicians. It helps physicians determine the suitable treatment options, gives hope to patients, and improves their psychological state. This paper aims to predict the survival period a patient can live after being diagnosed with cancer disease by surveying the performance of three different regression algorithms. The three regression algorithms used are Decision Tree Regression, Multilayer Perceptron Regression, and Support Vector Regression. The algorithms are trained and tested on nine cancer types selected from the SEER dataset. The prediction models of each regression algorithm are built using cross validation evaluation method and ensemble method. Our experimental results show that Decision Tree Regression outperforms the others in predicting the survival period in all the nine cancer types.

1 Introduction

Cancer is a vicious disease that is when diagnosed is considered, to many, as a death sentence. Indeed, it is one of the main causes of death and has claimed the life of millions around the world [25]. Therefore, predicting how long a cancer patient can survive has a tremendous impact on patients and physicians. It gives hope to patients and improves their psychological state. Also, it enables physicians to make a more accurate prognosis estimating the patient's chance of recovery, and choose the proper treatment options [21].

Nevertheless, the accurate prediction of cancer survivability remains a challenging task due to its complexity and heterogeneity. As a result, medical researchers have been increasingly using machine learning algorithms to effectively predict cancer prognosis. Machine learning algorithms can discover and identify patterns and detect relationships between features, they have been shown to be effective in predicting future outcomes of cancer survivability [14,15,19]. Different machine learning algorithms such as decision trees, neural

L. Barolli et al. (Eds.): EIDWT 2019, LNDECT 29, pp. 200–209, 2019.
https://doi.org/10.1007/978-3-030-12839-5_19

networks, and support vector machine have achieved varying success at predicting the survivability of cancer patients.

In this paper, we examine the performance of three regression algorithms in predicting how long, in months, a patient can survive after being diagnosed with cancer. The algorithms used in our study are: Decision Tree Regression (DTR), Multilayer Perceptron Regression (MLPR), and Support Vector Regression (SVR). The choice of these three regression algorithms is driven by their wide use and effectiveness in prediction tasks. Our experiments are conducted on the Surveillance, Epidemiology, and End Results program (SEER) dataset which includes cancer cases that have been collected in the United States since 1973. Thorough experiments are conducted to evaluate and compare the performance of the aforementioned algorithms. Our main contribution consists of the following:

1. Preprocessing the dataset which involves, among others, removing missing data, combining similar attributes, and data normalization.
2. Building the prediction models using cross validation and ensemble methods.
3. Evaluating and comparing the performance of the prediction models using mean absolute errors and correlation coefficient.

The rest of the paper is organized as follows: Sect. 2 reviews and discusses previous efforts related to predicting cancer survivability. In Sect. 3, we discuss the SEER dataset and present the propocessing methods used to prepare the data. Different methods to build the models are presented in Sect. 4. In Sect. 5, we discuss and analyze the experimental results and report our findings. Finally, Sect. 6 concludes the paper.

2 Related Work

Predicting the survival rates of cancer patients has been extensively researched in the literature. Machine learning algorithms are among the most common methods for the prediction of cancer survivability [14,15,19]. Of the early efforts, Frey et al. [12] develop ecological regression models based on country-level SEER incidence to estimate cancer incidence for the nation. Similarly, Mariotto et al. [17] apply hazard regression on 5-year breast, prostate, and colorectum relative survival rates from SEER program. The regression model is evaluated by comparing the predicted rates with the actual rates for countries not used in the process. They conclude that data from cancer registries could be used in ecological models to provide national survival rates.

Breast cancer survivability prediction has attracted much attention over the years. Delen et al. [9] compare three classification techniques: artificial neural networks, decision tree, and logistic regression to predict the survival rate of SEER breast cancer dataset. The artificial neural networks and decision tree achieve better results than logistic regression. In addition, Choi et al. [8] implement a hybrid model combining the artificial neural network and naïve bayesian to predict survival rate of breast cancer patients. The developed model achieves high area under the curve value and they conclude that the developed prediction model can be useful for clinicians in the medical fields. Support vector machine, artificial neural networks, and semi supervised learning model are compared in [19]. The authors employ these classification algorithms to predict the breast cancer survivability using SEER dataset. The experimental results show that the semi supervised learning model is the most accurate in predicting breast cancer survivability.

Kavith and Dorairangasamy [14] analyze the performance of naïve bayesian classifier and C4.5 algorithm in predicting the survival rate of breast cancer patients, they conclude that C4.5 shows better performance than naïve bayesian classifier. Similarly, Bellaachia and Guven [6] examine naïve bayes, back-propagation neural networks, and C4.5 decision tree algorithms on SEER breast cancer dataset. The prediction performance of the C4.5 decision tree algorithm has better performance than the other two techniques.

Alnemer et al. [2] investigate the performance of 4 classification algorithms: artificial neural networks, K-nearest neighbor, support vector machine, and decision tree to predict the survivability of breast cancer on SEER dataset. They apply conformal prediction algorithm to the classification results to improve the classification results.

Survivability in colon cancer has been also studied. In an effort by Al-Bahrani et al. [3], they analyze SEER colon cancer with the aim of developing accurate survival prediction models. They use synthetic minority oversampling technique (SMOTE) technique to solve the imbalance class label. Then the decision tree is applied, the experimental results show that the best accuracy is obtained when the SMOTE is used. Deep neural networks is applied to the SEER colon cancer data to build a survival prediction model [4]. Based on experiments, they conclude that a network with five hidden layers has the best result compared to other machine learning algorithms. They produce a colon cancer calculator based on the experimental models.

Lung cancer survivability is studied by Agrawal et al. [1]. They apply ensemble voting of five decision tree-based classifiers and meta-classifiers to SEER lung cancer dataset. Their effort results in developing an online lung cancer calculator for estimating the risk of mortality.

In comparison with previous efforts, we conduct an extensive study to investigate and compare the performance of decision tree regression, multilayer perceptron regression, and support vector regression on nine cancer types. Specifically, our prediction aims to learn how long, in months, cancer patients can survive. To the best of our knowledge, there is no similar studies that have been conducted in the literature.

3 Data Preprocessing

The Surveillance, Epidemiology, and End Results (SEER)[1] program of the National Cancer Institute (NCI) is an authorized source of cancer incidence and

[1] http://www.seer.cancer.gov.

survival in the United States. SEER program started collecting data about cancer patients in 1973. The SEER dataset contains various information about cancer patients including: cancer type, tumor size, cancer stage, cause of death, cancer site, and many other attributes collected from 1973 to 2014. Table 1 shows the list of cancer types, number of attributes and number of records used in our study.

Table 1. List of cancer types, number of attributes and number of records used in our study

Cancer type	No. of attributes	No. of records
Breast	43	447,426
Respiratory system	41	160,297
Colon	40	116,442
Rectum	41	101,272
Urinary	36	143,532
Female genital	40	61,820
Male genital	39	54,476
Leukemia	33	101,371
Lymphoma	40	110,996

Our preprocessing starts by separating each cancer type into its own dataset because we want to use them separately. The "Site Recode ICD-O-3/WHO 2008" attribute is used to filter out each cancer type. We further eliminate the missing records, remove the attributes that contain redundant values, and remove the unknown values. In addition, we combine the attributes that are essentially the same but named differently; for example, tumor size was named "EOD Tumor Size" cases in 1973–2003 and renamed to "CS Tumor Size" for cases in 2004–2014.

The *survival month* attribute is used in our comparative study as the target variable. It contains continuous values ranging from 0–503 months. We are going to use it to predict how many months a patient can survive after being diagnosed with cancer. The *survival month* is normalized to fall within [0–1] range using the min-max normalization. The motivation behind using this scaling is its robustness to small standard deviations of features and preserving zero entries in sparse data [23]. Equation 1 shows the min-max normalization.

$$x_{norm} = \frac{x_i - x_{min}}{x_{max} - x_{min}} \tag{1}$$

4 Building Prediction Models

We study and examine the performance of the three regression algorithms in predicting how many months a patient can survive after being diagnosed with cancer. The three regression algorithms are:

1. Decision Tree Regression (DTR) which is a simple and practical algorithm and has been shown to produce considerably higher cancer prediction results compared to other techniques [6,14].
2. Multilayer Perceptron Regression (MLPR) which has proved to be a competitive regression algorithm, it produces great prediction results in predicting time series compared to other methods [18].
3. Support Vector Regression (SVR) which has also shown to provide excellent performance results compared to other regression methods [22,24,26].

Each algorithm is trained, tested, and evaluated using two methods: cross validation and ensemble. In cross validation, we use 10-fold cross validation to evaluate the prediction models. Ensemble methods combine the outputs of prediction models and it is used to obtain better prediction performance. It has been proven experimentally by many researchers that ensemble methods produces higher performance results than a single model [5,16]. The ensemble methods have been used in many fields, such as image processing, biological engineering and medical predictions; and have enhanced the performance results [13]. Ensemble learning methods include Bagging, Boosting, and Random Forest.

Our study uses Bagging (Bootstrap Aggregating), which is one of the commonly used ensemble methods that was first proposed by Leo Breiman in 1996 [7] aiming to enhance the stability and performance of machine learning algorithms. Bagging works very well with non-stable learning algorithms and improves their performance results, moreover it minimizes the effect of overfitting the training data. Bagging can be used in both classification and regression problems. In bagging method, the original training set is sampled into multiple bootstrap samples; each bootstrap sample is selected randomly from the original training set and is the same size as the original training set [11]. Then the prediction model is built for each bootstrap sample. The last step in bagging method is to combine the predictions from the bootstrap samples by voting (in classification problems) or averaging (in regression problems). Figure 1 shows the steps of the ensemble method.

5 Experimental Results and Analysis

In our implementation, we use Scikit-learn library in Python [20]. For the evaluation of the regression prediction models we use the most frequently used regression evaluation methods: mean absolute error (MAE) and correlation coefficient between the actual value and predicted value.

Table 2 reports the experimental results obtained by the three regression algorithms using 10-fold cross validation approach. The results clearly show that decision tree regression outperforms the other two regression algorithms in all cancer types and in both measures MAE and correlation. In Fig. 2, we depict the correlation measure of the three algorithms on the nine cancer types. Correlation values range between -1 to 1, where -1 indicates perfect negative correlation between actual and predicted values and 1 indicates perfect positive correlation.

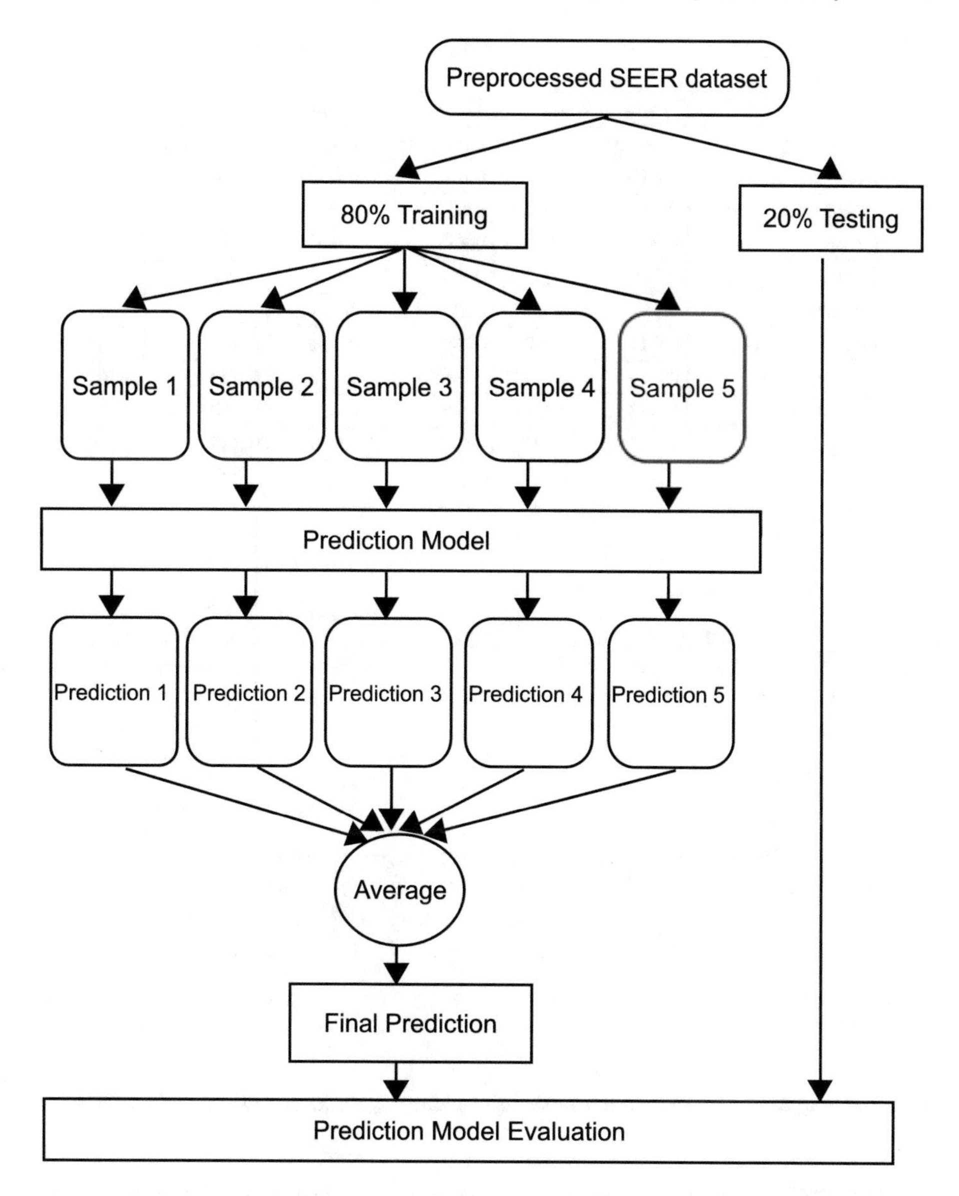

Fig. 1. Steps of building prediction model using ensemble method

The figure readily shows that all algorithms perform well (close to 1) and decision tree achieves the best performance. Multilayer perceptron comes second and performs nearly as good as decision tree on some cancer types. Support vector is the least performant and it does not help that it takes considerably more computational time to run on such large datasets.

Table 2. Results using 10-fold cross validation method

	Decision tree		Multilayer perceptron		Support vector	
Cancer type	Correlation	MAE	Correlation	MAE	Correlation	MAE
Breast	0.848	0.075	0.832	0.086	0.838	0.097
Respiratory system	0.842	0.054	0.840	0.057	0.714	0.124
Colon	0.839	0.083	0.830	0.096	0.816	0.106
Rectum	0.830	0.073	0.818	0.086	0.804	0.095
Urinary	0.854	0.059	0.800	0.084	0.832	0.081
Female genital	0.841	0.093	0.837	0.094	0.830	0.104
Male genital	0.864	0.064	0.860	0.067	0.823	0.104
Leukemia	0.853	0.040	0.820	0.059	0.810	0.067
Lymphoma	0.876	0.072	0.867	0.086	0.849	0.100

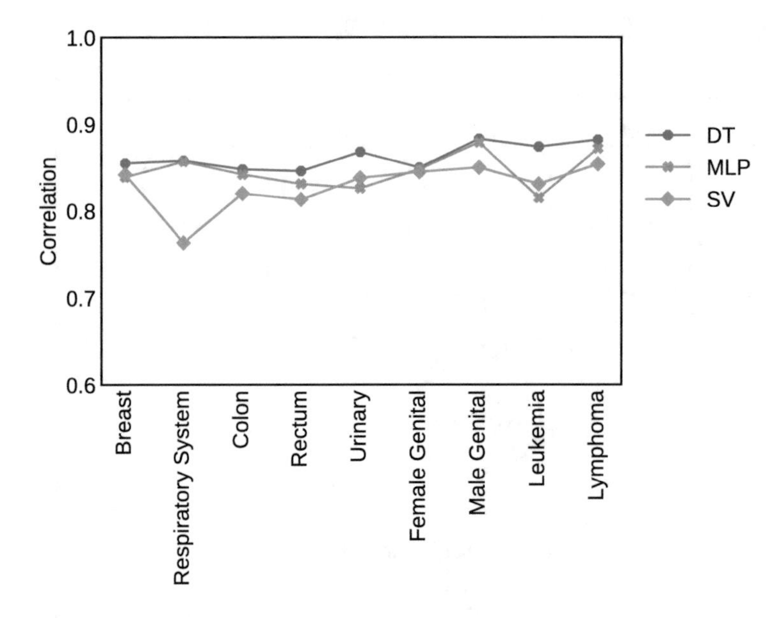

Fig. 2. Comparison of algorithms on the nine cancer types using 10-fold cross validation

The decision tree regression outperforms the other three regression algorithms because it performs a feature selection automatically when it is fitted to the training set. It partitions the attributes in which the most important attributes are at the top of the tree, it uses the most important attributes to predict the survival months of a cancer patient. Performing the automatic feature selection produced better prediction results compared to the other regression algorithms.

Table 3 reports the experimental results obtained by the three regression algorithms using ensemble method. The result shows that ensemble method enhances the prediction results when compared with the 10-fold cross validation. Similarly,

Table 3. Results using ensemble method

	Decision tree		Multilayer perceptron		Support vector	
Cancer type	Correlation	MAE	Correlation	MAE	Correlation	MAE
Breast	0.855	0.075	0.839	0.084	0.842	0.096
Respiratory system	0.858	0.052	0.857	0.056	0.763	0.122
Colon	0.848	0.082	0.842	0.091	0.820	0.105
Rectum	0.846	0.070	0.831	0.082	0.813	0.094
Urinary	0.868	0.057	0.826	0.077	0.838	0.080
Female genital	0.850	0.091	0.848	0.094	0.845	0.103
Male genital	0.883	0.063	0.879	0.066	0.85	0.103
Leukemia	0.874	0.038	0.815	0.056	0.831	0.063
Lymphoma	0.882	0.071	0.872	0.083	0.854	0.099

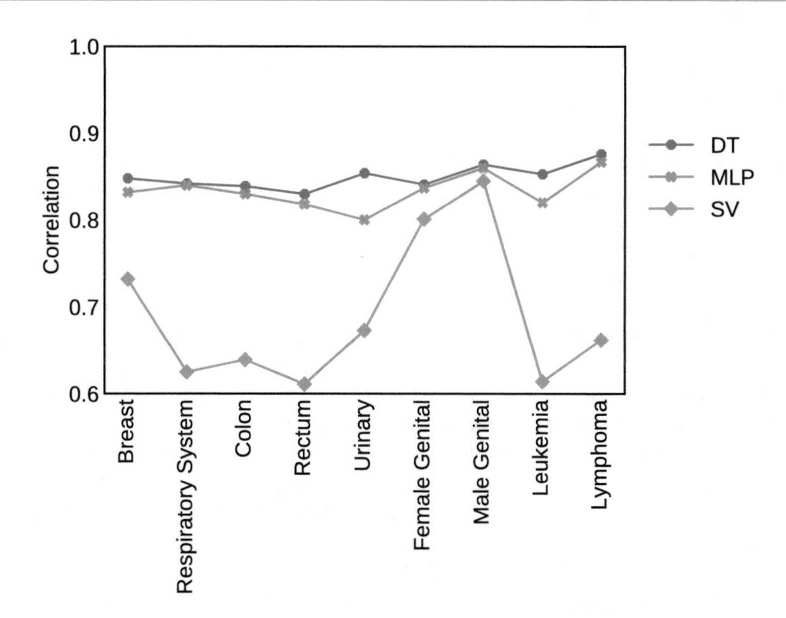

Fig. 3. Comparison of algorithms on the nine cancer types using ensemble method

decision tree outperforms the other two algorithms on all cancer types. Figure 3 shows the correlation measure performance for the three algorithms. The performance of support vector has considerably degraded and its time and space complexity makes it the least favorite. Decision tree and multilayer perceptron achieve higher performance with decision tree as the best performant.

6 Conclusion

Prediction of cancer survivability can positively impact the life of patients and help physicians improve prognosis. Therefore, we have conducted a comparative study to investigate the performance of three regression algorithms in predicting how long, in months, a cancer patient can survive after being diagnosed with cancer disease.

We study the performance of decision tree, multilayer perceptron, and support vector on nine cancer types obtained from SEER dataset. The models are built using k-fold cross validation and ensemble method. Our results have shown that the decision tree has the best performance in predicting the survival months of cancer patients compared to the other two regression algorithms. The decision tree outperforms the other regression algorithms on all the nine cancer datasets using both 10-folds cross validation evaluation method and ensemble method.

The multilayer perceptron performance results are very competitive to the decision tree. On the other hand, support vector comes last in performance and demands more time and space which makes it the least favourite algorithm to predict cancer survivability.

References

1. Agrawal, A., Misra, S., Narayanan, R., Polepeddi, L., Choudhary, A.: Lung cancer survival prediction using ensemble data mining on SEER data. Sci. Programm. **20**, 29–42 (2012)
2. Alnemer, L.M., Rajab, L., Aljarah, I.: Conformal prediction technique to predict breast cancer survivability. Int. J. Adv. Sci. Technol. **96**, 1–10 (2016)
3. Al-Bahrani, R., Agrawal, A., Choudhary, A.: Colon cancer survival prediction using ensemble data mining on SEER data. In: IEEE International Conference on Big Data (2013)
4. Al-Bahrani, R., Agrawal, A., Choudhary, A.: Survivability prediction of colon cancer patients using neural networks. Health Inform. J. (2017)
5. Bauer, E., Kohavi, R.: An empirical comparison of voting classification algorithms: bagging, boosting, and variants. Mach. Learn. **36**(1/2), 105–139 (1999)
6. Bellaachia A., Guven, E.: Predicting breast cancer survivability using data mining techniques. In: Scientific Data Mining Workshop, in conjunction with the 2006 SIAM Conference on Data Mining (2006)
7. Breiman, L.: Bagging predictors. Mach. Learn. **24**(2), 123–140 (1996)
8. Choi, J.P., Han, T.H., Park, R.W.: A hybrid bayesian network model for predicting breast cancer prognosis. J. Korean Soc. Med. Inform. **15**(1), 49 (2009)
9. Delen, D., Walker, G., Kadam, A.: Predicting breast cancer survivability: a comparison of three data mining methods. Artif. Intell. Med. **34**(2), 113–127 (2005)
10. Duda, R.O., Hart, P.E., Stork, D.G.: Pattern Classification, 2nd edn. Wiley-Interscience, New York (2001)
11. Fan, W., Zhang, K.: Bagging. Encyclopedia of Database Systems, pp. 1–5 (2016)
12. Frey, C.M., Feuer, E.J., Timmel, M.J.: Projection of incidence rates to a larger population using ecologic variables. Stat. Med. **13**(17), 1755–1770 (1994)
13. Ghodselahi, A.: A hybrid support vector machine ensemble model for credit scoring. Int. J. Comput. Appl. **17**(5), 1–5 (2011)
14. Kavitha, R., Dorairangasamy, D.: Predicting breast cancer survivability using Naïve bayesian classifier and C4.5 algorithm. Elysium J. Eng. Res. Manag. **1**(1), 61–63 (2014)

15. Kourou, K., Exarchos, T.P., Exarchos, K.P., Karamouzis, M.V., Fotiadis, D.I.: Machine learning applications in cancer prognosis and prediction. Comput. Struct. Biotechnol. J. **13**, 8–17 (2015)
16. Maclin, R., Opitz, D.: An empirical evaluation of bagging and boosting. In: Proceedings of the Fourteenth National Conference on Artificial Intelligence, pp. 546–551). AAAI Press/MIT Press, Cambridge (1997)
17. Mariotto, A., Capocaccia, R., Verdecchia, A., Micheli, A., Feuer, E., Pickle, L., Clegg, L.: Projecting SEER cancer survival rates to the US: an ecological regression approach. Cancer Causes Control **13**, 101–111 (2002)
18. Osowski, S., Siwek, K., Markiewicz, T.: MLP and SVM networks - a comparative study. In: Proceedings of the sixth Nordic Signal-Processing Symposium - NORSIG (2004)
19. Park, K., Ali, A., Kim, D., An, Y., Kim, M., Shin, H.: Robust predictive model for evaluating breast cancer survivability. Eng. Appl. Artif. Intell. **26**(9), 2194–2205 (2013)
20. Pedregosa, F., Varoquaux, G.: Scikit-learn: machine learning in Python (2011)
21. Rizzieri, D.A., Vredenburgh, J.J., Jones, R., Ross, M., Shpall, E.J., Hussein, A., Broadwater, G., Berry, D., Petros, W.P., Gilbert, C., Affronti, M.L., Coniglio, D., Rubin, P., Elkordy, M., Long, G.D., Chao, N.J., Peters, W.P.: Prognostic and predictive factors for patients with metastatic breast cancer undergoing aggressive induction therapy followed by high-dose chemotherapy with autologous stem-cell support. J. Clin. Oncol. **17**(10), 3064–3074 (1999)
22. Sangitab, P., Deshmukh, S.: Use of support vector machine for wind speed prediction. In: International Conference on Power and Energy Systems, pp. 1–8 (2011)
23. Shalabi, L.A., Shaaban, Z., Kasasbeh, B.: Data mining: a preprocessing engine. J. Comput. Sci. **2**(9), 735–739 (2006)
24. Sheta, A., Elsir, S., Faris, H.: A comparison between regression, artificial neural networks and support vector machines for predicting stock market index. Int. J. Adv. Res. Artif. Intell. **4**(7) (2015)
25. Stewart, W., Wild, P.: World Cancer Report. IARC, Geneva (2014)
26. Thongkam, J., Sukmak, V., Mayusiri, W.: A comparison of regression analysis for predicting the daily number of anxiety-related outpatient visits with different time series data mining. KKU Eng. J. **42**(3), 243–249 (2015)

Enhanced Classification of Sentiment Analysis of Arabic Reviews

Loai Alnemer[1], Bayan Alammouri[1], Jamal Alsakran[2](✉), and Omar El Ariss[3]

[1] The University of Jordan, Amman, Jordan
{l.nemer,b.ammouri}@ju.edu.jo
[2] Higher Colleges of Technology, Fujariah, UAE
jalsakran@hct.ac.ae
[3] Texas A&M University-Commerce, Commerce, TX, USA
Omar.El.Ariss@tamuc.edu

Abstract. Sentiment analysis is the process of mining textual data in order to extract the author's opinion, typically expressed as a positive, neutral, or negative attitude towards the written text. It is of great interest and has been extensively studied in the English language. However, sentiment analysis in the Arabic language has not received wide attention and most of the research done on Arabic either focuses on introducing new datasets or new sentiment lexicons. In this paper, we introduce a preprocessing suite that includes morphological processing, emoticon extraction, and negation processing to improve the sentiment analysis. Furthermore, we conduct experiments on sentiment analysis of hotel reviews that target two classification tasks: positive/negative and positive/negative/neutral. Our experimental results using various supervised learning algorithms, including deep learning algorithm, demonstrate the effectiveness of the proposed techniques.

1 Introduction

Until recently, people used to consult with family, friends, books, or newspapers when they wanted an opinion on a specific entity or service. Nowadays, with the advent of social media such as Twitter and Facebook, electronic news websites, and specialized websites such as tripadvisor.com, hotels.com and amazon.com you can find useful sources of people's opinions on almost everything. The volume of readily available data has greatly affected our daily decisions. This effect of online reviews has been investigated in various studies [10, 11]. As a result, mining these unstructured information is an important task for many researchers and e-commerce companies.

Sentiment analysis, which falls under the umbrella of opinion mining [17], is a topic of great interest and development since it has many practical applications such as collecting people's opinions and attitudes about entities and services, for example hotels, products, books, movies and many others [14]. There has been extensive research dedicated to advance the sentiment analysis techniques

L. Barolli et al. (Eds.): EIDWT 2019, LNDECT 29, pp. 210–220, 2019.
https://doi.org/10.1007/978-3-030-12839-5_20

towards a better understanding of raw text [18,20,21]. However, sentiment analysis in the Arabic language has not been adequately studied due to the limited available resources when compared to the English language. Moreover, the nature of the Arabic language and its morphological structure demands more manipulation and processing in order to get acceptable results.

Arabic is the official language for more than 20 countries and it is spoken by hundreds of millions. The written text used in official documents, newspapers, and books follows a standard form. It is the form of the language that is taught in schools and it is the same everywhere. Nevertheless, writings in less formal context such as forums or reviews rarely follow the standard form, rather, people more likely to write in their dialectic form of the language. Different forms of the Arabic language can be found on the internet, we here present the most popular forms:

- Classical Arabic is the old form of language used in ancient Arabic literature such as pre-Islamic poetry.
- Modern Standard Arabic (MSA) is the language that is taught in all Arab countries and is a modified version of the standard Arabic language. Some symbols such as Hamza, diacritics and tashkil have been added to improve the reading of the Arabic text.
- Slang or dialectic: in general are spoken languages and not written, but they began to grow on social networking sites.
- The Arabic chat alphabet (also known as Arabizi) where Arabic text, which can be either standard or colloquial, is written using the Latin alphabet. This format is now used extensively for writing on the Internet easily and without the need for an Arabic keyboard. The problem of different methods of writing is one of the most important problems in this type where everyone decides how to write an Arabic word. For example there are at least three Latin forms of the name لؤي in Arabic: Loai, Loay, Lo'ai. This form of writing requires more research and needs a large corpus so we decided not to target it in this work.

The existence of these forms dictates that when choosing a dataset for sentiment analysis it is crucial that the dataset supports multiple forms. The various forms of the Arabic language represent a challenge when choosing datasets for analysis. Any reviews dataset is likely to contain many forms especially MSA and colloquial dialects. In addition, some reviews are going to come with diacritics and others without. This is likely to inflate the vocabulary dictionary and subsequently make opinion mining harder to achieve. Finally, the presence of emoticons in the dataset provides useful emotional hints, yet it requires special treatments to extract such emotional clues.

Sentiment analysis in the Arabic language needs to deal with some particular challenges for instance large number of words are derivatives of the same root [23]. This problem is usually addressed using a stemmer which strips the words down to their roots and as a result reduce the number of distinct words in a text. Yet, the problem remains in the maturity of Arabic stemmers which still needs considerable improvements. In this paper, we introduce a preprocessing suite that includes filtering, segmentation, morphological processing specifically for the Arabic language, negation processing, and emoticon extraction. We evaluate the effectiveness of our proposed technique using five different classifiers; Bernoulli Naïve Bayes (BNB), Decision Tree (NB), Support Vector Machines (SVM), Neural Network (NN), and Convolutional Neural Network (CNN). The experimental results show an enhancement in the accuracy of classification of preprocessed dataset compared to the unprocessed dataset.

The rest of the paper is organized as follows. In Sect. 2, we discuss previous efforts in Arabic sentiment analysis. The preprocessing suite which includes emoticon handling and extraction, word normalization, morphological processing, and negation processing is presented in Sect. 3. In Sect. 4, we present our experimental study which employs five supervised machine learning techniques on the Arabic sentiment classification problem. Furthermore, we conduct two studies to investigate the effectiveness of the preprocessing steps. Finally in Sect. 5, we conclude the paper.

2 Related Work

Sentiment analysis of the Arabic language has not been extensively studied [7]. We can divide the previous work in this field into two main groups. The first group focuses on creating new datasets and sentiment lexicon using either reviews or tweets. The second group focuses on evaluating the techniques of analysis using different classification methods.

In an effort to introduce new Arabic datasets, Refaee and Rieser [22] present a dataset that contains around 8000 tweets for subjectivity and sentiment analysis. The authors manually annotate the dataset and analyze it. Their semi-supervised method that uses emoticons as noisy labels perform well for subjectivity analysis only. In [16], the authors introduce an Arabic social sentiment analysis dataset that contains around 10k tweets and investigate the properties of the dataset. Their experiment shows that the balanced dataset is more challenging than the unbalanced dataset and the support vector machine is the best classifier. On the other hand, Al-Ayyoub et al. [4] introduce their Arabic tweet corpus and a new sentiment lexicon for Arabic. The lexicon contains 120k terms and the corpus contains 900 tweets divided into three categories: positive, neutral and negative.

Another annotated dataset for book reviews is introduced by Al-Smadi et al. [5]. The dataset that has been designed to be as a reference benchmark for the aspect-based sentiment analysis. It contains 1513 books review selected from LABR [3]. SANA, a large-scale, multi-genre, multi-dialects lexicon for Arabic subjectivity and sentiment analysis, is introduced in [2]. The dataset contains 225K terms for Egypt and Levantine dialects and the entries are labelled without preprocessing and include useful information such as part of speech.

A standard Arabic sentiment lexicon is proposed be Eskander and Rambow [13]. The presented dataset contains 35k lemmas and it is based on linking the AraMorph with SentiWordNet. The authors conduct sentiment analysis based on evaluation that shows an improvement in the accuracy over the state-of-the-art lexicon.

In this work, we use the hotel reviews dataset that is introduced by ElSahar and El-Beltagy in [12]. In their effort, the authors present four different Arabic reviews dataset contains 33k reviews of hotels, movies, restaurants and products, in addition to new sentiment lexicon that contains 2K entries for these datasets. In order to evaluate their dataset, the authors perform an extensive sentiment analysis evaluation using different classification algorithms.

There have been many studies that focus on the classification of reviews as positive or negative, in which more than thirty techniques are used between 2009 and 2015 [4] and applied to datasets ranging from 625 [8] reviews to 63000 [3] reviews.

Mohammad et al. [15] study the effect of translation between Arabic and English in the sentiment analysis. They explore two approaches: (1) Translate the Arabic text to English to benefit from the wide range of resources of English language. (2) Translate English resources to Arabic and use them in the Arabic sentiment analysis. Their experiments show that the translation enhances the accuracy of the classification results specially in the case of translating resources. Subjectivity and sentiment analysis for modern standard Arabic using news articles is studied by Abdul-Mageed et al. [1]. Authors use a random graph walk approach and different features for subjectivity and sentiment analysis, their classification results show enhancement in accuracy comparing to different previous methods.

The largest Arabic book reviews corpus (LABR) proposed by Aly and Atiya [3] contains over 63,000 book reviews where each review is labelled from 1 to 5. The authors evaluate their proposed dataset using different classification techniques for two class classification problem and five class classification problem. Many researchers evaluate LABR using different methods. Al Shboul et al. [24] apply five different classification algorithms; decision tree, support vector machines, K-nearest neighbors, naïve based and ensemble of three classifiers for the five class problem. In [9], the authors evaluate different dataset including LABR using convolutional neural networks (CNN) and word Embeddings. Their results show increasing performance when CNN is used. Ariss and Alnemer [6] further introduce preprocessing that includes emoticon lexicons and morphological processing to enhance the accuracy of classification. Their method achieves higher accuracy when compared with the unprocessed dataset.

3 Preprocessing

The ubiquity of the internet has encouraged people from different backgrounds, geographical locations and age ranges to write comments and reviews. Unlike formal writings, reviews have no specific guidelines or standards that they need to adhere to and, more often than not, they contain a mixture of text and graphics (emoticons). This unstructured nature of reviews might drastically hinder the performance of our classifiers. Therefore, it is essential to preprocess and normalize the reviews in order to filter out redundant and unnecessary data. Our preprocessing involves many stages that are carried in an order to prevent the loss of sentiment clues. For example, removing punctuation marks prior to extracting emoticons can compromise the results of emoticon extraction. Here, we describe the preprocessing steps.

Emoticon Extraction are textual representations of facial expressions, and are very useful emotional cues in a text. We extract emoticons and construct a dictionary of 63 emoticons. We then categorize these emoticons into nine emotional categories. Finally, emoticons that we find in a review are replaced by the category label that they belong to. Table 1 shows a subset of the emoticons dictionary:

Table 1. Emoticon dictionary

Emoticon	Emotional category
;-) ;) *)	Wink
:'-(:'(	Cry
:-)) :-) :)	Smiley face
==3 -3	Laugh
:-|| :@ >:(	Angry
: [:[:-<	Sad
>: >:/ :-/	Annoyed
:* (')}{(')	Kisses
:$	Embarrassed

Word Normalization involves the tokenization of dates and numbers and the Arabic punctuations such as comma, semicolon and question mark are replaced by their English equivalents.

Furthermore, we need to normalize the same words that are written differently, for example if we are analyzing a corpus that contains both British and American English text then we need to normalize *colour* and *color* and *centre* and *center* since they are referring to the same concept. In Arabic this situation is even more prevalent due to tashkil: diacritics, hamzah and maddah. Therefore, the many variations of the same need to be unified into one. For example, here are four common variations of the word "felt" in Arabic: احسست أحسسِت أحسسُت أحسست

We also remove multiple repetitions of the same character, which are frequently added by user to stress a concept. For example, in English, if someone wants to emphasize that the weather is very cold, then he might excessively repeat the letters of the stressed syllable: verrrrrrrrrrrry (جددددددددددددددا).. Special caution must be taken not to remove valid repetitive characters. For example, the letter rā' (ر) in the word *decide* (قرر) is part of the word and should not be removed.

Morphological Processing focuses on simplifying the structure of words possibly by removing prefix/suffix, stemming the words or other techniques. Arabic is a morphological rich language, though, we perform a simple morphological processing, which is the removal of the definite article (ال), which is equivalent to *the* in English. Articles in Arabic are treated as prefixes and are part of the word. Figure 1 shows an example of Arabic word الأسئلة (in English *The questions*) before and after the removal of the article.

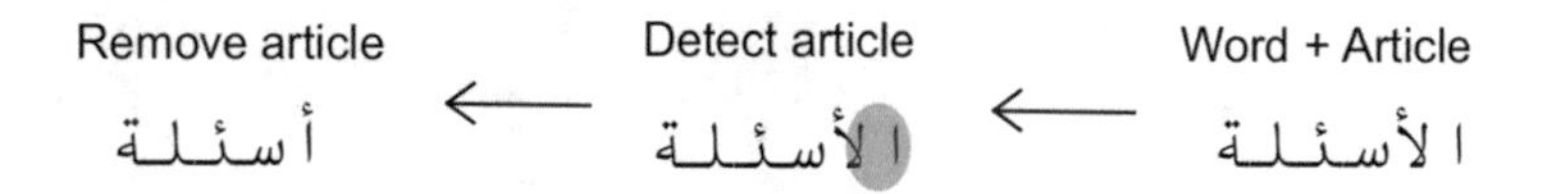

Fig. 1. Detection and removal of the Arabic article ال from the word الأسئلة

Negation Processing is tightly coupled with sentiment analysis because it represents ways to express negative polarity in a sentence. The presence of negation phrases such as ليس، غير، لا significantly impacts the sentiment conveyed in a sentence however when fed to a classifier they are treated like other words and their special contribution is undermined. For example, describing something as ليس مفيد , which means *not useful* in English, clearly indicates negative impression however the classifier reads ليس which suggests negative opinion followed by مفيد which is associated with positive impressions.

In our effort, we attach negation phrases to the words that come right after making them form one word, for example ليس مفيد is treated as one word. This concatenation aims to signify the impact of negation on all word count levels: Unigram, Bigram, and Trigram.

4 Experimental Setup and Evaluation

In this work, we use hotel reviews dataset collected from Tripadvisor.com [12], which contains 15,572 reviews with ratings that range from 1 to 5, where 1 is the lowest rating and 5 is the highest rating. Our sentiment classification focuses on two tasks:

1. Sentiment classification of two classes (Positive: 1 and Negative: 0) where ratings 1 and 2 represent Negative class while ratings 4 and 5 represent Positive class. Rating of 3 is excluded from our study in this task.
2. Sentiment classification of three classes (Positive: 1, Negative: –1, and Neutral: 0) where ratings 1 and 2 represent Negative class, rating 3 represents Neutral class and ratings 4 and 5 represents Positive class.

Table 2. Experimental results of the five classifiers when applied to two classes task

Feature	Classifier	Unprocessed Accuracy	Unprocessed f-measure	Preprocessed Accuracy	Preprocessed f-measure
Count unigram	BNB	0.812	0.799	0.826	0.825
	DT	0.896	0.896	0.918	0.910
	SVM	0.898	0.896	0.926	0.920
	NN	0.951	0.951	0.977	0.975
	CNN	0.952	0.952	0.972	0.981
Count bigram	BNB	0.725	0.636	0.752	0.662
	DT	0.868	0.859	0.883	0.884
	SVM	0.957	0.957	0.972	0.972
	NN	0.897	0.896	0.911	0.915
	CNN	0.958	0.958	0.984	0.986
Count trigram	BNB	0.698	0.574	0.726	0.593
	DT	0.699	0.576	0.723	0.606
	SVM	0.957	0.956	0.978	0.976
	NN	0.844	0.827	0.867	0.852
	CNN	0.902	0.902	0.924	0.920
Tfidf unigram	BNB	0.698	0.574	0.711	0.596
	DT	0.698	0.574	0.728	0.584
	SVM	0.956	0.956	0.966	0.966
	NN	0.721	0.627	0.742	0.650
	CNN	0.892	0.892	0.920	0.921
Tfidf bigram	BNB	0.698	0.574	0.723	0.586
	DT	0.701	0.580	0.714	0.596
	SVM	0.884	0.885	0.902	0.896
	NN	0.947	0.947	0.973	0.960
	CNN	0.948	0.949	0.978	0.977
Tfidf trigram	BNB	0.698	0.574	0.715	0.587
	DT	0.698	0.574	0.725	0.594
	SVM	0.699	0.577	0.715	0.589
	NN	0.869	0.870	0.888	0.898
	CNN	0.937	0.937	0.956	0.959

Several feature extraction methods are used for sentiment classification of texts have been presented in the literature. Word existence, word count, term frequency-inverse document frequency (Tfidf), and part of speech are among the most common ones. We employ both word count and Tfidf methods in our classification. In word count, we conduct experiments at three different levels: Unigram, Bigram and Trigram.

For classification, five well known and heavily used classifiers are used from the scikit-learn library in python [19]; Bernoulli Naïve Bayes (BNB), Decision Tree (NB), Support Vector Machines (SVM), Neural Network (NN), and Convolutional Neural Network (CNN). Our choice of classifiers is mainly because these classifiers are among the most commonly used for sentiment analysis in the literature. CNN is of particular interest to us because of its success in improving classification performance in other application domains while it has not been extensively studied on preprocessed Arabic sentiment analysis.

We report the accuracy and the f-measure for each classifiers and feature extraction methods. The comparison with the unprocessed datasets is also reported in terms of its accuracy for the two and three classes tasks.

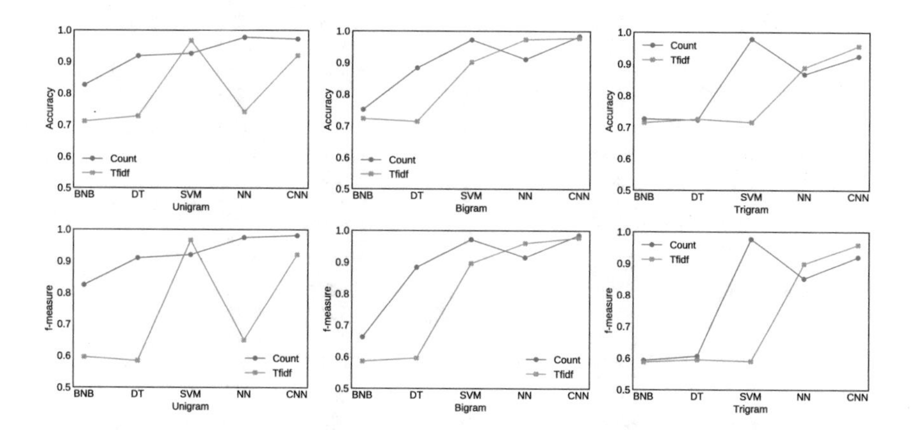

Fig. 2. Comparison of word count vs Tfidf on preprocessed dataset (2 classes)

Table 2 shows the experimental results of our proposed method for all classifiers and feature extraction methods when two classes dataset is used. Color is used to highlight the highest values (0.9 and above). The results show that the preprocessed dataset achieves higher accuracy in all classifiers and both features. This clearly proves the benefit of the proposed preprocessings in our approach.

Furthermore, Table 1 and Fig. 2 show that the Tfidf method provides a relatievly good improvement on the results of the preprocessed dataset while it does not have a significant effect on the processed datasets when compared to count, and in some cases the accuracy decreased with the use of Tfidf. This observation is different from the one reported in [3] which we believe happens because of the preprocessing of the dataset.

An observer of Fig. 2 can easily notice that CNN achieves the best performance while BNB and DT are the least performants. Count levels (Unigram vs Bigram vs Trigram) do not seem to have a clear cut winner even though we had the impression that Unigram should outperform the others. Our interpretation is that negation processing (see Sect. 3) has affected the word count of the original dataset.

Table 3 shows the corresponding results of our proposed method when a three classes dataset is used. Color is used to highlight the highest values (0.8 and above). Similar to our previous findings, The results show that the preprocessed dataset achieves higher accuracy in all classifiers and both features. However, the accuracy of three classes is considerably lower than what is achieved using two classes in Table 2. The degrading accuracy is likely to be attributed to the

Table 3. Experimental results of the five classifiers when applied to three classes task

Feature	Algorithm	Unprocessed Accuracy	Unprocessed f-measure	Preprocessed Accuracy	Preprocessed f-measure
Count unigram	BNB	0.641	0.501	0.667	0.548
	DT	0.679	0.654	0.713	0.694
	SVM	0.764	0.738	0.790	0.764
	NN	0.682	0.593	0.711	0.628
	CNN	0.733	0.649	0.757	0.670
Count bigram	BNB	0.641	0.501	0.686	0.541
	DT	0.642	0.504	0.675	0.514
	SVM	0.679	0.576	0.716	0.616
	NN	0.674	0.646	0.720	0.666
	CNN	0.779	0.745	0.826	0.790
Count trigram	BNB	0.641	0.501	0.683	0.519
	DT	0.680	0.650	0.719	0.680
	SVM	0.741	0.742	0.768	0.774
	NN	0.660	0.541	0.706	0.553
	CNN	0.776	0.740	0.788	0.786
Tfidf unigram	BNB	0.641	0.501	0.689	0.543
	DT	0.641	0.501	0.656	0.549
	SVM	0.658	0.640	0.680	0.670
	NN	0.641	0.501	0.681	0.532
	CNN	0.779	0.736	0.808	0.757
Tfidf bigram	BNB	0.641	0.501	0.686	0.521
	DT	0.641	0.501	0.677	0.525
	SVM	0.650	0.639	0.698	0.680
	NN	0.641	0.501	0.657	0.519
	CNN	0.775	0.723	0.807	0.760
Tfidf trigram	BNB	0.641	0.501	0.679	0.522
	DT	0.619	0.617	0.652	0.630
	SVM	0.641	0.501	0.669	0.536
	NN	0.641	0.501	0.672	0.534
	CNN	0.772	0.718	0.809	0.736

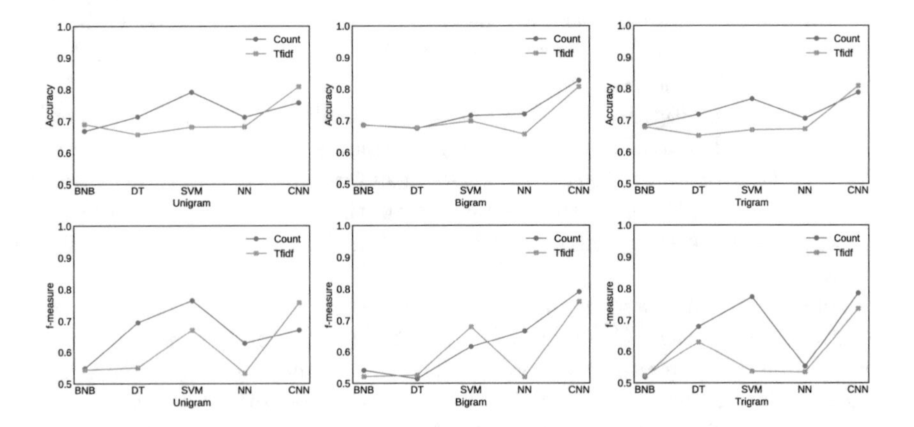

Fig. 3. Comparison of word count vs Tfidf on preprocessed dataset (3 classes)

inclusion of the Neutral class (rating 3) in the classification. Probably because people tend to choose a rating of 3 when they have mixed feelings about an entity.

In Fig. 3, we depict the accuracy of word count against Tfidf on the three classes task. The word count shows slightly better results which we believe happens because of the preprocessing of the dataset.

5 Conclusion

In this effort, we have presented a supervised sentiment analysis approach on a large Arabic hotel reviews dataset. The approach incorporates a suite of preprocesings which includes emoticon sentiment lexicon, word normalization, morphology processing, and negation processing. Five classification algorithms are used to evaluate the proposed method; bernoulli baïve bayes, decision tree, support vector machines, neural network, and convolutional neural network. The experimental results have shown that preprocessing the Arabic language enhances the classification of sentiment analysis. The results report improvements for both two classs (positive/negative) and three classes (positive/negative/neutral) classifications.

References

1. Abdul-Mageed, M., Diab, M.T., Korayem, M.: Subjectivity and sentiment analysis of modern standard Arabi0063. In: ACL–HLT 2011–Proceedings of 49th Annual Meet. Association for Computational Linguistics, vol. 2, pp. 587–591 (2011)
2. Abdul-Mageed, M., Diab, M.: SANA: a large scale multi-genre, multi-dialect Lexicon for Arabic subjectivity and sentiment analysis. In: Proceedings of the Language Resources and Evaluation Conference, pp. 1162–1169 (2014)
3. Aly, M., Atiya, A.: LABR: a large scale Arabic book reviews dataset. In: The 51st Annual Meeting of the Association for Computational Linguistics, pp. 494–498 (2013)
4. Al-Ayyoub, M., Essa, S.B., Alsmadi, I.: Lexicon-based sentiment analysis of Arabic tweets. Int. J. Soc. Netw. Min. **2**, 101–114 (2015)
5. Al-Smadi, M., Qawasmeh, O., Talafha, B., Quwaider, M.: Human annotated Arabic dataset of book reviews for aspect based sentiment analysis. In: Proceedings of 2015 International Conference on Future Internet of Things and Cloud, FiCloud 2015 and 2015 International Conference on Open and Big Data, OBD 2015, pp. 726–730 (2015)
6. Ariss, O.E., Alnemer, L.M.: Morphology based Arabic sentiment analysis of book reviews. In: Gelbukh, A. (eds.) Computational Linguistics and Intelligent Text Processing, CICLing 2017. Lecture Notes in Computer Science, vol. 10762. Springer, Cham (2018)
7. Biltawi, M., Etaiwi, W., Tedmori, S., Hudaib, A., Awajan, A.: Sentiment classification techniques for Arabic language: a survey. In: 2016 7th International Conference on Information and Communication Systems (ICICS), pp. 339–346 (2016)
8. Cherif, W., Madani, A., Kissi, M.: Towards an efficient opinion measurement in Arabic comments. Procedia Comput. Sci. **73**, 122–129 (2015)

9. Dahou, A., Xiong, S., Zhou, J., Haddoud, M.H., Duan, P.: Word embeddings and convolutional neural network for Arabic sentiment classification. In: Proceedings of COLING 2016, the 26th International Conference on Computational Linguistics: Technical Papers, pp. 2418–2427 (2016)
10. Dellarocas, C., (Michael) Zhang, X., Awad, N.F.: Exploring the value of online product reviews in forecasting sales: the case of motion pictures. J. Interact. Mark. **21**, 23–45 (2007)
11. Duan, W., Gu, B., Whinston, A.B.: Do online reviews matter?—an empirical investigation of panel data. Decis. Support Syst. **45**, 1007–1016 (2008)
12. ElSahar, H., El-Beltagy, S.R.: Building large Arabic multi-domain resources for sentiment analysis. In: Lecture Notes in Computer Science (including subseries Lecture Notes in Artificial Intelligence and Lecture Notes in Bioinformatics), pp. 23–34. Springer, Cham (2015)
13. Eskander, R., Rambow, O.: SLSA: a sentiment Lexicon for standard Arabic. In: Proceedings of the 2015 Conference on Empirical Methods in Natural Language Processing. Association for Computational Linguistics, pp. 2545–2550 (2015)
14. Liu, B.: Sentiment analysis and opinion mining. Synth. Lect. Hum. Lang. Technol. **5**, 1–167 (2012)
15. Mohammad, S.M., Salameh, M., Kiritchenko, S.: How translation alters sentiment. J. Artif. Intell. Res. **55**, 95–130 (2016)
16. Nabil, M., Aly, M., Atiya, A.: ASTD: Arabic sentiment tweets dataset. In: Proceedings of the 2015 Conference on Empirical Methods in Natural Language Processing. Association for Computational Linguistics, pp. 2515–2519 (2015)
17. Pang, B., Lee, L.: Opinion mining and sentiment analysis. Found. Trends Inf. Retr. **2**, 1–135 (2008)
18. Pang, B., Lee, L., Rd, H., Jose, S.: Thumbs up? sentiment classification using machine learning techniques. Lang. (Baltim), 79–86 (2002)
19. Pedregosa, F., Varoquaux, G.: Scikit-learn: machine learning in Python. J. Mach. Learn. Res. **12**, 2825–2830 (2011)
20. Poria, S., Cambria, E., Winterstein, G., Huang, G.: Bin: Sentic patterns: dependency-based rules for concept-level sentiment analysis. Knowl. Based Syst. **69**, 45–63 (2014)
21. Poria, S., Cambria, E., Gelbukh, A.: Aspect extraction for opinion mining with a deep convolutional neural network. Knowl. Based Syst. **108**, 42–49 (2016)
22. Refaee, E., Rieser, V.: An Arabic Twitter corpus for subjectivity and sentiment analysis. In: Proceedings of the Ninth International Conference on Language Resources and Evaluation, pp. 2268–2273 (2014)
23. Sarikaya, R., Kirchhoff, K., Schultz, T., Hakkani-Tur, D.: Introduction to the special issue on processing morphologically rich languages. IEEE Trans. Audio Speech Lang. Process. **17**, 861–862 (2009)
24. Al Shboul, B., Al-Ayyouby, M., Jararwehy, Y.: Multi-way sentiment classification of Arabic reviews. In: 2015 6th International Conference on Information and Communication Systems, ICICS 2015, pp. 206–211. IEEE (2015)

Load and Price Forecasting Based on Enhanced Logistic Regression in Smart Grid

Aroosa Tahir[1], Zahoor Ali Khan[2], Nadeem Javaid[3(✉)], Zeeshan Hussain[2], Aimen Rasool[2], and Syeda Aimal[2]

[1] Sardar Bhadur Khan Women University Quetta, Quetta 87300, Pakistan
[2] Computer Information Science, Higher Colleges of Technology, Fujairah 4114, UAE
[3] COMSATS University Islamabad, Islamabad 44000, Pakistan
nadeemjavaidqau@gmail.com
http://www.njavaid.com/

Abstract. Smart Grid (SG) is a modern electricity grid that enhance the efficiency and reliability of electricity generation, distribution and consumption. It plays an important role in modern energy infrastructure. Energy consumption and generation have fluctuating behaviour in SG. Load and price forecasting can decrease the variation between energy generation and consumption. In this paper, we proposed a model for forecasting, which consists of feature selection, extraction and classification. With the combination of Fast Correlation Based Filter (FCBF) and Recursive Feature Elimination (RFE) is used to perform feature selection to minimize the redundancy. Furthermore, Mutual Information technique is used for feature extraction. To forecast electricity load and price, we applied Naive Bayes (NB), Logistic Regression (LR) and Enhanced Logistic Regression (ELR) techniques. Our proposed technique ELR beats other techniques in term of forecasting accuracy of load and price. The load forecasting accuracy of ELR, LR and NB techniques are 80%, 82% and 85%, while price forecasting accuracy are 78%, 81% and 84%. Locational Marginal Price of Pennsylvania, Jersey, Maryland (LBM-PJM) market data used in our proposed model. Forecasting performance is assessed by using RMSE, MAPE, MAE and MSE. Simulation results show that our proposed technique ELR is perform better than other techniques.

Keywords: Smart Grid · Naive Bayes · Logistic Regression · Fast Correlation Based Filter · Recursive Feature Elimination · Mutual Information

1 Introduction

Traditional Grid (TG) is an interconnection of many elements such as transmission lines, distribution lines and different types of load. They are located far from power consumption areas, so power consumption transmitted from long

L. Barolli et al. (Eds.): EIDWT 2019, LNDECT 29, pp. 221–233, 2019.
https://doi.org/10.1007/978-3-030-12839-5_21

transmission lines. Traditional power grid is old version of Smart Grid (SG). SG precisely manage distribution, consumption and generation of energy which introduce communication and control of technology in power grid [1]. Smart grid helps the consumers to use electricity in secure manner. The new generation is attached against the SG due to extensive shortage of energy during summer. It suggest electricity consumers for load transferring and storage that can overcome their power consumption cost [2]. SG creates connection between consumer and utility. Customer consumes operation in SG for decreasing price by load shifting and energy storage. Consumer can manage their demand of energy on Demand Side Management (DSM) according to price variation [3]. Customer has choice to minimize their electricity cost through energy control and shifting the load. Consumer can switch the load and off depending on the electricity pricing. Competitive market takes advantage from load and price forecasting. Many valuable operational decisions are depends on load predication such as power reliability analysis, maintenance planning and generation scheduling. Consumer takes part in operation of SG that shifts the load from on-peak hours to off-eak hours.

In this paper, we highlight the electricity load and price problems, to solve these problems we used many techniques. Machine learning methods works excellent on data analytics, they categorize the data into training and testing step by step. Feature selection is an important operation for classification and selects important features. However, Feature selection have many common problems in the process is that redundancy of selected feature is not minimizing. Fast Correlation Based Filter (FCBF) and Recursive Feature Elimination (RFE) are used as a solution of feature selection problem. Mutual Information (MI) feature extraction method is used for feature reduction. We use two techniques Naive Bayes (NB) and Logistic Regression (LR) classifiers that have better capability of non-linear estimation and self learning. We proposed Enhance (ELR) classifier that suitable for load and price forecasting. These methods are capable for extract the complex data representation with better prediction. ELR has better computation power when comparing it with LR. The objective of our paper is to predict Load and price accurately using data analytics in the smart grid. For this problem we use NB, LR and ELR to predict the best results of electricity load and price forecasting.

1.1 Motivation

After reviewing papers [4,5], existing forecasting method have following motivations:

- Data analytics is not taken in the discussion of load and price forecasting method. Performance Matrix is taken on the electricity price data which is not enough large that minimize prediction.
- Intelligent methods such as ANN, SVM, DE and DWT have limited generalization capacity, they have fitting problem.
- Non-linear and other patterns of energy price is critical to predict with traditional data. The usage of data analytics make it feasible that generates the pattern of data and predicts accuracy.

- Automatically feature selection and extraction can efficiently extract and select useful hidden patterns in the data (Table 1).

Table 1. List of abbreviations

Abbreviation	Full form
ARMA	Auto Regressive Moving Average
AI	Artificial Intelligence
ANN	Artificial Neural Network
DSM	Demand Side Management
DE	Differential Evaluation
DWT	Discrete Wavelet Transformation
ELR	Enhanced Logistic Regression
FCBF	Fast Correlation based Filter
LR	Logistic Regression
MI	Mutual Information
MAPE	Mean Average Percentage Error
MSE	Mean Square Error
MAE	Mean Absolute Error
MLR	Multi Linear Regression
NB	Naive Bayes
RFE	Recursive Feature Elimination
RMSE	Root Mean Square Error
SG	Smart Grid
SVM	Support Vector Machine
SNN	Shallow Neural Network
SVR	Support Vector Regression
TG	Traditional Grid

1.2 Problem Statement

There is high variation between energy generation and consumption [6]. There is no proper strategy of energy generation, which leads to extra energy generation. To avoid spare energy generation, we perform forecasting. With the help of load and price forecasting, generation of energy can be controlled with in the limits.

1.3 Contributions

This paper have major contributions such as:

- We implement a framework for achieving accurate load and price forecasting by using the data analytics in SG.

- Feature selection, extraction and classification is proposed in our model to solve the addressed problem.
- In feature selection, we combine FCBF and RFE which gives us importance of feature selection. MI is used for feature extraction.
- Naive Bayes and LR classifiers are used for forecasting.
- However, we enhance LR to beat these two techniques.
- We have use the real world energy load and price data which performs better simulation that have better results.

2 Related Work

In the paper [1] Data Analytics Demand response technique propose for residential load. Data analytics scheme tested on PJM and Open Energy Information and show that it minimizing the load. In this paper author forecast day-ahead price forecasting, which contain six year long dataset and four auto regressive expert model transformed price [2]. ANN present for price forecasting, the focus is on the selection and preparation of fundamental data that has effect on electricity price [4]. In this paper [5], author admit the effect of data integrity attacks on the accuracy of four representative load forecasting model such as Multiple Linear regression (MLR), Support Vector Regression (SVR), ANN and Fuzzy Integration Regression (FIR). These models used to generate one-year-ahead forecasting to provide the comparison of their forecast error. The concepts of interaction for feature selection is introduce as Mutual Information (MI) and Interaction Gain (IG). These techniques measure the feature relevancy and redundancy, measures the load and price by merging them [6]. In [7] author reduce the energy purchase from wholesale market at the high price with operating Behind The Meter Storage System (BESS). MI use for feature selection and Intra Rolling Horizon (IRH) and Pre-Dispatch Price use for prediction. Forecasting the electricity price, three models are used to merge the Random Forest (RF) and Relief-F algorithm based on Gray Correlation Analysis (GCA) for feature selection to eliminate the feature redundancy [8].

Kernel Function and Principle Component Analysis (KPCA) used for feature extraction. Author use Differential Evaluation (DE) for price forecasting, which based on Support Vector Machine (SVM) classifier. Author hybrid the Discrete Wavelet Transformation (DWT), Empirical Mode Decomposition (EMD) and Random Vector Functional Link Network (RVFL) Techniques to forecast the short term load [9]. In paper [10], filter the feature by Stacked De-noising Auto-Encoder (SDAs) from energy load to train the SVR for prediction. The day-ahead load forecasting compare SVR and ANN, validates the performance for improvement. Author combine the Kernel Extreme Learning Machine (KELM) based on self-adapting (SAPSO) and (ARMA) [11]. Their experimental result show that developed method has more accurate prediction. Multi variation MI and SVR used for feature selection and price prediction [12].

These two methods also measure relation between price and load. Index bad sample matrix (IBSM), Optimize Algorithm (OA) and Dynamic Choosing (DCANN) are proposed for day-ahead price forecasting [13]. Proposed new feature strategy for short term load forecasting [17], uses for feature selection are MI and Neural Network (NN) which remove unnecessary and redundant candidates input. Data analysis is the way of predicting future value, if future stock price will decreases or increases [18]. To predict the future price used prediction

Table 2. Summary of related work

Techniques used	Objectives	Dataset	Achievement	Limitation
DA [1]	Peak load Reduction by analyzing the consumption data of SH	PJM and open energy information	Load forcasting	Trading of user satisfaction in response to the price for further boost the saving
MAP, SVT [2]	Price prediction	EPEX, PJM, Nord Pool	Electricity price forecasting	Redundancy in feature are not describe
ANN [4]	Develop ANN model for electricity forecasting from existing ANN model	European Power Exchange (EPEX)	Price forecasting	Measured on historical data and represented by connection weight
MLR, SVR, ANN [5]	Retrive load forecasting	Global electricity forcasting competition 2012	Load fecasting	Data intergrated attack in other similer forecasting feild
NN, SCA, MI [6]	Short term Price prediction	PJM	Price forecasting	Redundant feature not removed
IRH, MI [7]	Efficient for price forecasting	Ontario's electricity market	Price	Redundant feature not removed
DWT, EMI, RVFL [8]	Load prediction of time series	AEMO	Load forecasted	No good results for long term data
RF, Relief-f, GCA, KCPA, DE-SVM [9]	Price prediction	HSEC	Price forecasted	KPCA cannot model non-linear data
SDAs, SVR, ANN [10]	Compare SVR and ANN and forecast load	City or State of U.S	Load forecasting	Feature reduction do not remove
SAPSO, ARMA, KELM [11]	Price forecasting	PJM	Price forecasted	Big data is not used
IG, MI [12]	Best Feature selection	PJM	Price forecasted	More reduction of features, decreases the accuracy score
Multi variate MI, SVR [13]	Feature selection and price prediction	AEMO	Measure relation b/w price and load, Price forecasted	Redundancy in features not eliminated
IBSM, OA, DCANN [14]	Find the best parameter	PJM, AEM	Price forecasting	Limited for both dataset
NB, ANN [17]	Predict future Price using prediction concept	Logs from yahoo finance and store in stock market	Price forecasting	This method can not be use for big data
BR, NN, TB, GPR, MLR [18]	Predict future cooling load	WSHP	Load forecasting	Error arise because of weather change

concept. NB classifier use for prediction concept. Six performance indicators were made for use to assess the prediction performance of 6 models [19]. In simulation results display that precision of TB, GPR, NN, MLR and MAPE are compare for load forecasting (Table 2).

3 System Model

Our proposed system consist of many steps such as: preprocessing of data and splitting by training and testing, feature selection by RFE and FCBF, Feature extraction by MI, and forecasting of load and price oerformed by three techniques NB, LR and ELR shown in Fig. 1.

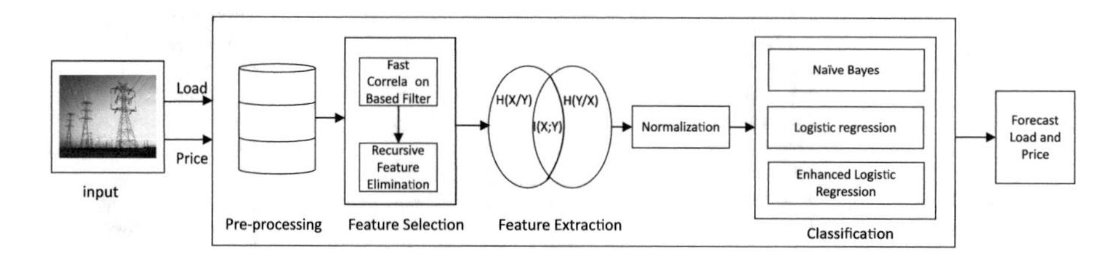

Fig. 1. System model

3.1 Preprocessing

For electricity load and price forecasting, it is necessary to collect load and price data that reflect the power usage of real world. In propose system model, Locational Marginal Price Pennsylvania, Jersey, Maryland (LBMP_PJM) 2015 to 2017 market dataset is used for load and price forecasting [20]. Three years data is used and divided month wise, i.e. January 2015, January 2016, January 2017 etc. this dataset is a regional transmission organization, which is managed by system operator, also responsible for whole sale energy market operations. Data is categorized into three steps as training, validation, and testing. First three weeks of month are used as training and last week for testing.

3.2 Methodology

We present two techniques RFE and FCBF method for feature selection. RFE removes the weakest feature until the specific number of feature is reached. FCBF works sequentially, where one feature is selected and then unusable features are removed by the selected one. FCBF is not use as single algorithm, however we combine other classifier with RFE for feature selection. MI that measure the dependency between two variable train for each data, extract the selected feature and remove the redundancy. After that Feature selection and extraction have normalized data. The output of all features are combined and classified by

NB and LR, when all weakest features are removed until the specified feature is received and then formulate final prediction. We enhance LR to get better result. Proposed model procedure can be concluded as:

- RFL and FCBF use for select the feature and MI to extract each feature into several inputs.
- NB and LG used as classifier for each obtained sub-series.
- ELR used to comparison with these classifiers.
- The load data is split and normalized into categories as train, validation and test.
- Training data is used for train the network and tested on validation data.
- Errors of prediction are calculated on data for validation.
- When network is tuned then update on new data.
- The network is tested on data, and predict the load and price. Prediction performance is performed by RMSE, MAPE, MAE, and MSE.

4 Description of Forecasting Techniques

Forecasting the load and price many techniques use such as RFE, MI, FCBF, NB and LR. Description of these techniques are as follows:

4.1 RFE

RFE is feature selection method that fits the model and remove the weakest feature until specific features are receive. RFE used to find best number of feature, cross validation and calculate the different feature subset and select the best collection of feature. Feature is selected in RFE by resourcefully considering smaller set of feature. Firstly RFE model fits on data, after that we have many features and its importance. We drop the feature with least importance, and then our model fits on the remaining feature. Process is repeated many times until we get best feature.

4.2 MI

MI used to detect the most applicable feature with less inessential information and measure the quality between two random variables [14]. Two random variable are a and b, which can be explained the information of b that we get by studying a. a and b denoted by (a; b) as continuous variable. Defined as joint probability distribution P (a; b) and individual probability distribution P(a) and P(b).

$$SC = a_1, a_2, a_3 ... a_n \tag{1}$$

Where SC is set of conditional value and b is variable for forecasting. Applicability of each input variable a_i with target variable b, which is denoted by Q(a_i).

$$Q(a_i) = |(a_i; b)| \tag{2}$$

For example, if a and b are not dependent to each other, then b does not give any information about a. so their mutual information can be zero.

4.3 FCBF

FCBF method is used for feature selection. Feature selection involves discrete expression before calculating the feature. FCBF have two categories filter and wrapper method. Filter method depend on the training data that select the features without including the training data. The wrapper methods depend upon the proposed algorithm in the feature selection and used to calculate and determine that which feature is selected. In filter method, use learning algorithm from feature selection is RFE which work with SVM and use efficiently because it show the dependency of them. In wrapped method, we use MI to guide the search process to weight the feature.

4.4 NB

Naive Bayes is classification technique depend on bayes theorem. NB classifier assume that the existence of specific feature in a class is separated to the existence of other feature. Bayes algorithm is used to calculate the posterior property. NB easy to construct and useful for very large dataset. NB also perform well in multi class forecasting. When assumption independent holds, NB perform better than other model such as logistic regression.

4.5 LR

Logistic Regression used for evaluate the dataset in which there are many variable that regulate an outcome. The outcome is measure with binary variable in which there are only two dependent variables on outcome (true or false).

Algorithm 1. Algorithm of Naive Bayes.

Require: Input: [Training data P]
1: $(F = (F_1, F_2, F_3......F_n))$ value of predict variation in the test data
2: **Output:** [Class of testing dataset]
3: Read the Training data P
4: Calculate the mean and Standard derivation
5: Predict the variable in each class
6: **for** $i = 1$ to *Population* **do**
7: Calculate the probability of Fi using distribution of each class;
8: Until the probability of all predictor variable;
9: $(F_1, F_2, ...F_n)$ has been calculated;
10: **end for**
11: **end for**
12: Calculated the probability of class
13: achieved the greatest result

LR generates combination of formulas to predict the logit transformation of the probability that being the characteristic of interest. LR algorithm use as performance baseline because easy to implement many task.

Algorithm 2. Algorithm of Logistic Regression.

Require: Input: [Initialize the prediction as $t = 0$ to $w = 0$]
1:
2: **for** $t = 1, 2, 3, ...n$ **do**
3: Compute the prediction gradient for (X_i);
4: For each example (X_i, Y_i);
5: For each non_Zero feature of Xi with index (X_j);
6: **if** j is not w, set w[j]=0 **then**
7: set $w[j] = w[j] + ((Y_i - P_i)X_j)$;
8: **end if**
9: **end if**
10: Update F
11: **end for**
12: **end for**
13: iterate to the next step until it is time to stop
14: **Output:** [of the feature is W.]

5 Results

Simulation results performed by python, daily electricity load data of LBM PJM are taken of three year as input data for simulator. To achieve the accuracy of load and price forecasting we are using three years electricity load and price data. FCBF and RFE are used for feature selection. We apply these two techniques which give us feature importance, control and grade of our feature. Every feature in feature selection have sequence like vector. After feature selection we apply MI for feature extraction. Mutual information measure the dependency between two variables that used for training the data. To remove the redundancy of selected feature we use MI. Feature that has effect on exact point are removed. After that load and price data is normalized, Then we apply NB, LR and ELR classifier on

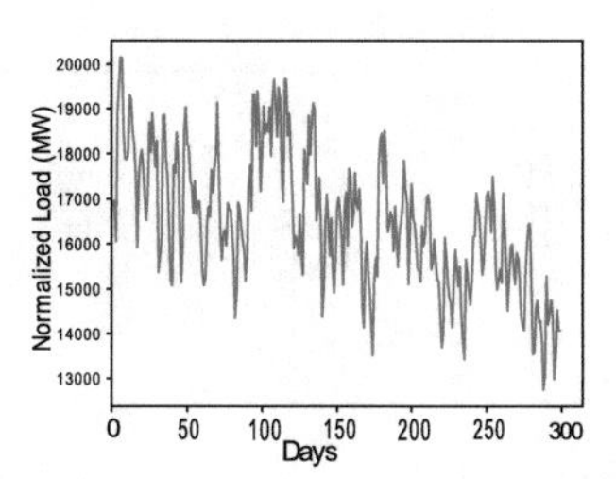

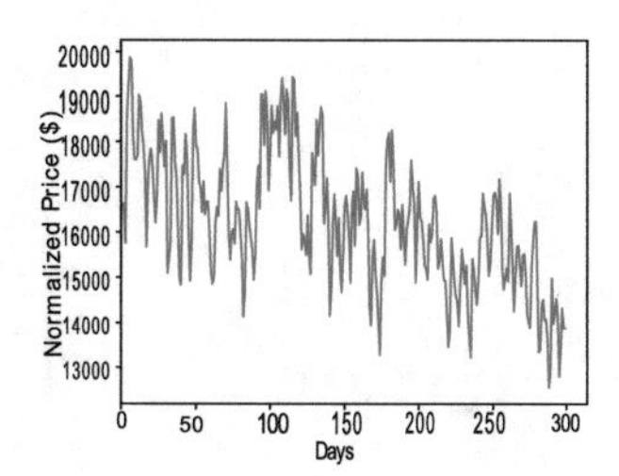

Fig. 2. Normalized load and price

load and price data. Data is divided into month wise and split into categories i.e. train, validate and test. After that network is tuned and validated, then network is tested the data. When we test the data then load and price is forecast.

5.1 Load Forecast

Firstly load data is used for training the forecast model. Data is splitting into training and testing in which training is 225% and testing is 75%. Figure 2 shows normalized load and price forecasting. Figure 3(a) Show one week prediction, Fig. 3(b) Show one month prediction and Fig. 3(c) show all nine months (January 2015 to March 2015, January 2016 to March 2016 and January 2017 to March 2017) prediction. All the similar months data are trained in the same pattern. we are comparing NB, LR with ELR for better prediction. Accuracy of NB is 80%, LG 82% and ELR 85% shows that our proposed technique ELR beats other two techniques.

5.2 Price Forecast

Price data is taken to forecast the price. Figure 4(a) Show one week prediction, Fig. 4(b) Show one month prediction and Fig. 4(c) show all nine months (January 2015 to March 2015 etc.) prediction. All the similar months of the data are trained in the same pattern. These graphs show that ELR performing better than NB and LR. Accuracy of these techniques show that NB has 78%, LG 81% and ELR 84%.

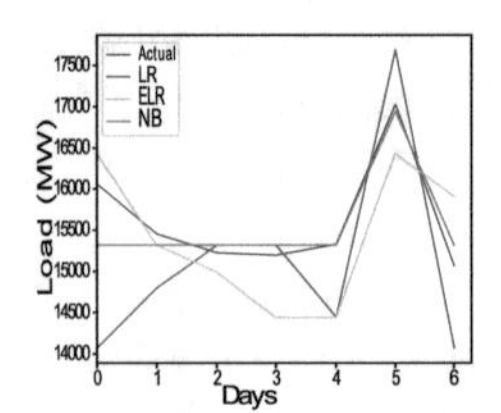

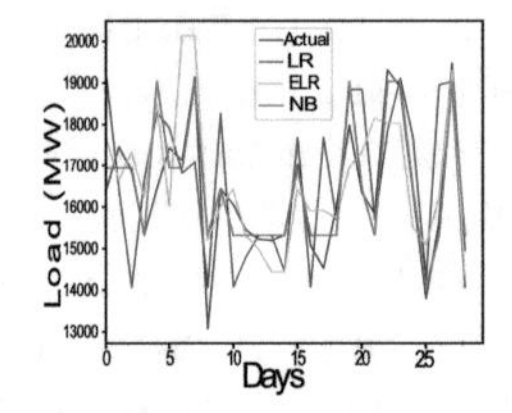

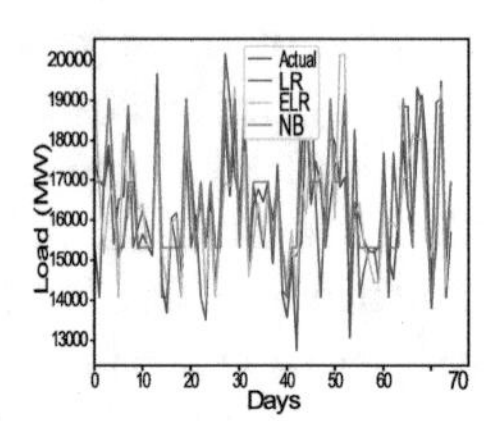

Fig. 3. Fig (a) shows one week prediction, fig (b) shows one month prediction, fig (c) shows nine months prediction

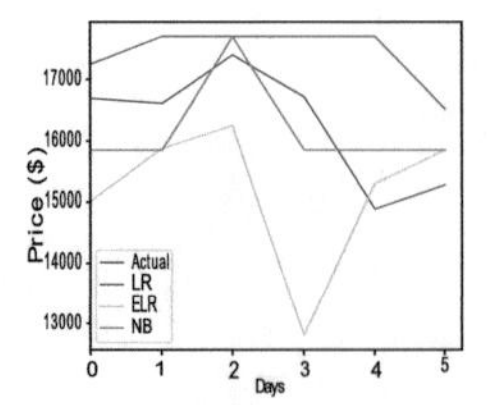

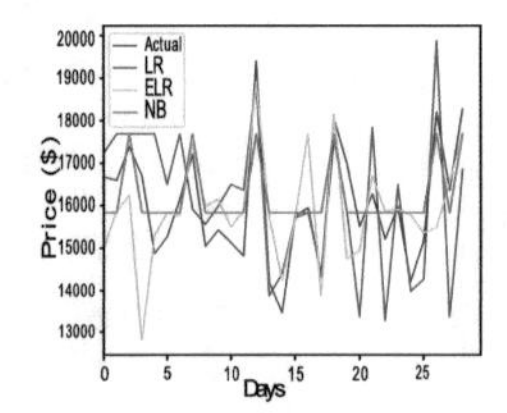

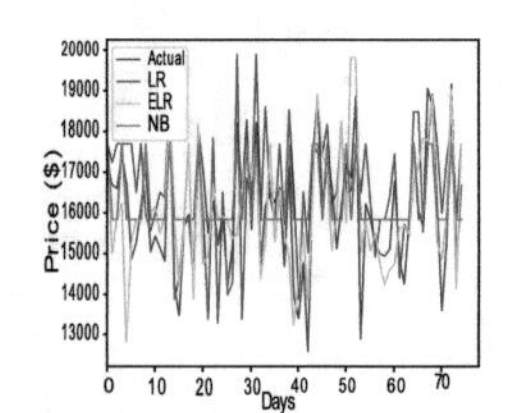

Fig. 4. Fig (a) shows one week prediction, fig (b) shows one month prediction, fig (c) shows nine month prediction

6 Performance Evaluation

To measure the performance of load and price four indicators are used as: Root Mean Square Error (RMSE), Mean Average Percentage Error (MAPE), Mean Square Error (MSE) and Mean Absolute Error (MAE). Figure 5(a) shows the comparison of load prediction and Fig. 5(b) shows the comparison of price prediction. MSE has the lowest error value in ELR i.e., 0.784 in price forecasting and 0.723 in load forecasting. Formulas of MAPE, RMSE MSE and MAE are given in Eqs. 3, 4, 5 and 6.

$$MAPE = \frac{1}{T}\sum_{n=1}^{T}|(s - Y_s)| \tag{3}$$

$$RMSE = \sqrt{\frac{1}{T}\sum_{tm=1}^{TM}(Av - Fv)^2} \tag{4}$$

$$MSE = \frac{1}{T}\sum_{tm=1}^{TM}(Av - Fv)^2 \tag{5}$$

$$MAE = \frac{\sum_{n=1}^{N}|(F_v - A_v)|}{N} \tag{6}$$

Where A_V is test value at time t and F_v is predicted value at time t. MSE has less error than MSE, MAE and RMSE. Experimental results show that ELR have better because it shows less error than other performance matrics.

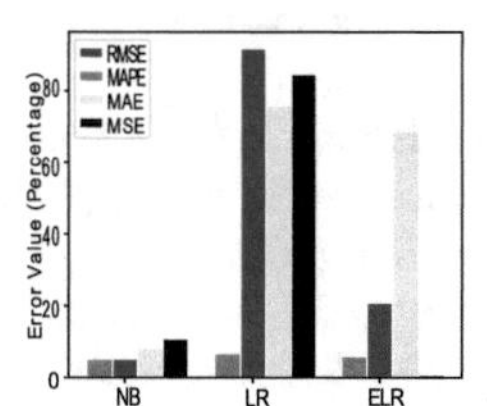

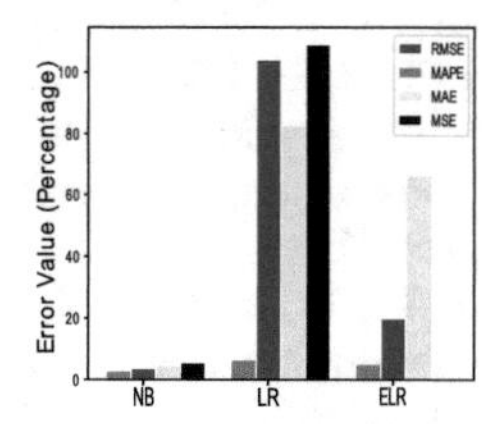

Fig. 5. Error value comparison of load and price

7 Conclusion

In this paper, three year LBM-PJM 2015 to 2017 market data is taken for forecating the load and price. The proposed model comprises form data preprocessing, selection, extraction and classification. We combine FCBF and RFE for feature selection and MI for Extraction. After selection of our selected feature, we forecast load and price with NB, LR and ELR techniques. We compare our proposed technique ELR with NB and LR to get better result. Experimental results prove the effectiveness of proposed ELG technique in forecasting. ELR beats other techniques in term of forecasting accuracy. ELR, LR and NB techniques accuracy is 84%, 82% and 80%. Numerical results show that ELG based forecasting model has lesser in MSE and MAPE than NB and LG. The feasibility of proposed ELR model is confined by its performance that is well known in our data.

References

1. Jindal, A., Singh, M., Kumar, N.: Consumption aware data analytical demand response scheme for peak load reduction in smart grid. IEEE Trans. Ind. Electron. **65**, 8993–9004 (2018)
2. Amjady, N., Keynia, F.: Day-ahead price forecasting of electricity markets by a new feature selection algorithm and cascaded neural network technique. Energy Convers. Manag. **50**(12), 2976–2982 (2009)
3. Huang, D., Zareipour, H., Rosehart, W.D., Amjady, N.: Data mining for electricity price classification and the application to demand-side management. IEEE Trans. Smart Grid **3**(2), 808–817 (2012)
4. Keles, D., Scelle, J., Paraschiv, F., Fichtner, W.: Extended forecast methods for day-ahead electricity spot prices applying artificial neural networks. Appl. Energy **162**, 218–230 (2016)
5. Luo, J., Hong, T., Fang, S.C.: Benchmarking robustness of load forecasting models under data integrity attacks. Int. J. Forecast. **34**(1), 89–104 (2018)
6. Wang, K., Xu, C., Zhang, Y., Guo, S., Zomaya, A.: Robust big data analytic for electricity price forecasting in the smart grid. IEEE Trans. Big Data (2017)
7. Chitsaz, H., Zamani-Dehkordi, P., Zareipour, H., Parikh, P.: Electricity price forecasting for operational scheduling of behind-the-meter storage systems. IEEE Trans. Smart Grid **9**, 6612–6622 (2017)
8. Qiu, X., Suganthan, P.N., Amaratunga, G.A.: Ensemble incremental learning random vector functional link network for short-term electric load forecasting. Knowl.-Based Syst. **145**, 182–196 (2018)
9. Ahmad, A., Javaid, N., Guizani, M., Alrajeh, N., Khan, Z.A.: An accurate and fast converging short-term load forecasting model for industrial applications in a smart grid. IEEE Trans. Industr. Inf. **13**(5), 2587–2596 (2017)
10. Tong, C., Li, J., Lang, C., Kong, F., Niu, J., Rodrigues, J.J.: An efficient deep model for day-ahead electricity load forecasting with stacked denoising auto-encoders. J. Parallel Distrib. Comput. **117**, 267–273 (2018)
11. Yang, Z., Ce, L., Lian, L.: Electricity price forecasting by a hybrid model, combining wavelet transform, ARMA and kernel-based extreme learning machine methods. Appl. Energy **190**, 291–305 (2017)
12. Abedinia, O., Amjady, N., Zareipour, H.: A new feature selection technique for load and price forecast of electrical power systems. IEEE Trans. Power Syst. **32**(1), 62–74 (2017)
13. Shi, H., Xu, M., Li, R.: Deep learning for household load forecasting—a novel pooling deep RNN. IEEE Trans. Smart Grid **9**(5), 5271–5280 (2018)

14. Wang, J., Liu, F., Song, Y., Zhao, J.: A novel model: dynamic choice artificial neural network (DCANN) for an electricity price forecasting system. Appl. Soft Comput. **48**, 281–297 (2016)
15. Wang, L., Zhang, Z., Chen, J.: Short-term electricity price forecasting with stacked denoising autoencoders. IEEE Trans. Power Syst. **32**(4), 2673–2681 (2017)
16. Ryu, S., Noh, J., Kim, H.: Deep neural network based demand side short term load forecasting. Energies **10**(1), 3 (2016)
17. Mahajan Shubhrata, D., Deshmukh Kaveri, V., Thite Pranit, R., Samel Bhavana, Y., Chate, P.J.: Stock market prediction and analysis using Naïve bayes. Int. J. Recent Innov. Trends Comput. Commun. **4**(11), 121–124 (2016)
18. Ahmad, T., Chen, H.: Short and medium-term forecasting of cooling and heating load demand in building environment with data-mining based approaches. Energy Build. **166**, 460–476 (2018)
19. Hawarah, L., Ploix, S., Jacomino, M.: User behavior prediction in energy consumption in housing using Bayesian networks, pp. 372–379. Springer, Heidelberg (2010). http://link.springer.com/10.1007/978-3-642-13208-747
20. Dataset. https://www.nyiso.com/public/marketsoperations/marketdata/custom-report/index.jsp

Effect of Peer Mobility for a Fuzzy-Based Peer Coordination Quality System in Mobile P2P Networks

Vladi Kolici[1](✉), Yi Liu[2], Keita Matsuo[3], and Leonard Barolli[3]

[1] Faculty of Information Technology, Polytechnic University of Tirana, Mother Theresa Square, No. 4, Tirana, Albania
vkolici@fti.edu.al

[2] Graduate School of Engineering, Fukuoka Institute of Technology (FIT), 3-30-1 Wajiro-Higashi, Higashi-Ku, Fukuoka 811-0295, Japan
ryuui1010@gmail.com

[3] Department of Information and Communication Engineering, Fukuoka Institute of Technology (FIT), 3-30-1 Wajiro-Higashi, Higashi-Ku, Fukuoka 811-0295, Japan
{kt-matsuo,barolli}@fit.ac.jp

Abstract. In this work, we present a distributed event-based awareness approach for P2P groupware systems. The awareness of collaboration will be achieved by using primitive operations and services that are integrated into the P2P middleware. We propose an abstract model for achieving these requirements and we discuss how this model can support awareness of collaboration in mobile teams. In this paper, we present a Fuzzy Peer Coordination Quality System (FPCQS) for P2P networks according to four parameters. We consider Peer Mobility (PM) as a new parameter. We evaluated the performance of proposed system by computer simulations. The simulation results show that when AA and GS are increased, the PCQ is increased. But, by increasing PCC, the PCQ is decreased. Comparing the complexity, by adding PM the proposed system is more complex than the system with three input parameters, but it has better peer coordination quality.

1 Introduction

Peer to Peer (P2P) technologies has been among most disruptive technologies after Internet. Indeed, the emergence of the P2P technologies changed drastically the concepts, paradigms and protocols of sharing and communication in large scale distributed systems. The nature of the sharing and the direct communication among peers in the system, being these machines or people, makes possible to overcome the limitations of the flat communications through email, newsgroups and other forum-based communication forms [1–5].

The usefulness of P2P technologies on one hand has been shown for the development of stand alone applications. On the other hand, P2P technologies, paradigms and protocols have penetrated other large scale distributed systems such as Mobile Ad hoc Networks (MANETs), Groupware systems, Mobile

L. Barolli et al. (Eds.): EIDWT 2019, LNDECT 29, pp. 234–246, 2019.
https://doi.org/10.1007/978-3-030-12839-5_22

Systems to achieve efficient sharing, communication, coordination, replication, awareness and synchronization. In fact, for every new form of Internet-based distributed systems, we are seeing how P2P concepts and paradigms again play an important role to enhance the efficiency and effectiveness of such systems or to enhance information sharing and online collaborative activities of groups of people. We briefly introduce below some common application scenarios that can benefit from P2P communications.

Awareness is a key feature of groupware systems. In its simplest terms, awareness can be defined as the system's ability to notify the members of a group of changes occurring in the group's workspace. Awareness systems for online collaborative work have been proposed since in early stages of Web technology. Such proposals started by approaching workspace awareness, aiming to inform users about changes occurring in the shared workspace. More recently, research has focussed on using new paradigms, such as P2P systems, to achieve fully decentralized, ubiquitous groupware systems and awareness in such systems. In P2P groupware systems group processes may be more efficient because peers can be aware of the status of other peers in the group, and can interact directly and share resources with peers in order to provide additional scaffolding or social support. Moreover, P2P systems are pervasive and ubiquitous in nature, thus enabling contextualized awareness.

Fuzzy Logic (FL) is the logic underlying modes of reasoning which are approximate rather then exact. The importance of FL derives from the fact that most modes of human reasoning and especially common sense reasoning are approximate in nature [6]. FL uses linguistic variables to describe the control parameters. By using relatively simple linguistic expressions it is possible to describe and grasp very complex problems. A very important property of the linguistic variables is the capability of describing imprecise parameters.

The concept of a fuzzy set deals with the representation of classes whose boundaries are not determined. It uses a characteristic function, taking values usually in the interval $[0, 1]$. The fuzzy sets are used for representing linguistic labels. This can be viewed as expressing an uncertainty about the clear-cut meaning of the label. But important point is that the valuation set is supposed to be common to the various linguistic labels that are involved in the given problem.

The fuzzy set theory uses the membership function to encode a preference among the possible interpretations of the corresponding label. A fuzzy set can be defined by exemplification, ranking elements according to their typicality with respect to the concept underlying the fuzzy set [7].

In this paper, we propose a Fuzzy Peer Coordination Quality System (FPCQS) considering four parameters: Activity Awareness (AA), Group Synchronization (GS), Peer Communication Cost (PCC) and Peer Mobility (PM) to decide the Peer Coordination Quality (PCQ). We evaluated the proposed system by simulations.

The structure of this paper is as follows. In Sect. 2, we introduce the group activity awareness model. In Sect. 3, we introduce FL used for control. In Sect. 4, we present the proposed fuzzy-based system. In Sect. 5, we discuss the simulation results. Finally, conclusions and future work are given in Sect. 6.

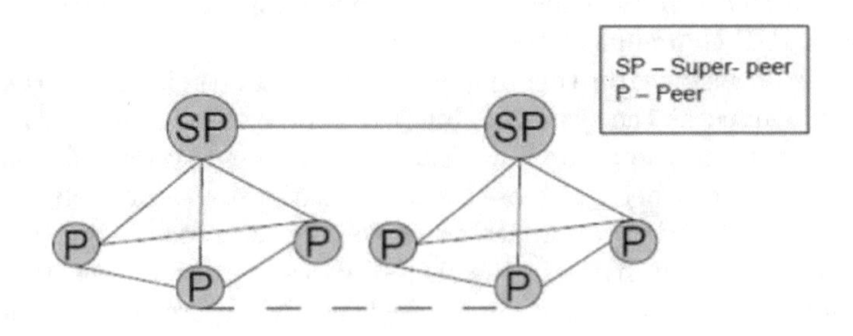

Fig. 1. Super-peer P2P group network.

2 Group Activity Awareness Model

The awareness model considered here focuses on supporting group activities so to accomplish a common group project, although it can also be used in a broader scope of teamwork [8–14]. The main building blocks of our model (see also [15,16] in the context of web-based groupware) are described below.

Activity Awareness: Activity awareness refers to awareness information about the project-related activities of group members. Project-based work is one of the most common methods of group working. Activity awareness aims to provide information about progress on the accomplishment of tasks by both individuals and the group as a whole. It comprises knowing about actions taken by members of the group according to the project schedule, and synchronization of activities with the project schedule. Activity awareness should therefore enable members to know about recent and past actions on the project's work by the group. As part of activity awareness, we also consider information on group artifacts such as documents and actions upon them (uploads, downloads, modifications, reading). Activity awareness is one of most important, and most complex, types of awareness. As well as the direct link to monitoring a group's progress on the work relating to a project, it also supports group communication and coordination processes.

Process Awareness: In project-based work, a project typically requires the enactment of a workflow. In such a case, the objective of the awareness is to track the state of the workflow and to inform users accordingly. We term this process awareness. The workflow is defined through a set of tasks and precedence relationships relating to their order of completion. Process awareness targets the information flow of the project, providing individuals and the group with a partial view (what they are each doing individually) and a complete view (what they are doing as a group), thus enabling the identification of past, current and next states of the workflow in order to move the collaboration process forward.

Communication Awareness: Another type of awareness considered in this work is that of communication awareness. We consider awareness information relating to message exchange, and synchronous and asynchronous discussion forums. The first is intended to support awareness of peer-to-peer communication (when some peer wants to establish a direct communication with another peer); the second is aimed at supporting awareness about chat room creation and lifetime (so that other peers can be aware of, and possibly eventually join, the chat room); the third refers to awareness of new messages posted at the discussion forum, replies, etc.

Availability Awareness: Availability awareness is useful for provide individuals and the group with information on members' and resources' availability. The former is necessary for establishing synchronous collaboration either in peer-to-peer mode or (sub)group mode. The later is useful for supporting members' tasks requiring available resources (e.g. a machine for running a software program). Groupware applications usually monitor availability of group members by simply looking at group workspaces. However, availability awareness encompasses not only knowing who is in the workspace at any given moment but also who is available when, via members' profiles (which include also personal calendars) and information explicitly provided by members. In the case of resources, awareness is achieved via the schedules of resources. Thus, both explicit and implicit forms of gathering availability awareness information should be supported.

3 Application of Fuzzy Logic for Control

The ability of fuzzy sets and possibility theory to model gradual properties or soft constraints whose satisfaction is matter of degree, as well as information pervaded with imprecision and uncertainty, makes them useful in a great variety of applications [17–23].

The most popular area of application is Fuzzy Control (FC), since the appearance, especially in Japan, of industrial applications in domestic appliances, process control, and automotive systems, among many other fields.

In the FC systems, expert knowledge is encoded in the form of fuzzy rules, which describe recommended actions for different classes of situations represented by fuzzy sets.

In fact, any kind of control law can be modeled by the FC methodology, provided that this law is expressible in terms of "if ... then ..." rules, just like in the case of expert systems. However, FL diverges from the standard expert system approach by providing an interpolation mechanism from several rules. In the contents of complex processes, it may turn out to be more practical to get knowledge from an expert operator than to calculate an optimal control, due to modeling costs or because a model is out of reach.

A concept that plays a central role in the application of FL is that of a linguistic variable. The linguistic variables may be viewed as a form of data compression. One linguistic variable may represent many numerical variables. It is suggestive to refer to this form of data compression as granulation.

The same effect can be achieved by conventional quantization, but in the case of quantization, the values are intervals, whereas in the case of granulation the values are overlapping fuzzy sets. The advantages of granulation over quantization are as follows:

- it is more general;
- it mimics the way in which humans interpret linguistic values;
- the transition from one linguistic value to a contiguous linguistic value is gradual rather than abrupt, resulting in continuity and robustness.

FC describes the algorithm for process control as a fuzzy relation between information about the conditions of the process to be controlled, x and y, and the output for the process z. The control algorithm is given in "if ... then ..." expression, such as:

If x is small and y is big, then z is medium;

If x is big and y is medium, then z is big.

These rules are called *FC rules*. The "if" clause of the rules is called the antecedent and the "then" clause is called consequent. In general, variables x and y are called the input and z the output. The "small" and"big" are fuzzy values for x and y, and they are expressed by fuzzy sets.

Fuzzy controllers are constructed of groups of these FC rules, and when an actual input is given, the output is calculated by means of fuzzy inference.

4 Proposed Fuzzy Peer Coordination Quality System

The P2P group-based model considered is that of a superpeer model as show in Fig. 1. In this model, the P2P network is fragmented into several disjoint peergroups (see Fig. 2). The peers of each peergroup are connected to a single superpeer. There is frequent local communication between peers in a peergroup, and less frequent global communication between superpeers.

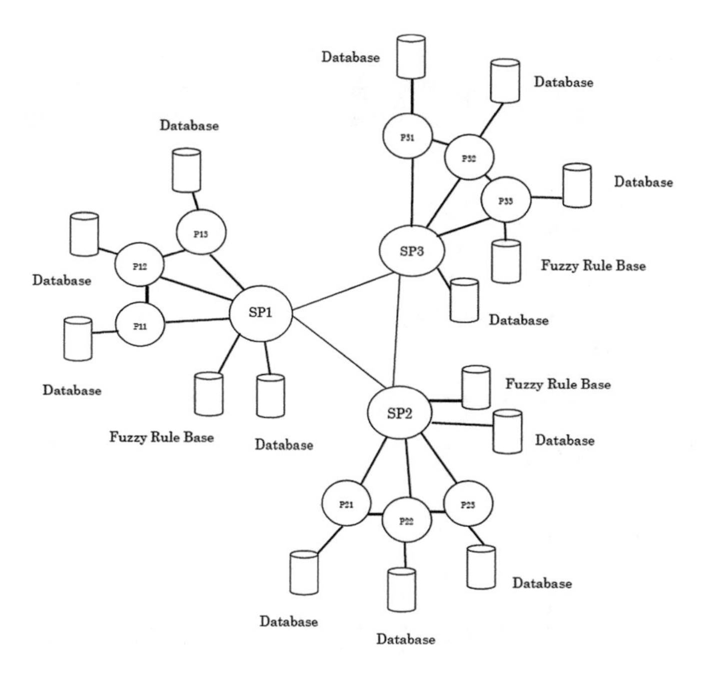

Fig. 2. P2P group-based model.

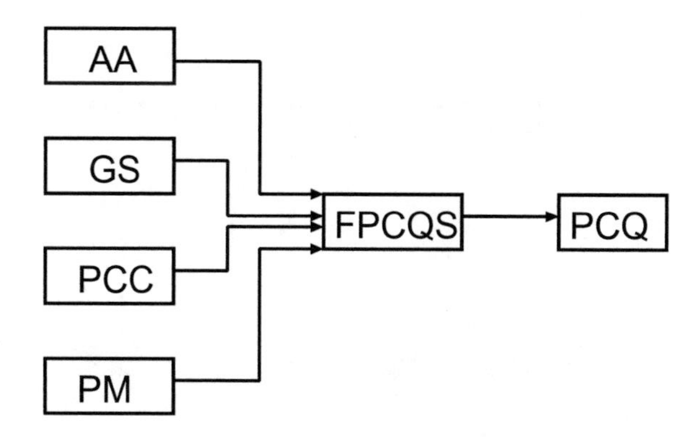

Fig. 3. Proposed system of structure.

To complete a certain task in P2P mobile collaborative team work, peers often have to in teract with unknow peers. Thus, it is important that group members must select reliable peers to interact.

In this work, we consider four parameters: Activity Awareness (AA), Group Synchronization (GS), Peer Communication Cost (PCC) and Peer Mobility (PM) to decide the Peer Coordination Quality (PCQ). The structure of FPCQS is shown in Fig. 3. These four parameters are fuzzified using fuzzy system, and based on the decision of fuzzy system the peer coordination quality is calculated. The membership functions for our system are shown in Fig. 4. In Table 1, we show the Fuzzy Rule Base (FRB) of our proposed system, which consists of 81 rules.

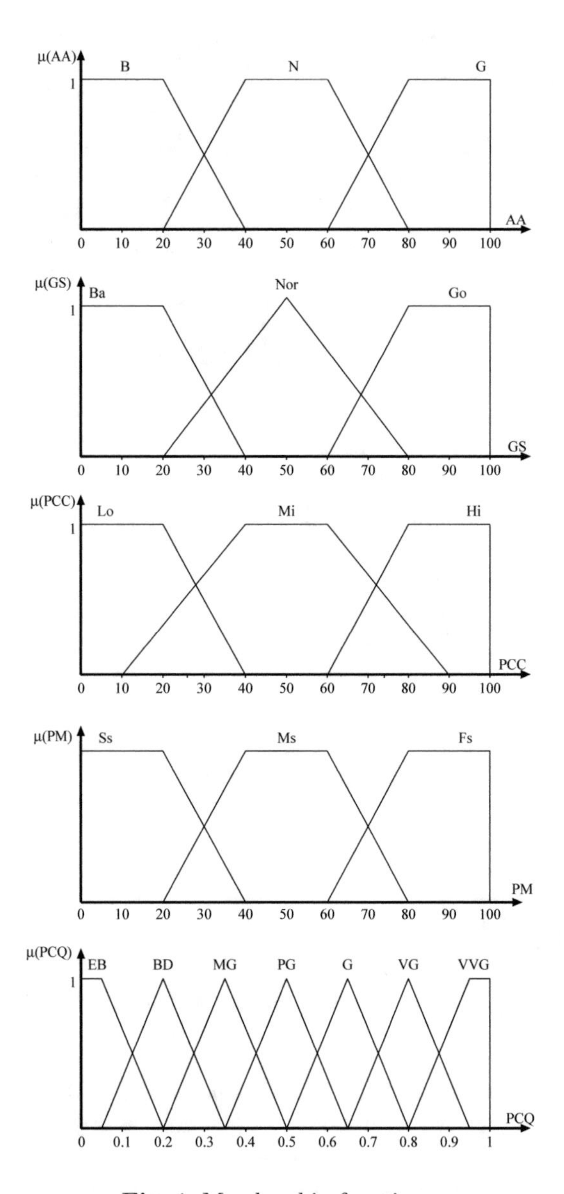

Fig. 4. Membership functions.

Table 1. FRB.

Rule	AA	GS	PCC	PM	PCQ	Rule	AA	GS	PCC	PM	PCQ	Rule	AA	GS	PCC	PM	PCQ
1	B	Ba	Lo	Ss	MG	28	N	Ba	Lo	Ss	PG	55	G	Ba	Lo	Ss	VG
2	B	Ba	Lo	Ms	G	29	N	Ba	Lo	Ms	VG	56	G	Ba	Lo	Ms	VVG
3	B	Ba	Lo	Fs	MG	30	N	Ba	Lo	Fs	PG	57	G	Ba	Lo	Fs	VG
4	B	Ba	Mi	Ss	EB	31	N	Ba	Mi	Ss	BD	58	G	Ba	Mi	Ss	PG
5	B	Ba	Mi	Ms	MG	32	N	Ba	Mi	Ms	PG	59	G	Ba	Mi	Ms	VG
6	B	Ba	Mi	Fs	EB	33	N	Ba	Mi	Fs	BD	60	G	Ba	Mi	Fs	PG
7	B	Ba	Hi	Ss	EB	34	N	Ba	Hi	Ss	EB	61	G	Ba	Hi	Ss	BD
8	B	Ba	Hi	Ms	EB	35	N	Ba	Hi	Ms	BD	62	G	Ba	Hi	Ms	PG
9	B	Ba	Hi	Fs	EB	36	N	Ba	Hi	Fs	EB	63	G	Ba	Hi	Fs	BD
10	B	Nor	Lo	Ss	PG	37	N	Nor	Lo	Ss	G	64	G	Nor	Lo	Ss	VG
11	B	Nor	Lo	Ms	VG	38	N	Nor	Lo	Ms	VVG	65	G	Nor	Lo	Ms	VVG
12	B	Nor	Lo	Fs	PG	39	N	Nor	Lo	Fs	G	66	G	Nor	Lo	Fs	VG
13	B	Nor	Mi	Ss	BD	40	N	Nor	Mi	Ss	MG	67	G	Nor	Mi	Ss	G
14	B	Nor	Mi	Ms	PG	41	N	Nor	Mi	Ms	G	68	G	Nor	Mi	Ms	VG
15	B	Nor	Mi	Fs	BD	42	N	Nor	Mi	Fs	MG	69	G	Nor	Mi	Fs	G
16	B	Nor	Hi	Ss	EB	43	N	Nor	Hi	Ss	BD	70	G	Nor	Hi	Ss	MG
17	B	Nor	Hi	Ms	BD	44	N	Nor	Hi	Ms	MG	71	G	Nor	Hi	Ms	G
18	B	Nor	Hi	Fs	EB	45	N	Nor	Hi	Fs	BD	72	G	Nor	Hi	Fs	MG
19	B	Go	Lo	Ss	G	46	N	Go	Lo	Ss	VG	73	G	Go	Lo	Ss	VVG
20	B	Go	Lo	Ms	VG	47	N	Go	Lo	Ms	VVG	74	G	Go	Lo	Ms	VVG
21	B	Go	Lo	Fs	G	48	N	Go	Lo	Fs	VG	75	G	Go	Lo	Fs	VVG
22	B	Go	Mi	Ss	MG	49	N	Go	Mi	Ss	G	76	G	Go	Mi	Ss	VG
23	B	Go	Mi	Ms	G	50	N	Go	Mi	Ms	VG	77	G	Go	Mi	Ms	VVG
24	B	Go	Mi	Fs	MG	51	N	Go	Mi	Fs	G	78	G	Go	Mi	Fs	VG
25	B	Go	Hi	Ss	EB	52	N	Go	Hi	Ss	MG	79	G	Go	Hi	Ss	PG
26	B	Go	Hi	Ms	MG	53	N	Go	Hi	Ms	G	80	G	Go	Hi	Ms	VG
27	B	Go	Hi	Fs	EB	54	N	Go	Hi	Fs	MG	81	G	Go	Hi	Fs	PG

The input parameters for FPCQS are: AA, GS, PCC and PM. The output linguistic parameter is PCQ. The term sets of AA, GS, PCC and PM are defined respectively as:

$$
\begin{aligned}
AA &= \{Bad,\ Normal,\ Good\} \\
&= \{B,\ N,\ G\}; \\
GS &= \{Bad,\ Normal,\ Good\} \\
&= \{Ba,\ Nor,\ Go\}; \\
PCC &= \{Low,\ Middle,\ High\} \\
&= \{Lo,\ Mi,\ Hi\}; \\
PM &= \{Slow\ Speed,\ Middle\ Speed,\ Fast\ Speed\} \\
&= \{Ss,\ Ms,\ Fs\}.
\end{aligned}
$$

and the term set for the output PCQ is defined as:

$$PCQ = \begin{pmatrix} Extremely\ Bad \\ Bad \\ Minimally\ Good \\ Partially\ Good \\ Good \\ Very\ Good \\ Very\ Very\ Good \end{pmatrix} = \begin{pmatrix} EB \\ BD \\ MG \\ PG \\ G \\ VG \\ VVG \end{pmatrix}$$

5 Simulation Results

In this section, we present the simulation results for our FPCQS system. In our system, we decided the number of term sets by carrying out many simulations.

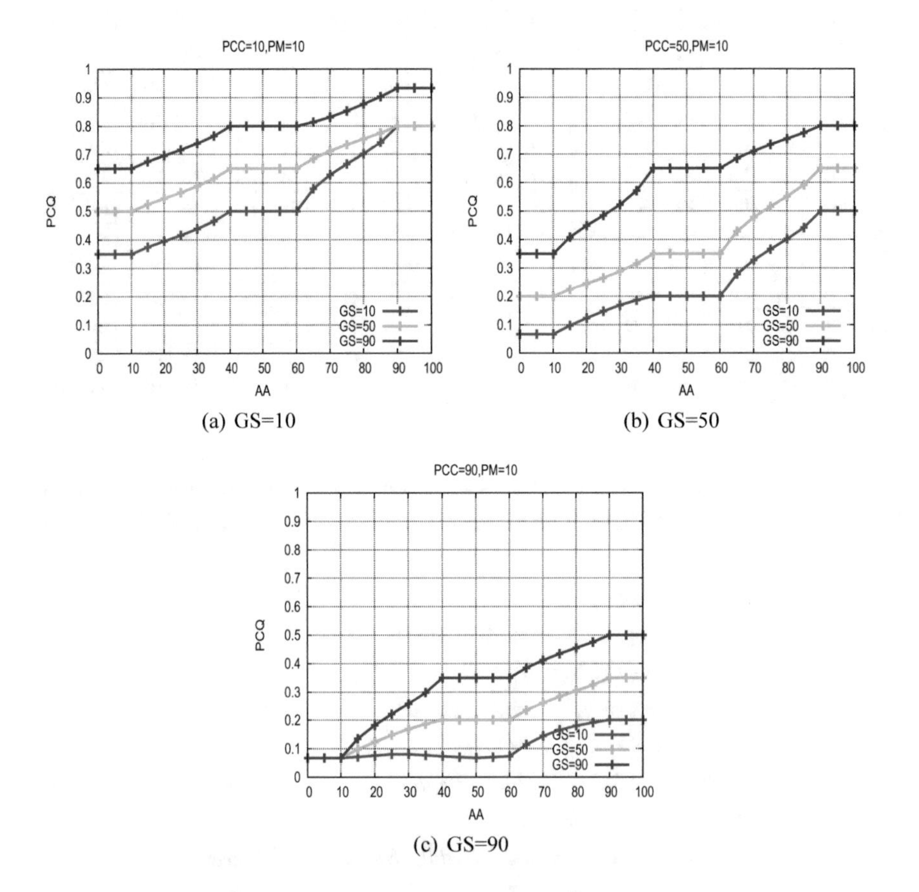

(a) GS=10

(b) GS=50

(c) GS=90

Fig. 5. Relation of PCQ with AA and GS for different PCC when PM = 10.

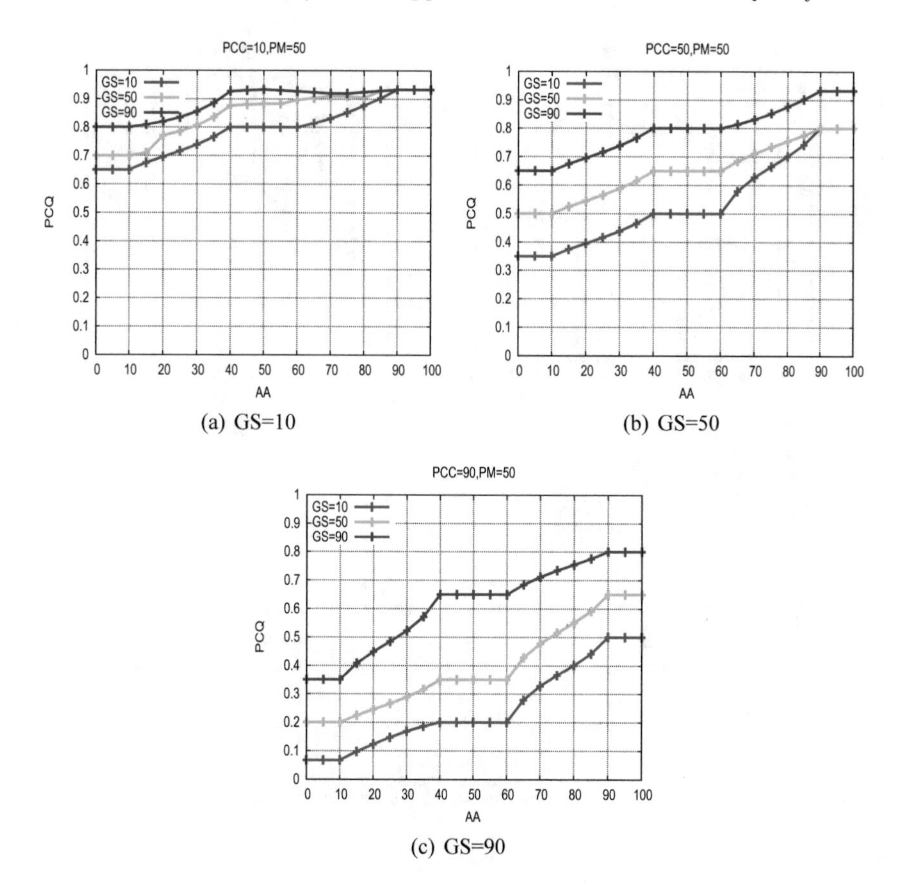

(a) GS=10

(b) GS=50

(c) GS=90

Fig. 6. Relation of PCQ with AA and GS for different PCC when PM = 50.

In Figs. 5, 6 and 7, we show the relation between PCQ and AA, GS, PCC and PM. In these simulations, we consider the PCC and PM as constant parameters. In Fig. 5, we consider the PCC value 10 units. We change the PCC value from 10 to 90 units. When the PCC increases, the PCQ is decreased. Also, when the AA and GS are increased, the PCQ value is high. In Figs. 6 and 7, we increase the PM values to 50 and 90 units, respectively. We see that, when the PM is 50 units, the PCQ is the best.

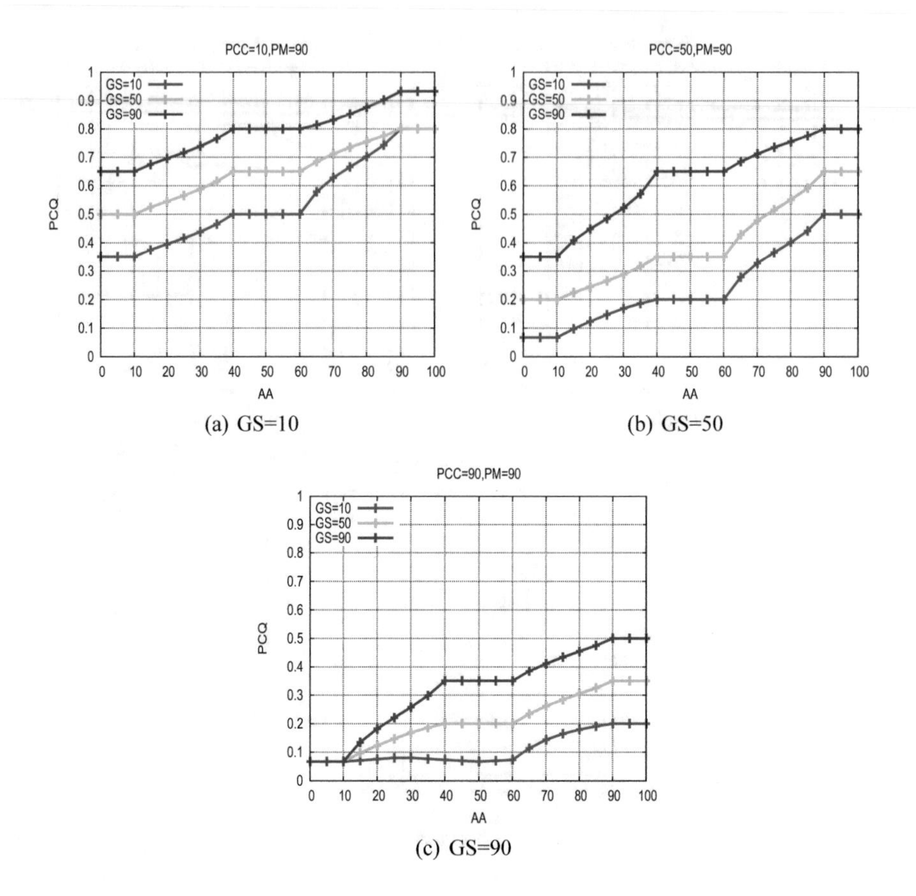

(a) GS=10

(b) GS=50

(c) GS=90

Fig. 7. Relation of PCQ with AA and GS for different PCC when PM = 90.

6 Conclusions and Future Work

In this paper, we proposed a FPCQS to decide the PCQ in P2P networks. We took into consideration four parameters: AA, GS, PCC and PM. We evaluated the performance of proposed system by computer simulations. From the simulations results, we conclude that when AA and GS are increased, the PCQ is increased. But, by increasing PCC, the PCQ is decreased. In [24], we considered three input parameters, while in this work we consider four input parameters. Comparing the complexity, by adding PM the system becomes more complex than the system with three input parameters, but it has better peer coordination quality.

In the future, we would like to make extensive simulations to evaluate the proposed systems and compare the performance with other systems.

References

1. Oram, A. (ed.): Peer-to-Peer: Harnessing the Power of Disruptive Technologies. O'Reilly and Associates, California (2001)
2. Sula, A., Spaho, E., Matsuo, K., Barolli, L., Xhafa, F., Miho, R.: A new system for supporting children with autism spectrum disorder based on IoT and P2P technology. Int. J. Space-Based Situated Comput. **4**(1), 55–64 (2014). https://doi.org/10.1504/IJSSC.2014.060688
3. Di Stefano, A., Morana, G., Zito, D.: QoS-aware services composition in P2PGrid environments. Int. J. Grid Util. Comput. **2**(2), 139–147 (2011). https://doi.org/10.1504/IJGUC.2011.040601
4. Sawamura, S., Barolli, A., Aikebaier, A., Takizawa, M., Enokido, T.: Design and evaluation of algorithms for obtaining objective trustworthiness on acquaintances in P2P overlay networks. Int. J. Grid Util. Comput. **2**(3), 196–203 (2011). https://doi.org/10.1504/IJGUC.2011.042042
5. Higashino, M., Hayakawa, T., Takahashi, K., Kawamura, T., Sugahara, K.: Management of streaming multimedia content using mobile agent technology on pure P2P-based distributed e-learning system. Int. J. Grid Util. Comput. **5**(3), 198–204 (2014). https://doi.org/10.1504/IJGUC.2014.062928
6. Inaba, T., Obukata, R., Sakamoto, S., Oda, T., Ikeda, M., Barolli, L.: Performance evaluation of a QoS-aware fuzzy-based CAC for LAN access. Int. J. Space-Based Situated Comput. **1**(1), 228–238 (2011). https://doi.org/10.1504/IJSSC.2016.082768
7. Terano, T., Asai, K., Sugeno, M.: Fuzzy Systems Theory and Its Applications. Academic Press, INC. Harcourt Brace Jovanovich, Publishers, Cambridge (1992)
8. Mori, T., Nakashima, M., Ito, T.: SpACCE: a sophisticated ad hoc cloud computing environment built by server migration to facilitate distributed collaboration. Int. J. Space-Based Situated Comput. **1**(1), 230–239 (2011). https://doi.org/10.1504/IJSSC.2012.050000
9. Xhafa, F., Poulovassilis, A.: Requirements for distributed event-based awareness in P2P groupware systems. In: Proceedings of AINA 2010, pp. 220–225, April 2010
10. Xhafa, F., Barolli, L., Caballé, S., Fernandez, R.: Supporting scenario-based online learning with P2P group-based systems. In: Proceedings of NBiS 2010, pp. 173–180, September 2010
11. Gupta, S., Kaiser, G.: P2P video synchronization in a collaborative virtual environment. In: Proceedings of the 4th International Conference on Advances in Web-Based Learning (ICWL 2005), pp. 86–98 (2005)
12. Martínez-Alemán, A.M., Wartman, K.L.: Online Social Networking on Campus Understanding What Matters in Student Culture. Taylor and Francis, Routledge, Abingdon (2008)
13. Puzar, M., Plagemann, T.: Data sharing in mobile ad-hoc networks - a study of replication and performance in the MIDAS data space. Int. J. Space-Based Situated Comput. **1**(1), 137–150 (2011). https://doi.org/10.1504/IJSSC.2011.040340
14. Spaho, E., Kulla, E., Xhafa, F., Barolli, L.: P2P solutions to efficient mobile peer collaboration in MANETs. In: Proceedings of 3PGCIC 2012, pp. 379–383, November 2012
15. Gutwin, C., Greenberg, S., Roseman, M.: Workspace awareness in real-time distributed groupware: framework, widgets, and evaluation. In: BCS HCI, pp. 281–298 (1996)

16. You, Y., Pekkola, S.: Meeting others - supporting situation awareness on the WWW. Decis. Support Syst. **32**(1), 71–82 (2001)
17. Kandel, A.: Fuzzy Expert Systems. CRC Press, Boca Raton (1992)
18. Zimmermann, H.J.: Fuzzy Set Theory and Its Applications. Kluwer Academic Publishers, Dordrecht (1991). Second Revised Edition
19. McNeill, F.M., Thro, E.: Fuzzy Logic. A Practical Approach. Academic Press, Inc., Cambridge (1994)
20. Zadeh, L.A., Kacprzyk, J.: Fuzzy Logic for the Management of Uncertainty. Wiley, Hoboken (1992)
21. Procyk, T.J., Mamdani, E.H.: A linguistic self-organizing process controller. Automatica **15**(1), 15–30 (1979)
22. Klir, G.J., Folger, T.A.: Fuzzy Sets, Uncertainty, and Information. Prentice Hall, Englewood Cliffs (1988)
23. Munakata, T., Jani, Y.: Fuzzy systems: an overview. Commun. ACM **37**(3), 69–76 (1994)
24. Liu, Y., Ozera, K., Matsuo, K., Ikeda, M., Barolli, L.: A fuzzy-based approach for improving peer coordination quality in MobilePeerDroid mobile system. In: Proceedings of IMIS 2018, pp. 60–73, July 2018

Load and Price Forecasting in Smart Grids Using Enhanced Support Vector Machine

Ammara Nayab[1], Tehreem Ashfaq[1], Syeda Aimal[1], Aimen Rasool[1], Nadeem Javaid[1(✉)], and Zahoor Ali Khan[2]

[1] COMSATS University Islamabad, Islamabad 44000, Pakistan
nadeemjavaidqau@gmail.com
[2] Computer Information Science, Higher College of Technology, Fujarah 4114, UAE
http://www.njavaid.com

Abstract. In this paper, an enhanced model for electricity load and price forecasting is proposed. This model consists of feature engineering and classification. Feature engineering consists of feature selection and extraction. For feature selection a hybrid feature selector is used which consists of Decision Tree (DT) and Recursive Feature Elimination (RFE) to remove redundancy. Furthermore, Singular Value Decomposition (SVD) is used for feature extraction to reduce the dimensionality of features. To forecast load and price, two classifiers Stochastic Gradient Descent (SGD) and Support Vector Machine (SVM) is used and for better accuracy an enhanced framework of SVM is proposed. Dataset is taken from NYISO and month wise forecasting is being conducted by proposed classifiers. To evaluate performance RMSE, MAPE, MAE, MSE is used.

1 Introduction

Accurate load and price forecasting are major issue in Smart Grids (SGs). However, accurate forecasting of electric load can improve the generation and consumption scheduling and decrease the peak load. With the growing demand of electricity, conventional power supply methods do not work effectively. A SG offers a promising solution to this rapid increase in electricity demand. Conventional power grids can only send electric power to one direction from a power plant to homes and offices however SGs enable information in two ways between suppliers and consumers. Through Demand-Response (DR) mechanism, user participation in energy savings and cooperation is encouraged. Another aim of the smart grid is to facilitate the public to participate in energy-saving programs through available information so as to control their energy consumption and costs.

Such actions also help utilities for estimating electric load and reducing peak load. DR can be referred as the changes made in the load/demand of users in response of policies adopted by utilities. Current DR schemes are implemented

L. Barolli et al. (Eds.): EIDWT 2019, LNDECT 29, pp. 247–258, 2019.
https://doi.org/10.1007/978-3-030-12839-5_23

with commercial as well as residential customers often through either time-based or incentive-based DR schemes [1,2]. In this paper Short Term Load Forecasting (STLF) is used. Which is essential for effective electricity management. STLF is used to forecast electric load from 1-hour to 1 week. A wide variety of models have been proposed for STLF due to its importance.

1.1 Applications of SGs

Basic applications of SGs are:

- Quick recovery after any sudden disturbance.
- Cost Reduction.
- Reduction in peak demand.
- Improvement in transmission lines.

1.2 Big Data Analytics

With the fast development of digital technology more and more data is being generated through digital equipments such as smart phones and computers. With the rapid increase in its generation the processing and analyzing of data has become much more complicated. Traditional electricity distribution system produces small data which can only be used for billing purposes. Now with the huge generated data because of the two way communication of smart grids needs advance data analytics to extract valuable information. There are several types of analytic models such as descriptive, diagnostic, predictive and prescriptive model that can be applied on SGs. The results of big data analytics can be used to predict customers behavior which helps them to improve network, to enhance security and to increase performance for future use.

The rest of this paper is organized as follows. In Sect. 2, related work is discussed. Section 3 covers proposed system model. Section 4, deals with the simulation results of proposed system model and In Sect. 5, performance is evaluated by using performance evaluators.

1.3 Motivation

After reviewing the past work done by many authors for electricity load and price forecasting [3,4] which we discussed in our related work we have the following motivation for this paper:

- For load and price forecasting, SGs should be used instead of conventional power grids.
- Because of exponential increase in data, data analytic techniques should be used for forecasting.
- A new technique should be made which resolve all the problems available in existing techniques.

1.4 Problem Statement

Due to increase in population, we are facing a huge gap between electricity demand and supply. To overcome this issue, it is necessary to migrate from Conventional Power grids to SGs. SGs involve the latest technical, control and communication technologies. SGs uses the Two way communication, which resulted in generation of huge amount of data. This huge amount of data, termed as big data, makes processing of data using conventional techniques, quite difficult. So we need to have efficient and enhanced techniques which could forecast load and price efficiently, resulting in cost minimization and efficient usage of electricity. In this paper we used SVM technique for load and price forecasting. However the basic limitation of SVM is higher computational complexity and tuning of parameters. To overcome this limitation we enhanced the SVM by tuning its parameters and simulation results show that our proposed technique beat the conventional techniques [5,6].

2 Related Work

For electricity load and price forecasting many techniques have been proposed. Forecasting techniques can be categorized in three groups:

- Data Driven
- Classical
- Artificially Intelligent

The statistical and mathematical models such as ARIMA, SARIMA, Random Forest etc. are classical models. These methods show good results in either load or price forecasting but do not consider the simultaneous forecasting of load and price. The existing forecasting methods mostly forecast only load or price. A forecasting method that can accurately forecast both load and price is required. Author in [7], used wavelet least square support vector machine (LSSVM) and sperm whale algorithm (SWM) with Discrete Wavelet Transform and inconsistency rate model (DWT-IR) to select the optimal features and to reduce the redundancy of input vectors. Dataset is taken from Commission for Energy Regulation (CER) Ireland to predict load. In this paper limitations was time complexity and only a day ahead forecasting can be done by default. In [8], SVR Support Vector Regression (SVR) is used on PJM dataset for day-ahead electricity forecasting. The basic limitation of SVR is, it requires a lot of data for training and requires retraining for any changes in decision model. In [9], proposes a very complex method containing multiple steps to predict the pattern of load, data is taken from an educational building in Hong Kong and multiple algorithms are used for this purpose. In [10], DLA technique is used for day-ahead electricity price and data set is taken from EPEX-Belgium and its limitation is over fitting problem. Author in [11], compared three different AI models for load and price prediction. Dataset is taken from PJM. The output results show that GRU (Gated Recurrent Unit) beats LSTM (Long Short Term Memory) and

DNN (Deep Neural Network) in terms of accuracy. In [12], Artificial Intelligent like CNN (Convolutional neural network) and LSTM (Long Short Term Memory) are being used for load and price prediction. The main problem with such models is that they are time and space complex models. Their working is just like the neurons in human brain i.e., they transfer information from one neuron to another and they also have the hidden layers, which work like the black box. In [13], two techniques ELM (Extreme Learning Machine) and WNN (Wavelet Neural Networks) are used for price prediction. In [14], for addressing the issue of peak load reduction author provides a DADR schemes. In [15], a neural network technique is used to predict the day ahead electricity load. The main problem with such model is that they are unnecessary complex. In [5], SVM (Support Vector Machine) is used for feature selection. The major drawback with SVM is that it is quite difficult to tune its parameters and the choice of kernel. Therefore many techniques have been used to solve this problem. In [16], Bayesian network framework is used to find out the price variation on the demand pattern. Author in [17], used ARIMA (Auto Regressive Integrated Moving Average) for price prediction in PJM region. Data is taken from residential sector and then ARIMA is used on that dataset. In [18], IWNN technique is used for load forecasting. In [19], Neural network based optimization approaches are used for energy demand prediction in smart grid. Result shows that proposed NNGA approach performs better for short term load forecasting and proposed NNPSO approach is more suitable for long term energy prediction. In [20], Artificial Intelligent model like RNN (Recurrent Neural Network) is used for load and price forecasting.

3 System Model

In this section, the proposed system model is described in detail as shown in Fig. 1:

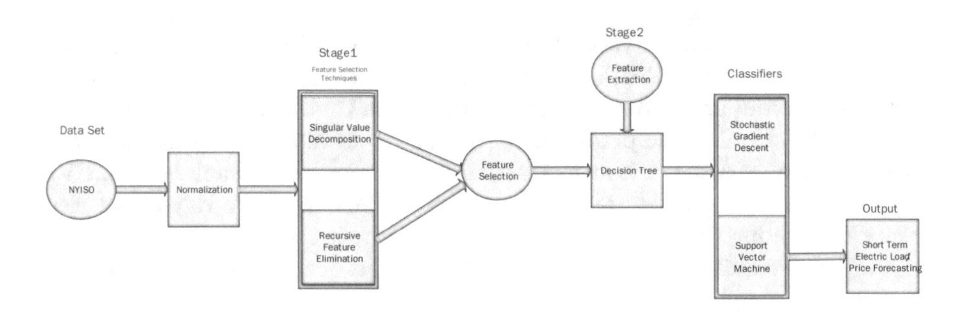

Fig. 1. System model

Preprocessing the Data. Data has been obtained from New York Independent System Operator (NYISO). Four years of data set from January 2014 to January 2017 are used in this paper. Data has been divided month-wise and similar months e.g. January 2014, January 2015, January 2016, and January 2017 data are used for training and testing. Data has been divided into three parts training, testing and validation of data. After that selection and extraction techniques are applied on this data.

Feature Selection Using DT, RFE. Feature selection is important when we have large set of data and we do not need to use every feature in our algorithm. Two feature selection techniques are used in this paper.

- Decision Tree (DT)
- Recursive Feature Elimination (RFE).

3.1 DT

Decision tree uses a tree like structure to build solutions. it breaks down a dataset into smaller subsets until it reaches on a base case. One advantage of DT is it does not require normalization of data.

Algorithm 1. DT

1: Calculate every attribute a of dataset S
2: Split the data S into subsets
3: Make a decision tree node using these attributes
4: Recur on subsets using remaining attributes

3.2 RFE

In recursive feature elimination technique the goal is to select features recursively considering smaller and smaller set of feature, at first by fitting the model on data we find feature importance of each feature. We drop the features with least importance, and then fits the model on remaining features. This process is repeated until our criteria is fulfilled. The stability of RFE depends upon the type of model that is used for feature ranking at each iteration.

3.3 Feature Extraction Using SVD

For dimensionality reduction feature selection and extraction techniques are used. Feature selection and feature extraction are two different approaches, in feature selection all original features are used but in feature extraction only subsets of feature extraction is used. Feature extraction basically is a process of deriving a new feature from the original features in order to reduce dimensionality and redundancy. In this paper SVD technique is used for feature extraction, which is used for dimensionality reduction.

Algorithm 2. RFE

1: 1.1 Tune/train the model on the training set using s predictors
2:
3: **for** 1.2 Each subset size A,i= 1 to A do **do**
4:
5: 1.3 Keep the A most important variables
6:
7: 1.4 [Optional] pre-process the data
8:
9: 1.5 calculate model performance
10:
11: 1.6 [optional] Recalculate the rankings for each predictors
12:
end

3.4 SVM and SGD for Forecasting

3.5 SGD

SGD is an optimization technique which is used to update the parameters of model. It is trying to find the best W by minimizing the error function Q. Advantages of SGD are: Efficiency and easy implementation. SGD is said to be efficient because of its linear time complexity (Fig. 2).

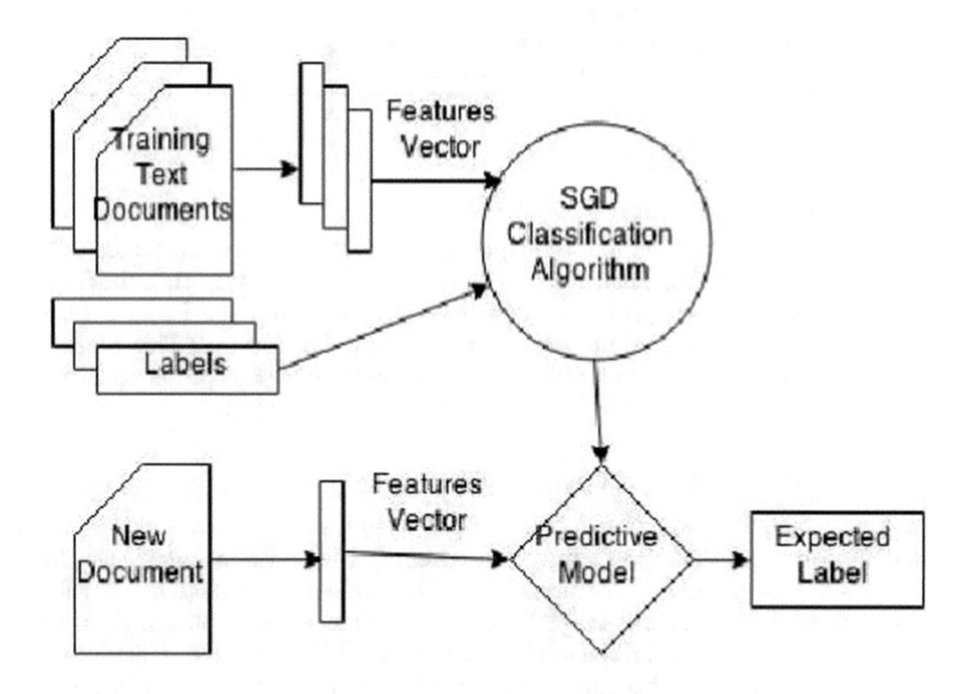

Fig. 2. Working of SGD classifier

Algorithm 3. SGD

1: initialization
2: input $z^0 \epsilon R^d$
3: **for** i=0 to T-1 **do**
4: uniformly randomly pick i from 1.....n
5: **while** not converged do **do**
6: **for** i=ϵ shuffle(1,2......N) **do**do
7: $z^0 + \lambda J^i(z)$ **return** z
8: return best possible solution

3.6 SVM

SVM is supervised learning that used for both classification and regression problems, however mostly it is use in classification problems. SVM searches for the closest path or points which it calls support vectors, once it has found the closest paths it draws a line for connecting them.

In this paper kernel parameter set as Radial Basic Function (RBF), cost penalty and gamma values are set to 27 and 38. These values are used for simulations and tuning the parameters. Steps involved in implementing model are listed below:

1. Data is normalized as (F/max(F)). After normalizaton, data is divided month wise and split into training, testing and validation.
2. This trained data is used to train model, testing and validation.
3. Now the model is tuned.
4. Model is tested on test data and month-wise load and price are forecasted.
5. To evaluate the performance forecasting, performance evaluation measurements RMSE, MAPE, MAE and MSE are used (Table 1).

Algorithm 4. SVM

1: initialization
2: training dataset D
3: define k attributes used in model
4: tune parameters
5: C for tuning margin and errors of SVM
6: **for** t=1 to T do **do**
7: sampling a training set using random attributes
8: SVM_t = SVM(D,k,t,C)
9: end
10: return SVM

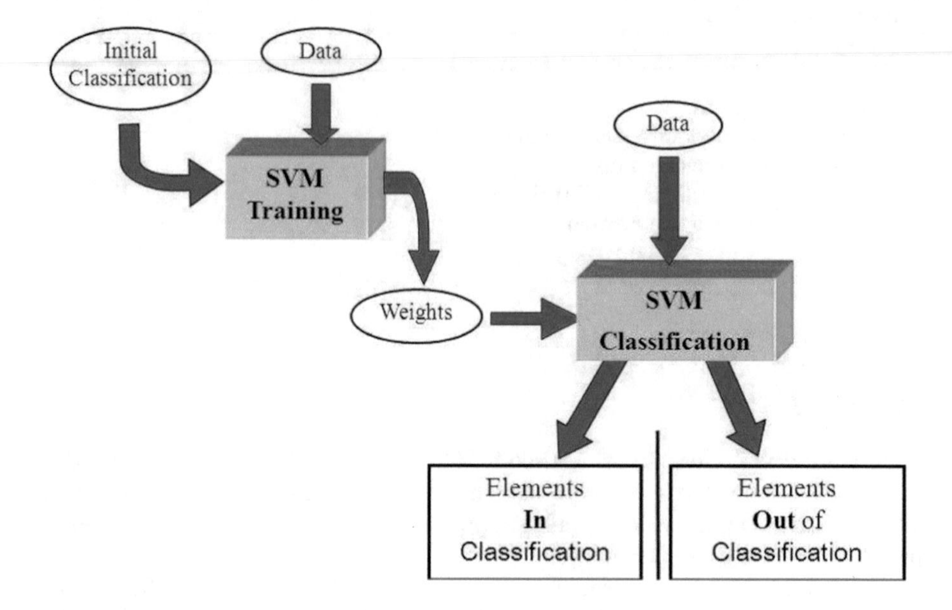

Fig. 3. SVM process overview

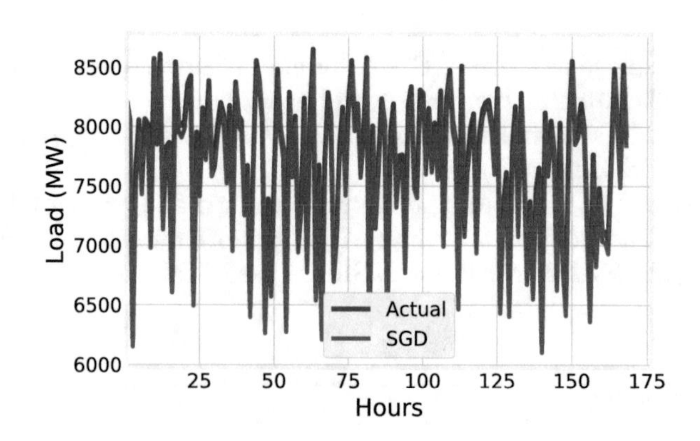

Fig. 4. Load using SGD

Table 1. Performance evaluators

Evaluator	Value
MSE	31.8
MAE	23.3
MAPE	31.1
RMSE	26.6

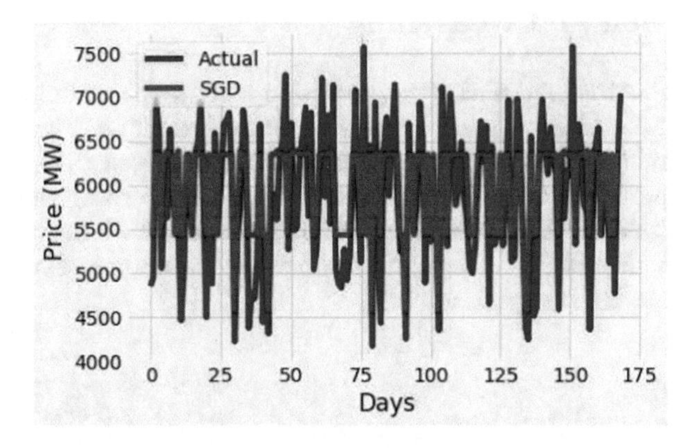

Fig. 5. Price using SGD

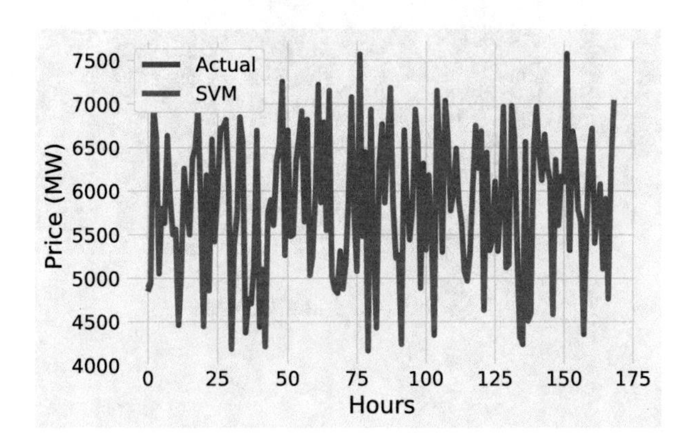

Fig. 6. Price using SVM

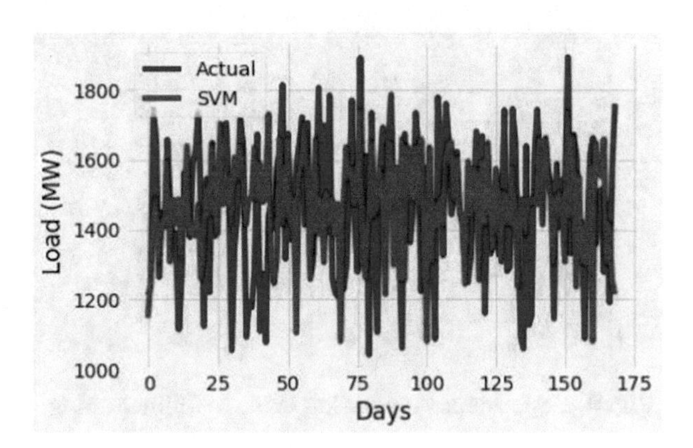

Fig. 7. Load using SVM

4 Simulation and Reasoning

Figures 3, 4 show the load and price prediction using SGD technique. Figures 5, 6 show the load and price prediction using SVM technique. Figure 7 show the enhanced technique of SVM. From the graph it is observed that our new technique beats the other two existing techniques. Figure 8 show the four performance evaluators for load forecasting. Figure 9 show the hourly load forecast comparison of models by using performance evaluator and Fig. 10 show the hourly price forecast comparison.

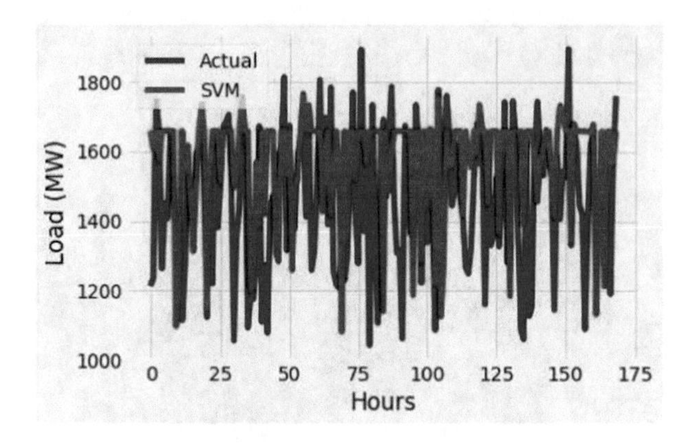

Fig. 8. Load using ESVM

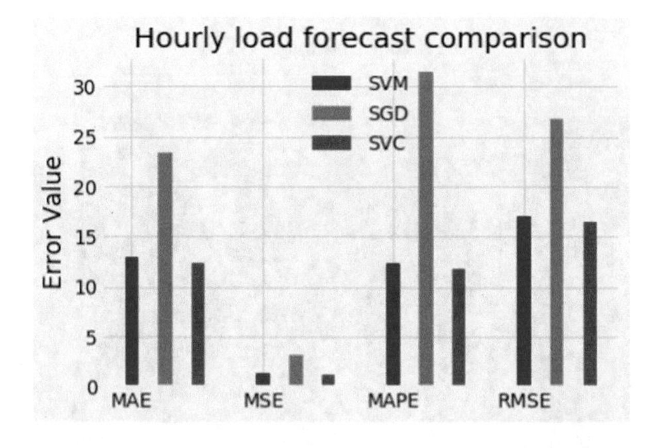

Fig. 9. Performance evaluation using load forecasting

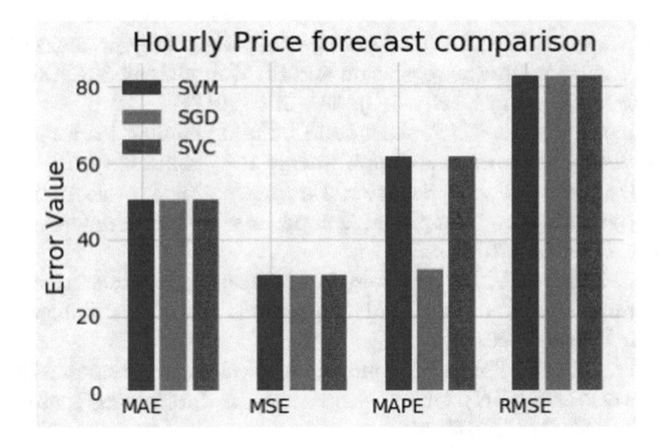

Fig. 10. Performance evaluation using price forecasting

5 Conclusion

In this paper, load and price forecasting in smart grids is done using data analytics technique. Data is taken from NYISO and categorized this data as month-wise for accurate load and price forecasting. Feature selection, extraction techniques are applied for normalizing data. For forecasting SGD and SVM classifiers are used. To increase the forecasting accuracy an enhanced form of SVM is used, which gives the better result compare to the conventional SVM technique. The performance evaluators RMSE, MAE, MSE and MAPE proves that the proposed technique is better than the conventional SVM technique.

References

1. Oren, S.S.: Demand response: a historical perspective and business models for load control aggregation. In: Presented at the Power System Engineering Research Center Public Webinar, 1 February 2011
2. Ghatikar, G., Mathieu, J.L., Piette, M.A., Liliccote, S.: Open automated demand response technologies for dynamic pricing and smartgrid. In: Presented at Grid-Interop Conference 2010, Chicago, IL, 13 December 2010. http://drrc.lbl.gov/sites/drrc.lbl.gov/files/lbnl-4028e.pdf
3. Liu, J.P., Li, C.L.: The short-term power load forecasting based on sperm whale algorithm and wavelet least square support vector machine with DWT-IR for feature selection. Sustainability **9**(7), 1188 (2017)
4. Vrablecov, P., Ezzeddine, A.B., Rozinajov, V., Slavomrrik, S., Sangaiah, A.K.: Smart grid load forecasting using online support vector regression. Comput. Electr. Eng. **65**, 102–117 (2017)
5. Wang, F., Li, K., Zhou, L., Ren, H., Contreras, J., Shafie-khah, M., Catalão, J.P.S.: Daily pattern prediction based classification modeling approach for dayahead electricity price forecasting. Int. J. Electr. Power Energy Syst. **105**, 529–540 (2019)
6. Wang, K., et al.: Robust big data analytics for electricity price forecasting in the smart grid. IEEE Trans. Big Data (2017)

7. Ziming, M.A., et al.: Month ahead average daily electricity price profile forecasting based on a hybrid nonlinear regression and SVM model: an ERCOT case study. J. Mod. Power Syst. Clean Energ. **6**(2), 281–291 (2018)
8. Fan, C., Xiao, F., Zhao, Y.: A short-term building cooling loadprediction method using deep learning algorithms. Appl. Energ. **195**, 222–233 (2017)
9. Jesus, L., De Ridder, F., De Schutter, B.: Forecasting spotelectricity prices: deep learning approaches and empirical comparison oftraditional algorithms. Appl. Energ. **221**, 386–405 (2018)
10. Lago, J., De Ridder, F., De Schutter, B.: Forecasting spotelectricity prices: Deep learning approaches and empirical comparison oftraditional algorithms. Appl. Energ. **221**, 386–405 (2018)
11. Jianzhou, W., Liu, F., Song, Y., Zhao, J.: A novel model: dynamic choice artificial neural network (DCANN) for an electricityprice forecasting system. Appl. Soft Comput. **48**, 281–297 (2016)
12. Jindal, A., Singh, M., Kumar, N.: Consumption-aware data analytical demand response scheme for peak load reduction in smart grid. IEEE Trans. Industr. Electron. (2018)
13. Marcjasz, G., Uniejewski, B., Weron, R.: On the importance of the long-term seasonal component in day-ahead electricity price forecasting with NARX neural networks. Int. J. Forecast. (2018)
14. Gao, W., Darvishan, A., Toghani, M., Mohammadi, M., Abedinia, O., Ghadimi, N.: Different states of multi-block based forecast engine for price and load prediction. Int. J. Electr. Power Energ. Syst. **104**, 423–435 (2019)
15. Long, W., Zhang, Z., Chen, J.: Short-term electricity price forecasting with stacked denoising autoencoders. IEEE Trans. Power Syst. **32**(4), 2673–2681 (2017)
16. Rafiei, M., et al.: Probabilistic load forecasting using an improved wavelet neural network trained by generalized extreme learning machine. IEEE Trans. Smart Grid (2018)
17. Muralitharan, K., Sakthivel, R., Vishnuvarthan, R.: Neural network based optimization approach for energy demand prediction in smart grid. Neurocomputing **273**, 199–208 (2018)
18. Shi, H., Xu, M., Li, R.: Deep learning for household load forecasting - a novelpooling deep RNN. IEEE Trans. Smart Grid **9**(5), 5271–5280 (2018)
19. Nguyen, H.K., Song, J.B., Han, Z.: Distributed demand side management with energy storage in smart grid. IEEE Trans. Parallel Distrib. Syst. **26**(12), 3346–3357 (2015)
20. Gorodetski, V.I., Popyack, L.J., Samoilov, V., Skormin, V.A.: SVD-based approach to transparent embedding data into digital images. In: Proceedings of International Workshop on Mathematical Methods, Models and Architecture for Computer Network Security 2052, pp. 263–274 (2001)

Data Analytics for Load and Price Forecasting via Enhanced Support Vector Regression

Tanzeela Sultana[1], Zahoor Ali Khan[2], Nadeem Javaid[1(✉)], Syeda Aimal[1], Aisha Fatima[1], and Shaista Shabbir[3]

[1] COMSATS University Islamabad, Islamabad 44000, Pakistan
nadeemjavaidqau@gmail.com
[2] Computer Information Science, Higher Colleges of Technology, Fujairah 4114, UAE
[3] Virtual University of Pakistan, Kotli Campus, Azad Kashmir 1100, Pakistan
http://www.njavaid.com

Abstract. In this paper, month-ahead electricity load and price forecasting is done to achieve accuracy. The data of electricity load is taken from the Smart Meter (SM) in London. Electricity load data of five months is taken from one block SM along with the weather data. Data Analytics (DA) techniques are used in the paper for month-ahead electricity load and price prediction. In this paper, forecasting is done in multiple stages. At first stage, feature extraction and selection is performed to make data suitable for efficient forecasting and to reduce complexity of data. After that, regression techniques are used for prediction. Singular Value Decomposition (SVD) is used for feature extraction afterwards; feature selection is done in two-stages, by using Random Forest (RF) and Recursive Feature Elimination (RFE). For electricity load and price forecasting Logistic Regression (LR), Support Vector Regression (SVR) is used. Moreover forecasting is done by the proposed technique Enhanced Support Vector Regression (EnSVR), which is modified from SVR. Simulation results show that the proposed system gives more accuracy in load and price prediction.

Keywords: Load forecasting · Price forecasting · Data Analytics · Logistic Regression · Support Vector Regression · Enhanced Support Vector Regression

1 Introduction

With the increasing demand of electricity along with the growth of advanced technologies, the energy markets are evolving consistently. The Traditional Grid (TG) systems are failed to incorporate with new technology. TG systems are centralized and often shows uncertainty. Only one-way transmission strategy is used in TGs. There were several issues that were rising in TGs. Due to increment in power demand, unstable voltage became major issue. Quality of power

L. Barolli et al. (Eds.): EIDWT 2019, LNDECT 29, pp. 259–270, 2019.
https://doi.org/10.1007/978-3-030-12839-5_24

also decreased due to voltage fluctuations. Resources were unpredictable, which causes low feasibility. To deal with the energy issues with the TGs, the grid systems are attracting toward the smart grid systems. The traditional approaches for power system are decreasing and the concept of Smart Meters (SMs) and Smart Grids (SGs) is getting high. SG systems are becoming more popular worldwide due to rise of technology. SG, can be defined as, an advanced development that gathers the Information Communication and Technology (ICT) and traditional energy systems to improve productivity and ability of the grid by making a relation between consumer and utility systems. SG provides the distributed forms of energy generation to minimize uncertainty. SG also provides efficient energy management, such as communication between utility store and consumer, security, generation cost saving, processing and storing energy data [1–4]. SG presents many new components like: SMs, Demand Response (DR) management scheme, growth in IoT devices, online interactions and billing systems, and integration of renewable energy. SG works on Information Technology (IT) systems, for performing several tasks such as processing and measurements, experiments and simulations, management [5–7] (Table 1).

As the power system is moving towards SG, more electrical data is generated from SMs. For efficient management of energy in utility markets, Data Analytics (DA) is used. DA is the process of collecting data and performs several operations on data such as, cleaning, transforming and modeling data, to provide solutions. Different DA techniques are used to perform multiple operations on data. DA is used almost in every field, for several purposes. However, in energy markets DA is mostly used for forecasting purposes. Forecasting is mostly done to estimate the electrical load and price of electricity in future and for efficient management of electricity. Also to take account of the consumption rate by user and generation of electricity. Several forecasting techniques are introduced till now, to provide accurate forecasting. The best forecasting algorithm gives accurate forecasting results that can minimizes the issues related to power [8] and it also helps to improve the electricity management system. As electric data generated from the smart meters is time series data, there are time series forecasting paradigms for different intervals: short term, medium term and long term [9]. Electricity grids are often affected by weather conditions. The PCR based method is used for prediction of temperature on power grids in [10], the conductor temperature is collected by the installation of a sensor over transmission lines of TaiPower grid in Taiwan. The main objective of the paper is to make improvement in transmission lines that are influenced by weather conditions. Many techniques are used in literature for forecasting both load and price in different horizon. However, in this paper, electricity load and price forecasting is done for month-ahead in three stages by regression techniques LR, SVR and EnSVR by taking the dataset of one block of smart meters in London. Load and weather data of that block is also used for month-ahead load and price forecasting.

Table 1. List of abbreviation

ABC	Artificial Bee Colony
AGG	Arian Golden Group
ARIMA	Autoregressive Integrated Moving Average
CMI	Conditional Mutual Information
DA	Data Analytics
DR	Demand Response
DTR	Dynamic Thermal Rating
DTW	Dynamic Time Warping
ELM	Extreme Learning Machine
EMD	Empirical Mode Decomposition
EnSVR	Enhanced Support Vector Regression
ERCOT	Electric Reliability Council of Texas
FWPT	Flexible Wavelet Packet Transform
GS	Greater Sydney
GTA	Grid Traverse Algorithm
ICT	Information Communication and Technology
LR	Logistic Regression
MA	Moving Average
MIMO	Multi-Input Multi-Output
MLE	Maximum Likelihood Estimation
MLR	Multiple Linear Regression
NA	North American
NLSSVM	Non-linear Least Square Support Vector Machine
NN	Neural Networks
NSW	New South Wales
PCR	Principle Component Regression
PLD	Power Load Decomposition
PJM	Pennsylvania-New Jersey-Maryland
PSO	Particle Swarm Optimization
QLD	QueensLand
RF	Random Forest
RFE	Recursive Feature Elimination
RVFL	Random Vector Functional Link Network
SG	Smart Grid
SM	Smart Meter
SMA	Simple Moving Average
SVC	Support Vector classification
SVD	Singular Value Decomposition
SVM	Support Vector Machine
SVR	Support Vector Regression
TG	Traditional Grid

1.1 Motivation

A lot of work is done in literature for efficient management of electricity. Many techniques are proposed to forecast electricity load and price in smart meters and smart grids for short-term, medium-term and long-term forecasting. Many papers proposed the hybrid model for prediction such as [2] proposes a hybrid of Grid Traverse Algorithm (GTA) and Particle Swarm Optimization (PSO) for short-term load forecasting, [4] does a three-stage prediction for both load and price through different methods at each stage, [6] mixes Non-Linear Regression with SVM for month-ahead price forecast. [2,5–7] uses SVR for forecasting. By taking all the studies into consideration, the proposed paper aims to forecast both load and price for month-ahead to achieve accuracy. The main objective of papers is to achieve more accuracy and performance in forecasting and to resolve peak load and price variation in smart grids.

1.2 Problem Statement

In paper [5] short-term load forecast is predicted to calculate Demand Response (DR) for an office building in eastern China. The model's objective is to achieve maximum forecasting accuracy under DR baseline using historical electric load and weather data. The paper uses SVR for accurate predictions. [7] predicts very short term half-hour ahead load forecasting, using on-line SVR where prediction is done to overcome the issues of memory and storage using advanced on-line methods. The paper used the dataset of Irish Commission for Energy Regulation (CER) utility company and achieves accuracy and minimizes the storage issues by using ensemble tree-based ensemble methods with advanced on-line methods. However, model outperforms for storage and memory issues, besides this model gives single output with only one-hour ahead prediction, a further methodology is required for the multiple outputs. Paper [7] also lacks in performance to predict separate day. Above papers forecast either load or price, these papers can only predict for few hours. However, the proposed model forecast both load and price by two existing schemes LR, SVR. Afterwards forecasting is done the scheme proposed by the paper.

1.3 Contribution

In this paper load and price forecast is performed in a new way, by making a combination of schemes that are never used before for forecasting. The major purpose is to achieve accuracy in terms of electricity load and price prediction.

The main contributions of this paper are:

- Only load data of one block smart meter, along with weather data is given in the dataset, the price is calculated from the given load data to predict price along with load,
- Data preprocessing is done by both feature extraction and feature selection. For extraction, SVD is used and for feature selection, first RF and after that RFE is used,

- Month-ahead forecasting for load and price is accomplished by executing multiple regressors such as LR, SVR. The EnSVR method is proposed to achieve more accuracy and for better electricity forecasting.

2 Related Work

With the generation of electricity in smart grids, the problem of electrical load and price is increasing. To tackle the electric load and price forecast issues, different schemes are proposed in the literature. Paper [1] founds the limitations in short-term forecasting for university buildings, because of diverse patterns. Paper uses a 2-stage forecast model for the short-term electric load prediction in the university campus. Five years data is taken from a university campus in Korea and 2-stage prediction is done. At the first stage, by using Moving Average (MA) method and secondly by using RF technique. Furthermore, the performance is evaluated by the time-series cross-validation process. Short-term load forecasting is also done by paper [2] over distributed systems using SVR and two-step hybrid optimization algorithm based on GTA and PSO. The data is calculated from the Partner utility's distribution feeder and the prediction is done on seasoned base. This paper overcome the issue for total load of small section of distributed feeder. However, this paper works effectively by using external factors. Load prediction is also done by taking external factors like temperature and weather data into consideration. Paper [3] determines the growth in household load forecast. The level of load in households is increasing due to which error rate is also increasing. The paper proposed a

Table 2. Summary of related work

Technique used	Features	Dataset	Region	Limitations
RF, ANN [1]	Academic year, load	University campus	Seoul, Korea	-
SVR, PSO [2]	Historical load	Partner's utility	US	Better perform only with external factors
MLR [3]	Calender effect, load, weather data	Ausgrid	NSW and GS	Minimize accuracy due to large dataset
3-Stage methodology [4]	Price, demand	NYISO and PJM	US	-
SVR [5]	Temperature, electricity load, weather	Office building	Eastern China	Predict few hours of one building only
Hybrid Non-Linear Regression, SVR [6]	Price, reading rate, average wages	ERCOT	Texas, USA	Model works well only for month-ahead forecast
On-line SVR [7]	Half-hour load, time, weekdays, weather data	Irish CER	Ireland	Model is not applicable for long term forecasting
EMD-ELM [8]	Half-hour electric load	NSW, Victoria, QLD	Australia	-
DWT, EMD, RVFL [9]	Year, max, median, mean, standard deviation	AEMO	Australia	-

forecasting model for scheduling and optimum planning by using a historical load with weather data and calendar effects of Australian grids of residential areas of New South Wales (NSW) and Greater Sydney (GS). Calendar effects are used for prediction in several periods of the calendar: days, months and years. Many forecasting techniques such as Multiple Linear Regression (MLR), Neural Networks (NN), regression trees and Support Vector Machine (SVM) are used. Accuracy of the paper is minimized due to large data. Paper [4] captures the limitation of forecasting algorithms, as they perform forecasting of load and price separately. So paper proposed hybrid algorithms based on demand-side management models to predict both price and load concurrently using data of New York Independent System Operator (NYISO), NSW and Pennsylvania New-Jersey Maryland (PJM). Forecasting in this paper is performed in three stages; at first stage, Flexible Wavelet Packet Transform (FWPT) and Conditional Mutual Information (CMI) algorithms are used for selection, then Multiple-Input Multiple-Output (MIMO) based Non-linear Least Square Support Vector Machine (NLSSVM) and Autoregressive Integrated Moving Average (ARIMA) is applied to define a relationship between load and price, at third stage enhances ABC on the basis of Time-Varying Coefficients and Stumble Artificial Bee Colony (TV-SABC) for parameter optimization. Different algorithms are used, that are complex to be used by the utility market. In paper [5] SVR model is employed reduce complexity and to minimize the risk of a load during the peak hours. Data is taken from four office buildings in China to examine DR. The model only predicts only few hours for one building efficiently. Electricity price is forecast in paper [6] to examine and overcome the abnormalities in the electric power industry. This paper also founds inaccuracy in month-ahead forecasting. Month-ahead average price forecast is done by hybrid Non-Linear Regression and SVM on data of Electric Reliability Council of Texas (ERCOT) electricity market in Texas, United States of America (USA). The limitation of this system is that it gives accurate forecasting for month-ahead only. With the enhancement of electricity grid market, most of the data is processed online. Paper [7] figures out the memory and storage problems in systems. Paper proposed a short-term load forecasting with online SVR to overcome the biasness of the Irish CER grid market. Ten models are compared in this paper, however, online SVR achieved more accuracy. The model is not applicable for long-term forecasting. For short-term load forecasting, paper [8] employed a novel method Empirical Mode Decomposition (EMD)-mixed-Extreme Learning Machine (ELM), where EMD is used for the decomposition of complex load features to smoothen the data. RBF and UKF kernels are used in EML method to handle the limitations of using one kernel. All the features of data are not captured by one kernel. In the paper load data of NSW, Victoria and Queensland (QLD) is used for forecasting. Another hybrid technique is used for short-term load prediction, which is comprised of EMD and Random Vector Functional Link Network (RVFL) in paper [9]. For better forecasting, ensemble method RVFL based on Dynamic Time Warping (DTW) and EMD is presented for accuracy and performance. Load data is collected from AEMO along with NSW, Tasmania (TAS), QLD, South Australia (SA) and Victoria. This paper targeted the insufficiency of the incremental methods, is the unnoticed change in the target value (Table 2).

3 System Model

In this section, the proposed system model of the paper is presented in detail. The flow of the system model is shown in Fig. 1.

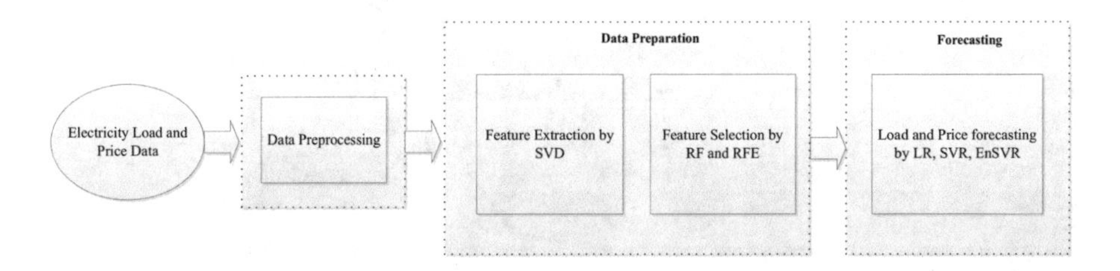

Fig. 1. Framework of proposed system

3.1 Dataset Description

The dataset of smart meters in London is used. In the whole dataset, data of household electric load of multiple blocks is given. Also, load data of half-hourly and hourly basis is presented along with the daily temperature data and hourly weather data of three years, from 2013 to 2015.

In this paper, for accurate forecasting of load and price, electrical load data of only one block smart meter is considered along with weather data of that block is used in this paper. Load and weather data of five months is used to forecast one-month ahead load and price of that block. The main focus of the paper is to forecast price and load. As the dataset does not have any price data of smart meter, the paper calculated price from the given load data to predict the future price for the smart meter of that block.

3.2 Feature Importance of Data

Feature importance of data is calculated for every feature in the dataset. It is examined by observing the effect of every independent feature in the data for the target feature. Figure 2, shows the feature importance for a load. For load forecasting, price feature shows more influence on the target than other features. However, the feature that does not affect the target is humidity.

Like load, feature importance for Price is also calculated. The Fig. 3, shows the feature importance of every feature to the target. For price, pressure shows more influence than other and humidity does not shows any effect also for price forecasting.

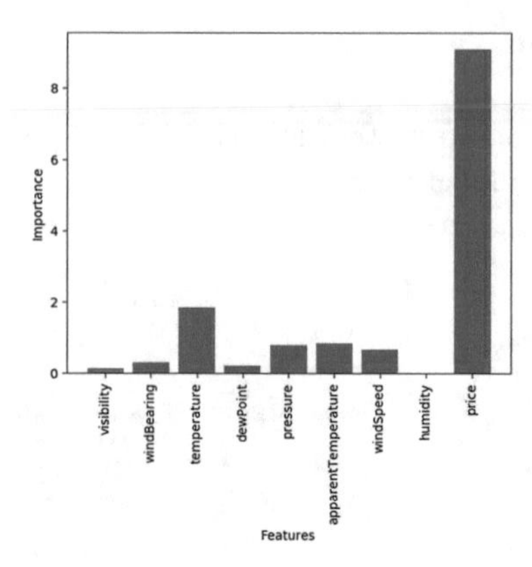

Fig. 2. Feature importance for load

3.3 Proposed System Model

Figure 1 demonstrates the flow of presented system model. In the proposed model, first, the load data of one block smart meter is taken along with the weather data of the block. Then the price is calculated for price forecasting and data is preprocessed. Feature extraction is done by SVD. After that feature selection is executed by two techniques, RF and RFE for getting the best features for accurate forecasting. In the proposed model, forecasting is done by LR, SVR and the SVR is modified to EnSVR.

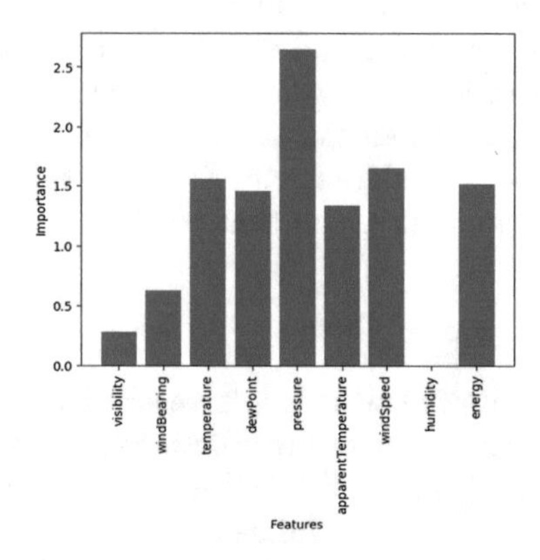

Fig. 3. Feature importance for price

4 Simulation Results and Reasoning

In this section, results of electricity load and price forecast are presented with reasoning. The proposed schemes are implemented by using five months data of load and price. The data is divided into training and testing as, 70% training data and 30% test data. Then feature and selection is done to extract meaningful features from the data. Forecasting is performed by two existing and one proposed scheme. Simulations are performed using a server of 2.50 GHz Intel Core i5 CPU and 4 GB RAM. LR, SVR and EnSVR are implemented in Python using spyder platform.

4.1 Load and Price Forecasting

In this paper, month-ahead load and price is predicted by taking five month historical data of smart meter in one block. Load and price forecast are shown in Figs. 4 and 5.

Electricity load and price forecasting is done by using multiple techniques. Feature extraction is done by SVD. Feature selection is performed by using two techniques, RF and RFE. Regression techniques are used for load and price forecasting. Existing techniques LR, SVR and proposed scheme EnSVR is used for prediction. At first step, LR is used and at second stage, SVR is used. At the final stage, EnSVR is used to achieve prediction accuracy.

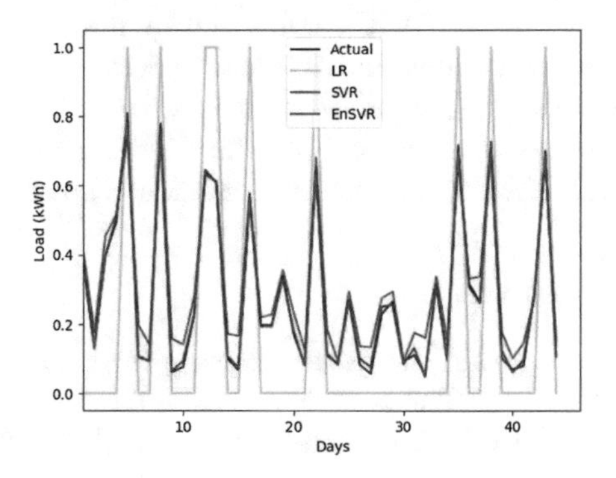

Fig. 4. Load forecast

Logistic Regression (LR). Logistic Regression (LR) is basically a statistical model and used as a classification technique in machine learning. LR shows nonlinear relationship of one dependent variable with multiple independent variables. Logistic or sigmoid function is used to show relationship by estimating probabilities and to transform the output. Logistic function can be defined as:

$$g(z) = 1/1 + e^{-z} \tag{1}$$

Support Vector Regression (SVR). Support Vector Regression (SVR) is the regression technique, based on Support Vector Machine (SVM) model. SVR learning is based on ratio of training and testing data. Input parameters of SVR are defined by all possible values.

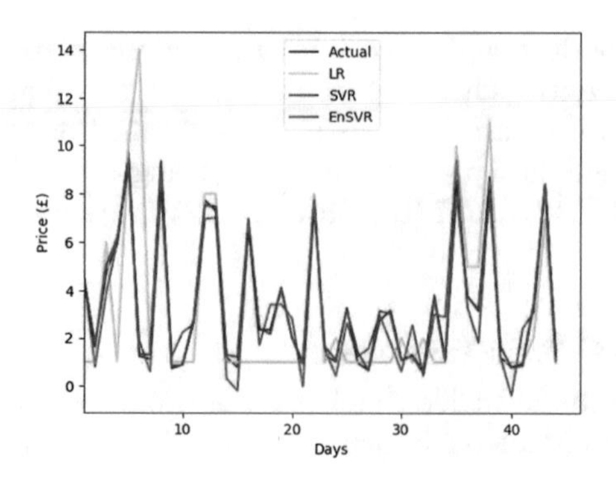

Fig. 5. Price forecast

Enhanced Support Vector Regression (EnSVR). To improve the performance of SVR, this paper proposed an Enhanced SVR technique named as EnSVR. EnSVR is modified from original SVR technique. To achieve more forecasting accuracy, best parameters are enhanced. Only those parameters are considered that give better result. SVR is enhanced by making changes in parameters such as cost, epsilon, gamma and tolerance. Best parameters are picked by cost = 1.0, epsilon = 0.001, gamma = 10, tolerance = 0.1. Forecasting results shows that EnSVR shows more accuracy than SVR for load and price as illustrated in accuracy table. Forecasting results are also shown in Figs. 4 and 5.

4.2 Accuracy Measurement

The goal of this paper is to achieve more accuracy for load and price forecasting on the data of smart meters. The accuracy is calculated for both load and price by using LR, SVR and EnSVR. Where LR has achieved 89% SVR has 60% accuracy and EnSVR got 94% accuracy. Accuracy is also examined for price on the base of every technique used. LR has achieved 74% and SVR has 60% accuracy for price. Whereas EnSVR achieves 91% accuracy that is more than other techniques. Proposed technique EnSVR achieves more accuracy more than other techniques used.

4.3 Performance Evaluation

To calculate the performance of forecasting techniques, performance metrics such as Mean Absolute Percentage Error (MAPE), Mean Absolute Error (MAE), Mean Squared Error (MSE) and Root Mean Square Error (RMSE) are used. These evaluators are mostly used to check the accuracy for forecasting. Table 3 shows the performance percentage for LR, SVR and EnSVR for load. Table 4 shows the error values for price.

Table 3. Performance evaluation for load

Method	RMSE%	MSE%	MAE%	MAPE%
LR	0.0024	0.0006	0.0021	inf
SVR	0.0005	2.704	0.0004	22.142
EnSVR	0.0001	1.065	8.3451	5.706

Table 4. Performance evaluation for price

Method	RMSE%	MSE%	MAE%	MAPE%
LR	0.0304	0.0927	0.0168	114.12
SVR	0.0071	0.0051	0.006	1893.3
EnSVR	0.002	0.0004	0.001	7.164

5 Conclusion

With the rapid growth of electricity it is becoming very difficult for utility market to handle the issues relating to forecast for future load and price. In this paper, short term load and price forecasting is done to achieve accuracy by taking load data of smart meters in London with weather data of about five months. There are many strategies proposed for forecasting. However in this paper, short-term load and price forecasting is done by using multiple techniques and with the proposed technique. For forecasting first LR and SVR is used. After that the forecasting is done by the proposed scheme EnSVR. Results show that the proposed EnSVR technique outperforms the existing techniques in both load and price forecasting.

References

1. Zhang, D., Li, S., Sun, M., O'Neill, Z.: An optimal and learning-based demand response and home energy management system. IEEE Trans. Smart Grid **7**(4), 1790–1801 (2016)
2. Shafie-khah, M., Siano, P.: A stochastic home energy management system considering satisfaction cost and response fatigue. IEEE Trans. Industr. Inf. **14**(2), 629–638 (2017)
3. Akhavan-Rezai, E., Shaaban, M.F., El-Saadany, E.F., Karray, F.: Online intelligent demand management of plug-in electric vehicles in future smart parking lots. IEEE Syst. J. **10**(2), 483–494 (2016)
4. Mirjalili, S., Mirjalili, S.M., Lewis, A.: Grey wolf optimizer. Adv. Eng. Softw. **69**, 46–61 (2014)
5. Jayabarathi, T., Raghunathan, T., Adarsh, B.R., Suganthan, P.N.: Economic dispatch using hybrid grey wolf optimizer. Energy **111**, 630–641 (2016)
6. Geem, Z.W., Yoon, Y.: Harmony search optimization of renewable energy charging with energy storage system. Int. J. Electr. Power Energy Syst. **86**, 120–126 (2017)

7. Ouyang, H.B., Gao, L.Q., Kong, X.Y., Li, S., Zou, D.X.: Hybrid harmony search particle swarm optimization with global dimension selection. Inf. Sci. **346**, 318–337 (2016)
8. Ambia, M.N., Hasanien, H.M., Al-Durra, A., Muyeen, S.M.: Harmony search algorithm-based controller parameters optimization for a distributed-generation system. IEEE Trans. Power Delivery **30**(1), 246–255 (2015)
9. Manzoor, A., Javaid, N., Ullah, I., Abdul, W., Almogren, A., Alamri, A.: An intelligent hybrid heuristic scheme for smart metering based demand side management in smart homes. Energies **10**(9), 1258 (2017)
10. Javaid, N., Javaid, S., Abdul, W., Ahmed, I., Almogren, A., Alamri, A., Niaz, I.A.: A hybrid genetic wind driven heuristic optimization algorithm for demand side management in smart grid. Energies **10**(3), 319 (2017)
11. Mahmood, D., Javaid, N., Ahmed, I., Alrajeh, N., Niaz, I.A., Khan, Z.A.: Multi-agent-based sharing power economy for a smart community. Int. J. Energy Res. **41**(14), 2074–2090 (2017)
12. Zhao, Z., Lee, W.C., Shin, Y., Song, K.B.: An optimal power scheduling method for demand response in home energy management system. IEEE Trans. Smart Grid **4**(3), 1391–1400 (2013)
13. Logenthiran, T., Srinivasan, D., Shun, T.Z.: Demand side management in smart grid using heuristic optimization. IEEE Trans. Smart Grid **3**(3), 1244–1252 (2012)
14. Rajalingam, S., Malathi, V.: HEM algorithm based smart controller for home power management system. Energy Buildings **131**, 184–192 (2016)
15. Ahmed, M.S., Mohamed, A., Khatib, T., Shareef, H., Homod, R.Z., Ali, J.A.: Real time optimal schedule controller for home energy management system using new binary backtracking search algorithm. Energy Buildings **138**, 215–227 (2017)
16. Zhang, Y., Li, C., Li, L.: Wavelet transform and Kernel-based extreme learning machine for electricity price forecasting. Energy Syst. **9**(1), 113–134 (2018)
17. Förderer, K., Ahrens, M., Bao, K., Mauser, I., Schmeck, H.: Towards the modeling of flexibility using artificial neural networks in energy management and smart grids: note. In: Proceedings of the Ninth International Conference on Future Energy Systems, pp. 85–90. ACM (2018)
18. Gao, W., Darvishan, A., Toghani, M., Mohammadi, M., Abedinia, O., Ghadimi, N.: Different states of multi-block based forecast engine for price and load prediction. Int. J. Electr. Power Energy Syst. **104**, 423–435 (2019)
19. Nowotarski, J., Weron, R.: Recent advances in electricity price forecasting: a review of probabilistic forecasting. Renew. Sustain. Energy Rev. (2017)
20. Bramer, L.M., Rounds, J., Burleyson, C.D., Fortin, D., Hathaway, J., Rice, J., Kraucunas, I.: Evaluating penalized logistic regression models to predict Heat-Related Electric grid stress days. Appl. Energy **205**, 1408–1418 (2017)

A Deep Learning Approach Towards Price Forecasting Using Enhanced Convolutional Neural Network in Smart Grid

Fahad Ahmed[1], Maheen Zahid[1], Nadeem Javaid[1](✉), Abdul Basit Majeed Khan[2], Zahoor Ali Khan[3], and Zain Murtaza[1]

[1] COMSATS University, Islamabad 44000, Pakistan
nadeemjavaidqau@gmail.com
[2] Abasyn University Islamabad Campus, Islamabad 44000, Pakistan
[3] Computer Information Science, Higher Colleges of Technology, Fujairah 4114, UAE
http://www.njavaid.com

Abstract. In this paper, we attempt to predict short term price forecasting in Smart Grid (SG) deep learning and data mining techniques. We proposed a model for price forecasting, which consists of three steps: feature engineering, tuning classifier and classification. A hybrid feature selector is propose by fusing XG-Boost (XGB) and Decision Tree (DT). To perform feature selection, threshold is defined to control selection. In addition, Recursive Feature Elimination (RFE) is used for to remove redundancy of data. In order, to tune the parameters of classifier dynamically according to dataset we adopt Grid Search (GS). Enhanced Convolutional Neural Network (ECNN) and Support Vector Regression (SVR) are used for classification. Lastly, to investigate the capability of proposed model, we compare proposed model with different benchmark scheme. The following performance metrics: MSE, RMSE, MAE, and MAPE are used to evaluate the performance of models.

1 Introduction

Nowadays, electricity plays an important role in economic and social development. Everything is dependent on electricity. Without electricity, our lives are imagined to be stuck. Electricity usage areas are divided into three categories: industrial, commercial and residential. According to [1], residential area consumes almost 65% of electricity from the whole generation. In the traditional grid, most of the electricity is wasted during generation, transmission and distribution. To solve this issue, SGs are introduced. A traditional grid is converted into SG when information, communication and technology (ICT) are integrated into the traditional grid. SG is an intelligent grid system that manages generation, consumption and distribution of energy more efficiently than traditional grid [2]. SG provides the facility of bidirectional communication between utility and consumer. As we know, energy is the most valuable asset of this world.

L. Barolli et al. (Eds.): EIDWT 2019, LNDECT 29, pp. 271–283, 2019.
https://doi.org/10.1007/978-3-030-12839-5_25

It is very necessary to utilize energy in an efficient way to increase productivity and to decrease losses and hazards. Energy crises are present everywhere, so industries are moving toward SG. The primary goal of SG is to keep balance between supply side (utility) and demand side (consumer) [3]. SG fulfills, all the demands from the consumer side and gives response to their requests. Consumers send their demands to the utility through Smart Meter (SM). Hence, a huge amount of data is collected via SM regarding the electricity consumption of consumers. Electricity usage may vary depending upon different factors such as: wind, temperature, humidity, seasons, holidays, working days, appliances usage and number of occupants. Utility must be aware of the usage pattern of consumer. [4], Data Analytics (DA) is a process of examining data. DA is basically used in business intelligence, for decision making. When data analyst wants to do an analysis of electricity load consumption and pricing trends, then they take dataset of any specific electricity company. To maintain the load of electricity consumption, many researchers are working on forecasting of electricity load and price [5]. There are three types of forecasting: Short Term Load Forecasting (STLF), Medium Term Load Forecasting (MTLF) and Long Term Load Forecasting (LTLF). STLF consists of time horizon from a few minutes to hours. Day ahead is considered in STLF. MTLF contains the horizon from one month to one year. LTLF consists of time horizon from one year to several years. Different researchers, used different types of time horizon for forecasting. STLF is mostly used for forecasting, because it gives better accurate prediction results as compared to others. Consumers can also take part in SG operations to reduce the cost of electricity by energy preservation and shifts their consumption load from on-peak hours to off-peak hours. Consumers can utilize energy according to their requirements. To manage supply and demand, both residential customers and industries require electricity price forecasting to cope with upcoming challenges [6]. Robustness, reliability, computational resources, complexity, cost of resources are some issues however, accurate price prediction is also an important issue [7]. When the utilization of electricity is maximum then prices are also high [8]. The price of electricity depends on various factors, such as renewable energy, fuel price and weather conditions etc. [9,10].

1.1 Motivation

In [11], they performed price forecasting of electricity through Hybrid Structured Deep Neural Network (HSDNN). This model is a combination of CNN and LSTM. In this model, batch normalization is used to increase the efficiency of training data. Authors in [12], proposed a model of Gated Recurrent Unit (GRU) in which LSTM is used as a base model for price forecasting accurately. In paper [13], authors predict load consumption using Back Propagation Neural Networks (BPNNs) model. Authors, used this model to reduce forecasting errors. In [14], authors proposed a combined model of Cuckoo Search, Singular Spectrum Analysis and Support Vector Machine (CS-SSA-SVM) to increase the accuracy of load forecasting. In [15], authors used data mining techniques such as k-mean and KNN algorithm for electricity price forecasting. They used k-mean algorithm to make three clusters for weekdays and also used KNN algorithm to divide the classified data into two patterns for the months of February to March and April to January. After classification, a price forecasting model is developed. The price data of 2014 is used as input and results are verified by 2015 data.

1.2 Problem Statement

We reviewed the related works in electricity price forecasting using deep learning techniques and feature engineering.

In [11] this paper proposes an electricity price forecasting system based on the combination of two deep neural networks, the Convolutional Neural Network (CNN) and the Long Short Term Memory (LSTM). However, they neglect the problem of over-fitting. Authors Ziming et al. [16], worked on price forecasting by using the hybrid model of nonlinear regression and SVM. However, the big data is not taken in consideration. Renewable resources, DR, and other factors are influenced on price and load [15,17]. The price of electricity changes frequently, that is why traditional methodologies and approaches are not suitable. We need some enhanced methods for price predictions.

1.3 Contributions

In this paper, main goal is to predict electricity price accurately by using data mining and deep learning techniques. To achieve this, we proposed model for price forecasting. In this work, SVR and CNN both classifiers are used for the prediction of price. Enhanced Convolutional Neural Network (ECNN) is used as a proposed classifier, its results are compared with different benchmark schemes. However, it is very hard to tune the parameters of these models according to dataset. The contributions of this paper are summarized as follows:

- Hybrid Feature Selector: Hybrid feature selector is proposed in this paper.
- Overfitting: Risk of overfitting is mitigated in this model.
- Grid Search and Cross Validation are used to tune the parameters of classifiers, by defining the subset of parameters,
- Enhance classifiers is used to increase the forecasting accuracy.

2 Related Work

Authors in [11], discussed price forecasting with the proposed model of Hybrid Structured Deep Neural Network (HSDNN) in which the combination of CNN and LSTM is used. The accuracy of this model is compared by performance evaluators i.e., MAE and RMSE with different models. In [12], authors described the prediction accuracy with the proposed model of LSTM and RNN named as Gated Recurrent Units (GRU) compared its accuracy with benchmark models: SARIMA, Markov chain and Naive Bayes. Rohit et al.

In [4], authors discussed the data pre-processing steps. They have worked on how to choose a technique for feature selection and feature extraction. These two phases are very important in data pre-processing. Feature selection and extraction techniques play very important role in forecasting. Pre-processing of data is a first step in every forecasting process. Normalized data provides better results for accuracy in forecasting. Data, which is present in raw form gives poor result in prediction. In this work, a meta learning approach is implemented and recommends the pre-processing technique, which shows better results.

In [17], authors proposed a model for price forecasting using Deep Learning approaches i.e. DNN as an extension of traditional MLP, hybrid LSTM-DNN structure, hybrid GRU-DNN structure and CNN model. Wang et al. [18], authors proposed a hybrid framework of feature selection, feature extraction and dimensionality reduction by GCA, KPCA and also predict the price of electricity through SVM. In [19], authors used Stacked Denoising Autoencoder (SDA) and DNN models. They also compared different models including SVM, classical Neural Network and multivariate regression. Lago et al. [20], worked on DNN to improve the predictive accuracy of a market, for feature selection. They used Bayesian optimization and functional analysis of variance. Also proposed, another model to perform price prediction of two markets simultaneously. Raviv et al. [21], used multivariate models for prediction hourly price instead of univariate, also mitigate the risk of overfitting by using dimensionality reduction techniques and forecast combination. Javaid et al. [22], proposed a deep-learning based model for the prediction of price, using DNN and LSTM. They worked on the prediction of both price and load. In [23], authors considered a probabilistic model for hourly price prediction. Generalize Extreme Learning Machine (GELM) is used for prediction. They used bootstrapping techniques, to increased the speed of model by reducing computational time. Abedinia et al. [24], focused on feature selection to performed better predictions. These proposed models, based on information theoretic criteria i.e. Mutual Information (MI) and Information Gain (IG) for feature select. Another contribution of this paper is a hybrid filter-wrapper approach.

In [25,26], proposed a hybrid algorithm for price and load forecasting. Also worked on new conditional feature selection, Least Square Support Vector Machine (LSSVM) and proposed a new modification for Artificial Bee Colony Optimization and Quasi-Oppositional Artificial Bee Colony (QOABC) algorithm. Keles et al. [27], proposed a method based on ANN. They also used different clustering algorithms to find optimal parameters for ANN. Wang et al. [28], proposed Dynamic Choice Artificial Neural Network (DCANN), this model is used for day-ahead price forecasting. This model is a combination of supervised and unsupervised learning, which deactivates the bad samples and search optimal inputs for a model to learn. In [29], developed a hybrid model based on Neural Network. Authors, in [30], used Multilayer Neural Network (MLNN) for electricity price forecasting.

3 Proposed Model

In this paper, a novel price prediction model is proposed. Figure 1 shows the proposed model for price prediction. Proposed model is divided into four modules. The modules of proposed models are:

1. Feature Selection,
2. Feature Extraction,
3. Grid Search and Cross Validation,
4. Price Prediction using SVR and CNN.

The individual module is further explained in the following subsections.

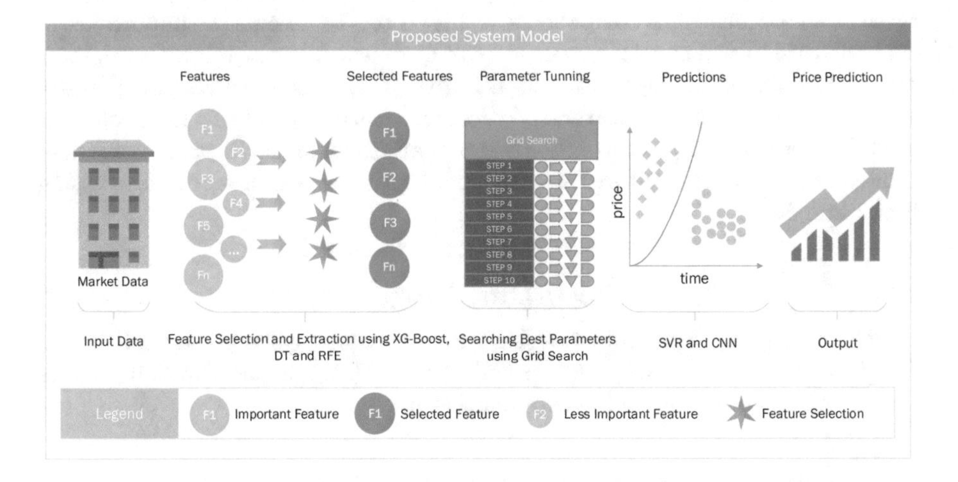

Fig. 1. Proposed model for price prediction

3.1 Model Overview

The accuracy of prediction is key issue in electricity price forecasting. As discussed earlier, the electricity price depends on various factors, which make training of classifiers difficult. To improve accuracy of price prediction, hybrid feature selector (i.e., DTC and XG-boost) is used to select most relevant features. At first, RFE is used to remove dimensionality and redundancy of data. In order to tune parameters of classifier, GS is used along with cross validation to select best subset of parameters. Finally, selected features and best parameters are used in classifiers to predict electricity price.

3.2 Feature Extraction Using RFE

RFE is used to select specified number of features from dataset. It removes weakest feature recursively, until the specified number of features is reached. RFE requires number of feature to select, however, it is difficult to decide in advance that how many features are most relevant. To address this issue, cross validation is used with RFE. Cross validation calculates accuracy of different subsets and select the subset with highest accuracy.

3.3 Feature Selection Using XG-Boost and DT

Using XG-boost and DT, importance of all features is calculated with respect to target, i.e., electricity price. These techniques calculate the importance of features in vector form. The components of this vector, represents the importance of every feature in sequence. However, we can drop features which have less importance. The fusion of Xg-boost and DT gives more accurate results. Figure 4 shows the importance of features. To control feature selection, threshold ϵ is used. Features having importance greater than or equal to threshold ϵ are considered and rest of the features are dropped. Feature selection is performed using Eqs. 1 and 2.

$$Fs = Reserve\ if\ I_{XG}[i] + I_{DT}[i] \geq \epsilon \tag{1}$$

$$Drop\ if\ I_{XG}[i] + I_{DT}[i] < \epsilon \tag{2}$$

Where, $I_{XG}[i]$ represents the feature importance calculate by XG-boost, $I_{DT}[i]$ is the feature importance calculated by DT. ϵ is the threshold values for the feature selection and i represent feature.

3.4 Tuning Hyper-parameters and Cross Validation

Tuning classifier is very important to do accurate and efficient forecasting. There is a strong relationship between hyper-parameter and results of classifier. GS is used to the tune parameters of classifier for higher accuracy. For this purpose, we define subset of hyper-parameters for SVM shown in Table 1.

Table 1. Subset of parameter for Grid Search

Parameter name	Parameter value(s)
kernel	['linear', 'rbf']
C	[3, 4, 5, 10, 15, 20, 30, 50]
gamma	['scale', 'auto', 5, 10, 20, 30, 50]
epsilon	[0.2, 0.02, 0.002, 0.0002]

3.5 Electricity Price Forecasting

After feature selection and parameter tuning, the processed data and the best subset of parameters are used in SVR and CNN to forecast electricity price. Hourly price data of two months (November and December 2016) are used to train classifier and predicts the price for first week of January 2017. We compared the results of first January 2017 and first week of January 2017 with actual price of electricity of NYISO. The results of SVR are shown in Fig. 5(a) and (b) whereas, results of CNN are shown in Fig. 5(c) and (d), respectively.

4 Simulation and Results

In this section, the simulation results are discussed in details.

4.1 Simulation Environment

For simulation purpose, we implement the proposed models by using the following python libraries i.e. Keras, Tensorflow, Sklearn, numpy and pandas. Models are implemented on a system with Intel core i3, 8 GB RAM and 500 GB storage capacity. Two different datasets are selected for simulation. Lastly, Dataset [31] is used as input in price prediction model, which is taken from New York Independent System Operator (NYISO). However, dataset 2 contains hourly data of price and electricity generation from 2016 to 2017.

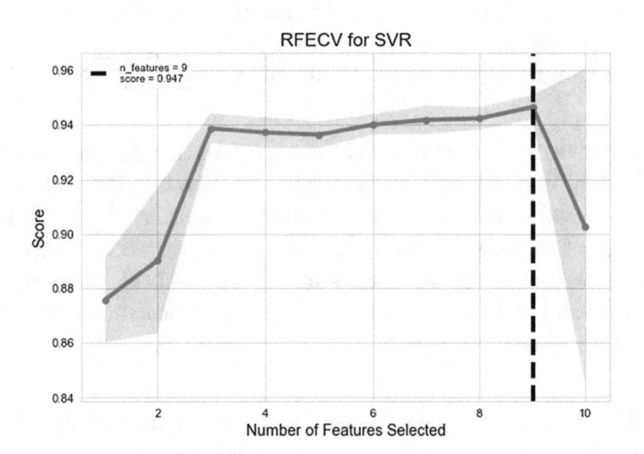

Fig. 2. Result of cross validation (RFE)

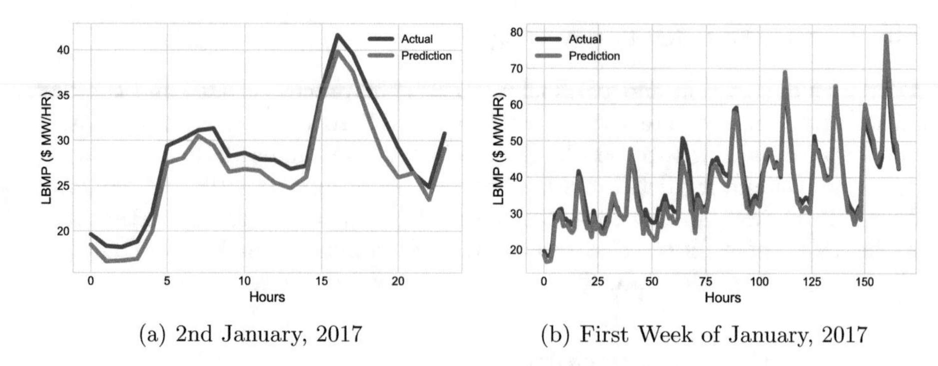

(a) 2nd January, 2017

(b) First Week of January, 2017

Fig. 3. Price prediction without parameter tuning.

4.2 Results of Price Prediction Model

The proposed model is shown in Fig. 1. NYISO dataset [31] is taken as input, which contains 9,314 real-world records. However, for the sake of demonstration, 75 days dataset are used to train model. This dataset invariably contain approximately 2000 h record, i.e., from 1st November, 2016 to 15th January, 2017. The whole simulation process is organized as:

1. Feature extraction using RFE
2. Feature selection by combining the attributes importance calculated by XG-boost and DT
3. Parameter tuning using cross validation and Grid Search
4. Prediction using SVR and CNN
5. Results and Comparison with real data of January 2017.

Feature Extraction: To remove redundancy and dimensionality of data, RFE is used. Although, it is difficult to determine in advance how many features set is required. To resolve this issue, cross validation is used with REF to select optimal number of features. Cross validation tests every combination of features and calculates the accuracy of each subset. The subset of features with the highest accuracy is used for prediction. Figure 2 shows the maximum accuracy score on seven number of features.

Feature Selection: Importance of selected features are calculated by both DT and XG-boost. By adding both importance, combined importance is calculated. For selection of features, a threshold value ϵ is defined. Features are selected with importance greater than or equal to threshold value. Figure 4 shows the importance of every feature. Some features have very high importance, i.e., TWI Zonal LBMP, RTC Zonal LBMP and Load. TWI Zonal Price Version shows very less importance as compared to others features. Most of the features have importance greater than 0.15 and that is why we set the values of threshold to 0.15. Those features whose values are less than threshold value are dropped.

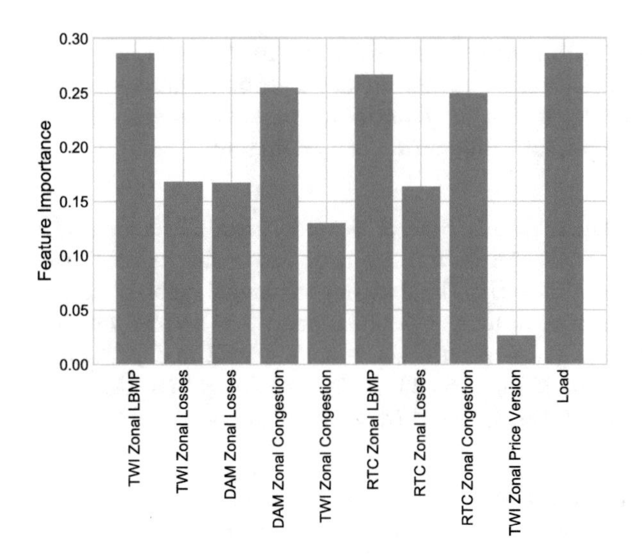

Fig. 4. Feature importance for price

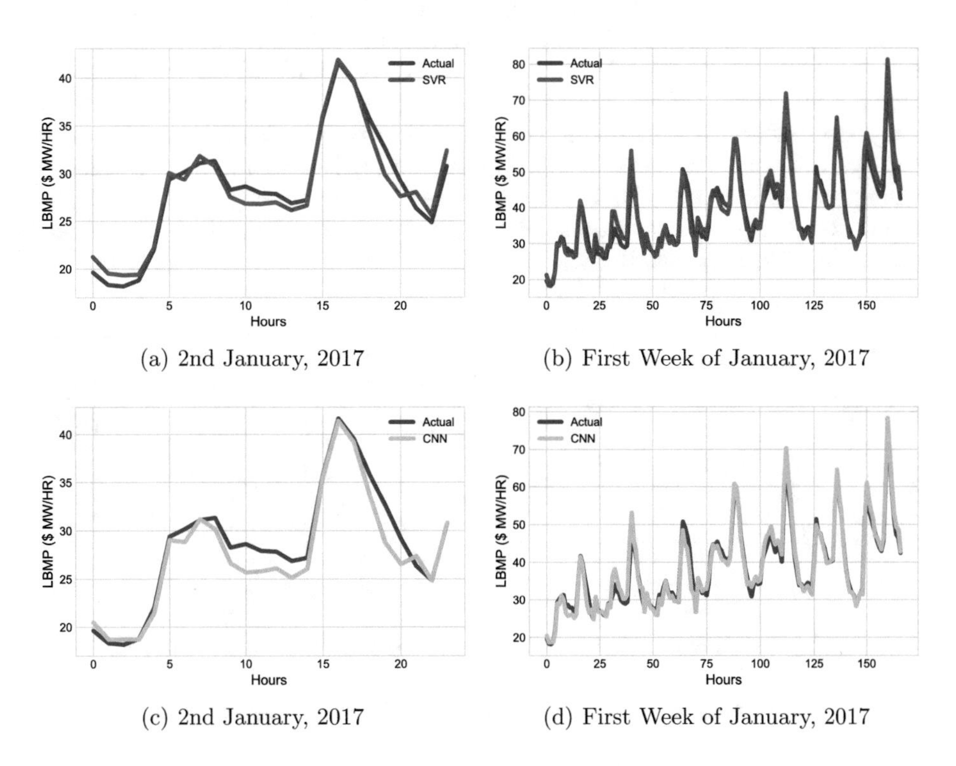

(a) 2nd January, 2017

(b) First Week of January, 2017

(c) 2nd January, 2017

(d) First Week of January, 2017

Fig. 5. Price prediction using SVR and CNN.

Parameter Tuning and Cross Validation: To find the optimal set of parameters for the classifiers, we use the defined a set of parameters as shown in Table 1. Using GS, every possible combination of parameters are tested by the propose model to find optimal combination of parameters.

Price Prediction: Hourly data of November and December 2016 is used to the train classifier. SVR and ECNN are used to predict price of electricity for first week of January. To verify the accuracy of model, predicted price is compared with the actual price of first week of January. The results are shown in Fig. 5(a), (b), (c) and (d). These figures show both actual and predicted price for first day and first week of January, 2017. Figure 5(a) and (b) shows the prediction of classifier SVR and Fig. 5(c) and (d) reports the result of CNN classifier.

Discussion of Results: As we know, the main goal of this proposed model is to improve the accuracy of classifier to predict price correctly. The results before parameter tuning of classifier are shown in Fig. 3(a) and (b) are less accurate. The MAE of prediction before parameter tuning is approximately equal to 2.83. After feature selection, extraction and parameter tuning through GS, the results are improved. The results after parameter tuning of SVR is shown in Fig. 5(a) and (b). The results of CNN is shown in Fig. 5(c) and (d). After parameter tuning, the accuracy of classifiers are improved, the MAE is reduced to 1.81 The comparison of actual values with before and after tuning classifiers are shown in Fig. 6(a) and (b). The value of MAE before tuning is 2.83 and after parameter tuning the value is 1.81. MAE value is decreased then it shows that results are improved after parameter tuning. Reducing 1% error values of MAE can save thousands of MW of electricity.

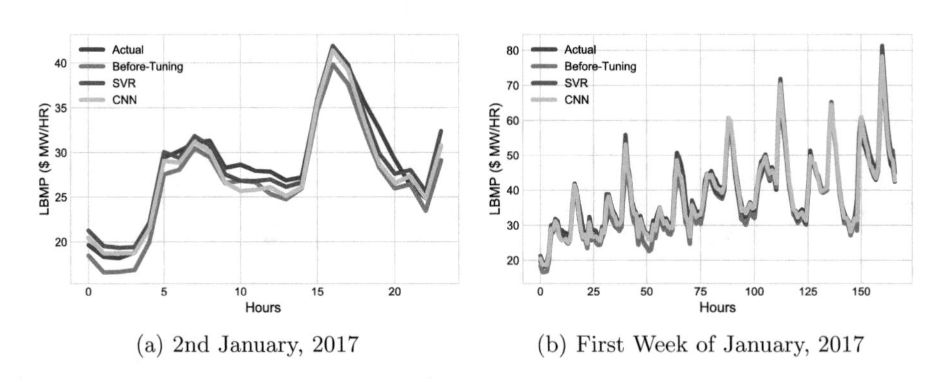

(a) 2nd January, 2017 (b) First Week of January, 2017

Fig. 6. Comparison of predictions before and after parameter tunning.

5 Conclusion

In this research study, a new model is established to predict the price of electricity efficiently and accurately. Proposed model is consist of feature selection, feature extraction, parameter tuning and classification. Hybrid feature selector (hybrid of DT and XG-boost) is used to select important features for prediction. For dimensionality reduction and feature extraction, RFE is used. In order to tune the parameters of classifiers, grid search is used, which boost the classifier's accuracy. Enhanced classifiers like CNN and SVR are used to predict price and load is proposed models for better accuracy. The results of classifiers is satisfactory and show better accuracy than benchmark scheme.

References

1. Masip-Bruin, X., Marin-Tordera, E., Jukan, A., Ren, G.-J.: Managing resources continuity from the edge to the cloud: architecture and performance. Future Gener. Comput. Syst. **79**, 777–785 (2018)
2. Osman, A.M.S.: A novel big data analytics framework for smart cities. Future Gener. Comput. Syst. (2018)
3. Chen, X., Zhou, Y., Wei, D., Tang, J., Guo, Y.: Design of intelligent Demand Side Management system respond to varieties of factors. In: 2010 China International Conference on Electricity Distribution (CICED), pp. 1–5. IEEE (2010)
4. Jindal, A., Singh, M., Kumar, N.: Consumption-aware data analytical demand response scheme for peak load reduction in smart grid. IEEE Trans. Industr. Electron. (2018)
5. Bilalli, B., Abell, A., Aluja-Banet, T., Wrembel, R.: Intelligent assistance for data pre-processing. Comput. Stand. Interfaces **57**, 101–109 (2018)
6. Mohsenian-Rad, A.-H., Leon-Garcia, A.: Optimal residential load control with price prediction in real-time electricity pricing environments. IEEE Trans. Smart Grid **1**(2), 120–133 (2010)
7. Chandrashekar, G., Sahin, F.: A survey on feature selection methods. Comput. Electr. Eng. **40**(1), 16–28 (2017)
8. Erol-Kantarci, M., Mouftah, H.T.: Energy-efficient information and communication infrastructures in the smart grid: a survey on interactions and open issues. IEEE Commun. Surv. Tutorials **17**(1), 179–197 (2015)
9. Wang, K., Li, H., Feng, Y., Tian, G.: Big data analytics for system stability evaluation strategy in the energy Internet. IEEE Trans. Industr. Inf. **13**(4), 1969–1978 (2017)
10. Wang, K., Ouyang, Z., Krishnan, R., Shu, L., He, L.: A game theory-based energy management system using price elasticity for smart grids. IEEE Trans. Industr. Inf. **11**(6), 1607–1616 (2015)
11. Kuo, P.-H., Huang, C.-J.: An electricity price forecasting model by hybrid structured deep neural networks. Sustainability **10**(4), 1280 (2018)
12. Ugurlu, U., Oksuz, I., Tas, O.: Electricity price forecasting using recurrent neural networks. Energies **11**(5), 1–23 (2018)

13. Eapen, R.R., Simon, S.P.: Performance analysis of combined similar day and day ahead short term electrical load forecasting using sequential hybrid neural networks. IETE J. Res. 1–11 (2018)
14. Chitsaz, H., Zamani-Dehkordi, P., Zareipour, H., Parikh, P.: Electricity price forecasting for operational scheduling of behind-the-meter storage systems. IEEE Trans. Smart Grid (2017)
15. Patil, M., Deshmukh, S.R., Agrawal, R.: Electric power price forecasting using data mining techniques. In: 2017 International Conference on Data Management, Analytics and Innovation (ICDMAI), pp. 217–223. IEEE (2017)
16. Ziming, M.A., Zhong, H., Xie, L., Xia, Q., Kang, C.: Month ahead average daily electricity price profile forecasting based on a hybrid nonlinear regression and SVM model: an ERCOT case study. J. Mod. Power Syst. Clean Energy **6**(2), 281–291 (2018)
17. Lago, J., De Ridder, F., De Schutter, B.: Forecasting spot electricity prices: deep learning approaches and empirical comparison of traditional algorithms. Appl. Energy **221**, 386–405 (2018)
18. Wang, K., Xu, C., Zhang, Y., Guo, S., Zomaya, A.: Robust big data analytics for electricity price forecasting in the smart grid. IEEE Trans. Big Data (2017)
19. Wang, L., Zhang, Z., Chen, J.: Short-term electricity price forecasting with stacked denoising autoencoders. IEEE Trans. Power Syst. **32**(4), 2673–2681 (2017)
20. Lago, J., De Ridder, F., Vrancx, P., De Schutter, B.: Forecasting day-ahead electricity prices in Europe: the importance of considering market integration. Appl. Energy **211**, 890–903 (2018)
21. Raviv, E., Bouwman, K.E., van Dijk, D.: Forecasting day-ahead electricity prices: utilizing hourly prices. Energy Econ. **50**, 227–239 (2015)
22. Mujeeb, S., Javaid, N., Akbar, M., Khalid, R., Nazeer, O., Khan, M.: Big data analytics for price and load forecasting in smart grids. In: International Conference on Broadband and Wireless Computing, Communication and Applications, pp. 77–87. Springer, Cham (2018)
23. Rafiei, M., Niknam, T., Khooban, M.-H.: Probabilistic forecasting of hourly electricity price by generalization of ELM for usage in improved wavelet neural network. IEEE Trans. Industr. Inf. **13**(1), 71–79 (2017)
24. Abedinia, O., Amjady, N., Zareipour, H.: A new feature selection technique for load and price forecast of electrical power systems. IEEE Trans. Power Syst. **32**(1), 62–74 (2017)
25. Ghasemi, A., Shayeghi, H., Moradzadeh, M., Nooshyar, M.: A novel hybrid algorithm for electricity price and load forecasting in smart grids with demand-side management. Appl. Energy **177**, 40–59 (2016)
26. Shayeghi, H., Ghasemi, A., Moradzadeh, M., Nooshyar, M.: Simultaneous day-ahead forecasting of electricity price and load in smart grids. Energy Convers. Manage. **95**, 371–384 (2015)
27. Keles, D., Scelle, J., Paraschiv, F., Fichtner, W.: Extended forecast methods for day-ahead electricity spot prices applying artificial neural networks. Appl. Energy **162**, 218–230 (2016)
28. Wang, J., Liu, F., Song, Y., Zhao, J.: A novel model: Dynamic Choice Artificial Neural Network (DCANN) for an electricity price forecasting system. Appl. Soft Comput. **48**, 281–297 (2016)

29. Varshney, H., Sharma, A., Kumar, R.: A hybrid approach to price forecasting incorporating exogenous variables for a day ahead electricity market. In: IEEE International Conference on Power Electronics, Intelligent Control and Energy Systems (ICPEICES), pp. 1–6. IEEE (2016)
30. Mosbah, H., El-Hawary, M.: Hourly electricity price forecasting for the next month using multilayer neural network. Can. J. Electr. Comput. Eng. **39**(4), 283–291 (2016)
31. NYISO Hourly SMD from 2011 to 2018. https://drive.google.com/open?id=1XO-ON8Jtmpsntw7GCeenBNFvw_AZCMlj

Semantic Multi Agent Architecture for Chronic Disease Monitoring and Management

Lina Nachabe[1,2(✉)], Bachar El Hassan[1], and Jean Taleb[2]

[1] Lebanse University, Beirut, Lebanon
bachar_elhassan@ul.edu.lb
[2] American University of Culture and Education, Beirut, Lebanon
linanachabe@auce.edu.lb, jhonny315.jt@gmail.com

Abstract. Population all over the world is experiencing an epidemic of chronic disease (such as diabetes, heart disease, etc.) which is considered as the leading cause of morbidity and mortality. These diseases are complex and require the intervention and interaction between different stakeholders (doctors, nurses, experts, dietitians). Moreover, patient self-monitoring and management can contribute in the diagnosis and treatment. With the evolution of medical sensors and m-health applications, vital signs monitoring is becoming easier. However, a new approach for chronic disease health care monitoring and management is needed in order to insure intelligence decision making, interoperability between existing systems, and interaction across stakeholders, as well as early diagnosis and self-monitoring. In this paper, we present a semantic multi agent architecture based on predefined ontology and using JADE framework. This architecture encompasses two main agents: contractor and manager in order to offer the adequate data to the requested parties (doctors/patients).

1 Introduction

In the last decade, the number of patients suffering from chronic disease is dramatically increasing. It has been estimated that, by 2020, 75% of worldwide deaths are caused by chronic diseases [20] which can be defined as illness that last for more than a year and can persist over time such as heart disease, cancer, diabetes, arthritis, stroke, chronic lower respiratory disease, etc. Unfortunately, these syndromes can progress and cannot be resolved impulsively [3]. It is worthy to note that these diseases are affecting poor as well as developed countries, thus, it is creating a major public health threat that distress all societies [20].

In addition to health burden, the economic impact of chronic disease is considered a major threat that should be solved. It is reported that by 2023, the total impact on the economy will increase to $4.2 trillion [6]. Furthermore, strong correlation and overlap exist between different chronic diseases, and require from caregivers to consider the interaction of multiple factors relevant for individual care plans. However, the actual healthcare systems lack for coordination and communication between medical staff of different care systems, patients and relatives [2].

L. Barolli et al. (Eds.): EIDWT 2019, LNDECT 29, pp. 284–294, 2019.
https://doi.org/10.1007/978-3-030-12839-5_26

Despite medical treatment, according to the report by the World Health Organization (WHO), patient's lifestyle and diet are considered as important adaptable elements of chronic diseases. Although many mobile applications and systems are providing nutrition consultancy and nutrients information, but it is built as stand-alone solution focusing on limited aspects.

As it can be seen, chronic diseases are complex syndromes that need new innovative approach for sustainable care to limit excessive cost and enhance the quality of life for the majority of patients worldwide. This new system should tackle the following requirements:

- Cooperation between different stakeholders using efficient data exchange and management.
- Patient involvement via real time monitoring and assistance.
- Proactive and reactive diagnosis.
- Advanced and intelligent decision making.
- Lack of expertise and medical staff.
- Security and privacy.
- Standardized solutions where government or other authorities can verify the reliability and correctness of the provided solutions.

For that purpose, we are proposing in this paper a semantic multi-agents architecture for chronic diseases (SMAS-CD) data monitoring, data management, patient's assistance and awareness alleviating the interoperability between different systems and enhancing decision making.

In Sect. 2 an explanation about semantic techniques and multi agents systems is depicted. Section 3 reviews existing multi agent systems in medical fields. In Sect. 4, we detail our proposed SMAS-CD architecture by describing the role of agents, the scenarios and the implementation. Finally, the paper is concluded in Sect. 5.

2 Background

Healthcare systems for chronic disease are generally characterized by being complex system, widely distributed where different actors should intervene for accurate diagnosis and assistance [19]. These systems require a general architecture that encompasses heterogeneous data about patients (vital signs, daily activity, diet, medical history, etc.), accessible by authorized parties (medical staff, patient, relatives, etc.), understandable between different systems (labs, mobile apps, etc.) and intelligent for advanced diagnosis and assistance. In other words, this architecture should resolve these main challenges: interoperability, effective data sharing, intelligence, context awareness, and cooperation. Thus, semantic techniques in conjunction with multi-agents methodology are promises techniques that fit these requirements.

2.1 Semantic Techniques

The term "semantic" refers to the meaning, so semantic techniques are used to give significance to the stored data. It can be used to represent generic knowledge, and has the ability to infer new information [5]. In particular, ontologies are "tools for specifying the semantics of terminology system in a well-defined and unambiguous manner" [11]. In 2004, W3C recommended the use of OWL language to describe ontologies. The main reason behind using ontologies in our proposed architecture resides in resolving the problem of interoperability. In such way, different e-health actors (experts, doctors, etc.) can share and exchange data and contribute in an efficient diagnosis and assistance. The semantic layer of our architecture is conform with the ETSI standard for BAN data management and representation [17] and extended in [12] to cover all aspects of the diabetes chronic disease. The proposed ontologies focused on food intake, vital signs monitoring (heart rate, sugar, level…), medication, family history, daily activities and symptoms. It insures the correlation between these different factors enabling an advanced data processing and decision making. Due to the use of semantic rules (SWRL) and inference agent (Pellet reasoner), complex inferences are made such the discovery of diabetes patients due to the family history parameter and symptoms. It offers the doctors the opportunities to add assumptions and infer advanced deduction to help patient's assistance and diagnosis. Moreover, it gives personalized services since each patient is fully described. Furthermore, where different systems suffer from lack of regulation and may not comply with medical evidence or may contain outdated formulas, (in particular mobile applications for chronic diseases monitoring and assistance) [16], the use of a common shared ontology can lessen this problem by giving authority to specialists like governments or FDA (Food and Drug Administration) to supervise medical annotated information and rules used for all e-health systems.

Besides semantic annotation, data sharing and inference, a distributed system of standalone modules offering services and able to communicate with each other and the surrounding environments, is required for chronic disease management architecture.

2.2 Multi-agents System

An agent is a software entity capable of choosing and performing a set of actions in order to achieve a goal required by the user [9]. Agents are characterized by [9, 10]:

- Autonomy: work dynamically and take own decisions.
- Pro-activity: answer to external request and adapt their actions without direct user's intervention.
- Negotiation/Sociability: communication between different agents and setting agreements.
- Intelligence/Learning/Context Awareness: improve their performance while working with the surrounding environment and learn from past actions.

Agents can perceive data, think and act accordingly. Multi agent system (MAS) is a "loosely coupled network" encompassing set of agents interacting with each other and the environment to take intelligent decisions or solve complex problems [13].

The use of multi agents systems in the medical field is not relatively new. As we have seen in the section above, MAS can be used to solve the complexity and interactivity requirements in the medical systems and especially when managing chronic diseases. Table 1 summarizes how our proposed MAS semantic architecture can tackle the challenges of chronic diseases management and monitoring systems mentioned in Sect. 1.

Table 1. Challenges and solution for diseases management and monitoring systems

Challenges	Solutions	Techniques used
Cooperation between stakeholders	Interaction between agents	MAS
Efficient data sharing	Semantic data annotation using ontologies	OWL language
Patient involvement	Agents interacting with patients and acting accordingly in intelligent manner	MAS
Proactive/Reactive diagnosis	Inference engine and semantic rule, in addition to the cooperation between agents	MAS/OWL/SWRL
Lack of expertise	Doctors/experts can act remotely by adding intelligent SWRL and accessing intelligent agents	MAS/SWRL
Security and privacy	Use of encryption and token	JADE-S/Token
Standardized solutions	Common ontology where authorized parties can define rules and supervise the semantic annotation process	OWL/SWRL

3 Existing MAS in the Medical Fields

Chakraborty et al. [4] summarized in their paper the existing multi-agent systems in medical or health care domain. They considered that multi-agent systems are suitable for healthcare paradigm. Attempt in this field started in 1998, with RETSINA agent, a BDI (belief desire intention) type agent for p1anning, scheduling, execution, information gathering, and coordination with other agents.

Moreno [9] argued why MAS should be used in health care systems. Moreover, some agents were proposed such as patient scheduling, organ transplant management, community care, information access, decision aid systems, and internal hospital tasks. The system AgentCities.NET, implemented using JADE-S (a security plug-in of JADE), has been depicted. The aim of the system is to provide agent-based services that improve the quality of life of the citizens and the visitors of a city. Moreover, Palliative Care Unit system adopted multi agent systems for health care. It includes Communication Manager, Data Base Wrapper, doctor agent, patient agent, and PC coordinator. The main purpose of this system is to facilitate the medical data access.

Omran et al. [13] suggested Bee-gent framework comprised of two main agents: Agent Wrappers used to model existing applications with agents, and mediation agents for coordination between applications. It uses the inference engine for advanced data analysis.

Tian et al. [18] proposed MAS for e-medicine that encompasses an interface agent representing the user graphical interface, broker agent which knows all the capabilities of the system, doctor agent for scheduling appointments, and admin agent for medical administration. The case study focused on diabetes patients. It proposes a monitoring, data processing, diagnosis, therapy and archival agents.

Li et al. [7] introduced the concept of agent based modeling for chronic diseases (diabetes, cardiovascular disease, and obesity). The existing systems model the patient as an agent where health factors such as smoking and physical activity and smoking are introduced. The authors suggest incorporating agents that deal with health behaviors, diet, and evidence from biological and medical researches, as well as multi-morbidity (effects and interactions of different chronic diseases).

In summary, all attempts confirm the importance of using MAS to solve the problem of complexity and communication in the domain of medical systems. However, none of these solutions provide a complete architecture for chronic diseases monitoring and management where a common framework is provided with semantic capability and distributed agents providing security, data accessing, data management, data discovery, inference, intelligent decision making, security/privacy, and user interaction. Thus, we are proposing our semantic MAS platform for chronic disease (SMAS-CD).

4 Our Semantic MAS Platform for Chronic Disease (SMAS-CD)

4.1 Agents Roles and Descriptions

SMAS-CD is basically divided into two main agents' types: Interface Agents provided by different application/user interface/developer and Service Agents that can cooperate, propose solutions for different problems, and deliver it to the Interface Agents. This segmentation is done in order to avoid overlapping between the platform's core where intelligence and decision making should be taken and the user interface for data insertion and retrieval. Thus, Interface Agents are accessed by users (patients/doctors/nurses) for data retrieval (e.g. heart rate monitoring); consultancy, and data insertion (e.g. add intake food or daily activity). Moreover, it can be used by doctors/experts to fill up patient's report or post decisions and diagnoses.

Service Agents are dedicated to deliver services to the medical staff/experts, patients and relatives. The service oriented paradigm is adopted [8, 14] as follow:

- Service requester is responsible for discovering a service by searching through the service descriptions given by the service broker. In our case, the service will be named as "Manager Agent" (as used in [21]).

- Service Provider creates description of the service and publishes it to the service broker. In our architecture the service will be named as "Contractor Agent" (as used in [21]).
- Service Broker contains a registry of service description in order to insure the link between the requestor and the provider. In our case, it is the Directory Facilitator "DF Agent" [1].

In general, the contractor agents related to same service will publish their services and register themselves in the yellow page (DF agent), while the manager service, when requested form an interface agent will search in the yellow page for the best contractor(s) and send back the data to the interface user.

For chronic diseases, the basic services that should be offered:

- Monitoring: a data collector agent is developed in order to collect requested data for patients' monitoring. It is a contactor agent that can take advantage of its social ability and coordinate with other agents to retrieve demanded data. The data collector agent is always invoked by interface agent when information retrieval is needed.
- Medication: a medicine tracker agent is used for each patient in order to record his/her medicines and track if he/she is taken it on time and appropriately. It is a contractor agent which gets information from an interface agent (it can be a mobile application for medicine tracking) and invoke, when necessary (e.g. reminder) the notification agent to inform the medical staff/relatives/patients for any intervention. Moreover, it can play the role of manager agent to retrieve the data and deliver it to an interface agent.
- Diet: as for the medication agent, diet tracker agent is developed as contractor agent for data entry and manager agent for data retrieval. Experts in food domain can interact by creating diet info agent for awareness and advanced food description.
- Notification: notification can be done using SMS, email, mobile notification and so on. It is the role of interface notification agent to precise the way of communication. Moreover, this agent can communicate with different parties such as hospital, ambulance, doctors, etc. In our proposed architecture, a notification manager agent will invoke the necessary notification contractor agent.
- Security: this agent is responsible of the encryption and the privacy of the data. It is neither a contractor nor a manager agent. It is used when data encryption is needed or when patient identity is requested.
- Diagnosis: the data collector agent, mediation agent, food agent, daily activity agent and intelligent agents cooperate together for efficient diagnosis. The diagnosis may involve one of more agents for better decision making. In fact, in SMAS-CD architecture, the intelligence resides in the ability to understand data and infer more complex information. Thus, a semantic agent is developed in order to interact with the predefined ontology for chronic disease management, in particular diabetes use case. (http://lina.ataouna.org/MySmartDiabetesOnto.owl) [12].

4.2 SMAS_CD Scenarios

The first step for each data contractor is to publish its service in the yellow page. To clarify the functionality of the architecture, we will detail the monitoring process because it includes the most used agents. First, the patient should have a token for identification requested from the security agent. This token is registered in the catalog with the unique patient ID (depicted in Fig. 1). Any interface agent requesting a service should send the patient ID and the token to grant access. In this way the privacy of information is provided.

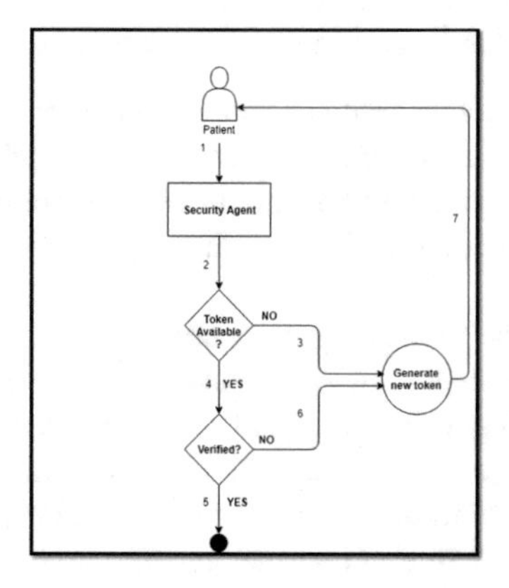

Fig. 1. Authentication process in SMAS-CD

Afterwards, the interface agent can invoke the manager agent based on its needs. For illustration, let us consider the case where a diabetic patient equipped with heart rate sensor felt extremely tired and need assistance. His/her doctor requests (via interface agent) the previous heart rate measurements (for e.g. from 2 days) and track the medicine intake during a week. The data collector agent as well as the medicine agent will be invoked. The data collector will check for heart rate measurements by sending the patient ID, the token, the measurement type and the date interval to the yellow page to inquiry the best contractor. The contractors can be a scanner agent (from sensors), a semantic agent (data annotated in the ontology) or data logger agent (from database or medical report). If many offers are available, the data collector will send these offers to the interface agent or choose the best offer. Figure 2 illustrates the steps for data collection. In such way, the doctor by receiving all needed information without having direct access to the data can make clear assumptions or request the intelligent agent for further assistance.

As depicted in Fig. 2 the data logger can be xml files, media, and database of a laboratory/hospital, medical report or others. The vital signs agent will read data from Bluetooth sensors, Wi-Fi or serial devices, etc.

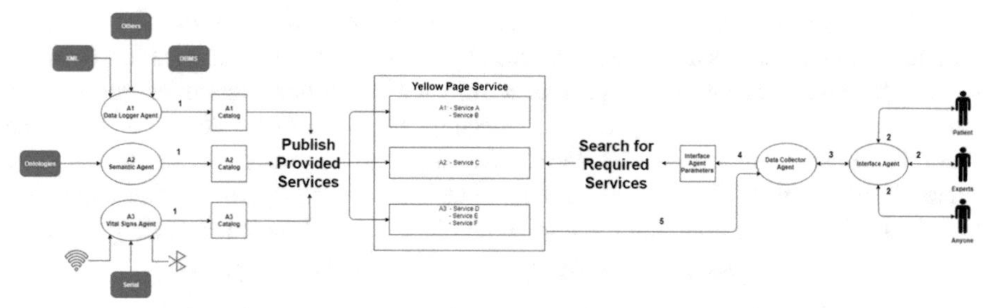

1. Ai agent adds data to the catalog and publish the service in the yellow page
2. Users send their request to the interface agent
3. The interface agent invokes the data collector agent and sends the patient ID and the token.
4. The data collector agent searches in the yellow page for the service matching with the sent parameters.
5. The Data collector sends back the found data to the interface agent.

Fig. 2. Monitoring service within SMAS-CD

4.3 SMAS-CD Implementation

To implement SMAS-CD architecture JADE middleware is used [15] for the following reasons:

- Java programming language which is an open source, hardware independent and portable language. Thus, these agents can be developed on the cloud side, local server, gateways, android applications, etc.
- Graphical tools that facilitates the administration and activity's monitoring of running agents.
- DF (Directory Facilitator) that provides a Yellow Pages service by means for agent discovery.
- Communication between agents due to asynchronous message passing implemented in ACLMESSAGE class.

Each agent is created in a container due to the fact that it can be implemented on different machine. Figure 3 depicts the vital sign agent, data logger agent, semantic agent and scanner agent. As illustration, the vital sign agent can be configured in the home gateway of the patient (e.g. raspberry pie), the data logger agent can be found in hospital or medical labs, and the semantic agent is a cloud agent connected to the ontology.

The patient (in our case the interface agent) will send the data to be fetched. As mentioned previously, the data collector agent will place these parameters (e.g. patient ID, measurement type, valid date) in the yellow page. Thus, all agents will be requested, the best offer (in terms of suitability or recent date) will be sent back to the patient.

Figure 3 depicts the GUI interface developed for agents to enter their data and publish it in the yellow page. Figure 4 illustrates the communication between agents. The data collector agent is trying to find offers based on requested data, and print out all the contractor agents available. In this example, two agents match with the patient request (the measurement type "Blood Pressure" and patient ID "1"). So, the agents are the semantic agent and vital signs but, each one of them has different date. As in Fig. 4, the semantic agent is chosen, because it has the newest date which is 29/6/2018 (where the vital signs agent has a date of 20/5/2017). At the end the communication between the patient and data collector agent is terminated to reduce system overhead. The testing was conducted on 50 patients requesting the same data collector at the same time. The average time taken to deliver data back to the patient is about 3 s. The average CPU usage is 40%. We should note that time needed to process requests varies based on the available contractor agents. It can be estimated by the following formula:

$$\text{Time\_back} = \text{Max}\,(n_1 * 20,\ n_2 * 5,\ n_3 * 10)\ \text{seconds}$$

Where n_1 is the number of semantic agent, n_2 is the number of vital signs agent and n_3 is the number of data logger agent. Moreover, for testing purposes, the agents where hosted on the same machine, thus the delay from network transmission has been ignored.

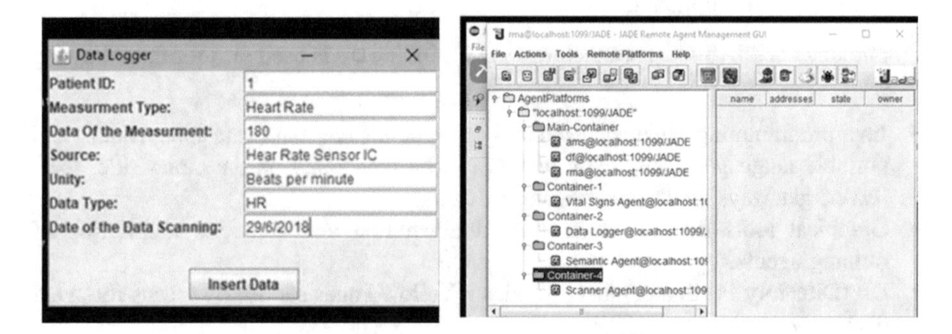

Fig. 3. GUI agents in SMAS-CD

```
Data Collector Agent ==> Data Collector Agent@localhost:1099/JADE
The Patient with ID: 1, newest information are successfully found in Contractor agent ==> Semantic Agent@localhost:1099/JADE.
With the best date, where Patient Information are:
1) Measurment Type: Blood Pressure
2) Data: 100/70
3) Source: Heart Rate Sensor IC
4) Unity: mm Hg
5) DataType: BP

=> BEST DATE = 29/6/2018

###########################

Data Collector - agent Data Collector Agent@localhost:1099/JADE terminating.

###########################
```

Fig. 4. The data collector finding the best offer for the patient's request

5 Conclusion

With the increasing number of patients suffering from chronic disease, and the high cost needed for monitoring and management, it is becoming primordial to propose new architecture for chronic disease monitoring and management. In this paper we suggested a semantic architecture to address the problem of standardization and interoperability, where multi intelligent agents can invoke each other's for patients/doctors/medical assistance and monitoring. The architecture was developed and tested suing JADE framework and relying on yellow page techniques. We are working now on developing a mobile application for data acquisition and notification that plays the role of interface agent and communicates with our proposed architecture. Advanced testing will be conducted to insure the reliability and effectiveness of our architecture.

References

1. Anandampilai, B.: Content-based multicasting using JADE. Int. J. Soft Comput. **2**, 422–425 (2007)
2. Brunner-La Rocca, H.P., Fleischhacker, L., Golubnitschaja, O., Heemskerk, F., Helms, T., Hoedemakers, T., Allianses, S.H., Jaarsma, T., Kinkorova, J., Ramaekers, J., Ruff, P., Schnur, I., Vanoli, E., Verdu, J., Zippel-Schultz, B.: Challenges in personalised management of chronic diseases-heart failure as prominent example to advance the care process. EPMA J. **7**, 2 (2016)
3. Van Buskirk, J.: 1 2, 3 1, 49, 2015–2017 (1993)
4. Chakraborty, S., Gupta, S.: Medical application using multi agent system - a literature survey. Int. J. Eng. Res. Appl. **4**, 528–546 (2014)
5. Gayathri, R., Uma, V.: Ontology based knowledge representation technique, domain modeling languages and planners for robotic path planning: a survey. ICT Express. **4**, 69–74 (2018)
6. Kovalenko, N.S.: The reconstruction of the linguistic world view on the basis of the old English substantives with the stem-building suffix-S, Vopr. Kognitivnoy Lingvistiki, pp. 109–114 (2014)
7. Li, Y., Lawley, M.A., Siscovick, D.S., Zhang, D., Pagán, J.A.: Agent-based modeling of chronic diseases: a narrative review and future research directions. Prev. Chronic Dis. **13**, 150561 (2016)
8. Mahmood, Z.: Service oriented architecture: potential benefits and challenges. In: Proceedings of the 11th WSEAS International Conference on Applied Computer Science, pp. 497–501 (2007)
9. Moreno, A.: Medical applications of multi-agent systems. In: Intelligent and Adaptive Systems in Medicine (2003)
10. Falah, T., Alwada, N.: Cloud computing and multi-agent system: monitoring and services, vol. 96, pp. 2435–2444 (2018)
11. Nachabe, L., Girod-Genet, M., El Hassan, B.: Unified data model for wireless sensor network. IEEE Sens. J. **15**, 3657–3667 (2015)
12. Nachabe, L., Girod-Genet, M., El Hassan, B., Al Mouhammad, D.: General semantic system for monitoring & assisting diabetes patient "MySmart diabetes OntoREFerence" ontology. In: 2018 IEEE Middle East North Africa Communications Conference, MENACOMM 2018, pp. 1–6 (2018)

13. Omran, S., Hassan, R.: Health Care Application Based on Intelligent Assistant Agent (2016)
14. Technical White Paper. Service Oriented Architecture (SOA) and Specialized Messaging Patterns, Architecture, vol. 345, pp. 1–15 (2007)
15. Jade Programming for Beginners, JADE Tutorial Jade Programming For Beginners (2003)
16. Ristau, R.A., Yang, J., White, J.R.: Evaluation and evolution of diabetes mobile applications: key factors for health care professionals seeking to guide patients. Diab. Spectr. **26**, 211–215 (2013)
17. Technical Specification, TS 103 378 - V1.1.1 - Smart Body Area Networks (SmartBAN) Unified data representation formats, semantic and open data model (2015)
18. Xiao, Z.X.: A multi-agent approach to the design of an programming ICAI system. Phys. Procedia **25**, 868–873 (2012)
19. Zhang, P., Bai, G., Carlsson, B., Johansson, S.J.: Applying multi-agent systems coordination to the diabetic healthcare collaboration. In: 2008 3rd International Conference on Information and Communication Technologies From Theory to Applications, ICTTA, pp. 1–6 (2008)
20. WHO, 2, 2 (n.d.)
21. Using Multi-Agent Systems for Healthcare and Social Services, pp. 1–20 (2003)

Big Data Analytics in E-procurement of a Chain Hotel

Elezabeth Mathew(✉)

RKW, HCT, Ras Al Khaimah, United Arab Emirates
ematthew@hct.ac.ae

Abstract. The hospitality industry is growing at a faster pace across the world which results in accumulating a huge amount of data in terms of employee details, property details, purchase details, vendor details and so on so forth. The industry is yet to fully benefit from these big data and the data has not been investigated to all extent for decision making or revenue/budget forecasting. Chae et al., 2014 has proposed in his paper to have descriptive, predictive and prescriptive data analysis as a good tool to analyze the big data collected in such means. In this paper, the author has tried to explore the data and produce some useful visual reports which is beneficial for the top management that give extra information about the inventoried data. The visual representation is limited to descriptive and predictive data analysis due to the time constraints. The author has used mainly R studio for the visual reporting. The author's vision is to do an extended study by further investigation with the information gathered for prediction and decision making.

1 Introduction

The spread of Information Technology (IT) generates openings and pressures in the hospitality industry that are aggressive and towing IT implementation. To enhance customer satisfaction, online data retrieval, online booking, feedbacks etc. has adopted by mostly all hospitality industry. Another major change is adopting electronic technology in supply chain management (SCM) apart from the implementation of digital technology in the business. Adoption of e-procurement is considered the latest trend and a lot of opportunities for big data analytics has been forecasted. Even though adoption has made substantial expansion of availability and transparency to information, big data analytics has experimented to very minimal level in e-procurement. This paper tries to put light on the e-procurement operation in running of a chain of hotels. Also to investigate the use of procured data and most importantly to do some applications with big data analysis.

In the pursuit to bring an awareness to practitioners in the supply chain management, this paper is mostly focused on (1) analyzing previously published papers on e-procurement (2) investigating the depth and width of data procured in a chain of hotel and its usage (3) apply big data analytics like predictive analysis or trend analysis on a sample of data to see the possibilities of further application of collated data. Furthermore, the study explores the hotel purchasing function in order to understand the purchase process and identify the limitation of usage of e-procured data. This is

L. Barolli et al. (Eds.): EIDWT 2019, LNDECT 29, pp. 295–308, 2019.
https://doi.org/10.1007/978-3-030-12839-5_27

accomplished through a case study methodology that integrates observations and interview with an executive of a hotel chain company with an analysis of the services offering in e-procurement solutions. Nowadays, the passionate race in industrial setting means that tourism and hospitality businesses have to toil rigidly to sustain and progress their competitiveness. Hence, digital or electronic technology supports organization to manage information dynamically and influences business competitiveness through assisting decision makers to make appropriate investments and decisions.

1.1 Statement of Problem

Several articles published has revealed the importance of adoption of e-procurement and most of the managers admit that there are more benefits in adopting e-procurement than barriers. The articles reviewed categories of consumers, technologies, and suppliers agrees on the fact that IT is increasingly becoming critical for the competitive operations of the tourism and hospitality organizations as well as for managing the distribution and marketing of organizations on a global scale.

- The papers reviewed in the study concluded that there is no much data analysis being done in the hospitality industry by which the managers can benefit from the big data.
- Various authors also mentioned that the managers do not have the correct knowledge to conduct a realistic assessment of the benefits and drawbacks of the system with the available data.

1.2 Purpose and Objective

E-procurement is considered one of the latest trend that has been digitalized. Hence, a lot of research can be done on the big data collected by e-procurement in marketing, procurement, warehouse operations, and transportation.

Hence, in this research paper, e-procurement data set from a chain of hotels in the UAE will be used and perform analysis like trend and graphical analysis to do prediction on the requirement of commodities in the hotel.

1.3 Research Questions

(a) Investigate most repeatedly published papers on e-procurement and SCM, and relevant topics being revealed in those studies.
(b) Find out the perception of project manager e-procurement on e-procured data.
(c) Do descriptive and predictive data analysis on e-procured data.

2 Literature Review

Several papers discuss about the benefits, barriers, critical success factor and organizational performance of e-procurement in the organization. The surveyed papers are categories into themes and clusters for various authors are detailed below.

2.1 BDBA and SCM

The team [1] did research and published a paper on big data analytics in logistics and supply chain management: Assured examinations for exploration and claims. As per their recent study, the quantity of data created and transferred over the internet is considerably growing. Thus, making challenges for the establishments that would like to earn the paybacks from evaluating this enormous inflow of big data. Since big data can offer exceptional comprehensions into market trends, customer buying patterns, and maintenance cycles, as well as into ways of diminishing the costs and allows to make better decisions in business. Realizing the importance of big data business analytics (BDBA), we review and classify the literature on the application of BDBA on logistics and supply chain management (LSCM) – that we define as supply chain analytics (SCA), based on the nature of analytics (descriptive, predictive, prescriptive) and the focus of the LSCM (strategy and operations). To gauge the level to which SCA is applied within LSCM, the authors recommend a maturity framework of SCA, based on four competence levels, that is, functional, process-based, collaborative, agile SCA, and sustainable SCA. [2] emphasize the role of SCA in LSCM and denote the use of procedures and skills to collect, disseminate, analyze, and use big data-driven information. Furthermore, [2] stress the need for managers to understand BDBA and SCA as deliberate resources that should be included through business deeds to allow combined enterprise business analytics.

2.2 Advanced Analytics

The study was done by a researcher on big data analytics in supply chain management: patterns and associated study describes on big data analytics propose vast views in today's business revolution [1]. Whilst big data have extraordinarily took the responsiveness of both professionals and explorers especially in the financial services and marketing sectors, there is an innumerable sites that big data analytics can produce better place in Supply Chain Management (SCM). Therefore, the authors intends to explore these premises. The inquiry arrays from the fundamentals of big data analytics, its nomenclature and the level of development of big data analytics solutions in each of them, to operational concerns and top performances.

Moreover, another researcher explains that advanced analytics is defined as the systematic procedure of changing data into intuition for manufacturing superior conclusions [2]. As a recognized field, advanced analytics have matured under the operational research domain. Since there are some fields that have significant overlap with analytics, [1] has proposed a grouping of advanced analytics in three main sub-types.

Descriptive Analytics: A previous business scenario data can be analyzed to make the trends, patterns, and exceptions obvious. This first degree of analytics is very helpful to answer the questions "what happened" and to get a clear comprehension from past that can be used for the future. Few techniques included are given in the table below.

Predictive Analytics: Predictive analytics (PA) analyses real-time and historical data to make predictions in the form of probabilities about future events. They encompass technology able to learn from data [3], based on the machine learning techniques and other computational algorithms of data mining. Predictive analytics are typically algorithmic-based techniques that include (but are not limited to).

Prescriptive Analytics: Prescriptive analytics use forecasting built on data to form information and later come up with an action plan that can be beneficial and hence to have a better outcome.

Prescriptive analysis also uses game theory or what-if analysis in analyzing unpredictable scenarios. Moreover, while the deterministic procedures are not suitable for a solution and rely on Monte Carlo techniques are also part of prescriptive technique.

This section below intends to provide some assistance to practitioners to understand where they could begin to incorporate Big Data Analytics across their supply chains, allowing them to potentially solve complex problems relevant for SCM.

2.3 E-procurement in the Chain of Hotels

[8] did an exploratory study on adopting e-Procurement technology in a chain hotel and the results are shown herewith. Hospitality industry are becoming more technology-based it is very important to understand e-procurement practices. In this study, the author attempted to find the challenges encountered by top management while adopting and implementing e-procurement. The major finding was that the company lacked standardization in purchasing across various properties and also, not all suppliers showed interest to be part of e-procurement. The study revealed a big ambiguity in the company's audit system. A centralized purchasing system would be a solution for this ambiguity. This system will control audit system from a corporate level and at the same time, simplify day to day communication between accounting and operations staff in different properties. Also, "adopting an e-Procurement system would enable The Hotel Company to more efficiently and accurately know how much they are spending corporate-wide in various purchasing product area, allowing them to use the leverage of their buying power to reduce costs" [9] narrated. The Hotel industry then can negotiate with the vendors for better prices and deals for bulk purchases.

The property under study being small had much variation in both what supplies procured and how it is purchased. Even the cost in various properties were different even if it was bought from the same vendor. Thus, it is obvious that there was a communication gap between corporate office and operations at the Hotel properties. Simultaneously, this is again this was resulting in increased operational cost at property level. The data being transferred to the corporate office was inconsistent which made the office difficult to track the progress of given property. The standardization lacked a common unit of measurement also.

Furthermore, in this study "these indicators argue for a standard centralized purchasing system that allows for a possibility of audit control at the corporate level and facilitation of regular communications between accounting and operations personnel at the corporate office and the various properties".

3 Methodology

3.1 An Exploratory Case Study

It is important to select a set of strategies that fit the research type. In this study, most of the questions are on how and why so it's better to use case studies, experiments of histories [14].

"A case study is an empirical inquiry that investigates a contemporary phenomenon within its real-life context; when the boundaries between phenomenon and context are not clearly evident; and in which multiple sources of evidence are used" [15].

A case study includes systematic interviewing and direct observations apart from historian's analysis. The key informants in this study have special knowledge or are considered experts with more than 10 years of experience in the relevant topic so as to clearly give comment on history and present. The interviewee in this study is the Project Manager for Middle East Africa region for e-procurement and Supply Chain Management. His role include decision making, report making, rolling out older software with newer e-procurement software in the properties that he is in charge. It can include already existing property or a brand new property. The H hotels is a worldwide branch having more than 2500 properties across the world. In this study, the H Hotels only in the UAE is considered. The name of the hotel company, its vendors and the interviewee has been disguised throughout the study for the sake of confidentiality. The chain hotel company studied is referred to as The H Hotels. The interview had semi-structured questions with few open-ended questions as well.

The data collected from the H Hotel is extracted, cleaned, analyzed for further reporting purposes which could be beneficial for the top management in this study. The data analytics is done using MS Excel and R studio mostly in this study. Two reports are generated using Oracle 11g. All descriptive reports are done using MS Excel. Due to time constraint, prescriptive analysis will be conducted in a later study.

3.2 Exploring the Setting

The H Hotels are managed hotels in this region which means it does not own property. There are 22 properties by around 10,000 employees with different brands ranging from 3 star to luxurious hotels. The interviewee has worked at both property level and corporate level. Moreover, he works with the chain for more than 15 years in various roles and in several properties. Now in his current role, he is in charge of rolling out of previous software for purchasing and coordinate to install and go live with the new e-procurement system. The H Hotels already have standardized rules and regulations on implementation and also for procuring items in all departments. The H Hotels have a database server to collect the big data which is collected for purchasing request. As this system is limited in reporting facility Project managers extract data from Oracle database into an excel sheet. In their opinion, Oracle is limited with data visualization and reporting together. Or in other words, their current system does not have the ability to display report as per the requirement of managers. So the project managers extract data from their database to make monthly report, expense records. A sample of extracted datasheet is displayed below (Fig. 1).

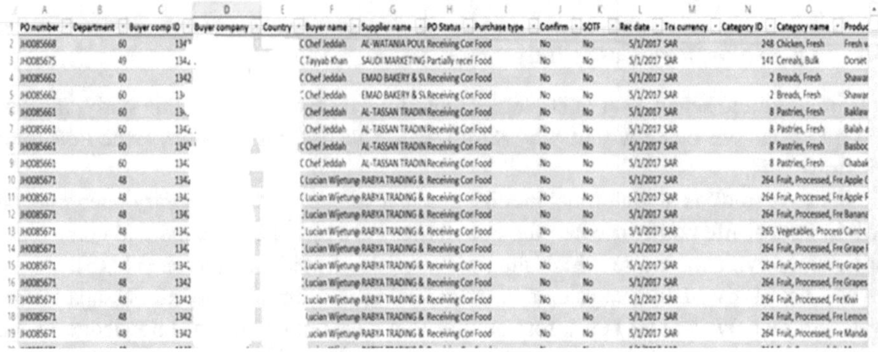

PO number	Department	Buyer comp ID	Buyer company	Country	Buyer name	Supplier name	PO Status	Purchase type	Confirm	SOTF	Rec date	Trx currency	Category ID	Category name	Produc
JH0085668	60	134			Chef Jeddah	AL-WATANIA POUL	Receiving Cor	Food	No	No	5/1/2017	SAR	248	Chicken, Fresh	Fresh v
JH0085675	49	134			Tayyab Khan	SAUDI MARKETING	Partially recei	Food	No	No	5/1/2017	SAR	141	Cereals, Bulk	Dorset
JH0085662	60	1342			Chef Jeddah	EMAD BAKERY & S	Receiving Cor	Food	No	No	5/1/2017	SAR	2	Breads, Fresh	Shawar
JH0085662	60	13			Chef Jeddah	EMAD BAKERY & S	Receiving Cor	Food	No	No	5/1/2017	SAR	2	Breads, Fresh	Shawar
JH0085661	60	13			Chef Jeddah	AL-TASSAN TRADIN	Receiving Cor	Food	No	No	5/1/2017	SAR	8	Pastries, Fresh	Baklav
JH0085661	60	134			Chef Jeddah	AL-TASSAN TRADIN	Receiving Cor	Food	No	No	5/1/2017	SAR	8	Pastries, Fresh	Balah e
JH0085661	60	134			Chef Jeddah	AL-TASSAN TRADIN	Receiving Cor	Food	No	No	5/1/2017	SAR	8	Pastries, Fresh	Basboc
JH0085661	60	134			Chef Jeddah	AL-TASSAN TRADIN	Receiving Cor	Food	No	No	5/1/2017	SAR	8	Pastries, Fresh	Chabak
JH0085671	48	134			Lucian Wijetung	RABYA TRADING &	Receiving Cor	Food	No	No	5/1/2017	SAR	264	Fruit, Processed, Fre	Apple (
JH0085671	48	134			Lucian Wijetung	RABYA TRADING &	Receiving Cor	Food	No	No	5/1/2017	SAR	264	Fruit, Processed, Fre	Apple F
JH0085671	48	134			Lucian Wijetung	RABYA TRADING &	Receiving Cor	Food	No	No	5/1/2017	SAR	264	Fruit, Processed, Fre	Banana
JH0085671	48	134			Lucian Wijetung	RABYA TRADING &	Receiving Cor	Food	No	No	5/1/2017	SAR	265	Vegetables, Process	Carrot
JH0085671	48	134			Lucian Wijetung	RABYA TRADING &	Receiving Cor	Food	No	No	5/1/2017	SAR	264	Fruit, Processed, Fre	Grape I
JH0085671	48	134			Lucian Wijetung	RABYA TRADING &	Receiving Cor	Food	No	No	5/1/2017	SAR	264	Fruit, Processed, Fre	Grapes
JH0085671	48	1342			Lucian Wijetung	RABYA TRADING &	Receiving Cor	Food	No	No	5/1/2017	SAR	264	Fruit, Processed, Fre	Grapes
JH0085671	48	1342			Lucian Wijetung	RABYA TRADING &	Receiving Cor	Food	No	No	5/1/2017	SAR	264	Fruit, Processed, Fre	Kiwi
JH0085671	48	1342			Lucian Wijetung	RABYA TRADING &	Receiving Cor	Food	No	No	5/1/2017	SAR	264	Fruit, Processed, Fre	Lemon
JH0085671	48	1342			Lucian Wijetung	RABYA TRADING &	Receiving Cor	Food	No	No	5/1/2017	SAR	264	Fruit, Processed, Fre	Manda

Fig. 1. Spreadsheet collecting information from various properties

3.3 Validity and Reliability of the Study

The interview conducted with the project manager could be considered valid and reliable as he is a seasoned expert in the field and also, thrive to do the best for e-procurement. Even then a further survey on team member to validate if their opinion matches with each other is recommended for triangulation purpose. This would be done in the next study. Surveying team members or other expert is not done in this study due to the time limitation.

3.4 Ethical Consideration

The interviewee was given with letter of permission and assurance to keep the participant's name confidential. Also, the rewritten codes of responses was given for the participant's review. Few names specified in the result section is with consent from the interviewee. *The company name, property names and the vendor names are hidden to maintain confidentiality* and so some pictures are blurred or distorted for the same purpose. There was not enough time for getting consent from each and every vendor in the study because each category is bought from probably different vendors.

4 Findings and Results

4.1 Summary of the Interview

An interview was conducted with the Project Manager (PM) for e-procurement at the H Hotels in the UAE to see the importance of adoption of e-procurement in the hospitality industry. There were only 2 main questions which was structured. The questions were (a) what are pros and cons in the present purchasing methods at a set of properties owned and managed by The H Hotel Company? (b) Why do you think adoption of e-procurement is important? The brief summary of interview is given below. The various examples given by the interviewee is omitted as part to keep the confidentiality of vendors.

As per the PM, the prices for commodities bought from same vendor at different properties vary drastically. The only way to make sure all properties have a standard rate for each commodity from a single vendor is to have a common system where vendors can put their fixed price which can be seen by the purchase managers. So there is no much negotiations required as the H Hotel will decide to buy from any of the vendors available. Thus, enabling a specific agreement with each vendor for their commodities supplied across the properties. Moreover, this way of dealings will definitely improve seller relations and also facilitates to play a part in chosen vendor fees and genuine discount plans as a result of their purchase history. Simultaneously, this new process will also give a new habit to the purchase managers that is to keep par stock- "Par Stock was defined as a level of inventory items that a buyer believes must be on hand to maintain a continuous supply of each item from one delivery date to the next" [10]. Furthermore, all properties will maintain a similar amount of information regarding the purchases and vendors. Any addition to the information will affect all properties at the same time. Thus, all properties of H Hotel can maintain a consistent data.

4.2 Data Analytics

The possibilities of descriptive, predictive and prescriptive data analytics are prevailing in the collected data. Few descriptive analytics are already been done in the H Hotel itself. In the next session, possible data analytics are explained for each category of data analytics. The analysis has been done using several software's like MS Excel, Oracle, R programming etc.

4.2.1 Descriptive Analytics

Some of the reports listed are already generated at the H hotel. Few possible and relevant descriptive data analytical reports are narrated below.

(1) Buyer summary: products bought from each buyer and the total amount paid. The report is generated in Excel using pivot table (Fig. 2).

Buyer company (All)

Supplier name	Supplier ID	Category name	Amount (USD)
BARAKAT VEGETABLES & FRUITS CO. LLC. (UAE)	3363	Fruits, Fresh	1,458,853
		Vegetables	1,269,760
		Herbs	31,755
		Veg, Frz	2,258
		Misc. Grocery, Dry	960
		Misc. Grocery, Refrigerated and Frozen	436
		Cakes, Fresh	293
		Fruit, Canned	236
		Vegetables, Processed, Fresh	97
		Veg, Cnd	66
		Vegetables, B & E	28
GHOLAMI & ABEDI FRUITS & VEGETABLES	3205	Fruits, Fresh	1,298,679
		Vegetables	1,249,363
		Misc. Grocery, Dry	113,807
		Herbs	27,978
		Egg Products, Processed	27,952
		Shell Eggs	12,323
		Vegetables, Processed, Fresh	354
		Butter, Blends and Margarine	240
		Cereals, Bulk	221
		Soups, Canned	120
		Fruit, Dried and Brined	23

Fig. 2. Summary of all items bought from a buyer

(2) Spend summary: Spend summary could be extracted for easy understanding it is made as pie chart as displayed below. It's useful and gives a quick idea on how much each property in the UAE has spent each month (Fig. 3).

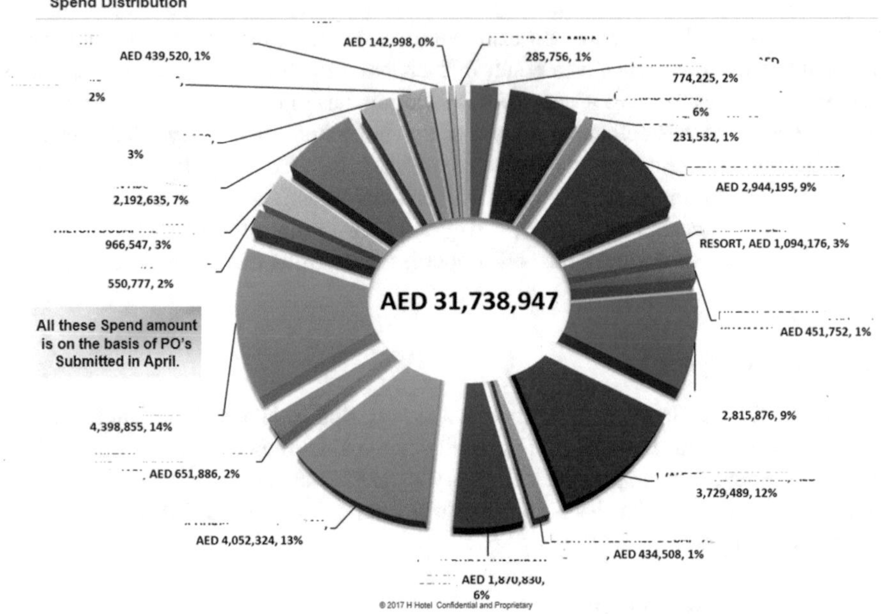

Fig. 3. Total spent in April is shown below for each category

(3) Spend by purchase type: Total purchased value is compared with previous months. Monthly expenses are extracted and can be used for a comparative study between months and even between years. Moreover, these data can still be used to find the variance in purchasing on monthly and yearly basis (Fig. 4).

PO UAE Total Purchased Value

2016-2017
2015-2016

Apr: AED 2,004,631; AED 2,289,443
Mar: AED 2,638,244; AED 2,382,571
Feb: AED 2,317,248; AED 2,067,156
Jan: AED 2,099,206; AED 2,215,155

Vs Previous Month -24%

0 500K 1,000K 1,500K 2,000K 2,500K 3,000K

Spend By Purchase Type

DUBA	Feb-17	Mar-17	Apr-17
Beverage	AED 1,116,268	AED 1,185,806	AED 624,831
Engineering Non Store	AED 39,160	AED 67,878	AED 145,553
Engineering Stores	AED 66,895	AED 69,047	AED 97,310
FF&E	AED -	AED -	AED -
Food	AED 811,748	AED 1,038,202	AED 786,472
General	AED 231,699	AED 230,340	AED 325,402
Miscellaneous Non Store	AED 1,520	AED 490	AED 4,160
Contracts and Prepaid	AED -	AED -	AED -
Operating Equipment	AED 47,271	AED 46,482	AED 12,543
Staff Canteen	AED 2,688	AED -	AED -
Service	AED -	AED -	AED 8,360
CAPEX	AED -	AED -	AED -
Grand Total	AED 2,317,248	AED 2,638,244	AED 2,004,631

Fig. 4. Spend by purchase type and total purchase value is shown below

(4) Top 25 in the list: to get the most purchased product and the top vendors can be extracted from the data as given below (Fig. 5).
   a. Suppliers
   b. Product

Spend Details April 2017

Top25 Suppliers

Suppliers Name	Total Spend		% of Total
Coastal Communities Distribution Fze	AED	1,184,124	29%
Gholami & Abedi Fruits & Vegetables	AED	216,113	6%
M H Enterprises Llc (uae)	AED	141,683	3%
Sysco Guest Supply - Uae - Ddp Uae (ua	AED	131,056	3%
Fresh Express (l.l.c.) (uae)	AED	113,625	3%
Elfab Co.llc (uae)	AED	92,989	2%
Indoguna Dubai Llc (uae)	AED	83,816	2%
Barakat Quality Plus Llc. (uae)	AED	81,534	2%
Repertoire Culinaire General Trading M	AED	81,122	2%
Gearhouse	AED	76,207	2%
Top Shelf Beverages	AED	67,785	2%
Mitras International Trading Llc-uae	AED	63,734	2%
Classic Fine Foods Em Llc (uae)	AED	62,348	2%
Chef Middle East L.l.c. (uae)	AED	57,534	1%
Dot N Drop Trading Fze	AED	46,900	1%
The Deep Seafood Co L L C - Uae	AED	45,715	1%
United Cool General Trading - (uae)	AED	42,885	1%
Family Meat Shop	AED	41,750	1%
Horeca Trade Llc. (uae)	AED	40,753	1%
Diversey Gulf Fze (uae)	AED	38,945	1%
Truebell Marketing & Trading (uae)	AED	36,340	1%
Farmfresh L.l.c. (uae)	AED	34,411	1%
Khoraswala & Al Basti Llc (masterbake	AED	33,011	1%
Bethel Foodstuff Trading Llc	AED	31,383	1%
Emirates Civil Defence Academy (part O	AED	31,280	1%

AED 4,052,324
Top 25 Product

Products Name	Total Spend		% of Total
Champagne, Dom Pérignon Rosé *, 75cl,	AED	217,588	6%
Whiskey Scotch, Blended, Johnnie Walke	AED	85,862	2%
Champagne, Laurent-perrier Brut Rosé N	AED	52,416	1%
Champagne, Dom Pérignon *, 75cl, Bot	AED	41,184	1%
Tenderloin Beef, Black Angus, Chilled,	AED	40,330	1%
Don Julio 1942 75 Cl	AED	38,182	1%
Hildon Mineral Water Still, 750ml, Gla	AED	38,000	1%
Av Rental For The Staff Party (gala Di	AED	33,145	1%
Grey Goose 75cl	AED	32,011	1%
Eggs, Whole With Shell, Pasteurized, 6	AED	31,383	1%
Training On Fire Safety Fundamentals A	AED	31,280	1%
Chivas Regal 18 Year Old 75 Cl	AED	30,537	1%
Zonin Prosecco Brut 75cl	AED	29,877	1%
Sperss, Gaja, Piemonte 75 Cl	AED	28,852	1%
Body Wash 60ml - Ferragamo	AED	28,289	1%
Heineken Bottle 33 Cl	AED	26,746	1%
Chilled Chicken Thigh Cubes/slice	AED	25,990	1%
Water Mineral Plastic Bottle, 0.6l, Al	AED	25,833	1%
Beer Draught, Stella Artois, Brand: St	AED	25,726	1%
Video Rental For The Adib Event On 22	AED	25,055	1%
Corona Bottle 33 Cl	AED	24,111	1%
Chicken Whole 1.3kgs Local Fresh Kgs	AED	23,606	1%
Lifeguard Services For The Month Of Ma	AED	23,000	1%
Gavi, La Luciana, Araldica, Piemonte 7	AED	22,780	1%
Shrimps Peeled Tail On 16-20, Brand: Se	AED	22,620	1%

© 2017 Hotel Confidential and Proprietary

Fig. 5. Top 25 suppliers is shown below

4.2.2 Predictive Analytics

Predictive analytics comprises mining data from present data sets with the aim of finding trends and patterns. These trends and patterns are then used to predict future outcomes and trends. While it's not an absolute science, predictive analytics does provide companies with the ability to reliably forecast future trends and behaviors.

The four predictive charts made a new to the Company and are very beneficial to the top management for futuristic purposes.

(1) Pie chart on category: This pie chart can be used to find which category has been used most and the range of spending. Moreover, this could be expanded to show which month there is maximum expenditure and further prediction could be done using the data obtained and that is represented in a bar diagram and displayed later in the study (Fig. 6).
   The data extracted in excel is imported to Oracle express edition 11g for querying and visualization of data. Since the chart looks similar the chart displayed below is form excel. The query in Oracle is shown in the appendix 9(a).

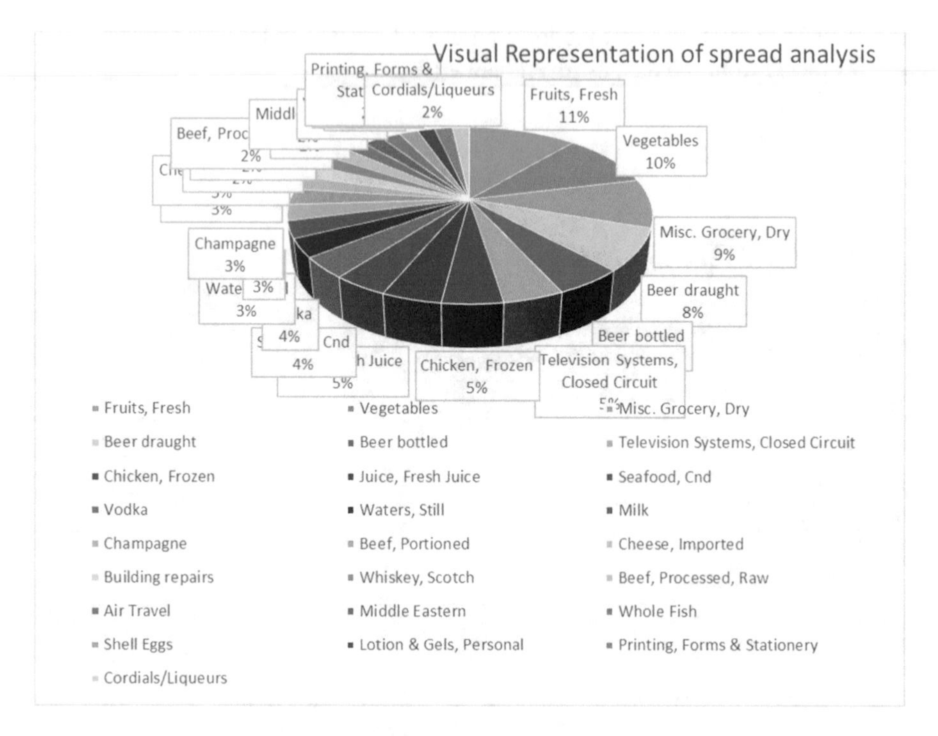

Fig. 6. Visual representation of spread analysis

(2) Time Series Graph:- There are lot of possibilities of time series graph from this big data. Splitting the received date field into date, month and year gives a huge number of graphs with respect to time. This study is only analyzing the possibilities out there rather than exploring it completely. The study could be considered as a preliminary study done for the thesis itself. Out of the thousands of commodities in the H Hotel, the methods to add value to this is by negotiating prices proactively could be from a decision making time series graph that shows price fluctuation during a period of year for each hotel and/or for each region and/or per supplier.

(3) Top 25 category for the year and their spend report in bar chart: The analysis is drawn using R Studio. The analysis shows a trend of top 25 category for the year and its expenditure for each month for the financial year. This graph is excellent for prediction using the trend and can also use to predict for coming years and month. One can visualize which item has been used the most in past one year and also in which month we spent the most for this item. The code is shown in the appendix 9(d).

The item that has purchased the most is calculated using total spend for each item which is displayed in 'Extension' column of the database. Then the items are grouped by category to find the grand total of each category. Later it is ordered by grand total in ascending order to select the top 25 from the list. In this finding, it is very clear that the H Hotel has spent the most in buying fruits and that too in March, October and November with maximum expense. Further detailing can be done in the next study to find trend for each property and for each department purchases which would further be analyzed for future forecasting (Fig. 7).

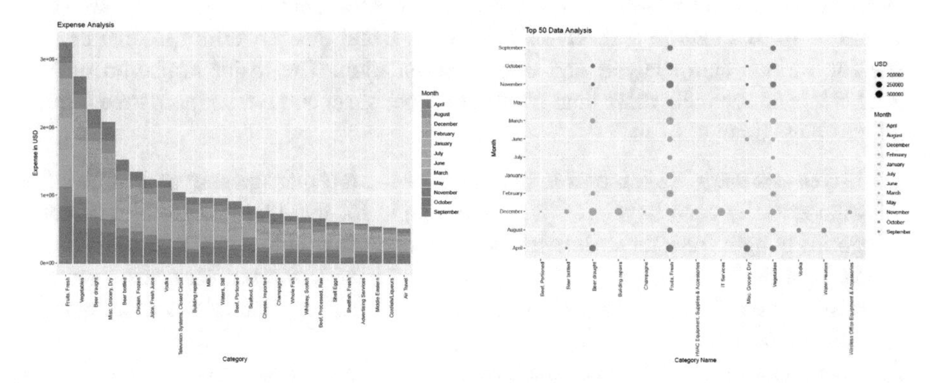

Fig. 7. Top 25 category for the year and their spend report in bar chart and scatter plot

(4) Scatter plot for the top 50 purchases per month: This visualization is good to know which month and which category has spent the most. The item that has used USD minimal, around 20000USD and around 40000USD is shown with points with varying diameter. The plot shows in few months "Building repairs" has spent huge amount overall and that makes sense as building materials are far more expensive than fruits and vegetables. But since fresh fruits are bought consistently every week/month the total spent is more for fresh fruits. The next category that shows huge spent on few months are champagne and advertising.

It would be interesting to do further study on why there is unexpected spent on some items in few months and is it recurring during similar period of time. A further study is recommended at this point for further clarification and that is considered by author in her next research.

The code for plotting it is given in appendix. Scatter plot gives an immediate and clearer idea about the data being currently analyzed. Fresh fruits and vegetables are bought in chunks on a monthly basis whereas other categories like beer draught and grocery items are also bought frequently in large quantity and that make sense for a hospitality industry. Another interesting thing to be noted is that December is the month in which most items are bought and a further analysis would give us extra details. For example, IT services and building repairs are other category which spend hugely in the year 2016–17 and in the month of December. Another thing to ponder is it the month of yearly maintenance, is it recurring in same month every year etc. For these, the authors need more time and further investigation which will be covered in the next study.

5 Discussions

In descriptive analysis, the author made an effort to display 4 report. Descriptive analysis is quite simple and easy representation of collected data and at the same time, it's pretty easy to understand. Descriptive analysis is this study is done in MS Excel as the author got the data in 3 sheets of MS Excel. Excel is a powerful tool that can be used for various data analysis and data representation. The major limitation of MS Excel is it cannot handle more than 65355 rows per sheet and it take more time to read and edit huge amount of data.

(1) Buyer summary: This report is important to the top management as it shows the total spent to each buyer for the items bought for them with the subset of categories bought from them. It helps H Hotel company with whom to negotiate and make a deal. Using analytical skills and experience the purchase manager can easily give the report on price, quality, speed of delivery and other factors when a situation arises.
(2) Spent summary: This report displayed in a pie chart gives management a clear picture of total money spend by each property in the UAE and what percentage is that amount among all properties. It helps the management to know which property spent the most during the month/quarter/year. Moreover, this report helps to do a comparative study with the budget given and money spent and forecast budget for coming year. Furthermore, if there was a noticeable fluctuation in budget allotted and money spent a further investigation would help the management to get more information on predicting next year's budget.
(3) Spend by purchase type: This report includes a comparative study of each category bought in the UAE properties and compare it with previous year same time. For example, 2017 quarter the properties spent an amount for beverages is compared with the same quarter previous year 2016. This would give management an idea on if H Hotels are getting more customers or on what rate the Hotel is progressing.
(4) Top 25: This report describes the top 25 suppliers and the top 25 category purchased in the UAE properties. This helps the management to know who are their close business partners with respect to procurement and who are the negotiators and deal makers. The tops 25 category would give the procurement manager a clear-cut idea on what items should be definitely be purchased during a period of time. Also, if it is not available with one supplier find another one.

Prediction analysis uses the current data and previous data and is used to do future predictions. The benefit of each of the report generated are explained below. All of these reports are new to H Hotel. The first 2 reports are made in Oracle 11g and the later 2 are made using R studio. Since Oracle 11g is student version it has limitation on number of rows to be added in it and so partial reports are generated.

(1) Category Analysis: The report gives top management an idea on various category bought with its total spent and its percentage spent with respect to other category is displayed. The category analysis helps the management to procure supplies more efficiently by giving time for well-organized and to have a planned procurement efforts with a better-dealt value and prediction, influencing the company's procurement power (Harrison 2013). A further study would be to find spends by subcategory.
(2) Time series graph: The change of values of total spent as time progresses for a quarter or for a year is displayed. It makes it easy for the management to know the trend in buying that particular product and fluctuation and use this for prediction purposes.
(3) Top 25 category for the year and their spend report in bar chart: Indulging in knowing how much is spent, on what and with whom is a significant baseline. These kind of visual representation can be used to investigate further on what value they have gained by these processes and decisions [4]. The chart depicts that "Fresh Fruits" is the category where the company spends the most throughout the year and it also shows that Fresh fruits are bought every month. Detailed data analysis shows it is bought on daily basis.
(4) Scatter plot for the top 50 purchases per month: In the earlier chart, top 25 category based on total spent is analyzed. In this graph, the total spend per month. The graphs show how much money in US dollars is spent for category per month. This graph shows little more in-depth detail on what categories are frequently required and which category consumed huge amount in one go. It's clear that fresh fruits and vegetables are bought every month and that makes sense, especially for hospitality industry. This could be further investigated for more prediction in coming months and years. It also shows that in December there was lot of purchase or spend on various categories. It would be interesting to do further interview or investigation to see if it was month for maintenance or is it because of guest complaint or was it due to climatic changes.

6 Conclusions

As Chae et al., 2014 has written in his paper descriptive and predictive analysis can be used for drawing many conclusions and decision making in top management level. Further analysis and investigation is required for prescriptive data analysis which involves decision tree making and other prediction tools. The author has constructed 4 main reports in descriptive analysis out of which 2 are already developed at the H Hotel. Furthermore, the author also constructed 4 different visual representation of big data from H Hotel by applying predictive analysis.

The rich data collected at H Hotel shows the flourishing or well-established business for them in the region of UAE. The H Hotel have thousands of properties across the world and there is a huge scope to study on the data being accumulated.

7 Recommendations

A similar study for 5 consecutive year will give more in-depth information on the spend analysis of each property or the region, which could be used of budget prediction for coming year will be a very useful information for the top management.

The study can be extended for each property to find which property is spending more and a time series analysis to predict the budget forecast of coming year. The study can also be furthered on vendors to see which vendor has fluctuation prices throughout the year or for which product, after that, a detailed investigation to find the reasons of these fluctuation and the effect of revenue in the property could also be done. These data can still be used for collaboration, innovation, integration, globalization, sustainability, and analytics as per [13].

References

1. Wang, G., Gunasekaran, A., Ngai, E., Papadopoulos, T.: Big data analytics in logistics and supply chain management: certain investigations for research and applications. Int. J. Prod. Econ. **176**, 98–110 (2016)
2. Varela Rozados, I., Tjahjono, B.: Big data analytics in supply chain management: trends and related research. In: International Conference on Operations and Supply Chain Management, vol. 6, no. 1 (2014)
3. Albano, G., Antellini Russo, F., Castaldi, G., Zampino, R.: Evaluating small businesses' performance in public e-Procurement: evidence from the Italian government's e-marketplace. J. Small Bus. Manag. **53**, 229–250 (2015)
4. Croom, S., Brandon-Jones, A.: Impact of e-procurement: experiences from implementation in the UK public sector. J. Purchasing Supply Manag. **13**(4), 294–303 (2007)
5. Johnson, M.: A study of e-market adoption barriers in the local government sector. J. Enterp. Inf. Manag. **25**(6), 509–536 (2012)
6. Kothari, T., Hu, C., Roehl, W.: Adopting e-Procurement technology in a chain hotel: an exploratory case study. Int. J. Hospitality Manag. **26**(4), 886–898 (2007). https://doi.org/10.1016/j.ijhm.2006.01.005
7. Kothari, T., Hu, C., Roehl, W.: Adopting e-Procurement technology in a chain hotel: An exploratory case study (2018). Accessed 4 June 2018
8. Piluso, J., Leimer, M., Zhang, K.: Empowered by analytics: procurement in 2025. EY: Building a better world, vol. 11, no. 14 (2016)
9. Shi, X., Liao, Z.: Managing supply chain relationships in the hospitality services: an empirical study of hotels and restaurants. Int. J. Hospitality Manag. **35**, 112–121 (2013)
10. Teo, T., Lin, S., Lai, K.: Adopters and non-adopters of e-procurement in Singapore: an empirical study. Omega **37**(5), 972–987 (2009)

Data Modeling and Visualization of Tax Strategies Employed by Overseas American Individuals and Firms

Alfred Howard Miller(✉)

Fujairah Women's College, Higher Colleges of Technology,
Box 1626, Fujairah, UAE
amiller@hct.ac.ae

Abstract. A study of the tax behavior of overseas American individuals and small firms, where the researcher models behavior, through text analysis, using data mining technologies of KH Coder, with data collected from a wide range of sources using interviews, surveys, blog and forum postings, published reports as well as personal communications, to demonstrate and inform using the pattern matching method. Text mining and modeling techniques, using unsupervised machine learning facilitate large-scale analysis of behavioral approaches to taxation to motivate a better understanding of the phenomenon tax avoidance and tax evasion. There are an estimated 9 million taxable overseas Americans corporations and business entities and estimated that as many as 100 billion U.S. dollars may go uncollected, due to tax evasion. A similar shortfall of 100 billion dollars is due to tax avoidance. The researcher proposes a model explaining tax avoidance behavior by the US taxable entities.

1 Introduction

This study targets the tax behavior of overseas Americans, both individuals and firms. The researcher proposes to discover and model their behavior through text analysis of data collected from a wide range of sources such as interviews, surveys, blog and forum postings, published reports as well as personal communications, to demonstrate and inform using the pattern matching method initially proposed by Trochim [1]. Advances in technology—modern methods such as unsupervised machine learning that utilize software and computing power to analyze text is motivated [2]. The text mining and modeling techniques have been widely deployed in language based studies and fall under the realm of human machine interaction. The aim is to examine tax behavior of United States of America (USA) overseas taxable entities.

Overseas Americans are taxable no matter where they reside globally, and are taxed on the basis American citizenship. In the case of non-citizens, those with a USA connection may also be subject to US taxes. Entities include corporations, individuals, estates and trusts, and flow-through entities filing under the individual code, namely S-corporations, sole proprietorships and partnerships. Through text mining, an understanding the behavioral approaches to taxation; via machine learning techniques, the phenomenon of tax avoidance and tax evasion, can be modeled and visualized.

L. Barolli et al. (Eds.): EIDWT 2019, LNDECT 29, pp. 309–321, 2019.
https://doi.org/10.1007/978-3-030-12839-5_28

2 Problem Statement

There is potentially, a 200–450 million-dollar tax shortfall problem, with an estimated 9 million overseas Americans and numerous corporations and business entities of all sizes responsible. The Congressional Research Service [3], reports that as many as 100 billion U.S. dollars may go uncollected due to tax evasion, while a similar figure of 100 billion dollars of estimated tax shortfall due to tax avoidance—mainly by U.S. origin, multinational corporations. The issue is exacerbated by changes to the 2018 tax code, which was proposed to encourage compliance through tax cuts to a fixed 21% for the corporate sector, and reduced taxes for individuals. However, changes in law may open new aggressive tax avoidance strategies. A gap in the literature centers around the uncertainty regarding changing of the U.S. tax code in 2018 and how it will effect tax filing by overseas American entities.

3 Literature Review

The literature review consists of three sections. One, a review of literature on taxation by overseas Americans, a second on the unsupervised machine learning method. Finally, a third section provides a context for the study. Some tax avoidance and evasion mechanisms are revealed and further explored as this study develops.

3.1 Tax Collection, Avoidance and Evasion

The US Department of State, Bureau of Consular Affairs, reports that the number of overseas Americans could be as high 9 million [4] from an overall size of the global diaspora of 244 million for 2017. While the US share of outgoing global diaspora is quite high, India, Mexico, Russian Federation, China, Bangladesh, Pakistan and Ukraine send more of their citizens abroad [5]. The U.S. taxpayer diaspora is one facet, while the corporate sector is another. There is some overlap of the individual and corporate taxpaying sectors as many individuals choose to incorporate or deploy pass-through entities, as part of their tax avoidance strategy. Corporate inversion is a phenomenon whereby a corporation moves overseas or merges with a foreign company as part of its tax avoidance strategy.

The various tax planning strategies results in less taxes being remitted to the United States government and less jobs for Americans. US tax rates are reported to average 35%, and although legislation has been introduced to make inversion more difficult, many US corporations still seek the tax advantages of a foreign domicile [6].

A range of controls were sought or introduced by the Obama administration in an attempt to collect more tax, by closing tax loopholes. Corporations were more closely aligned with tax avoidance whereas individuals were more closely associated with tax evasion. Tax avoidance is considered legal whereas tax evasion is illegal [3].

Tax avoidance schemes using shifting of income by multinational business entities is thought to result in losses of 100 billion dollars per year or more. By shifting debt to high tax jurisdictions profits can potentially be deferred indefinitely. Hybrid entities have been used to reduce passive income through having varied treatment in different jurisdictions. Individual tax evasion is estimated to be a significant amount at 40–70 billion per year, by some estimations but as high as 100 billion. The individual tax evasion strategies included failure to report passive income from capital gains, interest and dividends originating from foreign income sources [3].

A European study reported that tax avoidance was typically seen alongside tax evasion in and that the taxpayer actively considered consequences of both. Typically, the taxpayer deployed as much tax avoidance as possible, exhausting that opportunity, and then decided how much tax evasion they could get away with without significantly increasing the likelihood of being audited [7].

The strategy advocated by [3] for combatting individual tax evasion was to seek to collect more information about the foreign dealings overseas American taxpayers. The Foreign Account Tax Act (FATCA) requires the registration of accounts with US ownership affiliation. The effects are not known yet but FATCA is being enforced to identify and combat interest paid to foreign recipients not taxed as overseas Americans. Actors often channel interest and funds into shell corporations and trusts located in foreign jurisdictions. New provisions recommend limits or repeal of deferral, foreign tax credits, and introduction of an apportionment of income system. The Gravelle article [3], was a white paper produced by the Congressional Research Service. The purpose appeared to guidance for legislation strategy by those in United States government seeking to collect more tax by closing tax loopholes that could be exploited by overseas American taxable entities including, but limited to both individuals and corporations.

Apple avoided paying taxes on 44 billion in otherwise taxable off-shore income over 2009–2013. Apple's tax strategy included structuring of three foreign subsidiaries for tax purposes that were not tax residents of any sovereign state. Apple claimed its tax strategy was legal and proper—typical of similar firms such as Microsoft and Google [8].

In principal, the residential tax system of the US assesses taxes regardless of where profits are earned, provided the entity is identified as being a US taxpaying entity. Taxes paid by the entity to other countries, are subtracted from this tax calculation. Most countries employ the more typical, territorial tax system, upon which, taxes are paid on economic activities taking place within that country and is the more common system [8].

The US system does permit deferment of overseas income until that income is repatriated to the US. The system encourages US firms to incorporate in tax havens and shift their income using transfer pricing for goods, services and intellectual property that are sold overseas. The transfer pricing and income shifting can occur across multiple subsidiary entities. For example, one arrangement called the 'Double Irish' is two Irish firms that tax each other; both are subsidiaries of a large US multinational, using advantageous transfer pricing to shift income. Examples of firms that deploy the Double Irish such as Adobe Systems, Eli Lilly, Facebook, G.E., Microsoft, Oracle, Pfizer and Starbucks. A variation sandwiches a Dutch firm between the two Irish firms [8].

Overseas American firms tend to defer tax indefinitely and then lobby lawmakers and administrators seeking a tax holiday; returning the cash to the US tax-free. When this repatriation of funds occurs, there is a subsequent jump in earnings and boost in share prices. The construction of such tax havens is costly in terms of consultants, accountants and lawyers, but ultimately less costly for firms due to tax savings [8].

Analysts concede that US-based firms, competing globally are at a disadvantage given the 35% statutory tax rate. US firms operating abroad are forced to maintain complex tax entity structures in order to remain competitive globally. European's see these same tax structures employed by US multinational as reducing their ability to collect European tax as well. Critics are calling for US corporate tax reform to taxes, re-incentivize investing in the US economy. Recommendations range from more aggressive tax collection by closing loopholes and maintaining the 35% tax rate, to plans that promote business growth at home by cutting taxes to 20% yet eliminating deferment as a tax planning tool [8]. The Internal Revenue Service (IRS) takes the position that taxpayers are not allowed to evade taxes by shifting liability to foreign tax entities. The IRS recognizes a range of schemes, which follow here, used by individuals to abuse the tax code and pay less taxes [9].

Tension between financial versus tax reporting is a highly researched topic. However, characteristics of aggressive tax avoiding firms and the effect of changes in the tax law on financial reporting, is an area where a gap exists in the literature. Researchers [10] found a connection between mandatory financial disclosures and tax reporting and collections. They also found that multi-state active firms showed a greater propensity for tax avoidance. This partly explains the drop in state level corporate tax revenue. Stricter disclosure requirements such as the FIN 48 regulation, of 2007, which required greater certainty in income reporting, encouraged greater collection of state income tax. Furthermore, it generally takes a period of years for planners to maximize exploitation of changes to the tax code.

The U.S. Tax code has recently changed, and as the tax code changes so do the approaches taken by tax planners. These planning tips or so-called hacks are described in an article in the New York Times by their inclusion in a category ranging from easy to difficult. Tax avoidance strategies generally revolve around three approaches, income shifting, timing of income and changing the character of income. Under the easy category: Changes in the tax code to increase the Standard Deduction, by nearly doubling it provide an incentive to take charitable deductions now before the changes go into effect. A year one and two of a child's life parental leave tax credit will be offered to businesses allowing their employees to take this benefit [11].

More difficult is use is the revised estate tax rules which doubles the tax free transfer exclusion limit to 11 million dollars until 2025. A 40% tax kicks-in for larger inheritances [11]. Only $10,000 will be deductible for state and local taxes and this effect residents of high tax states such as New York, New Jersey and Connecticut, all with state and local taxes at over 12%, while helping those residents from lower tax states such as Alaska, Wyoming, South Dakota, Texas and New Hampshire, are taxed ranging from 6.5 to 7.9% [12]. This encourages people in high tax states to shift as much income as possible forward to this year as well as reorganize the Traditional IRA as the Roth model. If a person is considering moving, they should do it now as the moving expense tax credit is being repealed [15]. Relaxed rules for 1031 capital gains swaps which where funds from the sale of one asset are swiftly rolled over to a new asset are being tightened to apply to real estate, but not other capital assets such as private jets. As such, non-real fully depreciated assets should be swapped before the tax code changes [15].

The new tax bill will introduce a tax shield for private education expense of up to $10,000 paid after passing through a Section 529 account. The new tax bill also favors pass-through tax enterprises where employees may declare a themselves as freelance consultants or self-employed thereby qualifying for a 21% tax rate for single earners making up to $157,500 and $315,000 for married taxpayers. There more stringent limits on income qualification when operating as a pass-through enterprise.

Another option is file as a C-Corporation and avail the 21% tax rate on all earnings. This would allow the deduction of state and local taxes at expenses of profit distributions to the owner(s) being taxed as dividends. Taking this concept, a step further, if an incorporated business is split into several entities, one entity could provide capital while another provides labor, working a achieve an optimal split of which firm claims how much in profits. Further ideas include operating one's personal brand as a pass through enterprise, or splitting company into enough dependent enterprises that everyone working at the firm qualifies for the pass through rate. Exchanges of assets for shares is non-taxable, the transferee can sell the asset taxable at a 21% rate, while shares can be passed on to heirs tax free [15].

Changes in the US tax code with the resulting favoritism toward the 21% corporate rate, opens up tax research and tax employment opportunities. It is believed that the new tax regime was designed to encourage repatriation of overseas US holdings due to more advantageous corporate tax rate structure back home.

Some professors, [17] see tax reform entirely differently. They provided empirical evidence that criminal prosecution including jail time for offending executives, was the best deterrent to tax evasion. Their methodology noted that states with stricter penalties and longer criminal jail sentences for executives of firms convicted of tax evasion, tended to have a higher effective tax rate at the state level indicating greater compliance and a less aggressive tax avoidance strategy. The authors found that the only time civil penalties were an effective deterrent was when management had a high rate of share ownership. Finally, Nevada and Delaware were identified as Domestic tax havens [17].

3.2 Unsupervised Machine Learning Method

With the unsupervised machine learning method proposed for this study, human factors are divorced from the process, and data analysis is objective and performed by a computer algorithm, except to interpret the initial evaluation model. Computerized analysis methods are gaining in popularity due to high data volumes.

Trochim traces the advent of theory-driven approaches, and their application to program evaluation and creation of theoretical frameworks aiding conceptualization methodology [2]. Trochim has advocated establishing validity by matching the observational realm with the theoretical realm [1]. Pattern matching studies today can draw upon techniques borrowed from search engine optimization and search engine marketing. Trochim [2] has noted how widespread and interdisciplinary concept mapping methods had become.

Assessing work readiness in accounting graduates via the SCIL-based model [13, 14], is based on learners at Higher Colleges of Technology taking the Taxation course during either Semester 7 or 8 during Year 4 of the Bachelor's degree. The sample size collected was n = 23 (2015) and n = 28 (2016), n = 9 (2017) and n = 26 (2017) in the latest study with additional data points analyzed beyond the initial study. The researcher deployed a corpus-based approach to ethnographic research that is collaborative and contextual. The research strategy was to collect a body of business discourse, namely student reflections from the Year 4 Accounting, Taxation course, and then performed a corpus-based computer aided linguistic analysis [14–16].

One of the early advocates of analyzing the corpus or body of language was Susan Hunston [17]. "The main argument in favour of using a corpus is that it is a more reliable guide to language use than native speaker intuition is" (p. 20). Hunston nevertheless valued intuition stating: "[Intuition] is an essential tool for extrapolating important generalizations from a mass of specific information in a corpus" (p. 22). The method is therefore a mixture a computer analysis and researcher intuition.

[18] advocated the self-organizing map as being an excellent tool in exploratory data mining. They further explained the dendrogram supported nature of clustering using the self-organizing map (SOM). Furthermore, hierarchical cluster analysis uses dendrograms to specify clusters and as a technology can reinforce the SOM.

According to [19] data visualization has become a must-have skill for all managers. This is because a visual abstraction is often times the only way to process the volume and velocity of data that arrives for processing. Furthermore, decision-making increasingly relies upon the ability to make-sense from and interpret of this voluminous amount of what is also known as big-data. Due to open source programs, the internet and proprietary, yet affordable tools, visualization is becoming widely accessible. Access to tools without the deeper understanding of their application can result in producing charts that are inadequate or ineffective.

[19] proposed that data managers and decision makers ask two questions. "Is the information conceptual or data-driven? and is the statement about the topic declarative or exploratory?" [19], Page 1. [19] decision model identifies which of four types of visualization goals will be most effective, namely: "idea illustration, idea generation, visual discovery, or general data visualization" [19] Page 1. The implication of what has been stated by [19] is guidance to the researcher conducting a study that is both quantitative and qualitative in nature.

The method to conduct data visualization with multidimensional scaling was explained by [20]. [21] used word analysis, co-occurrence analysis, and categorization to examine a socio-cultural experience shared on an internet discussion forum. [21] set word analysis limits and reported high frequency words appearing 40 or more times. Using co-occurrence networks and the Jaccard coefficient, [21] was able to intuitively create categories.

The Jaccard coefficient is a word frequency algorithm. It divides the frequency of word intersection by the union of word appearance. For example, if the frequency of word a is 4, and frequency of word b is 3, then the frequency of words a and b is 2. As stated as a formula; $2/(4 + 3 - 2) = 0.4$ [22], page 2.

[21] reported that computer coding allowed a researcher to objectively handle large amounts of data, and qualitative information could be easily represented numerically which served to decontextualize the data. [21] considered de-contextualization to be an advantage via removing human factors from a crucial portion of data analysis.

[23] used KH Coder for content analysis of 35 student survey responses. They focused on 20 key responses to Question 11, "How Student's Attitude Influences on Learning Achievement? An Analysis of Attitude-Representing Words Appearing in Looking-Back Evaluation Texts". The author's explored a potential link between learner's attitude and achievement performance. The authors characterized their study as Education Data-mining (EDD) missed with Knowledge Discovery Data-mining (KDD). Low performers used language and words similar to high performers yet they were poorly implemented while the middle performance group had identifiable word differences. Their approach analyzing a limited number of student responses drew a parallel to the method deployed to analyze content—reflections in the case report on the SCIL model. Dissimilarities were use of a questionnaire, instead of discussion points and class focus on IT—information retrieval [23].

The Handbook of Business Discourse provided a reference for motivating content analysis in the context of business studies. Systematic replicable methods were described for working with large data sets of text, apportioning words into categories, motivated by uniformly applied computer coding rules. Intuition as well as objective means can be used to process the output. The approach has commonality with other data analysis methods, as a means to identify patterns, and trends. Content analysis methods are best applied when triangulated with other methods [24]. Researchers stated that a computer aided quantitative approach to text-mining of natural language, was reliable, consistent, and compatible with grounded theory, provided the researcher remained neutral, and allowed categories to emerge from the data. They justified content analysis because it could be used to validate evidence [24].

KH-Coder has been accepted in the U.S. Court of Law [25, 26]. Text mining and content analysis has been used in over 600 studies. "Ultimately Algorithms are a set of instructions followed by computers to solve problems" [27]. However, [27] challenges big data and says rogue algorithms are often with too small of sample sizes with only a few dozen students [27]. This study will ultimately use the approach of [28] to test for statistical power, where a sample size of approximately 500 respondents may likely be specified. KH Coder permits coding-based text analysis and allows the researcher to deploy a range of techniques which include; word frequency analysis, hierarchical cluster analysis, co-occurrence network, multidimensional scaling and self-organizing map. While KH Coder is open source and user friendly, RapidMiner, a similar program permits larger scale data mining. Times to run statistical procedures can be expected to take anywhere from a several minutes for factor analysis to several hours for self-organizing maps.

3.3 Context of Study

The scope of data collection for pattern matching using approached such as multi-dimensional scaling and concept analysis is ideally governed by contextualism and multiplism. In "Contextualism; we should articulate and observe the specifics of our object of interest, be it a program, measure, or participants in a context." [1] page 363. This statement implies that data collected for analysis must be collected from overseas American entities who have a requirement to file taxes.

Multiplism implies that a wide range of different types of entities such as sole proprietorships, partnerships, limited liability companies, limited liability partnerships, S-corporations, estates and trusts and C-corporations, will need to be polled, in order to articulate and engage these multiple manifestations of the gamut of contexts of tax avoidance and tax evasion. Relationalism, is observing and discovering the relationship between these multiple manifestation of tax behaviors by various types of taxpayer entities. Gradualism is to apply a gradient in the analysis to gauge the level of similarity between entities, tax behaviors employed and deployment of approaches. The contextual approach encourages researchers to avoid simple categorization and realize that tax behavior varies multi-dimensionally.

Dualism implies that patterns should be emergently identified at both the theoretical and observational level. While Parallelism, is the linkage between the theoretical and observed patterns. Pattern matching requires that there be correspondence between theoretical and observed structures. Finally, the degree of correspondence should be assessed using a statistical method [1].

Concept mapping helps develop construct validity, using pattern matching, the Theory of Conceptualization, multidimensional scaling, cluster analysis, and bridging analysis, to interpret the results of a concept map, where low values described a point location on the map while high values bridged between locations. Further methods include a go-zone plot, bibliometric analysis and multivariate statistics.

[29] in a 2016 master's thesis deployed factor analysis and explored two factors as they applied to tax evasion and tax avoidance. These factors were civic duty to pay tax and moral reasons to not pay tax. The fourteen item Likert scale survey was based on a previously validated instrument and later updated by [30] and certified as reliable at α .909 with a sample size of 71 banking employees. The survey considered inputs such as moral obligation, participation in the shadow economy, age, gender, religiosity, educational and occupational factors. To address this issue, a survey was designed, based on the tax ethics literature and previous studies, and administered to banking employees from a bank institution operating in Portugal. The survey addresses several issues like tax ethics, tax morals, tax evasion and tax compliance and the sample obtained consisted of 71 observations. The findings identified a strong shared ethical position against tax evasion but virtually no differences across the differentiating factors such as gender or religiosity [29].

An investigation into problems of the shadow economy, tax evasion and tax avoidance in Greece's economy was tackled by [31]. They sought to quantify the effectiveness of Elenxis a new algorithm for detecting audit candidates through an 80 sample factor analysis study, using principle component analysis with varimax rotation. The sample consisted of accountants, namely public, private and freelance, while the data collection instrument was a 12 question, 4 factor Likert scale survey besides eight additional on demographic factors. The survey measured attitudes of respondents in relation to shadow economy, tax evasion and tax. The general consensus by 92.5% of those sample favored use of information systems as a way to combat tax evasion. The authors concluded that increased taxation led to an increase in tax evasion, that accountants were jointly responsible for tax evasion with their customers; and that the tax mechanism is itself was responsible for tax evasion, and a laxity and a less constant attitude toward dealing with taxation. The authors identified that the sample was too small to be statistically meaningful, and findings limited, as it was confined to only one country. Men tended to focus more on the joint responsibility and women on the tax mechanism. The authors concluded with a recommendation for a more comprehensive confirmatory factor analysis [31].

4 Method

Hypothesis

Tax evasion is prevalent among overseas Americans as individuals
Tax avoidance is prevalent among overseas American business entities.

Research Question

RQ 1. What is the character of tax avoidance and tax evasion strategies deployed by overseas Americans and American entities?

4.1 Methodology

A mixed methods study, used text from the web a survey of established validity and reliability, coupled with interview collected data and from which keywords and concepts were distilled using human intelligence and computer interaction. [32] The method was based more on web-based mining of relevant texts as this data was readily available. The approach was implemented in conjunction with unsupervised machine learning, in a natural language processing context, coupled with content analysis, a quantitative method, to identify emergent themes toward understanding and modeling tax avoidance and tax evasion behavior, of overseas American taxable entities. The three modalities which are a sharing of both qualitative and quantitative analysis, support answering the research questions through a triangulation of methods.

Primary research will ultimately be expanded to include full triangulation of three methods. (1) Survey by questionnaire, of approximately 50 overseas taxpayers, using a previously validated instrument, from a sampling frame representative of overseas American taxpayers, (2) Interviewing people from this same sampling frame, but not those same people as who participated in the survey, with a prepared set of questions. Sufficient interviews will take place to achieve topic saturation, such that no more, new information is being introduced, and this is estimated to be 20 interviewees. (3) A third method will target this sampling frame with knowledge harvesting analysis of text-based data collected from online communities, such as forums, blogs, and other web-based communications. Again, a sufficient sample will be collected to achieve saturation of the tax topic. Tested here is primarily the third method of 52 text samples totaling 74,782 words. Text includes, minimal samples of interview and survey data and is primarily comprised of scholarly journals, cases, and expat tax blog entries from the World Wide Web.

Anonymity was preserved due to the ethical and moral nature of tax avoidance, evasion, and fraud which is a sensitive topic for those American citizens potentially in violation of the tax law, either involuntarily or cases actively working to circumvent the code, through exploitation of loopholes and other means. Instruments developed from the work of others will be used with permission by the original author. In cases where a new instrument is devised, field testing and pilot study shall take place as the study is developed with checks for validity and reliability. Since this research is mixed methods, being primarily qualitative with quantitative analysis in the form of the natural language processing of the text analysis a sufficient sample size will be collected to achieve saturation rather than a calculated statistical power.

Reliability was obtained through confirmability, by using a uniform and repeated coding regime. The principal researcher (PI) will be the only person working with coding the data set. Data preparation included removal of extraneous data such as section numbers, references, and fonts and symbols incompatible with KH Coder. Reliability is also be supported by triangulation across and within methods. Such as similarity of member responses, and established reliability of survey instruments. Verification of participant answers, uniformity in their answers across responses, and triangulation across the methods provides a construct to test reliability of the survey, interview questions and corpus collected text from the online sources. Similarity in

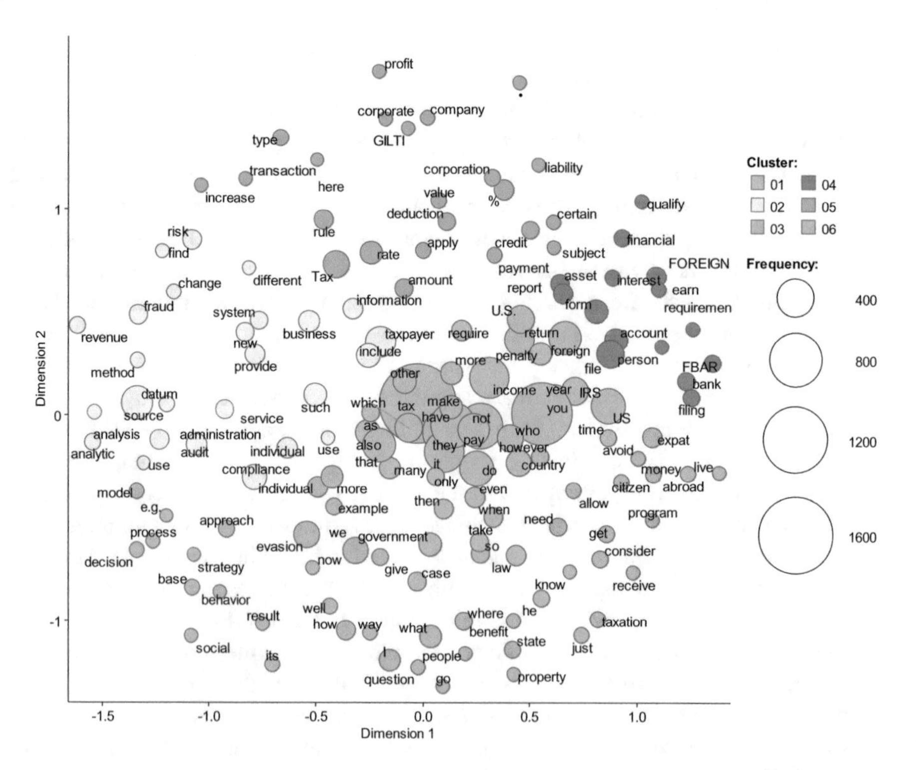

Fig. 1. Multidimensional Scaling, stress = .331, Kruskal method with Jaccard distance formula. yielded a six factor model, where 01 = Taxes Abroad, 02 = Data Analytics and Audit, 03 = Norms and Ethics, 04 = FBAR & FATCA, 05 = Impact of GILTI, and 06 = Concealed Assets Abroad. Dimension 1, X-axis = Analytics on the left and Regulations on the right. Dimension 2, Y-axis = New GILTI tax code at the top and Norms and Ethics at the bottom.

responses among the participants throughout the interview corroborates that research instrument and the accuracy of responses. A continuous member-checking feedback loop is reported by [33] as a way to boost accuracy of the reliability process.

Validity was assessed post analysis through web search of factor keywords on Google search of the three primary data collection methods of survey, interview and knowledge harvesting from the Web. Trochim's pattern matching approach as it aligns with grounded theory, achievement of data saturation, for internal validity and variation in means of text, and participant selection. Data subjected to a quantitative approach was examined using the output and interpretation of the machine learning algorithms of Kruskal method and Jaccard distance [2].

4.2 Results

To test the hypothesis, the researcher generated a comprehensive model that described the target group. Namely, output of tax strategies from text, covering overseas Americans and American corporations operating abroad. Results were analyzed using Multi-Dimensional Scaling, for a best fit of six factors at a reported Stress of .331, produced and analyzed through KH Coder an unsupervised machine learning technology, through graphical outputs to enable visualization of data. These findings will be disseminated in academic conferences, and scholarly journal publications and form the basis of further study on this topic. The six factor model and estimations of Dimensions 1 and 2 are given in Fig. 1 on the following page.

5 Conclusion

With a global diaspora of 9 million overseas Americans, potentially $100 billion dollars in tax evasion, by individuals and another $100 billion in tax avoidance, primarily by corporations, remain uncollected. Estimates in the coded material were as high as 450 million. The revised tax code of 2018 offers both an opportunity to collect additional revenue while at the same time opening up new avenues for pursuit of aggressive tax avoidance. The amount of uncollected revenue strongly supports the latest tax reform movement in the US and motivates exploration of further research into the various models of tax evasion, and aggressive tax avoidance. Outcomes of this study includes a graphic model that represents the tax environment based-on an emergent theory supported by analysis conducted.

5.1 Practical Application

This study contributes to the existing body of knowledge about taxation with a modeled understanding the problems of tax avoidance and tax evasion for overseas Americans and U.S. business entities. This problem is major in scope. An early stage model has been delivered, subject to further investigation and development. This study will be of use to those trying to understand how to minimize their tax and those seeking to combat fraud and collect full taxes owed. The academic community will appreciate the neutral perspective provided by the researcher.

Acknowledgments. The Higher Colleges of Technology, Fujairah Colleges, of Fujairah UAE, where the principle investigator is employed as an applied researcher.

References

1. Trochim, W.M.: Outcome program matching and pattern theory. Eval. Program Plan. **12**, 355–366 (1989). https://www.socialresearchmethods.net/research/Outcome%20Pattern%20Matching%20and%20Program%20Theory.pdf
2. Trochim, W.M.: Hindsight is 20/20: reflections on the evolution of concept mapping. Eval. Program Plan. (2016). https://doi.org/10.1016/j.evalprogplan.2016.08.009
3. Gravelle, J.G.: Tax havens: international tax avoidance and evasion. Congr. Res. Serv. (2015). https://fas.org/sgp/crs/misc/R40623.pdf
4. CA by the Numbers: Counsular Affairs by the Numbers. US Department of State, Bureau of Consular Affairs (2017). https://travel.state.gov/content/dam/travel/CA_By_the_Numbers.pdf
5. UN: Sustainable development goals (2017). http://www.un.org/sustainabledevelopment/blog/2016/01/244-million-international-migrants-living-abroad-worldwide-new-un-statistics-reveal/
6. Marshall, S.: Corporate inversion and the impact on the United States. Business, Lagrange.edu. (2016). http://www.lagrange.edu/resources/pdf/citations/2016/11_Marshall_Business.pdf
7. degl'Innocenti, D.G., Rablen, M.: Income tax avoidance and evasion: a narrow bracketing approach. Public Financ. Rev. **45**(6) (2017). https://doi.org/10.1177/1091142116676362
8. K@W: Corporate tax avoidance: can the system be fixed? Knowledge at Wharton (2013). http://knowledge.wharton.upenn.edu/article/corporate-tax-avoidance-can-the-system-be-fixed/#comments
9. IRS: Abusive offshore tax avoidance schemes-talking points (2017). https://www.irs.gov/businesses/small-businesses-self-employed/abusive-offshore-tax-avoidance-schemes-talking-points
10. Gupta, S., Mills, L.F., Towery, E.M.: The effect of mandatory financial disclosures of tax uncertainty on tax reporting and collection: the case of fin 48 and multistate tax avoidance. J. Am. Tax. Assoc. **36**(2), 231–239 (2014). https://doi.org/10.2308/atax-10402
11. Bui, Q., Sanger-Katz, M.: Hacking the tax plan: ways to profit off the republican tax bill. The Upshot, The New York Times (2017). https://www.nytimes.com/interactive/2017/12/12/upshot/tax-hacks.html. Accessed 23 Dec 2017
12. Tax Foundation: State-local tax burden rankings FY 2012 (2017). https://taxfoundation.org/publications/state-local-tax-burden-rankings/. Accessed Dec 30 2017
13. Miller, A.H.: Computer-aided content analysis of the corpus of business discourse: a comparison of accounting and HR learners, NETs 2016, Osaka, Japan, 25 July 2015 (2016)
14. Miller, A.H.: Preparing students for career success in accounting: the SCIL-based model with a focus on content analysis. Transnatl. J. Bus. (2017). http://www.acbsp.org/members/group.aspx?id=143359
15. Miller, A.H.: A corpus-based computer aided linguistic analysis of taxation learning outcomes. In: International Conference on Business, Big-Data, and Decision Sciences (2017 ICBBD), Chulalongkorn University, Bangkok, Thailand, 2–4 August 2017. http://icbbd2017.globalconf.org/site/page.aspx?pid=901&sid=1128&lang=en
16. Miller, A.H.: Modeling taxation learning outcomes using unsupervised machine learning. Int. J. Eng. Technol. **7**(2.4), 109–116 (2018). https://doi.org/10.14419/ijet.v7i2.4.13019. https://www.sciencepubco.com/index.php/ijet/article/view/13019/5200. Accessed 15 Aug 2018

17. Hunston, S.: Corpora in Applied Linguistics. Cambridge University Press, Cambridge (2010)
18. Vesanto, J., Alhoniemi, E.: Clustering of the self-organizing map. IEEE Trans. Neural Netw. **11**(3) (2000). http://ftp.it.murdoch.edu.au/units/ICT219/Papers%20for%20transfer/papers%20on%20Clustering/Clustering%20SOM.p
19. Berinato, S.: Visualizations that really work. Harv. Bus. Rev., June 2016. https://hbr.org/2016/06/visualizations-that-really-work. Accessed 2 Sept 2016
20. Buja, A., Swayne, D.F., Littman, M.L., Dean, N., Hofman, H., Chen, L.: Data visualization with multidimensional scaling. J. Comput. Graph. Stat. **17**(2), 444–472 (2008). https://doi.org/10.1198/106186008X318440
21. Tamura, T.: Application of text-mining methodology to sociological analysis of internet text in Japan, Japan and the internet: perspectives and practices (2011)
22. Mori, J., Matsuo, Y., Ishizuka, M., Faltings, B.: Keyword extraction from the web for FOAF metadata. In: Workshop on Friend of a Friend, Social Networking and the Semantic Web, Galway, Ireland. Google Scholar (2004)
23. Minami, A., Ohura, Y.: How student's attitude influences on learning achievement? An analysis of attitude representing words appearing in looking-back evaluation. In: ICIT Proceedings of the 8th International Conference on Information Technology and Applications, pp. 164–169 (2015)
24. Yu, C.H., Jannasch-Pennell, A., DiGangi, S.: Compatibility between text mining and qualitative research in the perspectives of grounded theory, content analysis, and reliability (2011)
25. Posner, R.: Opinion, United States of America, PlaintiffAppellee, v. Deanna L. Costello, Defendant-Appellant, No. 11-291 U.S. Court of Appeals Seventh Circuit Court (2012)
26. Smith, G.: Data and intuition. The Conglomerate (2014). http://www.theconglomerate.org/corpuslinguistics/
27. O'Neil, C.: 'Rogue Algorithms' and the dark side of big data. Knowledge@Wharton. Wharton, University of Pennsylvania (2016). http://knowledge.wharton.upenn.edu/article/roguealgorithms-dark-side-big-data/?utm_source=kw_newsletter&utm_medium=email&utm_campain=2016-09-22
28. Faul, F., Erdfelder, E., Buchner, A., Lang, A.G.: Statistical power analyses using G*Power 3.1: tests for correlation and regression analyses. Behav. Res. Methods **41**, 1149–1160 (2009). https://doi.org/10.3758/BRM.41.4.1149
29. Teixeira, D.P.: Attitudes on the ethics of tax evasion: a survey of banking employees (Master thesis), Lisbon school of economics and management (2016). https://www.iseg.ulisboa.pt/aquila/getFile.do?fileId=868711&method=getFile
30. Alm, J.: Measuring, explaining, and controlling tax evasion: lessons from theory, experiments, and field studies. Int. Tax Public Financ. **19**(1), 54–77 (2012)
31. Tenidoua, E., Valsamidisa, S., Petasakisa, I., Mandilasa, A.: Elenxis, an effective tool for the war against tax avoidance and evasion. Procedia Econ. Financ. **33**, 303–312 (2015). https://doi.org/10.1016/s2212-5671(15)01714-1. 7th International Conference, The Economies of Balkan and Eastern Europe Countries in the changed world, EBEEC 2015, 8–10 May 2015
32. Nie, L., Wang, M., Gao, Y., Zha, Z.J., Chua, T.S.: Beyond text QA: multimedia answer generation by harvesting web information. Trans. Multimed. **15**(2), 426–441 (2013). https://doi.org/10.1109/TMM.2012.2229971
33. Harvey, L: Beyond member-checking: a dialogic approach to the research interview. Int. J. Res. Method Educ., 1–16 (2014)

A New Transformation of 3D Models Using Chaotic Encryption Based on Arnold Cat Map

Benson Raj[1](✉), L. Jani Anbarasi[2], Modigari Narendra[3], and V. J. Subashini[4]

[1] Computer Information Science, Higher Colleges of Technology, Fujairah Women's Campus, Fujairah, United Arab Emirates
braj@hct.ac.ae

[2] School of Computing Science and Engineering, VIT University, Chennai, Tamil Nadu, India
janianbarasi.l@vit.ac.in

[3] Computer Science and Engineering, Madhav Institute of Technology and Science, Gwalior, India
narendramodigari@gmail.com

[4] Department of CSE, Jerusalem College of Engineering, Pallikaranai, Chennai, India
vjsubashini@yahoo.co.in

Abstract. In the emerging Virtual Reality era, 3D multimedia contents are popularized as images and videos. Encryption is a methodology that enhances the security by converting the original content into unintelligible content. The Arnold transform or Arnold cat map is a commonly used chaos-based encryption system that encrypts by shuffling the data. 3D models include vertices, faces and textures. An efficient secure symmetric chaotic cryptosystem is proposed for 3D mesh graphical models using 3D Arnold cat map. Arnold cat map is performed to encrypt the 3D mesh model using shuffling and substitution. The cryptosystem is proposed for vertices and faces separately and are composited together to form the final encrypted model. Each round introduces a good permutation and substitution through the confusion and diffusion generated by the 3D Arnold map. The chaotic function delivers more security for the 3D models through the shuffling and substitution. Simulation results show that the proposed scheme encrypts and decrypts the 3D mesh models and resists various attacks.

Keywords: 3D mesh models · Chaos · Image encryption · Arnold cat map · Arnold transform

1 Introduction

Security is an important issue in information technology, which is ruling the Internet world today. Recent technological advances in communication networks have turned the transmission of digital data into a popular task, and digital images are no exception. The necessity of transferring confidential multimedia data over open channels such as the Internet leads to security threats. With the advancement of multimedia technology

L. Barolli et al. (Eds.): EIDWT 2019, LNDECT 29, pp. 322–332, 2019.
https://doi.org/10.1007/978-3-030-12839-5_29

in the Internet, important information is increasingly becoming available in the form of digital images, videos and graphical 3D models. Rapid progress in the field of computer and information technology have augmented the usage of 3D mesh models in diverse areas like manufacturing, medical imaging, virtual reality, computer games, animation, etc., Public community is moving further towards securing their products digitally rather than physical views. So, protection of digital 3D models is an essential need in today's environment. Plenty of research work has been carried out to design robust and secure 2D image protection schemes, but 3D models have received less attention. Technical reports states that the worldwide shipment of 3D printers has increased around 49% in the year of 2013. Consequently, it shows that a huge advancement is expected in the forthcoming years due to their applications in real life. Techniques which successfully protected the 2D images, does not suite 3D solid models and 3D shell models. Graphical 3D models can be secured by means of encryption. Algorithms that are suitable for textual data and image data may not be appropriate for 3D models due to its nature of representation. Despite the sizeable number of encryption algorithms in existence, data scrambling is a common method employed to encrypt data so as to hide content from unauthorized users.

Chaotic maps are useful in ensuring the security of digital data by means of scrambling because they are easy to generate but deterministic and difficult to predict. Encryption is a technique used to enhance the security of a file by making incoherent content. The habitual transmission of digital 3D models necessitates that they may be made impervious to unauthorized access. It is, however, tedious to encrypt or decrypt 3D models directly. Chaos-based image encryption that works by shuffling the data is proposed to encrypt the graphical 3D models. A modified interleaving with shearing process is performed for encrypting the 3D models in the proposed scheme. Similar to Arnolds transform the original 3D model is reconstructed after certain number of iterations. The proposed method involves twice the number of cycles as Arnold's to bring back the original image. That is, on application, if observed, the iterations produce directional streaks and after a certain number of iterations then the 3D model returns to its initial state. The number of iterations taken by the image to return to its initial state is referred to as the period. Periodicity difference between the proposed scheme and various other chaotic schemes shows good encryption strength for the proposed scheme.

Chaotic map ensures the security of digital images by means of scrambling due to the easiness in generation and its prediction difficulty. The basic map is a combination of substitution and diffusion. In the substitution stage, chaotic map shuffles the image pixels and during diffusion the pixel values are altered. 2D cat map is generalized to 3D and are used for secure symmetric chaotic image encryption. 3D cat map permutes the image position during permutation stage and involves diffusion using logistic chaotic system during diffusion stage. Thamizhchelvy et al. [2] involved fractal image generation for hiding the image data where cracking this system was very difficult due to its chaotic behavior. Permutation was performed using chaotic map and image diffusion was done using logistic map. At the same time the logistic map that is used by Wang et al. [3] was cracked. Hyper chaotic sequence was proposed by Li et al. [4] for key stream generation and this scheme doesn't withstand chosen plain text attack. Wang et al. [5] performed image encryption through a perception model combined with

Lorenz map. Major drawback of Wang's algorithm lies with modification of a single bit which leads to single bit modification in encrypted image. Alvarez and Li [6] designed a generalized cryptosystem using chaotic map and identified the issues that occurs during implementation, key management and security analysis. Arnold cat map was combined with Chen map by Guan et al. [7] for improving the security purpose. Images have bulky data capacity and high correlation among the pixels, a high computational conventional algorithm is needed for encryption and decryption process.

Fridrich et al. [8] performed image encryption using a discrete 2D chaotic baker map. Properties like ergodicity, periodicity, sensitivity towards the initial conditions and control parameters, leads the chaotic maps for better security. Various chaotic maps have been analyzed by various researchers for digital image encryption. For real-time applications chaotic maps performs well due to simple iterative functions. Even small modification in the initial parameters leads to an enormous change on the resultant output. Continuous and discrete are the two types of chaotic maps. For designing complex image security algorithms, many researchers prefer chaotic transform due to its highly sensitive nature of the initial condition, along with confusion and diffusion. For image encryption process discrete chaotic maps are more appropriate. In most of the works as exemplified above, the basic scrambling is proposed using any transform and is generally used for various purposes like security enhancing and watermarking.

Rey [13] proposed an encryption for solid 3D models whereas Jolfaei et al. [14] and Jin et al. [15] used series of permutation and rotation for 3D point clouds. Few permutation based 3D mesh encryption was proposed in these research article. A transformation method proposed for 3D mesh models in this article is based on Arnold cat map and mainly deals with performing a cryptosystem using both 2D and 3D cat maps. Also, the proposed transformation is analyzed for both vertices and face shuffling, thus rendering transformation eminently appropriate for real-time applications.

The rest of this article is organized as follows. Section 2, describes the Arnold transformation and the pattern of 3D mesh model. The proposed transformation is detailed in Sect. 3. The investigations undertaken on the proposed method are given in Sect. 4. Finally, the conclusion on the findings is presented in Sect. 5.

2 Arnold Transform

The Russian mathematician, Vladimir Arnold proposed a chaotic transformation termed the Arnold transformation or Arnold cat map [1], while he was working on the ergodic theory. Arnold transform works in a very simple way: A Simple matrix is involved in transforming the coordinate position in the image, image is confused using increase in the number of iteration and retrieving back the original after reaching certain number of iteration. The main key in the Arnold transform is the number of iterations. The number of iteration that is required to regenerate the original is said to be the period in Arnold transform.

The confusion and diffusion property of the Arnold Cat Map makes it more suitable for image security. The General notation for image encryption using 2D cat map is used for image encryption as given in Eq. (1) where parameters 'a' and 'b' are the two important control parameters. A digital image can be represented as 2D matrix where (x, y) represents the pixel position in the image. After performing the 2D Arnold scrambling the pixels x and y becomes new pixel x′ and y′, represented as Eq. (2) where M is the size of the image.

$$ACM = \begin{bmatrix} 1 & a \\ b & ab+1 \end{bmatrix} \, or \quad ACM = \begin{bmatrix} ab+1 & a \\ b & 1 \end{bmatrix} \tag{1}$$

$$\begin{bmatrix} x' \\ y' \end{bmatrix} = ACM \begin{bmatrix} x \\ y \end{bmatrix} \mod M \tag{2}$$

Lian et al. [1] proposed a parametric 3D cat map in 2003 by extending 2D cat map in zx plane and yz plane. Keeping x plane unchanged and performing 2D cat map on y-z plane we achieve (3) and by Keeping y plane unchanged and performing 2D cat map on x-z plane we achieve (4) and Keeping z plane unchanged and performing 2D cat map on x-z plane we achieve (5).

$$ACM = \begin{bmatrix} 1 & 0 & 0 \\ 0 & 1 & a \\ 0 & b & ab+1 \end{bmatrix} \tag{3}$$

$$ACM = \begin{bmatrix} 1 & 0 & a \\ 0 & 1 & 0 \\ b & o & ab+1 \end{bmatrix} \tag{4}$$

$$ACM = \begin{bmatrix} 1 & a & 0 \\ b & ab+1 & 0 \\ 0 & 0 & 1 \end{bmatrix} \tag{5}$$

$$\begin{bmatrix} x' \\ y' \\ z' \end{bmatrix} = ACM * \begin{bmatrix} x \\ y \\ z \end{bmatrix} \mod M \tag{6}$$

The pixel position can be scrambled using a 3D cat map to a new location where [x′, y′, z′] is the scrambled position of the vector [x, y, z], and M refers to the size of the 3D models and is shown in Eq. (6).

2.1 3D Mesh Models

Geometrical 3D models include spatial relationships, particularly analytical geometry, where relationships are expressed in terms of algebraic formulas. 3D models are represented in the form of meshes or polygons where the mesh consists of edges, vertices and faces [9–12]. Vertices are represented as 'v' which includes $V = (V_1, V_2, \ldots V_m)$ and faces denote F_i € $V_i = 1 \ldots m$. Edges in the mesh refer to vertices which describes the intrinsic character of the 3D mesh and are invariant to translation, rotation, scaling and geometric distortions. The overall representation of 3D mesh model is shown in Fig. 1. Figure 2 shows the 3D mesh model of the piggy bank, its vertices and the faces.

$$V=\begin{bmatrix}V_1\\V_2\\\vdots\\V_m\end{bmatrix}=\begin{bmatrix}v_{1x} & v_{1y} & v_{1z}\\v_{2x} & v_{2y} & v_{2z}\\\vdots & \vdots & \vdots\\v_{mx} & v_{my} & v_{wz}\end{bmatrix}\quad F=\begin{bmatrix}F_1\\F_2\\\vdots\\F_n\end{bmatrix}=\begin{bmatrix}\alpha_x & \beta_y & \gamma_z\\\alpha_x & \beta_y & \gamma_z\\\vdots & \vdots & \vdots\\\alpha_x & \beta_y & \gamma_z\end{bmatrix}$$

Fig. 1. Representation of a 3D mesh model

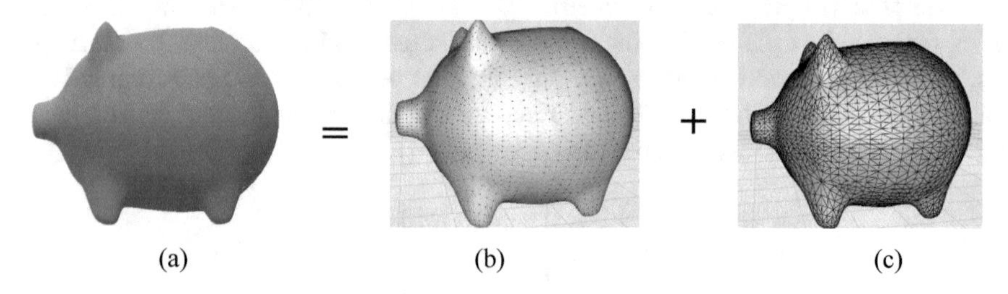

Fig. 2. (a) Original 3D model (b) vertices and (c) faces

3 3D Model Encryption

The proposed transformation matrix uses the basic principle behind Arnold's transformation matrix. Arnold's transformation matrix is based on the shearing transformation, both in the x, y and z directions, where the shearing factor is considered to be one. Image size always lies as MxN matrix whereas in the proposed work the vertices and faces of the 3D mesh models is always MX3. So altering x, y as well as x, y, z plane of 3D model is analysed in the proposed method.

The 3D model is decomposed into vertices and the face values and is scrambled using both 2D and 3D transform separately and finally combined to form an encrypted model. Every vertices and face values Vx, Vy, Vz, fx, fy, fz are scrambled to $Vx_i', Vy', Vz', fx', fy', fz'$. Since the vertices values and faces are modified, the 3D models are scrambled even during the earlier iterations. Figure 3 shows the input 3D mesh model used for experimental purpose. Table 2 shows the scrambling of 3D models after every iteration. Figure 4 shows the overall architecture of the proposed system. From Table 2, it is observed that the 342th iteration in Arnold's transform reconstructs the original 3D model for 3D cat map and reconstructed the original 3D model during 75th iteration for 2D cat map. Table 1 shows the characteristic of the mesh model used for experimentation. Figure 5 shows the proposed method of encryption and decryption

3.1 Vertex and Face Encryption

The vertices of the 3D mesh models are in the form of triplets that includes x, y, z plane.
$V = [V_{1,} V_2, \ldots \ldots \ldots V_m] = [\{v_{1x}\, v_{1y}\, v_{1z}\}, \{v_{2x}\, v_{2y}\, v_{2z}\}, \ldots \ldots \ldots \ldots, \{v_{mx}\, v_{my}\, v_{wz}\}]$
Where v_{mx}, v_{my}, v_{wz} the coordinates of the vertices and 'm' is number of vertices present in the 3D mesh model.

When applying the cat map, the element wise mapping of the coordinates is shown in

$$[v_{1x}\, v_{1y}\, v_{1z}] \begin{bmatrix} 1 & 0 & 0 \\ 0 & 1 & a \\ 0 & b & ab+1 \end{bmatrix} = [v_{1x}\, v_{1y} + v_{1zb}\, v_{1y} + v_{1z\,ab+1}]. \quad (7)$$

Similarly when shuffling the face values the triangular mesh gets modified to a new coordinates as shown below.

$$F = [F_1, F_{2,} \ldots F_n] = \left[\{\alpha_x\, \beta_y\, \gamma_z\}, \{\alpha_x\, \beta_y\, \gamma_z\}, \vdots \vdots \vdots, \{\alpha_x\, \beta_y\, \gamma_z\}\right]$$

Where

$$\begin{aligned} \alpha_x &=> v\alpha_x, v\alpha_y, v\alpha_z \\ \beta_y &=> v\beta_x, v\beta_y, v\beta_z \\ \gamma_z &=> v\gamma_{x,}\, v\gamma_{y,}\, v\gamma_{z,} \end{aligned}$$

$$[\alpha_x \beta_y \gamma_z] \begin{bmatrix} 1 & 0 & 0 \\ 0 & 1 & a \\ 0 & b & ab+1 \end{bmatrix} =>$$

$$[\{v\alpha_x, v\alpha_y, v\alpha_z\}, \{v\beta_x, v\beta_y, v\beta_z\}, \{v\gamma_{x,}\; v\gamma_{y,}\; v\gamma_{z,}\}] \begin{bmatrix} 1 & 0 & 0 \\ 0 & 1 & a \\ 0 & b & ab+1 \end{bmatrix} =>$$

$$\begin{aligned} &[\{v\alpha_x, v\alpha_y, v\alpha_z\}, \{\{v\beta_x, v\beta_y, v\beta_z\} + \{v\gamma_x b,\; v\gamma_y b\; v\gamma_z b\}\}, \\ &\{\{v\beta_x a, v\beta_y a, v\beta_z a\} + \{v\gamma_x ab + 1,\; v\gamma_y ab + 1\; v\gamma_z ab + 1\}\}] \end{aligned} \quad (8)$$

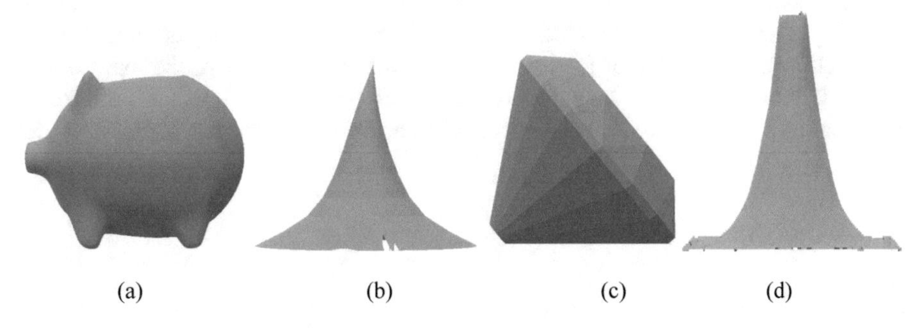

Fig. 3. Shows input 3D mesh model

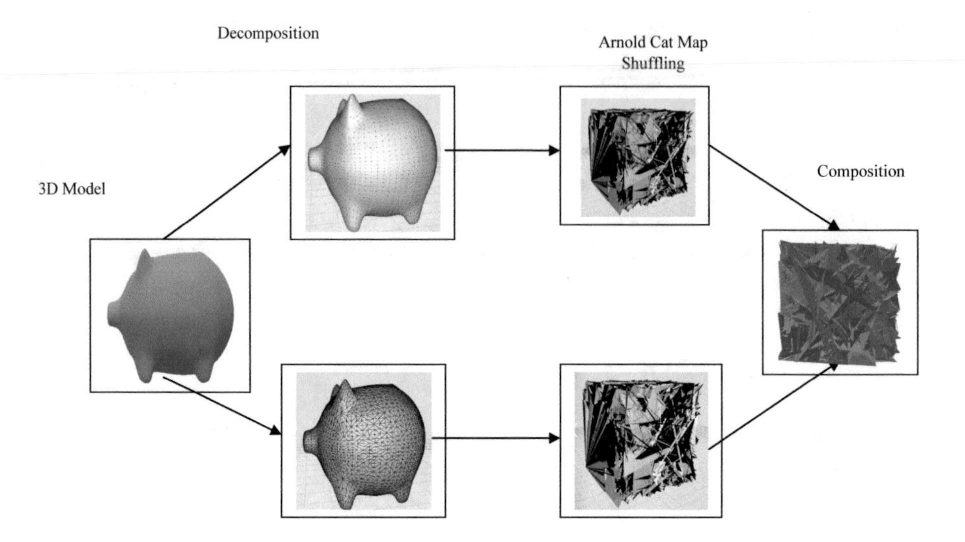

Fig. 4. Overall architecture of the proposed system

Table 1. Characteristics of 3D models used in our experiments

Object	Vertices	Faces
Diamond	228	110
Wizard-Hat	4788	4707
Piggybank	4040	4008
Valcono	15078	16756

4 Performance Evaluation

The proposed method uses chaotic transformation for securing 3D mesh models. That is, on application, if observed, the iterations produce directional streaks and after a certain number of iterations the models returns to its initial state. The number of iterations taken by the image to return to its initial state is referred to as the period. In this proposed work, only square mesh models of size 2n, ($n \geq 1$) are considered for evaluation. If the number of vertices and faces is not of that size then new vertices and faces are included with values 0. Each individual point in a mesh models is referred to as an x, y, z coordinates of face values and the xyz-plane where they are laid is called the pixel map. The x, y, z of the map describes the face position of the mesh model. Matlab is used to generate the experimental results. Tables 3 and 4 shows the Vertex Traversal of the mesh model for the transform and Periodicity of the 3D mesh model.

Table 2. Scrambling of 3D models after few iterations

No of Iterations	Arnold 3D Cat Map	Arnold 2D Cat Map
2		
45		
75		
189		
267		
345		

Table 3. Vertex traversal of the mesh model for the transform

Iterations		Initial position	16	32	75	96	112	345
Pixel position	Arnold 2D cat map	(35, 1)	(49, 3)	(44, 1)	(35, 1)	-	-	-
	Arnold 3D cat map	(35, 1)	(19, 4)	(5, 2)	(9, 3)	(53, 1)	(4, 4)	(35, 1)

Table 4. Periodicity of the 3D mesh model

N	2	4	8	16	32	64	128	256	512
Arnold 2D cat map	3	3	3	3	5	24	26	26	32
Arnold 3D cat map	3	3	6	12	24	48	96	192	345

4.1 Resistance to Statistic Attack

View point histogram (VFH) is used to describe the statistical feature point of the vertices and the faces of the 3D mesh model. The VFH is shown in Fig. 5 to show the statistical characteristics of the 3D mesh model before and after encryption. The figure shows that the VFH is completely different before and after encryption which leads to resistance to statistical attack.

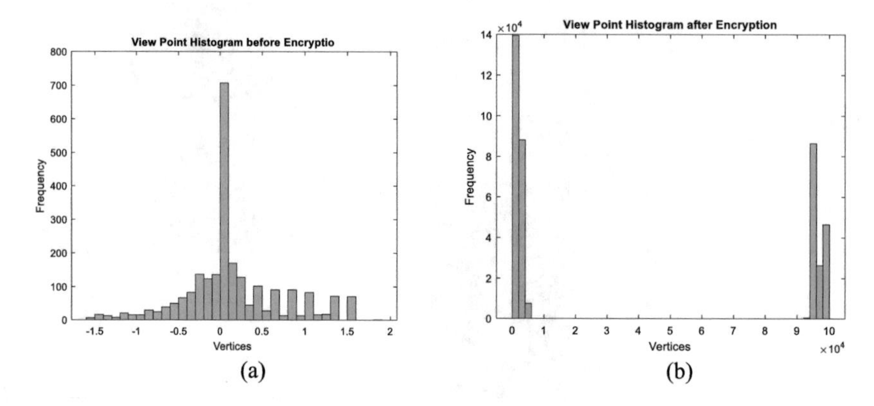

Fig. 5. VFH of 3D mesh models before and after encryption.

Quality of the proposed model is analysed using Peak Signal Noise Ratio (PSNR) and Normalised Coefficients (NC). Peak Signal Noise Ratio is computed to estimate the imperceptibility and NC is used to access robustness of the proposed crypto system.

$$\mathrm{PSNR} = 10\log_{10}\left(\max(\mathrm{I}(\mathrm{i},\mathrm{j}))^2/\mathrm{MSE}\right)$$

Where max(I(i, j)) is the maximum possible coordinate of the 3D mesh model where MSE is computed for the host 3D model 'I' before encryption and after encryption 3D model $\mathrm{W_m}$

$$MSE = \frac{1}{mxn}\sum\nolimits_{i=1}^{m}\sum\nolimits_{j=1}^{n}\left(I(i,j) - W_e(i,j)\right)^2$$

Where m, n is the dimension of the 3D model and normalised correlation is computed as Eq. (9) Where w and $\mathrm{w_e}$ indicates the original and the encrypted 3D modelmark, μ_e, μ_e indicates the mean of the original 3D mesh model and the mean of the encrypted 3D mesh model respectively. Tables 5 and 6 shows the PSNR and Normalized correlation values obtained between original and the encrypted 3D model The value shows that there is a huge deviation in values between the original and encrypted model.

$$NC(w, w_e) = \frac{\sum_{i=1}^{M}\sum_{j=1}^{N}(w(i,j) - \mu_w(w_e(i,j) - \mu_e}{\sqrt{\sum_{i=1}^{M}\sum_{j=1}^{N}(w(i,j) - \mu_w)^2}\sqrt{\sum_{i=1}^{M}\sum_{j=1}^{N}(w(i,j) - \mu_e)^2}} \quad (9)$$

Table 5. PSNR value between the original and the encrypted 3D model

Object	PSNR in dB
Pinetree	29.8
Piggybank	26.23
Golfball	20.3
Vase	24.7
Starfish	18.32

Table 6. Normalized correlation between the original and the encrypted 3D model

Object	Correlation value
Pinetree	0.5755
Piggybank	0.5437
Golfball	0.3451
Vase	0.6615

5 Conclusions

Multimedia models are widely used in the web, and ensuring the security of image data in terms of unauthorised access is crucial. A chaotic encryption method based on Arnold cat map, is proposed for the same. The shuffling, substitution and confusion and diffusion mechanism make it secure. Key size and 3D chaotic maps provide the additional security for the 3D mesh model. It is observed that the Arnold cat map gives greater scrambling choices for 3D mesh model. This leads to more chaos and better encryption by providing a larger key space.

References

1. Lian, S., Mao, Y., Wang, Z.: 3D extensions of some 2D chaotic maps and their usage in data encryption. In: 4th International Conference on Control and Automation, pp. 819–823, June 2003
2. Thamizhchelvy, K., Geetha, G.: Data hiding technique with fractal image generation method using chaos theory and watermarking. Indian J. Sci. Technol. **7**(9), 1271–1278 (2014)
3. Wang, X., Teng, L., Qin, X.: A novel colour image encryption algorithm based on chaos. Signal Process. **92**(4), 1101–1108 (2012)
4. Li, C., Zhang, L.Y., Ou, R., Wong, K.-W., Shu, S.: Breaking a novel colour image encryption algorithm based on chaos. Nonlinear Dyn. **70**(4), 2383–2388 (2012)
5. Wang, X.-Y., Yang, L., Liu, R., Kadir, A.: A chaotic image encryption algorithm based on perceptron model. Nonlinear Dyn. **62**(3), 615–621 (2010)
6. Alvarez, G., Li, S.: Some basic cryptographic requirements for chaos-based cryptosystems. Int. J. Bifurc. Chaos **16**(08), 2129–2151 (2006)
7. Guan, Z.-H., Huang, F., Guan, W.: Chaos-based image encryption algorithm. Phys. Lett. A **346**(1), 153–157 (2005)
8. Fridrich, J.: Secure image ciphering based on chaos. Final report for AFRL (1997)
9. Qi, D., Zou, J., Han, X.: Sci. China Ser. E: Technol. Sci. **43**(3), 304–312 (2000)
10. Jin, X., zhu, S., Xiao, C., Sun, H., Li, X., Zhao, G., Ge, S.: 3D textured model encryption via 3D Lu chaotic mapping. Science China Inf. Sci. **60**, 122107 (2017)
11. Jani Anbarasi, L., Anandha Mala, G.S.: Verifiable multi secret sharing scheme for 3D models. Int. Arab. J. Inf. Technol. (IAJIT) **14**(6), 1683–3198 (2015)
12. Jani Anbarasi, L., Narendra, M.: Robust watermarking scheme using Weber Law for 3D mesh models. Imaging Sci. J. **65**, 409–417 (2017)
13. Rey, A.M.D.: A method to encrypt 3D solid objects based on three-dimensional cellular automata. In: Proceedings of the 10th International Conference on Hybrid ArtiLcial Intelligent Systems, Bilbao, pp. 427–438 (2015)
14. Jolfaei, A., Wu, X.W., Muthukkumarasamy, V.: A 3D object encryption scheme which maintains dimensional and spatial stability. IEEE Trans. Inf. Foren. Secur. **10**, 409–422 (2015)
15. Jin, X., Wu, Z.X., Song, C.G., et al.: 3D point cloud encryption through chaotic mapping. In: Proceedings of the 17th Pacific Rim Conference on Multimedia Information Processing, Xi'an, pp. 119–129 (2016)

Feasibility Approach Based on SecMonet Framework to Protect Networks from Advanced Persistent Threat Attacks

Maher Salem(✉) and Moayyad Mohammed

Department of Computer Information and Sciences,
Higher Colleges of Technology, Al Ain, United Arab Emirates
{msaleml, mmohammed}@hct.ac.ae

Abstract. Advanced Persistent Threat (APT) principally steal data once the attacker gains unauthorized access to network resources. In this paper, we propose a detection and defense technique based on SecMonet framework to avoid this sophisticated attack. SecMonet is a security framework that can gather events and flows, normalize them, create a valuable dataset, train a classifier based neural networks, and detect and defend against APT attacks. In this regard, log data from logging servers or Firewall has been considered by SecMonet. In addition, a ranking criterion for detected suspicious activities has been also considered by the classifier to detect APT attack. The proposed method has been evaluated by a local simulated network and by a real network scenario. The result shows that the proposed technique can significantly detected APT attacks.

1 Introduction

Advanced Persistent Threat (APT) is a sophisticated cyber-attack that use multi stage techniques trying to exploit sources to target and compromise systems for long time undetected. These attacks are so advanced; that even cyber defense applications are unable to detect it. APT attack, namely Operation Aurora, targeted giant organizations such as Google or Adobe. In this attack, valuable intellectual property was stolen and the attack went undetected for six months [1]. The term "Advanced Persistent Threat" refers to a well-organized, malicious group of people who launch stealthy attacks against computer systems of specific targets, such as governments, companies or military. Main life cycle of APT is summarized by: research, preparation, intrusion, conquering the network, hiding presence, gathering data and gaining access [2].

Most of computer networks in any organization are exposed to this kind of attacks since they have covered or uncovered vulnerabilities. The attackers in this case keep trying to gain unauthorized access by exploiting any vulnerability and install the code in the target machine, usually it is encrypted code, to be launched based on certain trigger. The code intentionally seeks for any sensitive or personal information and steal them by communicating to a remote host. This is a major challenge and a problem for the networking as well as for security. To avoid this kind of attacks, this paper presents an advanced approach using SecMonet [3], which is an intrusion detection system based on machine learning and data mining.

L. Barolli et al. (Eds.): EIDWT 2019, LNDECT 29, pp. 333–343, 2019.
https://doi.org/10.1007/978-3-030-12839-5_30

SecMonet is an adaptive real-time anomaly-based intrusion detection system contains of three major parts, namely, OptiFilter, Normal Network Behavior (NNB) and Adaptive Classifier. OptiFilter aggregate the network flows and hosts events, preprocess them and generate datasets based on selected features [4]. Where the adaptive classifier is an improved Growing Hierarchical Self Organizing Map (EGHSOM) that classifies the abnormal and suspicious connections and activities [5].

If the adaptive classifier was not able to classify any connection as normal or abnormal, then it classifies it as unknown, which then redirected to the NNB model. The NNB model is a EGHSOM but trained only by clean connections, thus it represents the normal network behavior. Accordingly, NNB classifies the unknown traffic as abnormal if it deviates from the trained normal behavior [6]. For more illustration, Fig. 1 shows a general overview of SecMonet.

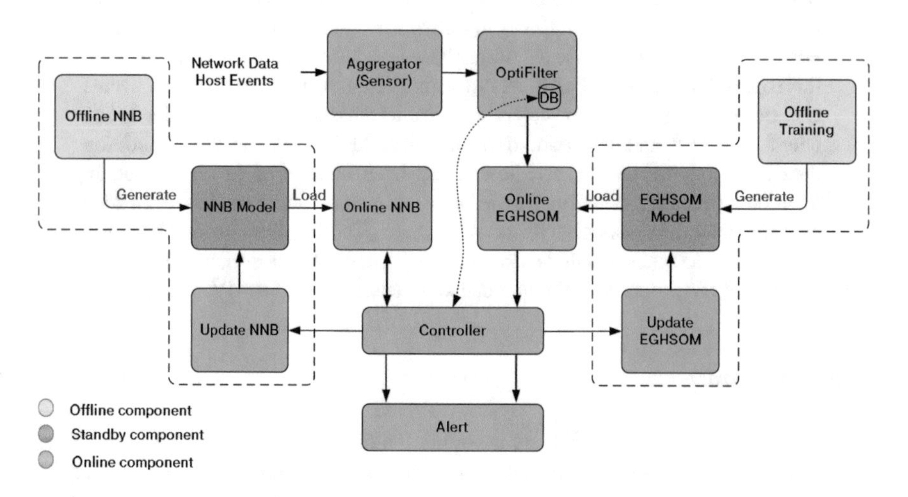

Fig. 1. General overview of SecMonet

In this article, we adopt the framework SecMonet but with further modifications to significantly detect APT attacks and defend against it.

In addition to flows and events, log data from logging servers or Firewall has been considered by OptiFilter. Moreover, a ranking criterion for detected suspicious activities has been also considered by the classifier to detect APT attack. Based on the ranking and the classified logs, a defense rule is created to avoid any exploit of network resources or any unauthorized access.

The rest of this article is organized as follows, section two discuss the recent and valuable works in detecting APT. Section three describes the modifications of SecMonet. An analysis and discussion of the modifications will be presented in section four. Finally, section five conclude the work.

2 Related Work

APT exists since long time and now has more attentions from the researcher due to its serious threat on the systems, specifically, the sensitive and personal data. Moreover, it has attacked as mentioned above some organizations and stole users' information. Therefore, a lot of significant researches and application are trying to detect and combat this kind of attacks. Chen et al. conducted a comprehensive study about the APT attack, characterizing its distinguishing characteristics and attack model, and analyzing techniques commonly seen in APT attacks [7]. They explained the main phase of APT attacks and the proper countermeasures to defense such kind of attacks. Almost the same but more comprehensive explanation and investigation of recent APT detection techniques, has been also presented by Ussath et al. [8].

Bhatt et al. have introduced a framework to detect APT using the following components: a multi-stage attack model, a layered security architecture and a security event collection and analysis system [9]. They use Intrusion Kill Chain (IKC) attack model, which allows a better tuning of security configuration and it has been used as a hypothesis model to improve the correlation of logs and thereby facilitate the identification of ongoing attacks.

On the other hand, Friedberg et al. [10] have treated the detection process of APT differently than the conventional signature based detection. They work on a whitelist approach and their anomaly detection technique learns the normal system behavior over time by keeping track of system events and their dependencies. The main input source of their work was utilizing the log events from several network components and process them accordingly. Then a system model has been created based on certain pattern, event classes and hypothesis. This approach is delivering a very promising result since it focuses on the log events. In our approach, we refer to their approach and utilize some functionalities. Quader et al. also use the network data collected by the intrusion detection to identify and detect persistence threat patterns [11]. They utilize Association Rule Mining (ARM) to detect persistent threat patterns on network data. They first capture the IDS log data and preprocess it. After refining the data, they bin the data into equal timespans. After that, they identify the frequent patterns using the Association Rule Mining and determine the overlapping and non-overlapping rules. Last step, they isolate the high priority unusual persistent threats. Finally, they correlate the time of the day when these persistent threats occur and validate the result. It is clear that they did not mentioned which type of IDS they are utilizing and if they consider the log files or activities in their data. From other perspective, Saranya et al. gathered the host data such as CPU usage, Memory and system files to classify APT patterns based on known Malware patterns [12]. The approach is using known malware to detect APT where emerged APT attacks will not be detected. Moreover, they use the host features which are not representing the entire network.

Some researchers have prepared the minimum requirements to establish an effective APT detection method as presented by Yang et al. [13]. One of the advanced model that detect also dynamic APT in the network is presented by Niu et al. [14]. Their model realise on Targeted Complex Attack Network (TCAN) model for APT attack process based on dynamic attack graph and network evolution. The experiment has been performed on a test environment and the result achieves high accuracy. Just the question is still open about the testing in real-time and online network. Another recent work for AbdElatif Mohamed et al. [15] focused again in the operating system by installing special login script to enhance the process of detecting APT. They uncover how organization ignore security part in their devices by minimal awareness and by the weakness in designing, implementing, and updating of information security policy. Many more topics investigates the severity of APT attacks and how to avoid it, we have just discussed few of them but more details can be also acquired from [16–19].

The previous discussed works showed a great interest in investigating the techniques of defensing APT attacks. But they have in some or several points some shortages such as not including the log files or events in a proper and sufficient way, or by focusing on irrelevant data source which leads to ineffectiveness, or build their model based on known malware patterns. In our approach, we study the feasibility of improving SecMonet to overcome all of the aforementioned shortages and then introduce a new ranking to detect suspicious logs on the network.

3 Proposed Approach

In the proposed approach, all logs from all intended network devices are considered for processing. Therefore, all logs will be correlated with the flows and event. Then a log ranking technique will weight every correlated log. Figure 2 shows a comprehensive structure of the proposed framework.

Just to mention again that the main components of SecMonet will be utilized in this article, hence, OptiFilter will now process more data received from network devices, that is, logs.

OptiFilter processes the main information of each log as follows:

1. It will recognize the source IP address and create an object for each IP.
2. Logs will be grouped based on their IP address and each log will be assigned a unique ID, timestamp, and contents for monitoring and tracking.
3. Then OptiFilter normalizes and correlates all logs with the aggregated flows and events as described in [18].
4. When the correlation process finishes, the classifier assigns each correlated input an initial score (default is 1).
5. As long as the ID or IP or even the content of an input (correlated logs) is appeared again during the classification, the classifier increase the score of it by one more point.
6. The classifier decides to classify certain input as APT based on a threshold value (when the score reaches certain value, we decided based on expertise to have the score threshold value 20).
7. Finally, an alert is launched and the classifier updates its model accordingly.

The structure of the framework as seen in Fig. 2.

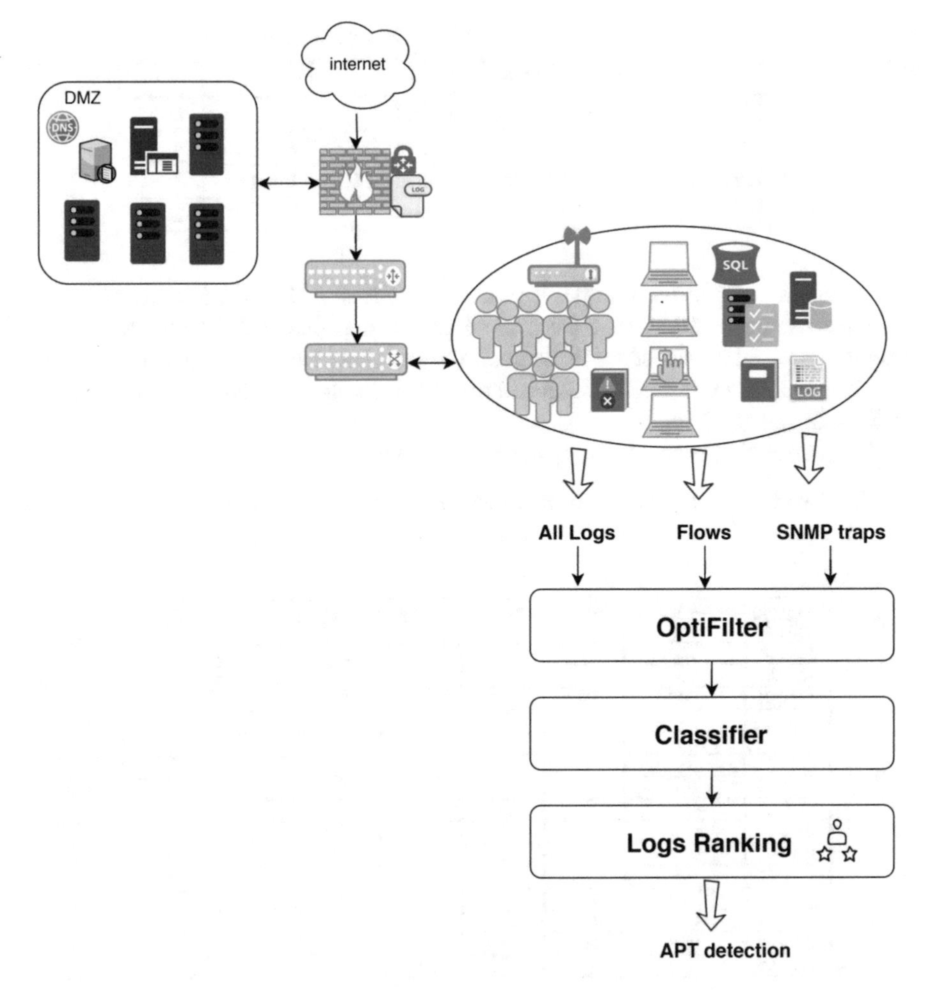

Fig. 2. General overview of the proposed framework

3.1 Correlation Process

We define three features for logs, namely, {*log_IP, log_Timestamp, log_content*}. Hence, the correlation process from OptiFilter in [18] in Fig. 3, is modified such that log features are also correlated with the connections and events in the same dataset.

However, for a ranking purpose, we also generate a separate dataset only from received logs, in which the dataset contains only the aforementioned three log features. In addition to these three features, we add a fourth feature called score, which indicates a weight value of the log.

The following dataset sample in Table 1 clarifies the idea of log dataset with score feature.

Table 1. Dataset sample of logs and their score value

ID	IP	Time	Contents	Score
1	10.13.10.4	12:33:45	Failed attempts	6
2	192.168.0.2	03:09:43	User1 has deactivated the firewall	4
3	33.44.2.3	16:44:22	User1 enables RPC and logout	3
4	13.2.44.3	22:15:53	Failed attempt	12

As an example in the first ID, it means, at 2:33:45 AM a failed attempt has been recorded on the IP 10.13.10.4 the score 6 means that this is the sixth time a failed attempt occurred on the same IP address. On the other hand, ID 3 means, user1 has enabled the RPC in the IP 33.44.2.3 and then logged out, user1 did this action last time at 6:44:22 and for 3 times.

The correlation process simply happens as follows:

- SecMonet generates dataset from connections and SNMP events, say each instance has almost 17 features (attributes). See Fig. 3 below.

Attributes

Instances

timestamp	protocol_type	service	src_byte	count	flag	... ...	logged_in	class
12:00:00	TCP	SSH	20	1	SF	... ...	0	Normal
12:00:01	TCP	FTP	5000	2	SF	... ...	0	Normal
12:00:01	UDP	DNS	33	5	S0	... ...	0	Normal
12:00:01	TCP	HTTP	65	3	SF	... ...	0	Normal
⋮	⋮	⋮	⋮	⋮	⋮	... ...	⋮	⋮
⋮	⋮	⋮	⋮	⋮	⋮	... ...	⋮	⋮
⋮	⋮	⋮	⋮	⋮	⋮	... ...	⋮	⋮
12:00:03	TCP	HTTP	152	2	S1	... ...	1	Normal
12:00:04	ICMP	0	150	1	S1	... ...	0	Anomaly
12:00:04	ICMP	0	150	1	REJ	... ...	0	Normal
12:00:04	UDP	TFTP	65241	1	S1	... ...	0	Normal
12:00:06	ARP	0	0	2	SF	... ...	1	Anomaly
12:00:07	TCP	HTTP	325	1	SF	... ...	0	Anomaly
12:00:08	TCP	HTTP	845	45	SF	... ...	0	Anomaly
12:00:09	UDP	DNS	26552	2	SF	... ...	0	Normal
12:00:09	TCP	FTP	548	5	SF	... ...	0	Normal

Fig. 3. Sample SecMonet dataset

- Each connection has a source and destination IP as features.
- A separate dataset is also created from the logs received from the firewall and logging server (or any other network device) as shown in Table 1.
- If the IP address of the log is the same IP in SecMonet dataset, then for that instance, we add all logs features in Table 1 so the dataset have now 21 features.
- If the time is also the same the log will be also amended to SecMonet dataset.
- The score is just for the classifier to monitor the logs and their occurrences.

Flow diagram in Fig. 4 shows an overview about the correlation process of logs with SecMonet dataset.

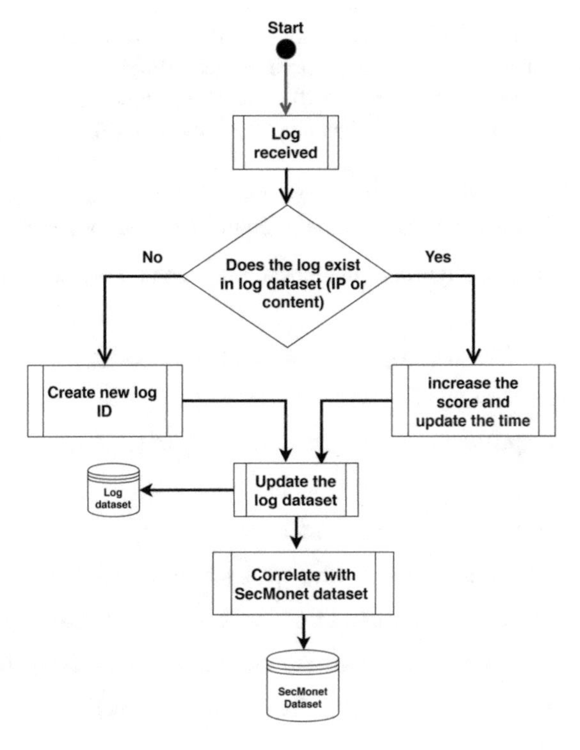

Fig. 4. Correlation process of logs with SecMonet dataset

3.2 Ranking Method

SecMonet classifier monitors the score for all correlated logs to detect any attempt for APT attacks.

To rank the logs accurately, a training phase of all legitimate logs has been conducted on the classifier of SecMonet, so that it can rank the logs based on the trained model.

The trained model is an EGHSOM [5] adaptive classifier that builds a map for all trained logs, the result then:

- The model has certain BMUs [3] representing legitimate logs.
- In real-time if a new log is received and correlated, then it should be related to one BMU based on a distance measure (Euclidean distance): The block distance between any two vectors $x, y \in \mathbb{R}^n$ in an n-dimensional space can be formally described as

$$d(x, y) = |x_1 - y_1| + |x_2 - y_2| + \ldots + |x_n - y_n| \equiv \sum_{i=1}^{n} |x_i - y_i|$$

- Based on the minimum distance, the log instance x is now belonging to one BMU y and the classifier monitors the log x for the future detection.

- Assume the log x is "failed attempt", then it will be correlated to the ID that contains the same content and the score will increase accordingly.
- Once this log is presented to the classifier and the distance to the BMU is increased significantly from the previous distance, then the classifier detects it as an APT attack.
- So, the significant deviation of the log x ranked it to be suspicious.
- SecMonet generates a rule to stop the attempt of gaining unauthorized access.

We describe the main algorithm behind ranking technique in the following pseudo code:

```
Collect legitimate logs in a dataset Log_D
Train EGHSOM using Log_D
          Output: Best Matching Units(BMUs)
Go Online
       New Log x is received
       Log x receives a score and stored in Log Dataset
          Log x is correlated with (SecMonet Dataset)
          Calculate distance to corresponding BMU
          If (Distance slightly deviates)
            Then normal
            Check log x score
            If (score > threshold)
            Then suspicious APT
          Else
            Then suspicious APT
            Issue proper alarm
Generate a rule to defend against APT
Goto Online
```

4 Analysis, Result and Discussion

The proposed approach studies the feasibility of utilizing an existed framework called SecMonet to detect APT attacks. Our assumption is that, we can correlate all logs with the dataset in SecMonet and classify them based on a ranking method.

We have evaluated the proposed model using synthetic and realistic data sources. In the university campus, we have installed a test network as seen in Fig. 5 and synthetically generated network traffic with a maximum bit rate of 200 Mbps. We have generated the traffic based on realistic network of our industrial partner. We capture the traffic at specific point (the bridge) and use it to evaluate the proposed model. There are several services running on the test network such as HTTP, SSH, FTP, SMTP, Rsync, Telnet [18].

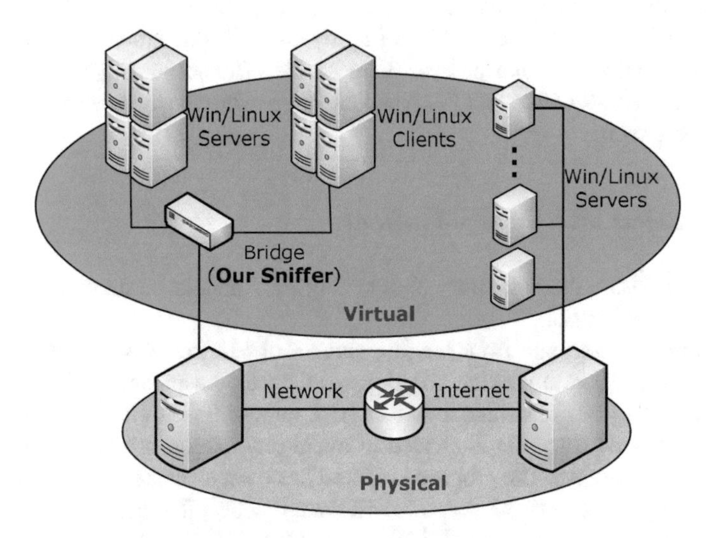

Fig. 5. Test network

We ran this network for longtime and attempted several times to login using wrong credential. Moreover, tried to connect remotely using SSH or Telnet. We have configured a logging server that sends logs to our SecMonet.

SecMoent delivers the following results as shown in Table 2.

Table 2. SecMonet results

Log_ID	Score	Detection rate	Rank	Class
00001	13	89%	<20	Normal
00002	4	92%	>20	APT
00003	112	90%	<20	Normal
...	...	...	...	...
...	...	...	...	...
00250	87	90%	<20	Normal

The result is just to indicate that the feasibility of the proposed approach is valid. SecMoent could detect up to 92% accuracy and therefore could detect an APT attack.

The challenge of this approach still if the correlation process as well as the ranking technique are scalable and sustainable.

Since SecMonet has been tested in a real life scenario in 2014 with one of the partners JUMO GmbH in Germany, we did not perform the same test in this article. The main contribution in this article is to study the feasibility of improving SecMonet to consider logs and detect APT attacks. The test network has been prepared just to verify that the improvement works and SecMonet can handle the logs together with Flows and SNMP traps. In addition, it can in real-time process and generates continuous datasets for further classifications.

But, this article has just verified that considering logs in SecMonet to detect APT attacks is feasible, hence, the next first improvement in this article is to handle large networks with large logs and process them online. The second improvement will be integrating SecMonet with a real life problem and perform a comprehensive functional and stress test on it.

5 Conclusion and Future Outlook

The article discusses the possibility of detecting APT attacks based on a previous IDS system called SecMonet.

In this regard, log data from logging servers or Firewall has been considered by SecMonet. In addition, a ranking criterion for detected suspicious activities has been also considered by the classifier to detect APT attack. Based on the ranking and the classified logs, a defense rule is created to avoid any exploit of network resources or any unauthorized access. The proposed method has been evaluated by a local simulated network and by a real network scenario. The result shows that the proposed technique can detect APT attacks. However, this is just a feasibility study about detecting and combating APT attacks. In the future, Logging and time dimension will be taken under consideration, SecMonet will be re-implemented to detect only APT attacks, and more features will be selected for training.

References

1. Vance, A.: Flow based analysis of advanced persistent threats. In: IEEE First International Scientific-Practical Conference: Problems of Info Communications. Science and Technology, Kharkov, Ukraine, pp. 173–176 (2014)
2. Vukalović, J., Delija, D.: Advanced persistent threats - detection and defense. In: International Convention on Information and Communication Technology, Electronics and Microelectronics (MIPRO), Opatija, Croatia, pp. 1324–1330 (2015)
3. Salem, M.: Adaptive Real-Time Anomaly-based Intrusion Detection Using Data Mining and Machine Learning Techniques. Kassel University, Kassel (2014)
4. Salem, M., Buehler, U.: Mining techniques in network security to enhance intrusion detection systems. Int. J. Netw. Secur. Appl. **4**(6), 51–66 (2012)
5. Salem, M., Buehler, U.: An enhanced GHSOM for IDS. In: IEEE International Conference on Systems, Man, and Cybernetics, Manchester, UK, pp. 1138–1143 (2014)
6. Salem, M.: Normal Network Behavior Model: In Adaptive Real-time Anomaly-based Intrusion Detection using Data Mining and Machine Learning Techniques, pp. 96–99. University Kassel, Kassel (2014)

7. Chen, P., Desmet, L., Huygens, C.: A study on advanced persistent threats. In: IFIP International Conference on Communications and Multimedia Security, Berlin, pp. 63–72 (2014)
8. Ussath, M., Jaeger, D., Cheng, F., Meinel, C.: Advanced persistent threats: behind the scenes. In: Annual Conference on Information Science and Systems (CISS), Princeton, NJ, USA, pp. 181–186 (2016)
9. Bhatt, P., Toshiro Yano, E., Gustavsson, P.M.: Towards a framework to detect multi-stage advanced persistent threats attacks. In: IEEE 8th International Symposium on Service Oriented System Engineering, Oxford, UK, pp. 390–395 (2014)
10. Friedberg, I., Skopik, F., Settanni, G., Fiedler, R.: Combating advanced persistent threats: from network event correlation to incident detection. Comput. Secur. **48**, 35–57 (2015)
11. Quader, F., Janeja, V., Stauffer, J.: Persistent threat pattern discovery. In: IEEE International Conference on Intelligence and Security Informatics (ISI), Baltimore, MD, USA, pp. 179–181 (2015)
12. Chandran, S., Hrudya, P., Poornachandran, P.: An efficient classification model for detecting advanced persistent threat. In: ICACCI, pp. 2001–2009. IEEE, India (2015)
13. Yang, L.X., Li, P., Yang, X.: Security evaluation of the cyber networks under advanced persistent threats. IEEE Access **5**, 20111–20123 (2017)
14. Niu, W., Zhang, X., Yang, G., Chen, R., Wang, D.: Modeling attack process of advanced persistent threat using network evolution. IEICE Trans. Inf. Syst. **E100.D**(10), 2275–2286 (2017)
15. AbdElatif Mohamed, N., Jantan, A., Isaac Abiodun, O.: An improved behaviour specification to stop advanced persistent threat on governments and organizations network. In: Proceedings of the International MultiConference of Engineers and Computer Scientists, pp. 219–224. International Association of Engineers (IAENG), Hong Kong (2018)
16. Hu, P., Li, H., Fu, H., Cansever, D., Mohapatra, P.: Dynamic defense strategy against advanced persistent threat with insiders. In: IEEE Conference on Computer Communications (INFOCOM), pp. 747–755. IEEE, Kowloon (2015)
17. Rass, S., Koenig, S., Schauer, S.: Defending against advanced persistent threats using game-theory. PLOS ONE **12**(1), e0168675 (2017)
18. Salem, M., Buehler, U.: A comprehensive model for revealing anomaly in network data flow. In: Lecture Notes in Informatics Proceedings, pp. 913–924. Gesellschaft für Informatik e.V., Bonn (2014)
19. Marchetti, M., Pierazzi, F., Colajanni, M., Guido, A.: Analysis of high volumes of network traffic for Advanced Persistent Threat detection. Comput. Netw. **109**(2), 127–141 (2016)
20. Salem, M., Buehler, U.: Transforming voluminous data flow into continuous connection vectors for IDS. Int. J. Internet Technol. Secur. Trans. **5**(4), 307–326 (2014)

Trust Analysis for Information Concerning Food-Related Risks

Alessandra Amato(✉) and Giovanni Cozzolino

University of Naples Federico II, Naples, Italy
alessandra.amato@studenti.unina.it, giovanni.cozzolino@unina.it

Abstract. In last years many business activities, scientific researches and applications exploit social networks as important sources for gathering data with different aims. Knowing the habits and preferences of user can be useful for different purposes, firstly to build marketing and advertising campaigns, but also to analyse other social phenomena for statistics, demography or security reasons. Thanks to their wide adoption among people, social networks are becoming the first media adopted to publish and share real-time news about happening events and, consequently, also the main media to retrieve information on what happens around you. Taking into account this consideration, in this paper we investigate a methodology for *semantic* analysis of textual information obtained from social media streams, in order to perform an early identification of food contaminations. As a case study, we consider a set of reviews gathered from the social network Yelp [26], on which we perform the textual analysis foreseen in the proposed methodology.

1 Introduction

Social networks user's trends and preferences are used for many application fields, above all in "social media marketing" field: analysing users' behaviour, companies can infer their tastes and habits and, consequently, can predict users' needs and preferences [4]. Such information can be used for commercial purposes in order to influence a great amount of customers and implement more effective advertising campaigns.

In the last decade, communities and search engines - such as TripAdvisor [25] or Yelp [26] - has started to be also a valuable source for many others *social phenomena* in order to analyse behaviours that influence or is influenced by individuals who relate each another. The early identification of food diseases is an hot trend among scientific communities, and in the near future social network analysis will likely acquire more and more weight to reach this aim.

This is also the conclusion of a study [11] conducted by a team of bioinformatics researchers of the Computer Science Department of Columbia University (New York) and published on the Journal of the American Medical Informatics Association. In fact, in 2012 the team developed and launched a program able to identify, among the reviews of Yelp community users (also present in

L. Barolli et al. (Eds.): EIDWT 2019, LNDECT 29, pp. 344–354, 2019.
https://doi.org/10.1007/978-3-030-12839-5_31

Italy), reports on food-borne outbreaks of single or groups of customers, among restaurants of New York. From its activation, the system has analysed over 13 thousand of posts, identifying 8,523 cases of infections and 10 outbreaks of potentially dangerous crises. But above all, what matters most, the social network analysis approach has been proved to be faster and more reliable than the traditional ones, which include a report to the Department of Health and Mental Hygiene. In fact, despite the number of food contamination deaths or infections in the United States each year, people, especially young, feel more practical and comfortable to post messages on social media rather then to report any accidents to the public health institutions. This trend has been proven by the fact that, in the first year of activity, only 3% of the toxins detected by the Columbia program were reported to competent institution through the classical systems.

This trend is occurring not only in the case of food infections: according to a recent estimate, nowadays the first reports of epidemics, for example influenza, are made in 70% of cases via social media.

On these basis, in this paper we propose a methodology, based on *semantic* technologies, aiming to extract topics from posts gathered from social network streams in order to perform an early identification of food contamination events. As a case study, we consider a set of reviews gathered from the social network Yelp [26], on which we perform the textual analysis foreseen in the proposed methodology. Advanced approach can be further exploited to improve performances through application-specific design methodologies [7, 13] or dedicated hardware [6].

Text Mining

Information retrieval and knowledge management activities on data gathered form social network requires to adopt Big Data techniques and efficient processing algorithms [5]. Big Data approaches adopt a graph model and parallel and distributed architectures that make use of data-intensive distributed platforms like Apache Hadoop [21], Spark and Flink [16]. These platforms implement popular programming models, such as MapReduce, in which a master node divides input datasets mapping each subset into parallel and independent worker nodes. Their outputs are reduced to calculate the final result.

Many *morpho-semantic* approaches have been proposed in medical domain for text analysis. Pratt [20] dealt with the identification and the transformation of terminal morphemes in the English medical dictionary; Wolff [27] worked on the classification of the medical lexicon based on formative elements of Latin and Greek origin; Pacak et al. [20] classified the diseases words ending in *-itis*; Norton and Pacak [19] classified the surgical operation words ending in *-ectomy* or *-stomy*; Dujols *et al.* treated the suffix *-osis*.

Lovis *et al.* [17], proposed an approach to solve the problem of automatic population of thesauri, deriving the meaning of the words from the morphemes that compose them; Hahn et *al.* [15] segmented the sub-words in order to recognise medical documents; in [14] authors exploit SNOMED thesaurus applying machine learning methods on it to analyse morphological data.

Many other works deal with the definition and the implementation of an architecture for information structuring, by mapping, through the exploitation of semantic technologies, the entities extracted from different corpora on a set of shared concepts [18,28], in order to ensure the interoperability.

Many studies have been conducted for the enhancement of semi-automatic medical information extraction from structured repositories or unstructured texts. In [12] author proposes a clinical information system, that, through Information Extraction techniques, transform medical unstructured information retrieved on clinical records into a structured format. Rink [23] proposes a method that exploits machine learning algorithms, in conjunction with some language resources (such as Wikipedia, WordNet, etc.) and processing resources (POS tagger, chunk parser, etc.) to perform a semi-automatic automatic feature extraction from electronic medical reports. In [10] the authors adopt NLP tools to automatically extract medications and related information from discharge summaries. Finally, in [2] authors propose a knowledge Based collaborative framework for e-health.

The rest of the paper is organised as follows. In Sect. 2, we discuss the graph model, used in social network analysis, and we introduce Text Analysis Tools adopted in our methodology. Then, in Sect. 3, we report the design of performed experimentation. Computational results are summarised in Sect. 4. Conclusions and remarks are discussed in Sect. 5.

2 Methodology

Nowadays, social networks analysis techniques mostly exploit the graph theory to model the network. Through graph theory, the users of a social network can be represented as a nodes of the graph and their relationships as edges. In this way, network data can be represented in a matrix form, directly analysable without the need of drawing the graph, that could be unfeasible for large datasets [3,4].

2.1 Text Analysis Tools

2.1.1 TaLTac

TaLTaC (*Trattamento automatico Lessicale e Testuale per l'analisi del Contenuto di un Corpus*) [24] is a software for natural language processing (Lexical and Textual), able to analyse textual content of a set of document contained in a Corpus. It can perform many different operations on documents like Text Analysis and Text Mining. Text Mining is a features that extracts relevant information from the Corpus, by dividing the document in fragments and associating them a set of variables, in order to identify logically related fractions of the corpus. Other resources made available by TaLTaC allows to identify peculiar language of the text by computing frequency of a term or lexical unit in a corpus.

2.1.2 GATE

GATE (General Architecture for Text Engineering) [22] is a family of software built around a common specific framework (GATE Embedded) with which it is possible to implement text analysis and information extraction applications. GATE comes with many language and processing resources that can be combined into running applications. For example, ANNIE is one of GATE's built-in application, that is an information extraction system constituted by a set of language resources and processing resources structured in a pipeline specialised for information extraction tasks.

2.2 Methodology

In our methodology we foresee multiple phases performing different textual elaboration, in order to achieve a fine grain analysis of different data sources.

2.2.1 Text Pre-treatment

A text pre-treatment has to be performed in order to obtain better results in the subsequent phases. It consists of two steps: normalization and correction of Spelling errors

Normalization aims at writing in a unique "normal" form all the names, acronyms and other entities by executing the following operations:

- changing apostrophes into stresses;
- annotate words or sequence of words with labels reporting their right meaning (i.e. 'Rose' has a different meaning of 'rose');
- capitalize letters, according to the eventual labels associated with the word (i.e. the term 'Rose' must have the capital letter if it is a proper name).

Correction of spelling errors compares misspelled words with the system dictionary in order to correct them.

2.2.2 Lexical Analysis

Lexical analysis performs the segmentation of the Corpus (i.e. the division of the text into segments, sequences of word separated by a divider like punctuation) and the estimation of various lexical parameter of each segment, starting from the words that compose them. Other two operations are Tagging and Lexation: Tagging links a description of the grammatical or semantic characteristics [1,8,9] to each word, while Lexation identifies sequence of close words tagged as a unique concept during the pretreatment. Finally, we extract the Corpus' keywords by comparing their frequency, assuming more relevant those having a appreciable standard deviation.

2.2.3 Textual Analysis

Textual Analysis is mainly composed by the study of the Concordances and by co-occurrences identification.

Concordances require to take into consideration the context of every word. We adopt TF-IDF index to rank the results by computing the frequency and distribution of the query keywords in the documents provided, according to Eq. (1):

$$\text{TF-IDF} = f_{t,d} \cdot \log \frac{N}{D_t} \tag{1}$$

where:

$f_{t,d}$ is the frequency of term t in the resource d;
N is the number of resources of the corpus;
D_t is the number of resources containing the term t.

Through the identification of co-occurrences, the set of near elements that repeats in the text are treated as single entities, thus enhancing the definition of primary concepts contained in the Corpus.

3 Experimental Design

We design an experimental campaign in order to analyse on-line social networks data sets and to derive information regarding contamination and food-related risks. The data set is constituted of tuples collected from the social network Yelp, where users publish reviews about local businesses [26]. As a simplified experimentation, we extracted from collected tuples an anonymized subset of fields, i.e. user name, reviewed place, date of the review and the textual review.

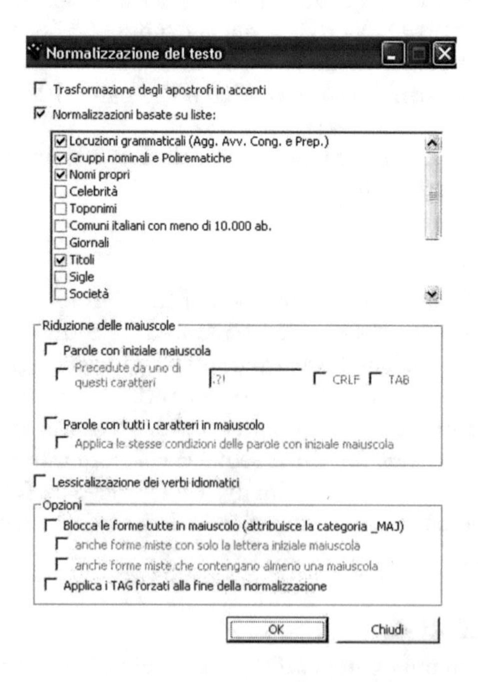

Fig. 1. Text normalization window

3.1 TaLTac Analysis

Firstly, we perform a text pre-treatment, in order to normalize and compute the sub-occurrences in the Corpus (see Fig. 1).

Forma pivot			
and	the	is	s
the	the	a	s
and	a	a	good
the	of	is	place
is	the	is	The
the	in	place	this
the	was	that	to
and	to	and	good
I	the	the	for
the	to	that	is
I	to	and	here
a	to	the	area
a	is	are	the
I	a	that	of
a	of	a	the
on	the	and	you
I	and	the	all
and	was	is	of
and	is	of	out
and	of	a	place
and	for	and	beer
s	to	beer	selectior
the	but	t	don'
it	to	but	was
that	the	and	that
the	place	I	t
I	it	a	that
and	and	and	are
the	food	and	with
and	s	and	were
a	for	the	from
on	a	have	a
with	the	the	s
the	bar	get	to
I	was	ve	I'
a	a	I	of
for	to	the	were
you	to	the	games
it	was	It'	s
to	to	the	they
place	to	good	was
it	and	the	beer
have	the	and	games
with	a	I	would
on	and	a	great
The	was	s	it'
a	bar	a	there
the	t	it	the
a	you	and	Dave
I	have	it	a

Fig. 2. Co-occurrences window

Through the Textual Analysis we perform the identification of the segments, that are saved in two files:

1. the "List of Segments" which contains the segments with their number of occurrences, the number of elements forming the segment and the IS index;
2. the "List of Significant Segments" containing significant segments.

Then, we compute TF-IDF index and we analyse the specificity of our Corpus. Last, we compute the concordances and co-occurrences (see Fig. 2).

3.2 GATE Analysis

We exploited the functionalities offered by a Gate plugin for Italian language available at https://github.com/sercanto/GATE-plugin-Lang_Italian, that provides a complete application, named *Italian NE*, with a set of processing resources for the Italian language treatment: Italian Grammar, Italian

Gazetteer, Italian Tokeniser, Adapt Tokeniser to Tagger. Moreover, in our processing pipeline we add the *Italian POS Tagger* and *ANNIE Sentence Splitter* plugins.

Then we proceed with the composition and initialization of the Information Retrieval pipeline, customizing the configuration on the basis of ANNIE pipeline (see Fig. 3):

- **Document Reset PR**: remove all the annotation sets and their contents made on document, resetting it to its original state;
- **ANNIE Italian Tokeniser**: splits the text into very simple tokens such as numbers, punctuation and words of different types;
- **ANNIE Sentence Splitter**: segments the text into sentences through a cascade of finite-state transducers;
- **Adapt Tokeniser to Tagger**
- **Italian POS Tagger**: assigns a part-of-speech tag to each word or symbol;
- **Italian Gazetteer**: identifies entity names in the text exploiting a set of predefined lists (such as Organizations, Cities, Names, etc.);
- **Italian Grammar**: ANNIE named entity grammar;

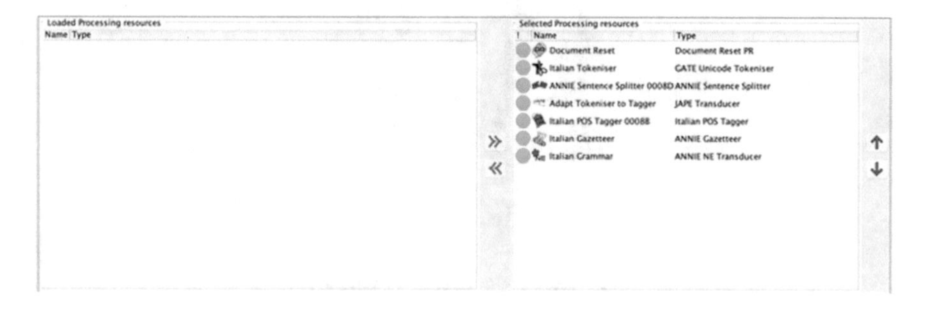

Fig. 3. ANNIE pipeline for Italian language

4 Computational Results

We show the computational results obtained on a subset of 50 tuples taken from the initial data set, made up of more than 5000 tuples collected from Yelp social network, regarding a specific place reviewed by the users.

4.1 TaLTac Results

Through TaLTac we obtain the peculiar lexicon (shown in Fig. 4), that is the list of words with highest occurrences and TF-IDF. Since the lexicon is composed by words like "games", "beer", "good", "great", "food", "place", we can assume that the analysed tuples mostly contain reviews about a place where you can eat, drink beer or play some games.

The co-occurrences analysis, reported in Fig. 2, strengthen the good sentiment about the place reviewed, with expressions such as "a good", "the bar", "beer selection", "a place", "the games". Last, to improve the contextualization of extracted tuples, we studied the concordances of a meaningful word, "food", through all the Corpus (see Fig. 5), in order to understand the context of the lemma.

Forma grafica	Occorrenze totali	Lunghezza	TFIDF
was	83	03	3,37217
s	62	01	2,72497
for	71	03	2,61815
games	34	05	2,46866
is	97	02	2,36757
beer	24	04	2,19451
are	34	03	2,10028
to	132	02	2,07886
t	37	01	2,06400
good	37	04	2,06279
selection	15	09	2,06025
you	46	03	2,04468
were	32	04	1,98688
there	29	05	1,97350
that	57	04	1,95741
-	25	01	1,93592
Great	7	05	1,92191
in	56	02	1,91502
on	47	02	1,90879
food	39	04	1,87488
1	26	01	1,86714
of	91	02	1,84201
1qCuOcks5HRv67OHovA	26	22	1,58947
all	18	03	1,58031
and	194	03	1,57490

Forma grafica	Occorrenze totali	Lunghezza	TFIDF
they	38	04	1,82534
here	21	04	1,82342
area	15	04	1,79898
bar	24	03	1,79825
get	20	03	1,78191
great	20	05	1,77350
as	17	02	1,77066
with	37	04	1,76399
had	27	03	1,75748
it	57	02	1,75409
place	44	05	1,75382
out	27	03	1,72490
I	131	01	1,71641
fun	13	03	1,70398
4.0	15	03	1,65914
It	18	02	1,65544
The	52	03	1,65391
from	24	04	1,64308
wings	9	05	1,64125
be	18	02	1,62180
&	20	01	1,59426
have	46	04	1,59094
but	48	03	1,57259
can	12	03	1,55776
5.0	10	03	1,55417

Fig. 4. Peculiar lexicon by occurrences

ID Fr...	Intorno sinistro	Forma grafica	Intorno destro
fragm...	, and authentic . If you' re looking for good Irish	food	and a cold pint , you can' t go wrong at the Pour House
fragm...	worth seeking out . They have some of the best Irish	food	I' ve had in Pittsburgh- the colcannon is awesome and
fragm...	out of this world . If you' re not looking for Irish	food	, then try the grilled cheese- and make sure you ask
fragm...	there on a Saturday night with a mind to try the Irish	food	. Apparently , we were out of luck . I' ve always thought
fragm...	secrets of restaurant success is to actually stock	food	for people to eat . He told us before we ordered that
fragm...	were out " . At that point , realizing that the only	food	to be had in the place was what was crusted on the
fragm...	heoc96QXrTbecWVw933qhQ | 2011-08-20 Best Irish	food	in the Burgh . Great bar food too . The service is
fragm...	2011-08-20 Best Irish food in the Burgh . Great bar	food	too . The service is maybe a bit surly and it' s not
fragm...	Excellent wings and sandwiches , generally good	food	otherwise , and fair prices-- nice casual place . The
fragm...	1qCu0cks5HRv670HovAVpg tPUGLIDZLF7HrOC46NqT...	food	here can actually be a little hit and Miss , but I
fragm...	wonderful character and ambiance of this place , but the	food	was average at best . We were there on a Pens play-off
fragm...	2 people working all the tables in both rooms . Our	food	came to us cold and unimpressive at that . Our orders
fragm...	were pretty good , also . If you are looking for good	food	, Homestead better choices at Blue Dust or Tin Front
fragm...	beer selection , but it won' t be for a while . The	food	has always been average at best , and the pizza sub
fragm...	op2Gve4sAMQ4qEzq2Tad0g | 2013-09-15 I' m reading o...	food	is really hit or Miss . I' ve only gone to Duke' s
fragm...	Miss . I' ve only gone to Duke' s one time , but the	food	was good . It wasn' t too busy so our service was very
fragm...	very attentive and it didn' t take long to get our	food	at all . The place is separated between a bar area

Fig. 5. Concordances of the word "food"

4.2 GATE Results

By using GATE processing resources (Fig. 6) we can deduce even more meaning respect to the analysis described in Subsect. 4.1, thanks to an advanced tagging system that annotate segments of text with a semantic description, according to the context in which they are sited.

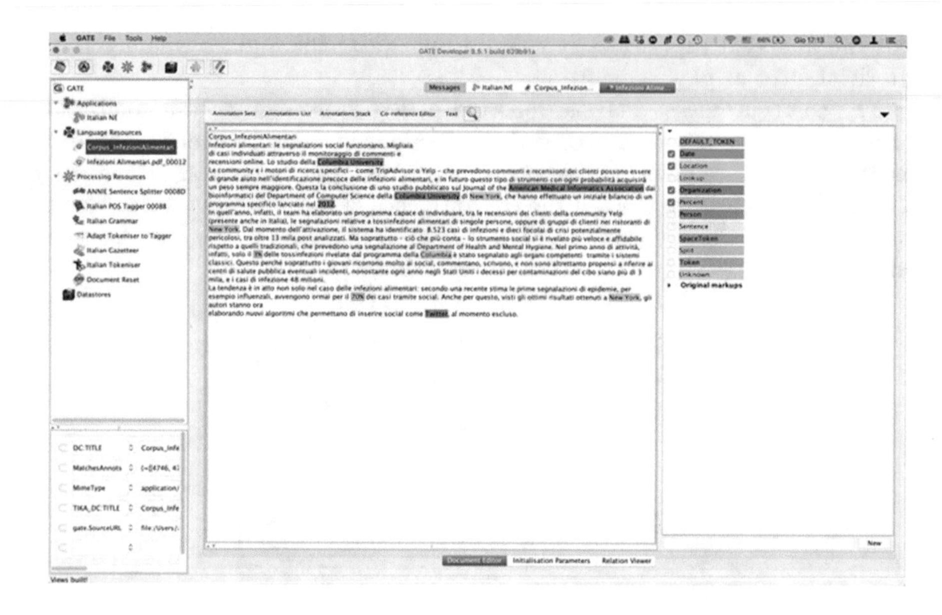

Fig. 6. Annotation sets produced by GATE for an Italian corpus.

5 Conclusions

In this work, we have proposed a methodology aiming to analyse the content of textual information gathered from messages posted on social networks. More in detail, we aimed to design a methodology for early detection of food diseases by analysing the reviews of restaurants, or other places where you can eat. From these basis, it can be performed further analysis in health domain (i.e. the first reports of epidemics, like influenza), by crawling information from social networks data streams and processing them with specific textual analysis, that make use of a peculiar lexicon.

Acknowledgment. This work is supported by CREA European Project: Conflict Resolution with Equitative Algorithms, Grant Agreement number: 766463, CREA, JUST-AG-2016/JUST-AG-2016-05.

References

1. Albanese, M., D'acierno, A., Moscato, V., Persia, F., Picariello, A.: Modeling recommendation as a social choice problem. In: Proceedings of the 4th ACM Conference on Recommender Systems, RecSys 2010, pp. 329–332 (2010)
2. Amato, F., Cozzolino, G., Maisto, A., Mazzeo, A., Moscato, V., Pelosi, S., Picariello, A., Romano, S., Sansone, C.: ABC: a knowledge based collaborative framework for E-health, pp. 258–263 (2015)
3. Amato, F., Cozzolino, G., Mazzeo, A., Pizzata, A.: Sentiment analysis on Yelp social network, pp. 92–98 (2017)

4. Amato, F., Cozzolino, G., Moscato, V., Picariello, A., Sperlí, G.: Automatic personalization of visiting path based on users behaviour, pp. 692–697 (2017)
5. Balzano, W., Murano, A., Stranieri, S.: Logic-based clustering approach for management and improvement of VANETs. J. High Speed Netw. **23**(3), 225–236 (2017)
6. Cilardo, A.: Exploring the potential of threshold logic for cryptography-related operations. IEEE Trans. Comput. **60**(4), 452–462 (2011)
7. Cilardo, A., Durante, P., Lofiego, C., Mazzeo, A.: Early prediction of hardware complexity in HLL-to-HDL translation. In: International Conference on Field Programmable Logic and Applications (FPL), pp. 483–488. IEEE (2010)
8. Cozzolino, G.: Using semantic tools to represent data extracted from mobile devices, pp. 530–536 (2018)
9. D'Acierno, A., Moscato, V., Persia, F., Picariello, A., Penta, A.: iWIN: a summarizer system based on a semantic analysis of web documents. In: Proceedings of the IEEE 6th International Conference on Semantic Computing, ICSC 2012, pp. 162–169 (2012)
10. Doan, S., Bastarache, L., Klimkowski, S., Denny, J.C., Xu, H.: Integrating existing natural language processing tools for medication extraction from discharge summaries. J. Am. Med. Inform. Assoc. **17**(5), 528–531 (2010)
11. Effland, T., Lawson, A., Balter, S., Devinney, K., Reddy, V., Waechter, H., Gravano, L., Hsu, D.: Discovering foodborne illness in online restaurant reviews. J. Am. Med. Inform. Assoc. (2018). https://doi.org/10.1093/jamia/ocx093
12. Fette, G., Ertl, M., Wörner, A., Kluegl, P., Störk, S., Puppe, F.: Information extraction from unstructured electronic health records and integration into a data warehouse. In: GI-Jahrestagung, pp. 1237–1251 (2012)
13. Fusella, E., Cilardo, A.: Minimizing power loss in optical networks-on-chip through application-specific mapping. Microprocess. Microsyst. **43**, 4–13 (2016)
14. Grabar, N., Zweigenbaum, P.: Automatic acquisition of domain-specific morphological resources from thesauri. In: Proceedings of RIAO, pp. 765–784. Citeseer (2000)
15. Hahn, U., Honeck, M., Piotrowski, M., Schulz, S.: Subword segmentation–leveling out morphological variations for medical document retrieval. In: Proceedings of the AMIA Symposium, p. 229. American Medical Informatics Association (2001)
16. Javanmardi, S., Shojafar, M., Shariatmadari, S., Ahrabi, S.S.: FR trust: a fuzzy reputation-based model for trust management in semantic P2P grids. Int. J. Grid Utility Comput. **6**(1), 57–66 (2015)
17. Lovis, C., Baud, R., Rassinoux, A.-M., Michel, P.-A., Scherrer, J.-R.: Medical dictionaries for patient encoding systems: a methodology. Artif. Intell. Med. **14**(1), 201–214 (1998)
18. Moore, P., Xhafa, F., Barolli, L.: Semantic valence modeling: emotion recognition and affective states in context-aware systems. In: Proceedings of the IEEE 28th International Conference on Advanced Information Networking and Applications Workshops, WAINA 2014, pp. 536–541. IEEE (2014)
19. Norton, L.M., Pacak, M.G.: Morphosemantic analysis of compound word forms denoting surgical procedures. Methods Inf. Med. **22**(1), 29–36 (1983)
20. Pratt, A.W., Pacak, M.: Identification and transformation of terminal morphemes in medical English. Methods Inf. Med. **8**(2), 84–90 (1969)
21. The Apache Hadoop Project: Apache Hadoop
22. The GATE Project Team: Gate
23. Rink, B., Harabagiu, S., Roberts, K.: Automatic extraction of relations between medical concepts in clinical texts. J. Am. Med. Inform. Assoc. **18**(5), 594–600 (2011)

24. Morrone, A., Bolasco, S., Baiocchi, F.: TaLTac
25. The Free Encyclopedia Wikipedia: TripAdvisor
26. The Free Encyclopedia Wikipedia: Yelp
27. Wolff, S.: The use of morphosemantic regularities in the medical vocabulary for automatic lexical coding. Methods Inf. Med. **23**(4), 195–203 (1984)
28. Xhafa, F., Barolli, L.: Semantics, intelligent processing and services for big data. Fut. Gener. Comput. Syst. **37**, 201–202 (2014)

Outgoing Data Filtration for Detecting Spyware on Personal Computers

Aishwarya Afzulpurkar[1(✉)], Mouza Alshemaili[2(✉)], and Khalid Samara[2(✉)]

[1] RIT Dubai, Dubai, UAE
afzul.purkar.87@gmail.com
[2] CIS Division, HCT, Ras Al Khaimah, UAE
{malahemaili,ksamara}@hct.ac.ae

Abstract. One of the most critical issues emerging from the Internet is the diverse number of spyware and bots. When a spyware is installed in your PC then it will be difficult to detect, mainly because it deploys covert channels to communicate with outbound data transmissions. These attacks are usually sent from PCs infected with a bot that communicates with malicious controllers over an encrypted channel. However, the available pattern-based intrusion detection system (IDS) and antivirus systems (AVs) are unable to detect the infected PC. This paper presents a Monitoring and Filtering method (SMF) for outgoing packets based on machine learning and behavioral-based methods that can help in the protection of PCs. In addition, this paper presents recent research contributions and emerging tools in the field of spyware detection and identifies existing gaps in the literature. The paper then presents a High-level Architecture to inspect the outgoing packet from the hardware and the software installed in PCs as a solution.

1 Introduction

As the number of spyware products has intensified, better protection of personal and confidential information is needed. This paper presents a Monitoring and Filtering [MF] method for outgoing packets that not only support PC protection but also aids in the detection of botnets and subsequent prevention of the uncontrolled spread of malware.

A number of researchers [1–4] have stated that bot can be controlled through the command and control (C&C) servers to distribute attack packets. However, it is complex to detect the bots using the pattern-based IDS or pattern-based AV since the communication is usually controlled by a herder over an encrypted channel or a normal communication protocol.

A spyware is a program that monitors a computer user's activities and captures data about the user, storing the information to a third party to access it [5]. The prevention of these attacks is necessary. However, current methods and emerging detection methods for spyware and bots have proven not to provide efficient detection, especially when dealing with zero-day attacks. Thus, this study presents machine learning techniques that can be used to detect the spyware and bots using software and hardware systems.

L. Barolli et al. (Eds.): EIDWT 2019, LNDECT 29, pp. 355–362, 2019.
https://doi.org/10.1007/978-3-030-12839-5_32

The hardware device is connected to the network interface controller (NIC) in the PC to detect all the packets that reach the PC at the early stages. Also, the software solution is based on the machine learning methods. The smart detection system analyzes the outgoing packet on PCs from the data link layer until it reaches the application layer. The next section provides a review of the available spyware detection systems. Then, the paper explains the proposed solution. The implementation and evaluation of the proposed solution will be discussed in the upcoming paper on this topic. Finally, the paper presents future work and extensions.

2 Literature Review

The current spyware and bots detection systems are divided into two types. One is signature-based and the other is a behavioral-based method. The signature-based analysis is a static method that relies on pre-defined signatures such as a fingerprint, e.g. MD5 or SHA1 hashes, static strings, and file metadata. Thus, all files in the PC are statically analyzed by the anti-virus software. If any of the signatures are matched, an alert is triggered, stating that this file is suspicious [6]. This method is effective if the AVs software is updated with a well-known malware sample. However, nowadays attackers have started to develop spyware in a way that it can change its signature. This spyware feature is referred to as polymorphism. Most important, such spyware cannot be detected using a signature-based detection method. Indeed, this is the reason as to why the second type of detection system is proposed the behavior-based method [9]. In this method, the actual behavior of malware is observed and detected during its execution, observing the signs of malicious behavior such as modifying host files, registry keys, and establishing suspicious connections. All of these actions cannot be considered as a reasonable sign of spyware, but can certainly increase the suspiciousness of the file. This method is based in creating a threshold for each file based on suspicious actions and behaviors, thus in case the spyware exceeds the level of the threshold then it raises an alert that the file can be considered as a spyware [7].

Most previous research focuses on incoming malware prevention rather than outgoing communication monitoring because once the virus is inside a system, a virus could easily manipulate software and attempt to detect malicious outgoing packet data. Current incoming packet filtration mechanisms rely on a combination of static pattern matching (fast but unreliable) and dynamic, run-time monitoring (slow, but reliable) [8, 9]. Achieving both speed and accuracy is impossible as it has to detect vulnerabilities in the entire system, which makes incoming packet filtration insufficient for confident spyware detection. Outgoing packet filtration (PF) only has to detect user and system activity and compare outgoing packets to packets 'officially generated' by the system [10, 11]. Because the amount of outgoing data issued by a system is less than the amount of incoming data, the state space for doing outgoing PF is considerably less. The specifics on the proposed method are discussed further in the paper. This paper presents a novel solution that can increase the accuracy of behavioral-based methods by combining the software solution that depends on machine learning by defining the spyware malicious behaviors then set the triggers for these behaviors to call the file spyware, and most importantly focusing on analyzing the outgoing packets from the PC.

The basis of machine learning was formed from rapid development and tried and tested techniques including pattern recognition and data mining approaches which can be considered as one field of Artificial Intelligence. A fundamental understanding of machine learning is to train the model based on some algorithm so it can perform a specific task such as classification, clusterization, regression, etc. From a machine learning perspective, spyware detection can be considered as either classification or clusterization. The idea is that an unknown type of spyware should be clusterized into several clusters, based on certain properties, which is identified by the algorithm. Therefore this study proposes to apply various Machine Learning Algorithms in order to explore the most suitable one for our proposed solution. The deployment of Machine Learning is not new given that several studies were carried out in this field, aiming to ascertain the accuracy of various methods, such as [12–15]. The next section describes the proposed solution in more details.

3 The Proposed Solution

The increase in open source apps and code has helped us tremendously in understanding the issues of data monitoring. However, there are still too many default and proprietary apps that are not open source and thus imposes app-level encryptions and encapsulation on code logic sent over the network [16].

Currently, tools that expose raw packet data to the user, and trace application data information are software-based and have been subject to hacks on several occasions [17]. Flow analysis software that links network data to specific app-level or kernel level flows is confusing [18], as it leaves it to the user to link one activity to the other.

The objective of this research is to keep an isolated, un-hackable hardware area that 'checks' the packet trail from the data link layer up to the application layer. The 'check' matches 'first in queue' packet data in NIC, which is no longer vulnerable to hacks, to packet data generators (i.e. apps, services, system ops) in the hard disk. The 'verification' is based on historic, current and scheduled system and UI activity, tracking is performed by a software component and continuously verified by a hardware component. Because more and more enterprises are offloading simple network-based monitoring and sniffing activity to their NIC cards and other hardware components, mimicking more of the Level 2 and 3 functionality in the hardware component seemed to be a feasible option to explore.

3.1 High-Level Architecture

In this section, we present a High-Level Architecture (Figs. 1 and 2) that can effectively identify outgoing packet from the hardware and the software installed in PCs. Outgoing packets are communicated to a proposed external packet checker (EPC) before being sent out. Trail information is contained within the packet (additional headers) and in selected data from incoming NIC queue. To handle encryption and user interface (UI) level checks, the proposed app-level functionality and lower OSI layers of the system have to communicate with the EPC, hence the connection from the hard disk to the EPC is needed. Mismatches trigger warnings or further software-level checks are initiated by the EPC and sent to the proposed software on the hard drive (HD). Binary verification and memory access checks of the software is conducted on a random basis on machine code and initiated by the EPC. Log space is allocated by the user. In the two figures below, we show the communication channels of packets going from the NIC to the HD as well as to the EPC.

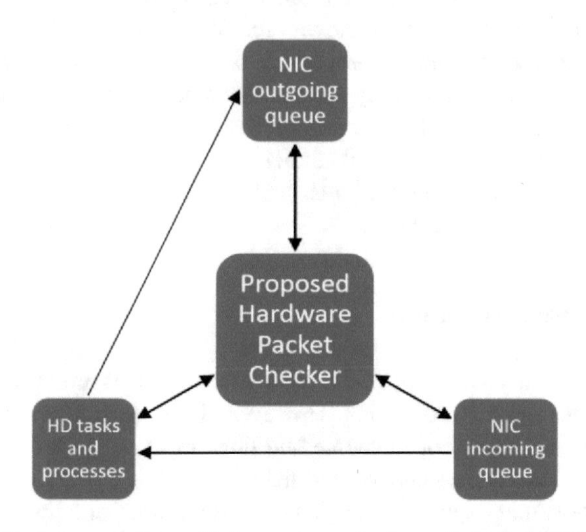

Fig. 1. Isolated Hardware solution with two communications channels to NIC and Hard drive.

Hardware-level functionality includes:

1. Matching information provided by software-level functionality to packets.
2. Maintaining a log of outgoing + incoming data.
3. Mimicking TCP/IP and DL layer functionality to keep a log of packet transfer rate, size, retransmission, protocols, etc.
4. Lower level Encryption.

Software level functionality includes:

1. UI analysis and logging.
2. Trace of packet data to app/system calls.
3. Ability to query the HD for data.
4. App and network activity monitoring for flagged apps or processes, including run-time behavior, user settings, data access, and induced patterns in network data transfer based on the machine learning algorithm.

3.2 App-Level Encryption

This subsection explained the Detailed Functionality of the application level.

I. UI component and data packet inspection: Freely available machine learning toolkits can be used to extract the 'functionality' being seen on the user's display screen. Lightweight tools to monitor screens (ie. ScreenMonitor) can be coupled

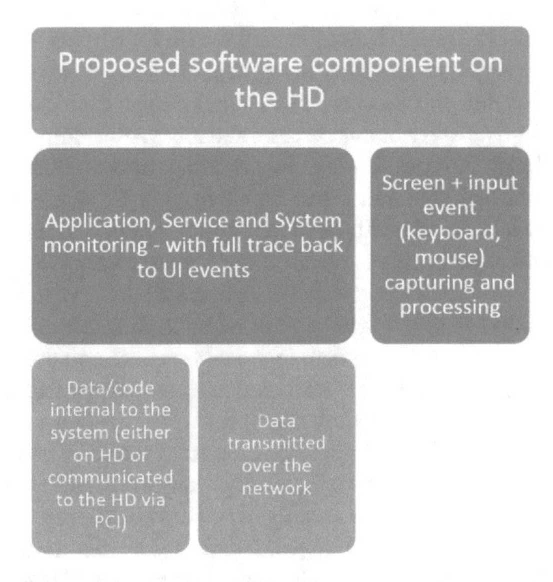

Fig. 2. Software solution with access to keyboard + mouse events, screen visibility, and user level services and processes.

with object recognition and OCR software to detect the running apps, services and user inputs. Every download/upload is checked based on whether the user has visibly scheduled the activity, what apps/system tools were used, and to what endpoint the connection was visibly established. If extraneous endpoint URLs are detected, or private data is queued for transmission, that has no history of being entered or seen on the UI, then the data transmission is stopped. Malware can also be in the form of a seemingly unharmful app – so the users view it, and use it for its assumed purpose, not knowing about its hidden intentions. In order to remain hidden, the app only uploads content to the malicious server when the user sends the content over the network to a verified endpoint. For example, if the user uploads the content on a verified endpoint, the content may simultaneously be sent to a different endpoint. A hacker could also alter the content entered by the user, either in the app itself or during the data transfer process, such that the content sent to the verified server ends up hacking the server, so the hacker could steal the content from that location. In that case, the content itself is checked (deep packet inspection) for malicious endpoints and sensitive user or system information, based on content that was visibly entered on the screen. However, if the user connects and visibly sends data to an already hacked server, there is no way to tell the server is hacked.

II. Encryption and Encapsulation: If encrypted/encapsulated packets are being sent in the background from a legitimate app/system level resource, the encrypted entity can either be code (i.e. browser sends JavaScript code) or data. If encrypted code exists on the app, on the system, (i.e. explicitly listed change in a file) or comes from outside (i.e. request carry-forward), the proposed component is able to search the encrypted code in the system or in its log of incoming encrypted data. If a match is not found, the code for the app may be encapsulated in a different way than the code being sent out (i.e. packaged machine code vs higher level code). In this case, every downloaded app, including browsers, is monitored for its history of data access, and any visible outputs generated for the purpose of being sent out. If possible, the component monitors variations in data outgoing from that app, and based on changes to its commonly accessed files, evaluates the likelihood and type of data being sent out. There have been several initiatives and experiments to match assembly code to process and kernel traces, to uncover the call stack and data being accessed by applications [19]. If time-based encryption is applied, the app is required to submit either the timestamp used, or its encryption mechanism to the component, and the encryption will be applied by the component before being sent out. Apps that perform two-way encryptions or public/private key encryptions are required to provide data prior to encryption and well as the public key of destination. The component ideally takes responsibility of performing encryption and sending out the new data packets because encryption is usually only performed prior to sending out data on a network. So, if there is any ambiguous data that is not explicitly from outside, or internal to the system, it is assumed that it had been generated, either by the user or taken from elsewhere on the computer, and the measures outlined above are taken to ensure security. An additional check for the number of packets sent and marked as retransmission/dropped compared to the amount of information required to be sent is done – so that an excess of retransmissions/drops can be evaluated against network bandwidth or other network conditions, and an excess of packet transfer can be evaluated based on the amount of raw data to be sent from the user's side.

III. Browser: An analysis of the components of the visible webpage (request + response) is conducted as it has links to several servers/URLs that may not all be safe. The user is informed about the endpoint URLs and server addresses of all components of the website, so the user can decide which endpoints to send data to, and which to ignore. Browsing data settings and extensions are kept track of incoming push notifications are not 'run' until explicitly allowed to do so. Webpage components are kept track of and request code pertaining only to visible user input fields are allowed. Responses only pertaining to allowed incoming request code are allowed to be sent out.

4 Conclusion

This paper presents a novel hardware and software level solution to the problem of spyware communication on PCs. An overview of the architecture, as well as details on targeted situations and solutions, are given. The models presented Isolated Hardware and Software solution shows the relationships between the defined components and the functionality for each component. The solution is comprehensive in terms of monitoring all outgoing data, encrypted or not. Thus, based on the previous research and available methods and tools, this work fills a significant gap in the existing literature.

5 Further Work

Tests need to be done on common apps and system level services that communicate via networks to determine bottlenecks and efficiency measures. While some functionality can perform with minimal delay, some may restrict network activity of the user. The solution can be eventually extended to detect man-in-the-middle attacks. Further research is required on whether current packet filtration functionality (i.e. Snort) is able to detect these attacks. If this is the case, then the solution presented in this paper will be able to mimic MITM detection functionality and perform in a more secure way, using binary verification and memory access checks to verify the software on a hardware level. This paper presents the proposed solution and we are currently working on the implementation to prove the accuracy of our solution.

References

1. Zou, C.C., Cunningham, R.: Honeypot-aware advanced botnet construction and maintenance. In: Proceeding of the DSN 2006, pp. 199–208, June 2006
2. Sudo, T., Fujiwara, K.: The evaluation of the botnet analysis system based on the virtual Internet environment. In: Proceeding of the CSS 2006, pp. 513–158. IPSJ, October 2006
3. Miwa, S., Miyachi, T., Miyachi, T., Eto, M., Yoshizumi, M., Shinoda, Y.: Design issues of isolated sandbox for analyzing. In: Proceeding of the IWSEC 2007, pp. 13–27. IPSJ, October 2007
4. Kondo, S., Sato, N.: Botnet traffic detection techniques by C&C session classification using SVM. In: Proceeding of the IWSEC 2007, pp. 91–104. IPSJ, October 2007
5. Chien, E.: Techniques of Adware and Spyware. WWW document (2005). https://www.symantec.com/avcenter/reference/techniques.of.adware.and.spyware.pdf. Accessed 15 Feb 2017
6. Jang-Jaccard, J., Nepal, S.: A survey of emerging threats in cybersecurity. J. Comput. Syst. Sci. **80**(5), 973–993 (2014). ISSN 0022-0000
7. Konrad, R., Trinius, P., Willems, C., Holz, T.: Automatic analysis of malware behavior using machine learning. J. Comput. Secur. **19**, 639–668 (2011)
8. Harley, D., Lee, A.: Heuristic Analysis—Detecting Unknown Viruses (2009)
9. Kaleem Awan, M.S., Burnap, P., Rana, O.: Identifying cyber risk hotspots: a framework for measuring temporal variance in computer network risk. Comput. Secur. **57**, 31–46 (2016). ISSN 0167-4048
10. Sultan, K., Ali, H., Zhang, Z.: Call detail records driven anomaly detection and traffic prediction in mobile cellular networks. IEEE Access **6**, 41728–41737 (2018)
11. Takemori, K., Nishigaki, M., Takami, T., Miyake, Y.: Detection of Bot infected PCs using destination-based IP and domain whitelists during a non-operating term. In: IEEE GLOBECOM 2008 - 2008 IEEE Global Telecommunications Conference, pp. 1–6 (2008)
12. Dragos, G., Cimpoesu, M., Anton, D., Ciortuz, L.: Malware detection using machine learning. In: Proceedings of the International Multiconference on Computer Science and Information Technology, pp. 735–741 (2009)
13. Priyank, S., Raul, N.: Malware Detection Module using Machine Learning Algorithms to Assist in Centralized Security in Enterprise Networks (2015)

14. Usukhbayar, B., Jambaljav, N., Horng, S.: A Static Malware Detection System Using Data Mining Methods. Cornell University (2013)
15. Mamoun, A., Venkatraman, S., Watters, P., Alazab, M.: Zero-day malware detection based on supervised learning algorithms of API call signatures. In: Proceedings of the 9-th Australasian Data Mining Conference, pp. 171–181 (2011)
16. Forte, D.: Spyware: more than a costly annoyance. Netw. Secur. **2005**(12), 8–10 (2005). ISSN 1353-4858
17. Caballero, A.: Information security essentials for information technology managers. In: Computer and Information Security Handbook, pp. 393–419 (2017)
18. Chen, T.M., Walsh, P.J.: Guarding against network intrusions. In: Network and System Security, pp. 57–82 (2014)
19. Arasteh, A.R., Debbabi, M.: Forensic memory analysis: from stack and code to execution history. Sci. Direct Digital Invest. **4**(Supplement), 114–125 (2017)

Formulation of Information Hiding Model for One-Time Authentication Methods Using the Merkle Tree

Yuji Suga(✉)

Internet Initiative Japan Inc., Iidabashi Grand Bloom, 2-10-2, Chiyoda-ku, Fujimi 102-0071, Japan
suga@iij.ad.jp

Abstract. We consider the extension of the Lamport-like one-time password scheme proposed at BWCCA2017. In our new method, the act of authentication is performed by disclosing the digest values linked to the nodes located higher than the digests of the leaf node of the hash chain, just like as the Lamports' authentication method. In the tree structure like as Merkle tree in which there are multiple nodes disclosed at authentication phase, the prover can transmit secret data to the verifier by changing the disclosure order. This paper adopts a model that embeds information in the "edge having a node to be disclosed", sets up a kind of optimization problem, and discusses efficiency from concrete toy case examples with small depths of Merkle tree.

1 Introduction

ID (Identity) management technology involves more than technology related to the lifecycle, such as the issuing, usage, and disposal of IDs. It also encompasses an extremely diverse range of approaches, such as user authentication, issuing and usage of credentials, management and circulation of attribute information, access control for various resources, and delegation of authority, as well as technology for coordinating this information between different entities. Consequently, there are a wide variety of specifications related to the management and coordination of IDs used on the Internet. Because each is developed based on different concepts, it is necessary to select the one most appropriate for the environment it will be used in. However, due to the different technical terms used and the increasingly broad range that specifications cover, this is unfortunately becoming an extremely difficult area to understand.

1.1 Approaches to IDs - Entities and Identifiers

Using IDs in the digital world enables us to identify different entities. However, IDs are public information, and it would pose a problem if simply anyone could claim they were a particular entity. Consequently, an authentication system is

L. Barolli et al. (Eds.): EIDWT 2019, LNDECT 29, pp. 363–373, 2019.
https://doi.org/10.1007/978-3-030-12839-5_33

required for verifying an entity that is requesting access really is the entity that ID was assigned to. For this, private information such as a password that matches the corresponding ID is needed. Here, in line with the NIST SP800-63 [1] definitions, we will explain credentials and tokens separately. Tokens indicate information that should be kept secret held by users assigned the corresponding ID. They fall into the categories of "something you know," "something you have," or "something you are." Examples of tokens include passwords or the private keys used in public key cryptosystems, physical media such as IC cards or dongles, and body characteristics used for biometric authentication such as fingerprints or irises. Meanwhile, credentials are public information that indicate associations between tokens and IDs. In some cases credentials are signed by a trusted entity, and serve as information that a third party can verify the validity of. For password-type tokens, IDs can be thought of as credentials. Considering public key cryptosystem authentication methods used with protocols such as SSL/TLS, the X.509 certificate [2] is the credential, and the private key that matches the public key included in the certificate corresponds to the token. SSL/TLS server certificates also include the FQDN, and this can be thought of as an ID. Bitcoin addresses can be regarded as IDs, but because the address itself is a public key, and transaction signature verification can be performed using only the address, you can also regard them as credentials. An authentication method using multiple [token/credential] pairs instead of a single pair to authenticate an entity with a certain ID assigned is called multi-factor authentication. The most widely-known method is parallel multi-factor authentication, in which independent tokens are used for each authentication method. Meanwhile, some opt for a cascading approach to multi-factor authentication. For example, when using a private key as a token in public key cryptosystems, the private key file is generally encrypted, so the password must be entered to decrypt it. The use of both the password and private key as tokens at this time can be regarded as cascading multi-factor authentication. The same applies to IC cards and PIN numbers. Additionally, for hardware tokens that fall into the "something you have" token category, a one-time password that is shown on the physical media and periodically updated is typically presented as a token, and used as a part of multifactor authentication. For network environments, a one-time password system using Lamport's hash chains [3] is known, and has been drawn up in specifications [4] such as S/Key. One-time password systems show promise as a proposed countermeasure [5] for the list-based attacks of recent years, through replacement of or combined use with conventional ID/password systems.

In a realm there is an IdP (identity provider) that performs authentication-related tasks, and an entity in the said realm is assigned a unique ID. This can be thought of as the registration screen you input the necessary details into when using an online service for the first time. When registering, there is a "user ID" or similar field for entering an ID, and a check is performed to ensure that the ID is not already in use by another user. On the other hand, there are cases in which a random ID is assigned on the service side. In either case, personal data such as an email address or phone number is normally also entered upon user registration. Registration is only provisional at this stage, so notification is sent along with a "challenge" by way of an email to the email address entered or a short message to the phone number, and the user's identity is confirmed by having them input the corresponding challenge. During provisional registration or input of the challenge, a password as a type of token is generally chosen. In this way, use of the service is possible once registration of the ID and password pair is complete. There are now cases in which an email address is substituted rather than assigning an ID unique to a realm. This is done to reduce the cost for managing IDs within a realm, and because it has the added benefit of avoiding situations where an ID selected by the user, or in particular a random ID assigned by the IdP, is forgotten. When assigning IDs unique to a realm, it is necessary to operate a Web page for providing ID reminders or resetting the password by prompting users to enter the email address or phone number used to confirm their identity, and this increases the burden on the service side. Meanwhile, when using an email address as the ID, you must consider the fact that it may be targeted in list-based attacks. It could be said that there is the same risk when identical IDs are reused on each realm. Of course, not reusing the same password is a fundamental countermeasure against this, but the name binding or collation of leaked data due to IDs being identical is also a risk. Additionally, there have been cases in the past where the leak of private information has been caused by entering the email address of an acquaintance on a reminder page for the input of IDs on an SNS or shopping site due to email addresses being used as IDs. In light of these circumstances, we are now seeing systems that issue IDs unique to a realm, but also give users the right to choose whether or not to use their email address during the login process.

As far as the actual purpose of this goes, one possible use could be saving derivative tokens for each application on a smartphone (this could include cases in which tokens are encrypted using a master password and stored, rather than using the actual token itself). This would be a useful function in environments such as smartphones where password input is difficult. This also has the benefit of limiting the damage caused if a password were to leak from a specific application and be disclosed. Another advantage is that you only have to disable the derivative password, so the base password does not need to be changed. From this perspective, you can think of it as delegating authority, and limit the scope of authorization when a derivative token is used in comparison to the scope of authorization for the base token. This enables you to minimize the impact if the token leaks.

2 One-Time Password Authentication Schemes

In the previous section we discussed variations regarding the use of IDs and tokens. Next we will examine one-time passwords, which are a type of token that is used once and then discarded. One-time passwords have mainly been used in Internet banking systems up to this point, and they were seen as a secure authentication method. However, in February 2015, Japan's National

Police Agency published case studies regarding illegal remittances in Internet banking for 2014 [6], indicating that over 100 financial institutions had incurred losses, and that damages were also on the increase. Particular attention was given to attacks targeting corporate accounts that involve comparatively large remittance amounts, and an alert was issued. In addition to methods that simply use a password, those that incorporate client authentication based on X.509 certificates (strong authentication) have also been introduced for corporate Internet banking authentication, and these are known to provide more security. However, when these digital certificates are used there have still been cases disclosed in which automatic transfers have been made from PCs or browsers infected with malware [7]. This demonstrates that even authentication methods using SSL/TLS client certificates are not foolproof.

Based on these circumstances, an effort is being made to improve the authentication methods used in Internet banking systems. In the past, random number tables listed on paper or card were used, along with hardware devices that display a one-time password. This latter case is an authentication method in which a temporary password is input at the same time as the primary PIN number (a 4-digit number) that corresponds to the password used for authentication at ATMs, etc. Even if the primary PIN number leaked, use of a one-time password that is discarded each time reinforces the identity verification with particularly important processes, such as address changes or large remittances. However, as shown by Man-in-the-Browser attacks and MITM attacks, if banking system transaction details such as the destination account number or transfer amount are re-written, illegal remittances are possible even when a disposable one-time password is used. This demonstrates the problem of it being impossible for a user to explicitly confirm whether or not a transaction is legitimate based on the information shown in the browser alone, even if authentication methods are improved. There is an undeniable possibility that transactions may have actually been rewritten by an attacker. In response to this problem, progress is being made towards migrating to the use of hardware devices equipped with an input device. In the past, measures featuring the combined use of hardware devices during authentication have also been adopted at a number of banking systems. However, because this hardware device was simply a one-time password generator with no input interface, it could only be used to identify whether the user has the correct token, which has nothing to do with the transaction at hand. In addition to the abovementioned use of X.509 certificates, corporate banking systems also incorporate secondary measures such as only accepting transactions from specific IP addresses and PCs. However, because it is possible that the transactions a user sees have been rewritten, there is no fundamental countermeasure for Man-in-the-Browser attacks. In other words, even with improved identity verification the countermeasures were ineffective. In response, there were announcements from financial institutions in 2015 that they would start using one-time password cards. Unlike previous devices, these new hardware devices have a keypad input interface, enabling identity verification while also incorporating techniques that enable users to confirm the legitimacy of transactions that may have been rewritten. The devices are not merely for inputting the one-time password that is output from the hardware device during authentication like before. They feature functions for generating and displaying one-time passwords that guarantee the correctness of account numbers, by having users themselves input the account number they want to transfer money to. This prevents money being sent to the account intended by an attacker, and by also recording the transaction log generated at this time, it is possible to automatically create blacklists for the attacker's accounts. Additionally, for user convenience, they can also be used as one-time password generators with no input device. The technique is based on the premise that accounts registered by users in advance are safe, and omits the input of account numbers for transfers to these pre-registered accounts.

Finally, we introduce standard related to One-Time Password published in IETF community. In 1998, RFC 2289: "A One-Time Password System" [8] was published. After that, HMAC-Based One-Time Password (RFC4226) [9] in 2005, Time-based One-time Password (RFC6238) [10] in 2011 were standardized. These specifications are usually used as one of two factor/two step authentication schemes, and there are software implementations for smartphones.

3 Sausage-Style Original Schemes in BWCCA2017

A previous scheme proposed at BWCCA2017 [11] is one of extensions about Lamport-like one-time password schemes [3]. In this proposed method, it is possible to send arbitrary data from the client (prover) to the server (verifier) at the same time that the authentication is performed. By reveling parent node value for authentication similar to Lamports' scheme, however the user can freely select a path in the Merkle tree.

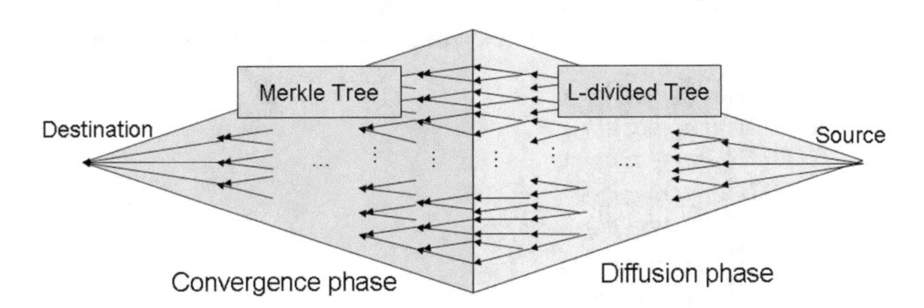

Fig. 1. Concept of sausage-style one-time authentication schemes

3.1 Concept

Figure 1 shows the concept of the authentication method proposed at BWCCA2017. Since it is in the form of sausage as a whole, it was named sausage-style one-time authentication scheme. In the graph of Fig. 1, the right half is an

L branch tree, a Merkle tree [12,13] is constructed on the basis of the leaf, and its route is called an end point (destination). Each node has a value (hash digest given from parent node) and is connected by directed edges. It can be considered that the value allocated to a node using a cryptographic hash function (needs onewayness functionality) changes as it passes through an edge. Also note that since it has onewayness, it can not be reverted back to the base node from the edge node.

3.2 Diffusion Phase

First, nodes of the L branch tree are calculated from the parent node as follows:

$$p_i := H(p||q_i)$$

where $||$ means concatenation of data, $i = 1, \ldots, L$, p is a value of the parent node, and q_i is a pre-shared secrets between the client and the server.

These parameters q_i mean that we make L distinct hash functions virtually from just one hash function $H()$.

3.3 Convergence Phase

In the Merkle tree, we can calculate the child node p from the L parent nodes $p_i (i = 1, \ldots, L)$ straightforwardly as follows:

$$p := H(p_1|| \ldots ||p_L)$$

3.4 Example with $L = 2, d = 3$

There are 8 boundary nodes between the diffusion part and the convergence part. If they are $r_1, \ldots, r_8$, they are calculated from the origin s as follows:

- $r_1 := H(H(H(s||q_1)||q_1)||q_1)$
- $r_2 := H(H(H(s||q_1)||q_1)||q_2)$
- $r_3 := H(H(H(s||q_1)||q_2)||q_1)$
- $r_4 := H(H(H(s||q_1)||q_2)||q_2)$
- $r_5 := H(H(H(s||q_2)||q_1)||q_1)$
- $r_6 := H(H(H(s||q_2)||q_1)||q_2)$
- $r_7 := H(H(H(s||q_2)||q_2)||q_1)$
- $r_8 := H(H(H(s||q_2)||q_2)||q_2)$

Finally, if we construct a Merkle tree from here, the end point t will be as follows:

$$t := H(H(H(r1||r2)||H(r3||r4))||H(H(r5||r6)||H(r7||r8)))$$

Note that there are just 2^3 candidate routes that client can decide only one path, so this means that client can send 3-bit message to server in authentication phase simultaneously (Fig. 2).

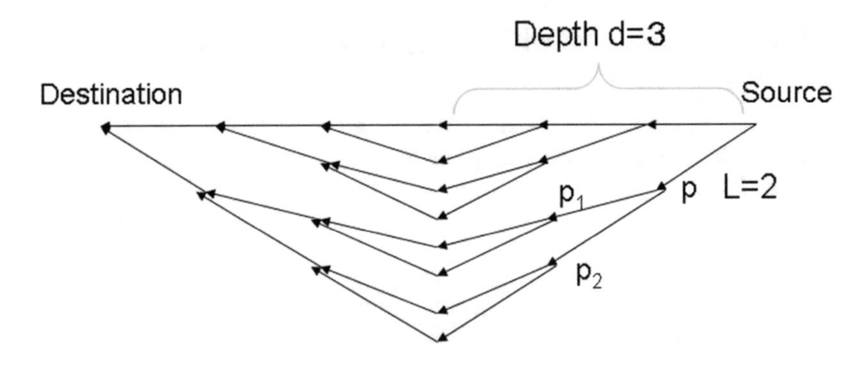

Fig. 2. One sausage example with $L = 2, d = 3$

3.5 Evaluation of Sausage-Style Original Schemes

In the subgraph of the convergence phase area (left side), cost of reveling the parent nodes (authentication cost) is very high when L is greater than 2, because the counter nodes for checking correctness of hash chain is $L - 1$, namely it is wasteful. So we recognize that L should be 2.

Moreover this previous scheme is inefficient: (1) there are a lot of unused edges (unrevealed values), (2) the weight of one edge in a given graph is just only 1 bit, namely the amount of secret between client and server by revealing a value is 1 bit. So we can extend the concept of sausage-style original schemes.

4 A New Model About Information Hiding

We proposed new model about information hiding. The key idea is that depth-d edge has d bits volume in the left part of sausage, Fig. 3 is an example with $L = 2, d = 3$. Each edge on left part has from 1 to 3 bits.

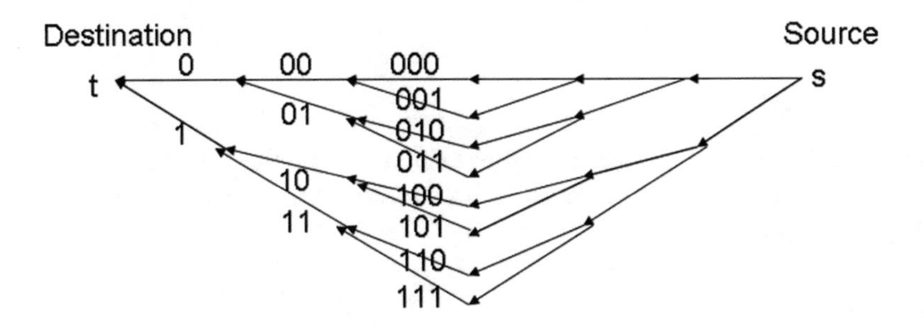

Fig. 3. One sausage example including assigned 1–3 bits with $L = 2, d = 3$

4.1 Toy Case Examples with Small Depths of Merkle Tree

4.1.1 $d = 1$

d is very small, so the disclosure options are limited, It can be seen that only one bit of information is given depending on which one of the two nodes is disclosed (Fig. 4 and Table 1).

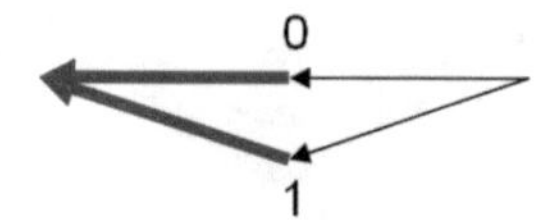

Fig. 4. In the case of $d = 1$

Table 1. 1-bit disclosure case

Transferable secret	1st
0	0
1	1

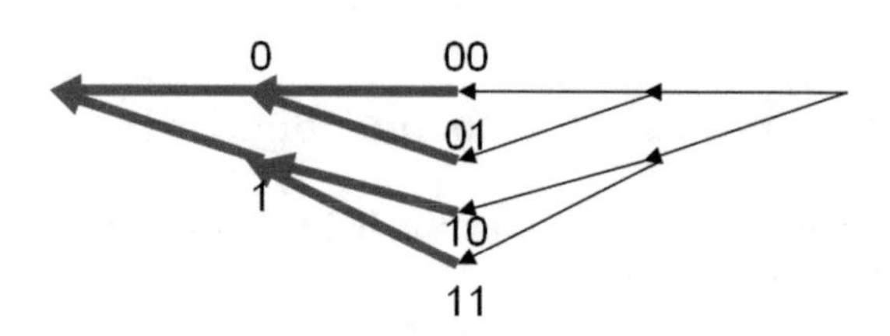

Fig. 5. In the case of $d = 2$

4.1.2 $d = 2$

The mark ? indicates an arbitrary one bit. In this method, it can be seen that transmission of 2.5 bits on average is possible by 2 node revelings (authentications) (Fig. 5).

4.1.3 $d = 3$

The mark $*$ indicates an arbitrary zero or one bit, so the third disclose of 0?$*$ in 0000 indicates the candidates as follows: 01, 000, 001. Similar to 0000, the third disclose of 0?$*$ in 0010 indicates the candidates as follows: 00, 010, 011 (Fig. 6).

Note that the complement of $(1x_2x_3x4)_2$ is $(0\overline{x_2}\overline{x_3}\overline{x4})_2$ where $\overline{0} = 1$ and $\overline{1} = 0$. So we can omit the half of bottom of Table 3.

Table 2. 2-bits disclosure case

Transferable secret	1st	2nd
00	0	0?
01	0	1
10	1	0
11	1	1?

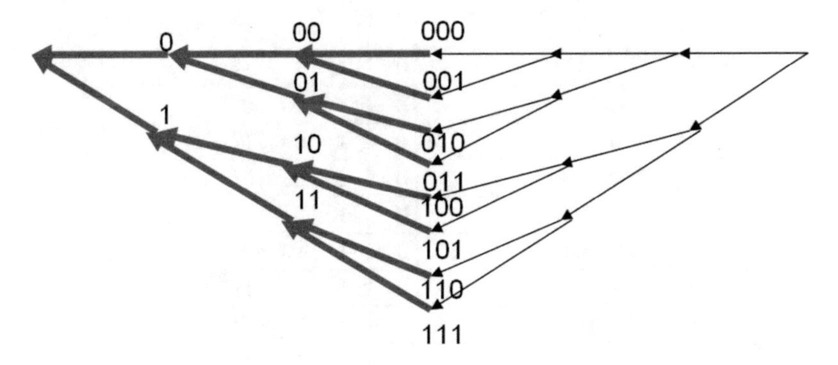

Fig. 6. In the case of $d = 3$

Table 3. 4-bits disclosure case

Transferable secret	1st	2nd	3rd
0000	0	00	0?*
0001	0	00	1
0010	0	01	0?*
0011	0	01	1
0100	0	1	00
0101	0	1	01
0110	0	1	10
0111	0	1	11
1000	1	0	00
1001	1	0	01
1010	1	0	10
1011	1	0	11
1100	1	10	0
1101	1	10	1?*
1110	1	11	0
1111	1	11	1?*

Table 4. 5-bits disclosure case

Transferable secret	1st	2nd	3rd	4th
00000	0	00	00?	
00001	0	00	01	
00010	0	00	1	0?*
00011	0	00	1	1?
00100	0	01	00	
00101	0	01	01?	
00110	0	01	1	0?*
00111	0	01	1	1?
01000	0	1	00	0?*
01001	0	1	00	1?
01010	0	1	01	0?*
01011	0	1	01	1?
01100	0	1	10	0?
01101	0	1	10	1?*
01110	0	1	11	0?
01111	0	1	11	1?*

4.2 Discussions

In this paper, we assume that abilities of attackers is just observations of one-time authentication protocol, so we could assume that flows have preserved order of revealing nodes. If attacker can modify the flows, so we need much more countermeasures (Table 4). For example, 2-bits disclosure case in the Table 2, 1st and 2nd nodes are changeable when transferable secret is 01 or 10.

5 Conclusions and Future Work

This paper shows the new model about information hiding on one-time password protocols. This proposal uses jointed hash chain with Merkle tree and binary tree, and we assigned 1-d bits to edges on sausage, so client can send several bits message to server in authentication phase securely. We construct concrete tables for parameters $L = 2, d = 3$, and we checked soundness in our new model for small cases.

We assume that abilities of attackers is just observations of one-time authentication protocol, however if attacker can modify the flows, so we need much more countermeasures. In the future work, we would like to clarify that attackers' model and try to implement on actual protocols standardized in IETF.

References

1. NIST Special Publication 800-63-3, Digital Identity Guidelines. http://nvlpubs.nist.gov/nistpubs/SpecialPublications/NIST.SP.800-63-3.pdf
2. ITU-T Recommendation X.509 — ISO/IEC 9594-8, Information Technology - Open Systems Interconnection - The Directory: Public-key and attribute certificate frameworks (2016)
3. Lamport, L.: Password authentication with insecure communication. Commun. ACM **24**(11), 770–772 (1981)
4. Haller, N.: The S/KEY One-Time Password System. http://tools.ietf.org/html/rfc1760
5. IIJ, Internet Infrastructure Review Vol.25, 1.4.3 The Status of List-Based Attacks and Their Countermeasures. https://www.iij.ad.jp/en/company/development/iir/pdf/iir_vol25_infra_EN.pdf
6. National Police Agency, Status of Incidents of Illegal Remittance Related to Internet Banking in 2014, February 2015 (in Japanese). http://www.npa.go.jp/cyber/pdf/H270212_banking.pdf
7. Trend Micro Security Blog, Analyzing digital certificate theft attacks targeting corporate net banking. http://blog.trendmicro.co.jp/archives/9417
8. Haller, N., et al.: A One-Time Password System. http://tools.ietf.org/html/rfc2289
9. M'Raihi, D., et al.: HOTP: An HMAC-Based One-Time Password Algorithm. http://tools.ietf.org/html/rfc4226
10. M'Raihi, D., et al.: TOTP: Time-Based One-Time Password Algorithm. http://tools.ietf.org/html/rfc6238
11. Suga, Y.: Sausage-style one-time authentication schemes. In: Proceedings of the 12th International Conference on Broad-Band Wireless Computing, Communication and Applications (BWCCA-2017), pp. 658–667 (2017)
12. Merkle, R.: Secrecy, authentication and public key systems. A certified digital signature. Ph.D. dissertation, Dept. of Electrical Engineering, Stanford University (1979)
13. Szydlo, M.: Merkle Tree Traversal in Log Space and Time. In: EUROCRYPT 2004 (2004)

An Approach to Discover Malicious Online Users in Collaborative Systems

Ossama Embarak(✉), Maryam Khaleifah, and Alya Ali

Department of Computer Sciences, Higher Colleges of Technology, Fujairah, UAE
{oembarak,H00357078,H00356166}@hct.ac.ae

Abstract. Collaborative filtering systems employ numerous techniques to provide recommendations for online users. These techniques depend on the collected data about online users' preferences or ratings of provided items, topics, news, etc. Such systems are vulnerable to malicious users' attacks in which malicious users carefully target specific profiles in order to boost up or diminish the predictions of some targeted items. In this paper, we suggested an approach to find malicious attacks and remove attackers' profiles. This leads to placing the user ratings in the region of rejection and thereby affecting his level of trustiness, the impact of specific user ratings is affected by the user calculated trustiness level, where completely untrusted user's ratings will be neglected by recommendation system. We used a Movie Lens of 1 M rating dataset to perform the required training and test the suggested framework. The suggested method distinguished perfectly between Normal, Excess, Inferiority, and completely dishonest users.

1 Introduction

The presences of noise ratings violate recommendation systems and lead to inaccurate predictions/recommendations. Vicious users deliberately insert attack profiles (pus/nuke attacks) to the system to bias generated recommendations to suit their benefits. Web collaborative systems depend on the collected users' preferences in forms of ratings or spend time. To predict and estimate items of recommendation for visitors. However, these systems malfunctioned due to the huge amount of noisy data which deliberately directs the system computations towards a specific direction [1]. The reputation of many systems has been affected, and a lot of money and users have been lost due to such behaviors. This paper addresses a few of these attacks and provides a proper approach for solution and preventions.

2 Literature Review

A well-known rating attack is called the Shilling attacks, which aims to inject some profiles in order to affect the performance of the targeted system [2], this attack is divided into two types *push attack* which inserts malicious profiles in order to rate a particular item(s) highly; and *nuke attack* which injects malicious profiles in order to

L. Barolli et al. (Eds.): EIDWT 2019, LNDECT 29, pp. 374–382, 2019.
https://doi.org/10.1007/978-3-030-12839-5_34

downgrade the popularity of a specific item(s). Playbooks attack is one of these attacks which reflect the situation where a sequence of actions has been taken to maximize the importance of a specific item and to increase its profits [3]. Some systems provide low-quality items based on previously created good reputations, as soon as the users discover such attacks, the system loses the respect and loyalty of its user [4]. Unfair rating attack which reflects the situation when a system receives biased ratings and depends upon comparison ratings about the same service entity provided by different agents, and hence use ratings set by trusted agents as a scale. This attack leads to inaccurate predictions which do not reflect the actual ratings of site visitors [5]. Re-entry attacks which reflect the situation when an agent with low ratings renames itself with a different identity to avoid his associated low ratings and to create a refreshing start. Sybil attacks which reflect the use of multiple identifiers by the same user in order to provide different ratings [6].

Nowadays, the robustness of any recommendation systems is an important event if it will cause an additional cost in the development and maintenance of the system. Robustness aims to measure the ability of a specific algorithm to create good predictions with the existence of noisy data. As soon as a user enters new ratings, the system should update its datasets; we should mention here that there is no guarantee that these ratings reflect the user's real favorites. Sometimes rating entry becomes a vapid process [7], then users might be careless, and errors in the data could be possible. Also, malicious agents might attack as well. There are two main aspects in robustness, the first is the accuracy of recommendation, which are the products recommended after the attack occurred, and the second is the stability - does the system recommend different products with the existence of the attack? [8]. Whatever the attack type, it should be discovered before storing users' ratings into the system and hence before affecting the system performance and to keep its good reputation. Profile injection another example of malicious users who aim to inject forged data into recommendation systems to manipulate recommendation ranking. Both user-based and item-based collaborative filtering recommendation systems are vulnerable to profile attacks. In this attacking, attackers try to rate both target and non-target items according to the attack type and to make it looks like normal ones by forging rating profiles and inject them into the rating of recommendation systems [9]. Studies show that group attack profiles are used to manipulate recommendation ranking of target items. Where many attackers work together to perform an attack on specific target items in a particular time frame to quickly push the particular item to a preferred list [10]. The following section demonstrates the suggested approach and experimental results.

3 Suggested Method

A user-item rating matrix is composed of three components, including users, items and ratings.

$$U = \{u_1, u_2, \ldots., u_{m-1}, u_m\}$$

Where U is users set, m is the number of users in the dataset.

$$I = \{I_1, I_2, \ldots, I_{n-1}, n\}$$

Where I is all items set, n is the number of items in a web recommendation system. Each user rate all items or a set of items, all generated matrix for user-items ratings is used to find recommendations

$$\begin{bmatrix} I_{u_1}^1 & \cdots & I_{u_1}^n \\ \vdots & \ddots & \vdots \\ I_{u_m}^1 & \cdots & I_{u_m}^n \end{bmatrix}$$

Where $I_{u_1}^1$ is the rating of item 1, user 1, while $I_{u_m}^n$ is the rating of item n, user m. Therefore, for each item n, we have all rating of all users who rate the item. Similarly, for each user m, we have all rating of items conducted by that user and by all other users. This paper aims to detect any anomalies ratings of a particular user compared to the other ratings by all other users who rate the considered item. The suggested method aims to find out if user ratings fall in rejection or non-rejection regions compared with others who rate the same item. This leads to a decrease or an increase in a user level of trustness. Subsequently, the confidence level is calculated for each user. Users' standard deviation is considered, and the items standard deviation of ratings for each domain ratings provided by all users in that domain. The following sections clearly demonstrate the suggested approach.

3.1 User Confidence Level

Each user has a percentage of honest in his/her ratings, lets us consider the dishonest part of his ratings as β, let κ be the number of times β has occurred for specific user $\mathring{u}$. Then the user ů confidence level α can be calculated using the following formula.

$$\alpha(\mathring{u}) = 1 - 1/\kappa; \quad \text{and } \kappa \geq 2. \tag{1}$$

The user $\mathring{u}$ confidence level might be increased or decreased based on the occurrence of β in his ratings. With the increase in the user ů confidence level, the region of non-rejection increased too for that user, while the decrease in his confidence level leads to increase in rejection region and a decrease in the non-rejection region, until it reaches the peak point where the system should refuse all his ratings.

3.2 Domains Region of Acceptance

Since web applications contain multiple domains of different interest levels, we calculated non-rejection regions for each domain. Therefore, each user ratings related to a specific domain may, or may not, fall in the non-rejection regions. The user level of confidence affected accordingly with the changes happen to regions of each domain. Let Ð be a set of domains, and Ð = $\{đ_1, đ_2, \ldots đ_n\}$, and each $đ_i$ has a mean value μ which reflect the average ratings for this domain, and the domain standard deviation is σ. Vicious users deliberately insert inaccurate ratings which are very high or very low to the domain average ratings. Let R_{V+} be the excess part, and let R_{V-} be the inferiority part. Therefore, we can find differences between users' ratings in a specific domain and then find mean differences of domain ratings as shown by Eq. 2.

$$\mu_{d_i} = \frac{\sum_{i=1}^{m} \mathrm{I}_i}{m}$$

$$d_i = \left|\mu_{d_i} - \mathrm{I}_i\right| \tag{2}$$

$$\overline{d_i} = \frac{d_i}{n}$$

Where $\overline{d_i}$ is the absolute differences of ratings for specific domain (items) i by all users m, and $m >= 2$. We computed the standard deviation of the differences for the domain d as shown by Eq. 3.

$$\sigma(d_i) = \sqrt{\frac{\sum_{j=1}^{l} \left(I_{u_j}^{i} - \overline{d_\imath}\right)^2}{n}} \tag{3}$$

Where $I_{u_j}^{i}$ is the rating of item I buy the user u_j, $\overline{d_\imath}$ is the calculated average differences of domain i ratings. The domain calculated standard deviation $\sigma(d_i)$ increase with the high vulnerability of the system. Users' ratings for specific domain (item) could be within the normal range, R_{V+}, or R_{V-}, if the user rating was R_{V+}, or R_{V-} then increment the user κ value, which in turn affects the user confidence level.

$$\check{\mathrm{D}} = \overline{d_\imath} \pm \sigma(d_i) \tag{4}$$

Where the expected maximum point is $\check{\mathrm{D}}_{+}$, and the expected lowest point is $\check{\mathrm{D}}_{-}$, and both points are calculated as shown in Eqs. 5 and 6.

$$\check{\mathrm{D}}_{+} = \overline{d_\imath} + \sigma(d_i) \tag{5}$$

$$\check{\mathrm{D}}_{-} = \overline{d_\imath} - \sigma(d_i) \tag{6}$$

Figure 1 shows the rejection regions, if specific user standard deviation of differences found in this region, then his κ value increases, and accordingly, his confidence level is decreased, and his domain interval is affected as well, in reflection to the dishonest part of his ratings.

3.3 Calculation of Users Confidence Interval

A user-domain average rating difference value is calculated using Eq. 7, as well as, the standard deviation of differences for specific users on specific domains is calculated as shown by Eq. 8.

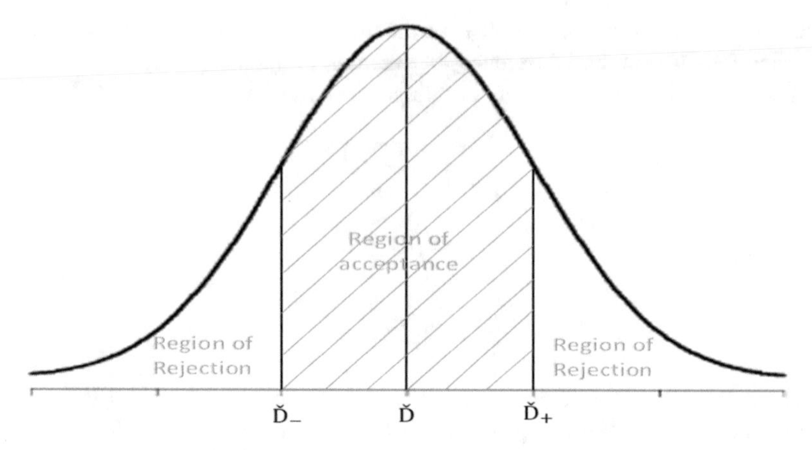

Fig. 1. Rejection and non-rejection regions.

$$\overline{d^{i}_{u_j}} = \frac{\sum_{i=1}^{l} \left| I^{i}_{u_j} - \overline{d}_{i} \right|}{n} \quad (7)$$

Where $I^{i}_{u_j}$ is the rating conducted by particular user u_j *for the item* I^i *in the domain* d^i. If the calculated $d^{i}_{u_j} > \check{D}_{+}$, then the user rating is R_{V+}, else if the calculated $d^{i}_{u_j} < \check{D}_{-}$, then the user rating is R_{V-}.

Where n is the total number of ratings done by the specific user in items of the domain đ, and n >=2. We computed the stander deviation of differences for ratings done by a user ů in the domain as shown by Eq. 8.

$$\sigma\left(d^{i}_{u_j}\right) = \sqrt{\frac{\sum_{j=1}^{l} \left(I^{i}_{u_j} - \overline{d^{i}_{u_j}}\right)^2}{n}} \quad (8)$$

The calculated user standard deviation of differences reflects the variation of the user-domain ratings done by a particular user.

The acceptable user interval shows the calculated acceptable interval for a specific user based on his ratings on different domains. Equation (9) is used to find the user acceptable interval points. The user acceptable domain interval is affected by the user confidence level α. We calculated the acceptable interval which determines the regions of rejection and non-rejection for each user-domain, the following formula is used to find each user-domain acceptable interval.

$$\tilde{I} = \alpha\left(\overline{d^{i}_{u_j}} \pm \sigma\left(d^{i}_{u_j}\right)\right) \quad (9)$$

Where the expected particular user highest point is $\tilde{I}_{+}$, and the expected lowest point is $\tilde{I}_{-}$.

$$\tilde{\mathrm{I}}_{+} = \alpha\left(\overline{d^{i}_{u_J}} + \sigma\left(d^{i}_{u_j}\right)\right) \quad (10)$$

$$\tilde{\mathrm{I}}_{-} = \alpha\left(\overline{d^{i}_{u_J}} - \sigma\left(d^{i}_{u_j}\right)\right) \quad (11)$$

The acceptable user interval shows the calculated acceptable interval which the user ratings should be within in the related domain. Each time we found a vicious user on a specific domain we increased κ value for that user, which reflects the number of times a user rating was found to be dishonest i.e. the user rating was outside the domain interval found in the previous section.

3.4 Vicious Users Detection

The user is considered as malicious if the calculated user rating was greater than $\tilde{\mathrm{I}}_{+}$ or less than $\tilde{\mathrm{I}}_{-}$. The following Table 1 summarizes different prone situations.

Table 1. A user prone description.

Situation	Prone description (Prone to)
$\check{\mathrm{D}} \leq \tilde{\mathrm{I}}_{+} \leq \check{\mathrm{D}}_{+}$, and $\check{\mathrm{D}}_{-} \leq \tilde{\mathrm{I}}_{-} \leq \check{\mathrm{D}}$	Normal
$\tilde{\mathrm{I}}_{+} > \check{\mathrm{D}}_{+}$ and $\tilde{\mathrm{I}}_{-} > \check{\mathrm{D}}_{-}$	Excess
$\tilde{\mathrm{I}}_{-} < \check{\mathrm{D}}_{-}$ and $\tilde{\mathrm{I}}_{+} < \check{\mathrm{D}}_{+}$	Inferiority
$\tilde{\mathrm{I}}_{-} < \check{\mathrm{D}}_{-}$ and $\tilde{\mathrm{I}}_{+} > \check{\mathrm{D}}_{+}$	Fully dishonest

A normal user is the one with an acceptable interval which falls within the calculated acceptable domain interval, while Excess user reflects the situation when there are tend to increase the items rates to deliberately excess its rates, and therefore affect the systems selections for recommendations. Inferiority reflects the situation when the user tries to downgrade the domain items in order to affect the system and leads to unselect specific items in recommendations, while Fully-dishonest reflects the situation where the user calculated maximum point, exceeds the domain maximum point and his minimum point is lower than the domain minimum point.

4 Experimental Results

We used the online available Movie Lens 1 M dataset [11], the dataset contains 1,000,209 anonymous ratings of approximately 3,900 movies made by 6,040 Movie Lens users. Data processing is conducted using 2.8 GHz PC, 16.0 GB RAM, the data cleaning and processing takes approximately two weeks. The dataset ratings are integer values between the lowest 1 and the highest 5. The dataset has 18 different domains, as well as we included only all users who rated at least 20 movies. We calculated each domain average value and then each domain average difference of its ratings $\bar{d}_{i}$, each domain standard deviation of differences $\sigma(d_i)$, and then we found the domain acceptable interval $\check{\mathrm{D}}$.

Although we planned to involve only users who rate items in all domains in the training set to show the fully dishonest users we involved some users who did not rate items in all domains. We calculated each user average differences of domain items' ratings $\bar{d}^{i}_{u_j}$, and we calculated each user standard deviation of differences $\sigma\left(d^{i}_{u_j}\right)$, and then we found the user acceptable interval points $\tilde{I}$ on each domain. Spiteful user prone is discovered using different situations shown in Table 1. We should mention here that every user κ value in the domain start with the value 1, and then this value increased every time we found R_{V+}, or R_{V-} in the user ratings. All domains acceptable intervals have been calculated, and then the users domains prone has been calculated based on the detection of users' acceptable interval against each domains' acceptable intervals, as well as, based on the user vicious prone shown in Table 1. All users/domains prone are summarized in Table 2.

Table 2. Users-domains malicious prone results.

Domain name	Normal	Excess	Inferiority	Fully dishonest
Action	167	121	148	2
Adventure	159	124	146	0
Animation	160	100	101	4
Children	133	131	117	0
Comedy	175	129	133	3
Crime	173	114	112	0
Documentary	71	37	32	0
Drama	240	96	101	15
Fantasy	123	109	104	0
Film-Noir	140	81	57	3
Horror	157	126	88	0
Musical	125	103	105	7
Mystery	151	110	112	0
Romance	177	117	135	0
Sci-Fi	173	109	143	11
Thriller	176	125	134	0
War	225	105	85	14
Western	109	88	87	0

Figure 2 shows the users' tendencies on different domains of the movie lens data set; Normal users represent the highest numbers while Excess and Inferiority users alternate the second rank. Our experimental results showed that many users rated items in random, and these random rating led to inaccurate recommendations. Therefore, recommendation systems should consider such situations by controlling the ratings and accept only normal users.

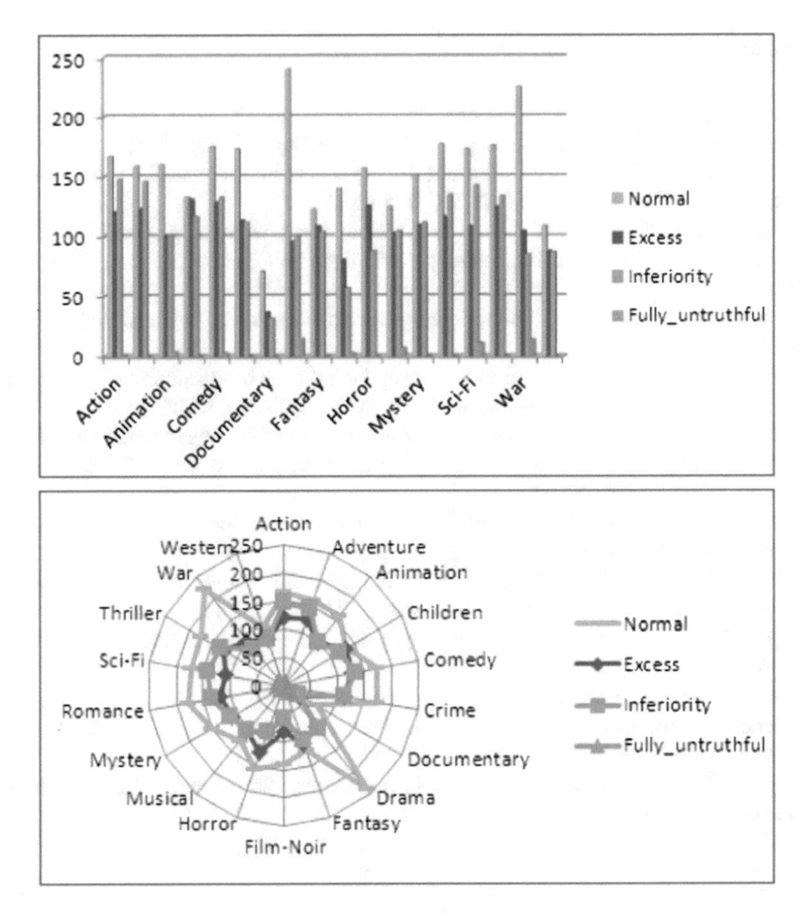

Fig. 2. Domain – users tendency.

The suggested method is valid for large-scale collaborative systems. It should be mentioned here that this method needs sufficient ratings for processing, we could not expect accurate detection of vicious users for a newly released system with insufficient ratings.

5 Conclusion and Future Works

The suggested method provides a practical solution to protect recommendation systems from vicious users' attacks, as well as, to protect users from being rejected by recommendation systems. Users are affected by the confidence level calculated from his/her ratings against the calculated domain ratings conducted by all users. The system will not allow malicious users to affect the system predictions for recommendations. In the future, we want to evaluate the performance of recommendation systems with a massive number of users and domains using this approach.

References

1. Burke, R., O'Mahony, M.P., Hurley, N.J.: Robust collaborative recommendation. In: Recommender Systems Handbook, pp. 961–995. Springer, Boston (2015)
2. Yang, L., Huang, W., Niu, X.: Defending shilling attacks in recommender systems using soft co-clustering. IET Inf. Secur. **11**(6), 319–325 (2017)
3. Embarak, O.H., Corne, D.W.: Detecting vicious users in recommendation systems. In: Developments in E-systems Engineering (DeSE), pp. 339–344. IEEE, December 2011
4. Panagopoulos, A., Koutrouli, E., Tsalgatidou, A.: Modeling and evaluating a robust feedback-based reputation system for e-commerce platforms. ACM Trans. Web (TWEB) **11**(3), 18 (2017)
5. Ricci, F., Rokach, L., Shapira, B.: Recommender systems: introduction and challenges. In: Recommender Systems Handbook, pp. 1–34. Springer, Boston (2015)
6. Wang, S., Zheng, Z., Wu, Z., Lyu, M.R., Yang, F.: Reputation measurement and malicious feedback rating prevention in web service recommendation systems. IEEE Trans. Serv. Comput. **8**(5), 755–767 (2015)
7. Moradi, P., Ahmadian, S.: A reliability-based recommendation method to improve trust-aware recommender systems. Expert Syst. Appl. **42**(21), 7386–7398 (2015)
8. Gunes, I., Kaleli, C., Bilge, A., Polat, H.: Shilling attacks against recommender systems: a comprehensive survey. Artif. Intell. Rev. **42**(4), 767–799 (2014)
9. Aggarwal, C.C.: Recommender Systems, pp. 1–28. Springer International Publishing, Cham (2016)
10. Zhou, W., Wen, J., Koh, Y.S., Xiong, Q., Gao, M., Dobbie, G., Alam, S.: Shilling attacks detection in recommender systems based on target item analysis. PLoS ONE **10**(7), e0130968 (2015)
11. Harper, F.M., Konstan, J.A.: The movielens datasets: History and context. ACM Trans. Interact. Intell. Syst. (TIIS) **5**(4), 19 (2016)

Understanding Students Personality to Detect Their Learning Differences

Ossama Embarak[1(✉)], Zahoor Khan[1], and Binod Gurung[2]

[1] Department of Computer Sciences, Higher Colleges of Technology, Fujairah, UAE
{oembarak, zkhanl}@hct.ac.ae

[2] Department of Education, Higher Colleges of Technology, Fujairah, UAE
bgurung@hct.ac.ae

Abstract. Students have different levels of intellectual capabilities and learning styles which affect their understanding of specific academic concepts and gaining specific skills. Educational institutions dedicate much efforts to support at-risk students. However, this support usually comes as a reaction of students' low performance, while learners need proactive support to keep their academic performance high, this needs a deep understanding of students learning capabilities and the real support they need accordingly. This research tries to investigate students' personalities and the impact of learners' personalities on their academic capabilities. A sample of 180 students was involved in this pilot study to evaluate the impact of their personalities on their academic standing. The sample collected from three different computing majors which are security forensics, Networking, and Application development majors. Myers-Briggs Type Indicator (MBTI) test is conducted twice using two different platforms and in different periods to figure out any random answers might happen by the participants. We excluded anomalies found and only quit fair data are kept for processing. This paper implements t-test to check if learners' personalities affect their academic standing which might drop them into at-risk category. The experimental results showed that students' personalities have a direct impact on their knowledge acquisition.

1 Introduction

Learners' academic performance is key for all educational institutions, it does not only affect students' study period, but also it has a direct impact on educational cost and increases the burden on the limited available resources. Discover learners personality type can be used to understand learning patterns and give proactive support learners might need to keep good standing. The Myers-Briggs test is used to discover personality patterns. Finding HCT learners' patterns will help to discover leaning differences and provide personality based advising. MBTI acts as a useful reference point to understand individuals' personality type. There are four pairs of patterns which help to identify individual personality. Firstly, the energy directions, Where learners prefer to get and focus their energy or attention?, persons who prefer to direct their energy to deal with situations, people, and things (i.e., the outer world) are known as extraversion

L. Barolli et al. (Eds.): EIDWT 2019, LNDECT 29, pp. 383–390, 2019.
https://doi.org/10.1007/978-3-030-12839-5_35

(E), compared to their complement who prefer to direct their energy to deal with information, beliefs, ideas, and explanations (i.e., the inner world) are Introversion (I). Secondly, the type of information and things a person prefers to process, what kind of information learner prefer to gather and trust? People who prefer to deal with facts to have clarity and clearly describe situations are Sensing (S), compared to their complement who prefer to deal with ideas, anticipations and deal with unknown factors which are not obvious, i.e., prefer to depend on Intuition (N). Thirdly, the type of decision making, what process learners prefer to use in their decisions? Thinker people prefer to make decisions based on analytics and logic compared with people who take their decisions based on their feelings, i.e. based on their values and believes. Finally, the lifestyle people adapt, how learners prefer to deal with the world around, learners 'lifestyle'? People who prefer planned, well-structured lifestyle are Judging (J), compared to those who prefer to maintain flexibility, goes with the flow, and response to things as they arise, i.e., Perception (P).

The combination of these four pairs produces 16 alternative types of personality which by measuring help researchers to understand individuals and their abilities for progression in a specific field(s). It should be clear that being Judging doesn't mean a person is not Perception, but it means his or her personality is Judging tends. Extraversion personality favors more interaction with people than involving in research activities, while Introversion personality advantage the research activities, this justifies the preferences of introverted students to involve in detail work rather than being in a too loud environment with many interactions. Therefore, they might prefer to select concentrations with less activity and interference. Therefore, this research examines personality patterns that fit each CIS concentration (Application development, Networking, Security and Forensics) in colleges, and how important the use of personality test to understand learners' learning academic capabilities and performance, hence provide right proactive advising considering individuality.

2 Literature Review

Many faculties believe that students with low performance are very weak and unqualified to pursue in the computing field. This believe entirely wrong and destroy the main fact of differences between learners' capabilities. We should admit that not all students in the same major have the same perception level as well as not all students in different majors are entirely different in their learning capabilities. Students have different backgrounds, interests, belief system, motivations, strengths, weakness, senses of responsibilities, and approach of study. Learners can be described by their intellectual learning style and how they perceive and processes different type of information [1]. Numerous models are developed to classify learners based on their learning style, one of the more trusted models is the Myers-Briggs Type Indicator (MBTI). This test assesses personality types where each candidate MBTI profile is detected to have strong learning style implications [2]. This instrument was the basis for numerous studies [3–7]. A study at the University of Western Ontario found that students with personality extroverted, sensing, feeling, and perceiving (ESFP) achieved low academic performance in the first year of the engineering curriculum than those with personality type introverts, intuition, thinkers, and judges (INTJ) [8].

Another study investigated the impact of personality on five subsequent courses in the engineering program, for this purpose they ran MBTI to a group of 116 students. Thinker personality outperformed feelers who were more likely to drop out of the curriculum regardless of their academic performance [9]. Sensors personality performance were significantly lower than the Intuitions in courses with high abstraction and practical ones. Extraverted were positively reacted in teamwork than introverted personalities. Intuitions show a remarkable rating for creative problem-solving abilities comparing to sensors personalities. Another model to understand students learning difference developed by Felder and Silverman depend on answering four questions. What type of information the learners perceive differently? Leaners with intuitive personality prefer to memories and are more comfortable with theories and abstractions models compared with learners with sensory personality who prefer practical, hands-on activities, and concrete facts [10]. What are the type of sensory data learners perceived? i.e., visual personality or verbal personality. How the learners intellectually like to process received data? Reflectively by meditation or actively by engagement in physical activities. How the learners naturally progress toward understanding? Using sequential logical progression of incremental steps where learners tend to think linearly and can function with only partial understanding. Alternatively, globally where students think in a system-oriented manner and won't be able to apply their knowledge until they fully grasp the full picture and having a holistic perspective [11].

Kolb mode [12–15] is applied extensively, also Felder and Silverman [16, 17] is applied massively in many research studies. This model presumes that in all classes we should expect various learners preferences and personalities. As a result, we could provide effective support and adaptive teaching for students' better performance. The main question is how to detect the proper learning approach for the new learners where we have a lack of information about them. Numerous methods are suggested and used to solve the new state case which is called the cold start problem, [18] implement mind thinker route which finds the closed thinker to the new leaner. Understand learners intellectual style is key for better advising and adaptive learning approach; an additional study found a direct correlation between leaners performance in technology majors and their personality type where computing students tended to be more ISFJ, ESFP, INTJ. While management information system students tend to be more ESTJ, and ISTJ. The correlation between students' personality and their academic progression depend on the matches between students' intellectual style and the study field contents, they found that System Analysts are more extroverted and intuitive, computer designers are more thinkers and introverted, and computer programmers tend to be more introverted, thinker and judging [19].

3 Research Problem

This research aims to examine the intellectual differences between learners in the same major and between different majors as well. Examine if there is a correlation between student personality and their abilities to maintain good standing in specific major and digest specific concepts. Hence the null hypothesis stated that there are no significant differences in students' academic performance due to their personality differences.

The null hypothesis: H_0: *In computing majors, students' academic performance doesn't affected by their personality types.*

The alternative hypothesis: H_1: *In computing majors, students' academic performance are affected by their personality types.*

4 Research Procedures

To proceed with this study, we involved 180 students in the study; each student has been requested to take the Myers-Briggs Type test twice using different platforms. All anomalies have been removed, and only accurate and cleaned data are kept for processing. We collected data about learners' different personalities weights, Extraversion (E), Introversion (I), Sensing (S), INtuition (N), Thinking (T), Feeling (F), Judging (J), and Perceiving (P) in the three majors (App Development, Networking, Security and Forensics).

Variables
This research deals with four independent pairs of variables which are Extraverted and Introverted, Sensing and Intuitive, Thinking and Feeling, Judging and Prospecting. For the three CIS concentration Networking, Security and Forensics, and Application.

5 Analyses

Firstly, we cleaned data from all NA cases, then calculate the mean, standard deviation, and range for each personality pattern. We have used python libraries [20] to calculate the mean and stander deviation and quantiles of the GPA variable for all concentrations as shown in Table 1.

Table 1. Basic GPA descriptive statists of the cis three concentrations.

Concentration	Mean	Std	Min	25%	50%	75%	Max
Applications development	3.10	0.40	2.15	3.00	3.20	3.25	3.90
Networking	2.69	0.61	1.40	2.27	2.56	3.20	3.74
Security and Forensics	2.69	0.54	1.30	2.40	2.68	3.00	3.68

Application Development students have the highest GPA mean values as well as, app development has the lowest standard deviation compared to security and networking majors.

Table 2 shows that application learners are more introverted, intuitive, feeling and perceiving. Networking students are more extroverted, sensing, thinking, and perceiving. While major security students are more introverted, sensing, thinking and judging (Fig. 1).

Table 2. Basic GPA descriptive statists of the cis three concentrations and for all personality patterns.

Concentration	Extraverted	Introverted	Sensing	Intuitive	Thinking	Feeling	Judging	Perceiving
Applications development	32.50	67.50	40.47	59.53	39.88	60.13	33.81	66.19
Networking	51.43	48.57	57.34	42.66	63.04	36.96	36.59	63.41
Security and Forensics	37.55	62.45	57.06	42.94	58.13	41.87	63.35	36.65

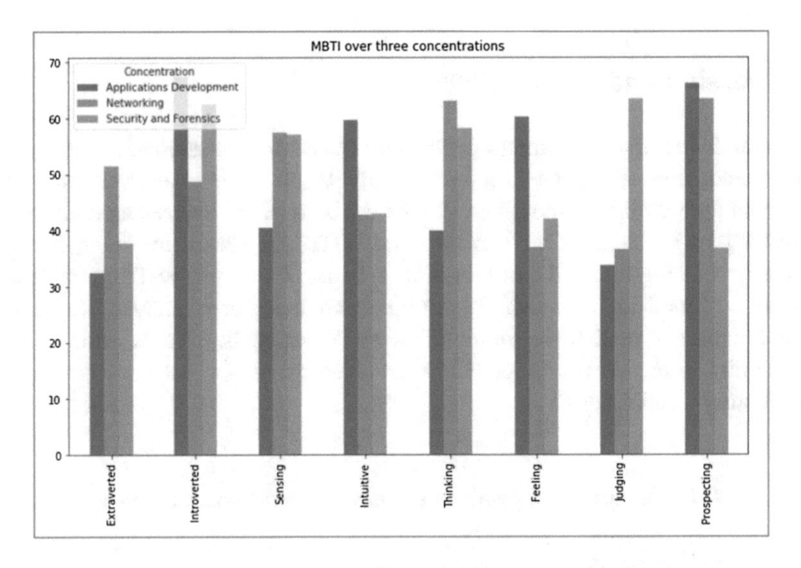

Fig. 1. Learners' personality weights per concentrations.

Methodology: we are going to accept the null hypothesis if the calculated t-test value greater than or equal $\alpha = 0.05$ significance level. This study use t-test Eq. (1) to determine if there is a significant relationship between personalities and academic standing.

$$t = \frac{\bar{x}_1 - \bar{x}_2}{\sqrt{\frac{var_1}{n} + \frac{var_2}{n}}} \tag{1}$$

Test Results

Alternative hypotheses supported when the null hypothesis rejected. According to the calculated t-test value, the null hypothesis should be rejected as there is statistical significant difference between the proportions represented by $p < .05$. The calculated t-test values for personalities INFP (0.00000083), ISTJ (0.00002313) and ISTP (0.02244996) for both App development vs. Security majors. The calculated t-test values for different personalities (INFP, ISTJ, and ISTP) against the academic standing was less than the $\alpha = 0.5$. Similarly, the calculated t-test values for personalities are INFP (0.00005234), ISTJ (0.00341639) and ISTP (0.04031650) for both App development vs. Networking majors. The calculated t-test value for the above personalities were less than the $\alpha = 0.5$. As a result, we should reject the null hypothesis and accept the alternative hypothesis. There is a significant impact of personality in learner's academic performance. We found that the personality ISTP t-test value is 0.90 which is greater than $\alpha = 0.5$ for both Networking and Security majors, in this case, we can accept the null hypothesis and reject the alternate one i.e., there is no significant difference in the academic performance between students with personality ISTP in both Networking and Security majors.

6 Discussion and Conclusions

This research examines if learners personality affects their academic standing. Application development learners tend to be more INFP and usually have very well-formed ideas about the way things should be and are not wary about expressing their opinions. Networking major students tend to be more ESTP, they learn by doing better than learning by reading about the concepts in a book. Therefore, ESTP personality not likely to be found in fields which require much theoretical analysis. Security and Forensics students tend to be more ISTJ; such personality feel a strong sense of responsibility, interested in serving others, have more sense of space and functions, and value traditions and security.

Table 3. Suggested algorithm for detecting performance in courses.

If Programming Courses

```
If "INFP" THEN high performance
ELSE IF "ISTJ"|"ESTP" THEN middle performance
ELSE low performance
```

Else if Networking Courses

```
If "ESTP" THEN high performance
ELSE IF "ENTP"|"ENTJ"|"ISTP"|"ESFP" THEN middle
     performance
ELSE low performance
```

Else if Security & forensics Courses

```
If "ISTJ"|"ESFJ"|"INTJ" THEN high performance
ELSE IF "ISFJ"|"ISTP"|"ESTJ"|"INFP"|"ESTP"|"ISFP" THEN
     middle performance
ELSE low performance
```

The conclusion that being drawn from this study is that students who are Introverted and perceiving digest programming and development concepts better than the networking and security concepts, while students who are extroverted and perceiving digest networking concepts and do better in networking practical comparing to security concepts. Students who are introverted and judging perform better in security concepts. The analysis concluded that we could implement the algorithms shown in Table 3 to predict students' course performance based on their discovered personality.

7 Future Work

This research emphasizes the importance of deeply study learners personality and learning style. The research finds that the learning style affects students' performance and the differences between students not based on their selected majors but based on their intellectual capabilities. We see that collecting data of a large number of learners from different majors will help to dig deeper in the curricula and provide direct and precise advising for learners based on their learning style. In the future, we aim to collect data about learners from different majors not only the computing major. As well as, we want to consider the other factors might affect learners' performance such as teaching methodology, teachers' skills, and the available learning logistics.

Acknowledgments. This work was supported by HCT Research Grants (HRG) [Fund No: 103108]. We would also like to show our gratitude to Higher Colleges of Technology for the financial grant.

References

1. Entwistle, N., Ramsden, P.: Understanding student learning (Routledge revivals). Routledge (2015)
2. Furnham, A.: Myers-Briggs Type Indicator (MBTI). In: Encyclopedia of Personality and Individual Differences, pp. 1–4. Springer International Publishing (2017)
3. Rashid, G.J., Duys, D.K.: Counselor cognitive complexity: correlating and comparing the myers–briggs type indicator with the role category questionnaire. J. Employment Couns. **52** (2), 77–86 (2015)
4. Furnham, A., Crump, J.: The myers-briggs type indicator (MBTI) and promotion at work. Psychology **6**(12), 1510 (2015)
5. Sample, J.: A Review of the myers-briggs type indicator in public affairs education. J. Public Aff. Educ. **23**(4), 979–992 (2017)
6. Murphy, L., Eduljee, N.B., Croteau, K., Parkman, S.: Extraversion and introversion personality type and preferred teaching and classroom participation: a pilot study. J. Psychosoc. Res. **12**(2), 437–450 (2017)
7. Boghikian-Whitby, S., Mortagy, Y.: Student preferences and performance in online and face-to-face classes using Myers-Briggs Indicator: a longitudinal quasi-experimental study. Issues Inform. Sci. Inf. Technol. **13**, 89–109 (2016)
8. Prince, M.: Does active learning work? a review of the research. J. Eng. Educ. **93**(3), 223–231 (2004)

9. Felder, R.M., Felder, G.N., Dietz, E.J.: The effects of personality type on engineering student performance and attitudes. J. Eng. Educ. **91**(1), 3–17 (2002)
10. Yip, M.C.: Learning strategies and their relationships to academic performance of high school students in Hong Kong. Educ. Psychol. **33**(7), 817–827 (2013)
11. Zuffianò, A., Alessandri, G., Gerbino, M., Kanacri, B.P.L., Di Giunta, L., Milioni, M., Caprara, G.V.: Academic achievement: the unique contribution of self-efficacy beliefs in self-regulated learning beyond intelligence, personality traits, and self-esteem. Learn. Individ. Differ. **23**, 158–162 (2013)
12. Truong, H.M.: Integrating learning styles and adaptive e-learning system: current developments, problems and opportunities. Comput. Hum. Behav. **55**, 1185–1193 (2016)
13. Faria, A.R., Almeida, A., Martins, C., Gonçalves, R., Martins, J., Branco, F.: A global perspective on an emotional learning model proposal. Telemat. Inform. **34**(6), 824–837 (2017)
14. Basheer, G.S., Tang, A.Y., Ahmad, M.S.: Designing teachers' observation questionnaire based on curry's onion model for students' learning styles detection. TEM J. **5**(4), 515 (2016)
15. Ramírez-Correa, P.E., Rondan-Cataluña, F.J., Arenas-Gaitán, J., Alfaro-Perez, J.L.: Moderating effect of learning styles on a learning management system's success. Telemat. Inform. **34**(1), 272–286 (2017)
16. Fatahi, S., Moradi, H., Kashani-Vahid, L.: A of personality and learning styles models applied in virtual environments with emphasis on e-learning environments. Artif. Intell. Rev. **46**(3), 413–429 (2016)
17. Normadhi, N.B.A., Shuib, L., Nasir, H.N.M., Bimba, A., Idris, N., Balakrishnan, V.: Identification of personal traits in adaptive learning environment: systematic literature review. Comput. Educ. **130**, 168–190 (2019)
18. Embarak, O.H.: A method for solving the cold start problem in recommendation systems. In: 2011 International Conference on Paper presented at the Innovations in Information Technology (IIT) (2011)
19. Ayoubi Rami, M., Ustwani, B.: The relationship between student's MBTI, preferences and academic performance at a Syrian university. Educ. Training **56**(1), 78–90 (2014)
20. Embarak, O.: Data analysis and visualization using python. In: Data Analysis and Visualization Using Python, pp. 205–241. Apress, Berkeley (2018)

A Novel Simulation Based Classifier Using Random Tree and Reinforcement Learning

Israar Ahmed(✉), Munir Naveed(✉), and Mohammed Adnan(✉)

Higher Colleges of Technology, Abu Dhabi, United Arab Emirates
{iahmed,mnaveed,madnan}@hct.ac.ae

Abstract. In this work, we present a new classification model to solve Human Activity Recognition (HAR) problem. The new classifier is a hybrid of Random Tree and Monte-Carlo simulations where Random Tree is used to select random samples for each simulation. The simulation use a generative model to train a value function that predicts a activity depending on sensor values. The classifier trains in an unsupervised learning style and does not require a training example dataset. It builds value function depending on response from environment. The experiments are performed on HAR dataset and compared with the start-of-the-art rival techniques. The performance is measure using precision, recall, f-Score and accuracy rate. The results show the new algorithm performs better than its rival techniques in f-score and accuracy. The classifier is also scalable and can also generalize non-deterministic behaviours.

1 Introduction

Smartphones are of vital importance in our daily lives. Working with electronic devices is becoming the requirement of all professions. With these devices, many experiments were made to improve the quality of life.

Human Activity Recognition (HAR) is a field committed to observing the daily activities of people by means of computational techniques. Different classification techniques (AAL, HMM, ANNs, NB, etc.) have been connected to the theme, with varying degrees of success. Human intervention is needed in many systems for decision making, usually by means of interaction through traditional devices such as the mouse, keyboards, switches, touch pens, or touchscreens.

Human Activity Recognition has been adopted by many researchers in which methods for understanding human behavior are acquired by reading characteristics derived from motion, locality, physiological signals and environmental information, etc. This field is the first element (sensing) of the series for achieving smarter interactive cognitive environments jointly with data analysis, judgment making and taking action [1, 2], and the focus of this research paper. In this research Meta Bagging and Random tree algorithms are mainly focused on and compared based on their performance.

Bagging or Bootstrap Aggregating is a method that generates distinct samples of the training dataset and generates a classifier for each sample. The results of these multiple classifiers are then joined together. When each sample of the training dataset is dissimilar, each classifier that is trained is given a subtly different concentration and

L. Barolli et al. (Eds.): EIDWT 2019, LNDECT 29, pp. 391–400, 2019.
https://doi.org/10.1007/978-3-030-12839-5_36

evaluation on the problem. On a large data bagging might be too computationally concentrated. However, when data is small, bagging is computationally feasible and gives performance gains. In this work we present a hybrid of Random Tree and Monte-Carlo simulations to classify HAR activities.

2 Literature Review

Li et al. [3] presents an unofficial learning technique to generalize the Human Activity Recognition mechanism. The authors used Principal Component Analysis (PCA) along with sparse auto-encoding based learning methods to adapt to the changes in the activity. To improve the issues of hand-held feature, they introduced the component origin system which uses diverse informal component learning systems to learn valuable figure interpretation from two sensors accelerometer and gyro meter for human movement detection. The informal learning systems they explored incorporate inadequate automatic encoder, demising PCA, and auto-encoder. They assessed the execution on an open human movement acknowledgment dataset and furthermore contrast the technique and conventional highlights and another method for unsupervised element learning.

HAR is looking for the sensor information during a particular time period and recording the information based on the motion or action taken by the object. There are three stages or phases taken into account by HAR: dividing device information streams, extricating highlights & characterization. The information received through HAR is mostly using the sliding window strategy, in which recording of the data is done with settled time window with certain corresponding breaks to frames. The research on Human activity recognition done by many researchers, is elected through hand. The handmade highlights depend on specific space learning & possibly cause in a deficit of data in the wake of obtaining attributes.

Multi-layered neural network is explored by [4] to generalize Human Activity Recognition (HAR) activities. The authors in [4] used a variation of multi-layered networks called convolution neural networks (ConvNet). The work investigates different layers of ConvNet. The outcomes demonstrated that expanding the number of convolutional layers, the neural system execution is enhanced too. The execution of their work is estimated by utilizing forecast exactness. Time-arrangement data has unavoidable neighborhood dependence properties. In addition, exercises tend to be dynamic and translation invariant in nature. In this manner, convolutional neural frameworks (convnet) misuse these qualities, which make it appropriate in overseeing time-course of action sensor data. They proposed an outline of convnets with sensor data gathered from mobile phone sensors to see exercises. Analyses showed that growing the number of convolutional layers assembles execution, in any case, the complexity of the decided features decreases with every additional layer. Also, sparing the information from layer to layer is more basic, as opposed to aimlessly extending the hyperparameters to upgrade execution.

The structure of convnet can in like manner benefit from a broader channel amount and lesser size of pooling setting. The network stores the great weights between the layers rather than adding randomness for re-structuring in an evolutionary episode.

The results are compared with different variations of Support Vector Machines (SVM). The results show that ConVNet based learning approach performs better than SVM. The convNet takes benefit of using restructuring features e.g. dynamic settings for pooling size and filter-size.

Reyes-Ortiz et al. [5] present a tight-time bond classifier to categorize the human activities for HAR dataset. Their work uses the transition activities e.g. transition from Running to Walking as a category. Since there could be huge set transition activities, their proposal adds additional category so-called "unknown activities" for new or exceptional transition activities. The main intuition behind mapping the transition activities is to refine the rules for identification of actual activity at a given time.

A variation of convolutional neural systems is investigated by [6] for HAR dataset. The creators investigated diverse layers of neural systems to outline input highlights to the yield classes given in HAR. The main feature of convolutional neural networks is their adaptive nature with respect to the temporal variations in data. In the paper, they recommended a convnet as the programmed include extractor for considering human activities using mobile phone sensors. The convolution task adequately exploits the impermanent neighborhood dependence of the time-plan signals and the absorption movement crosses out the effect of little understandings. The results demonstrate that lower pooling size e.g. 1×2–1×3 can give the calculation a chance to take in the mapping in a quicker speed and with high accuracy.

Using Hidden Markov Models Ronao [7] explored HAR by utilizing cell phone sensors. HAR was the main focus of their research using multiple sensors. Human activity recognition are primary activities human beings perform on an everyday basis, like sitting, walking, standing, and laying. They proposed an approach by the usage of Random Forests (RF) variable measures for characteristic choice and two-stage hidden Markov model to diagnose human activities. RF variable measures unbroken and possibly highly-associated variables, whilst two-stage CHMMs are suitable for the traits of sensory information and the normal hierarchical shape of events. The results showed that two-stage continuous hidden Markov Model is an achievable method for common action recognition on smart mobile phones. Continuous hidden Markov model is able to switch time series data and use smaller numbers as values applying the most applicable approach.

Use of HAR data set to train the machine learning algorithms to detect motion disorder has been presented in [8] where authors propose the use of transition duration between two activities as an indicator to detect the motion disorders. The authors mentioned a new task as an outcome of the simple and easy approach to substantial amounts of facts gathered from different sources like wearable sensors, concurrent database, and compact computing gadgets. They focused mainly on factual world trouble especially helping those who need immediate help and movement related action of human sicknesses. The candidate solution for classification are K-nearest Neighbour (KNN), Artificial Neural Networks (ANN), Naïve Bayes Classifier and SVM.

HAR dataset has also been explored by authors in [9] for its application in wearable devices. Their aim was to improve a correct and tremendous method for pastime consciousness using wearable devices. They assembled signals of gyroscopes and accelerometers into a diverse activity image, which permitted Deep Convolutional

Neural Networks to routinely research the ideal facets from the features for the recreation identification job. They applied SVM to categorize uncertain labels. The input was generated from sensor data and DCNN was used as output. The movement data is transformed to a model image and then Discrete Fourier Transform (DFT) is applied on the model construct image. The results showed that the model outperforms rivals without using transformations.

Ortiz et al. [10] explores the use of machine learning algorithms to classify a motion into different categories of transitory movement e.g. postural transitions. Like many HAR projects, they use the same wearable devices to get user-related signal report. The authors presented a smartphone-based available Human Activity Recognition procedure for the division of actions, which proceeds into concern the effect of PTs in method completion. The Human Activity Recognition methods used in past ignored evolutions among methods since the during between events is usually very minute. The work uses SVM to train the classifier to detect transitory movements from the sensor values of the gyroscope of a smartphone. The result shows that SVM can detect these movements with high accuracy.

3 Problem Definition

This research work addresses the problem of classification of the motion-based human activities where activities are sensed via smartphones. One of the possible way to solve such problem is to represent the human motions by using the motion-based contexts. These solutions are context-aware and can align each motion with a well-defined activity such as Walking, Running, Moving upstairs etc. The main challenge for context-aware solutions is to balance the quality of the solution and speed of finding solution. The traditional classifier takes huge time to converge but generally produce a quality solution. In this work, we propose a solution that finds the quality solution in a much shorter time as compared to the traditional classifiers.

4 Activity Recognition Framework

Human activity recognition (HAR) manages the coordination of detecting and reasoning so as to better comprehend individuals activities. Research identified with human action acknowledgment has turned out to be significant in unavoidable and versatile computing, observation-based security, setting mindful processing, and surrounding assistive living [8]. HAR research, mainly involves the use of different sensing technologies. The most ordinarily utilized sensors are accelerometer and gyroscope in HAR utilizing a smartphone. Most of the features used are preferred by hand regarding HAR. These hand-crafted features may result in a shortfall of information as they rely on the particular domain knowledge. Activity recognition is a vital innovation in unavoidable processing since it can be connected to many real-life, human-driven issues, for example, eldercare and healthcare services.

5 Justification

The requirement for fast convergence of classifier is required to provide solutions in real-time e.g. mobile devices has limited resources (CPU and memory) and cannot afford to run machine learning solutions for a longer time period. Although more focus is on more speed and better performance of these devices but still more enhancements are required as desired by most users. Day by day the need for electronic devices are growing and particularly mobile devices are more in demand. Therefore, we focus on optimizing the converging time of the classifiers without comprising on the quality of the solution.

6 Propose Model

We propose a hybrid of Random Tree and Monte-Carlo simulation based reinforcement learning model called RBCM [25, 26]. The generative model in RBCM is updated by using Random Tree. The value function in RBCM remains intact. The architecture of the proposed model shown in the following figure (Fig. 1).

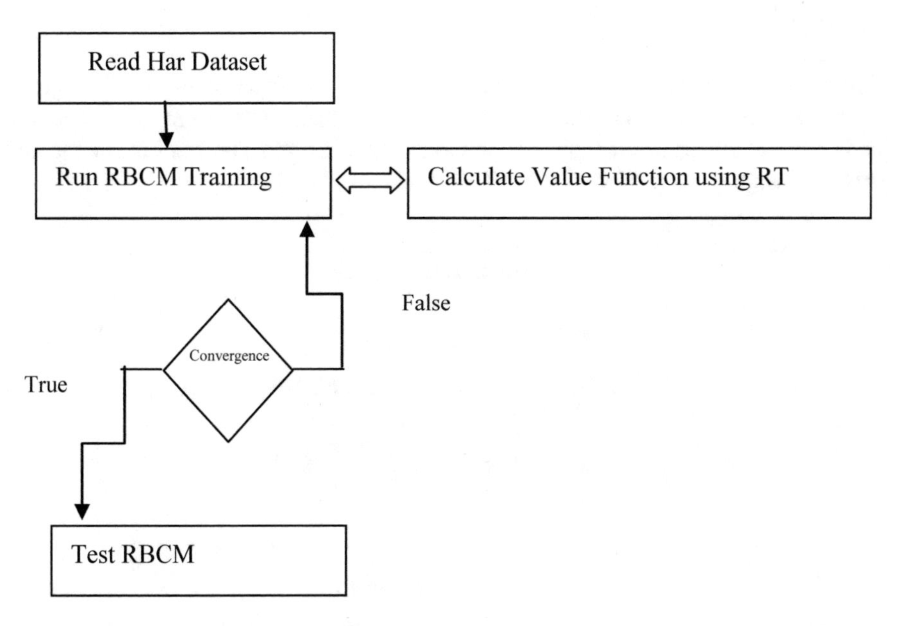

Fig. 1. Proposed model

7 Experiment Setup

Like other machine learning, activity recognition has two stages, training and testing. The wearable sensors attached to the subject body to measure different attributes has a wide range like accelerometer, GPS, heart monitor, ECG, thermometer, light sensor etc.

There can be different attributes measured like movement [19], temperature [20], location [21] and ECG [22]. The sensor sends the signal to the integration device that can be a smartphone, laptop or any other embedded system. The communication protocol can be user datagram protocol or transport control protocol depends upon the security requirement of the data set. In order to perform experiments a laptop with Microsoft Windows 8.1 (64 bit), having 8 GB RAM, 500 GB Hard disk drive and Intel Core i7 processor is used. All experiments are performed using WEKA.

7.1 Performance Criteria

The models are compared using the following parameters:

1. Mean absolute Error: Model will lower value of mean absolute error is considered better than the one with higher.
2. Time to build model: Smaller time is better than long time.
3. Accuracy: higher accuracy is better than model with lower accuracy
4. Precision, Recall and F-Score: higher values are better than lower values.

8 Results

Figure 2 shows the results of all models with respect to mean absolute error. The graph shows that there is a minor difference regarding the Mean absolute error. The bagging mean error is 3% while Random tree is having error as 4%. The Root mean squared error for bagging has a value of 0.115 (11%) and Random tree result is 0.1994 (20%).

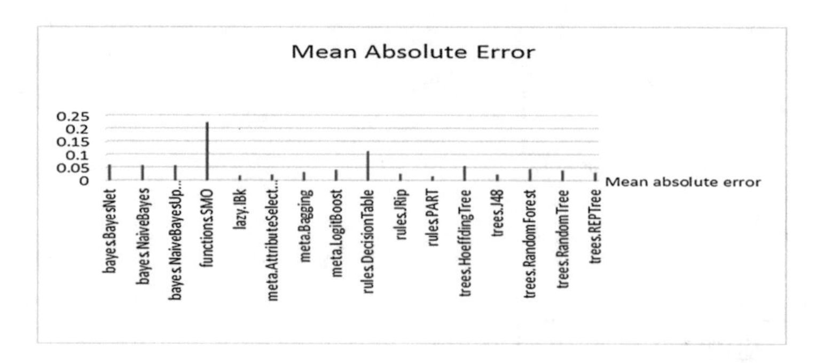

Fig. 2. Mean absolute error

The time taken by each model to be built is shown in Fig. 3. In other words, it is time taken be each model to converge. Meta bagging takes 4.15 s to build a model, while Random tree consumes only 0.13 s. RBCM take 0.56 s. Random tree and decision tree work almost in a similar way with one exemption that decision tree use the irregular subset for each split. Decision tree takes in decision trees from both numerical and ostensible data.

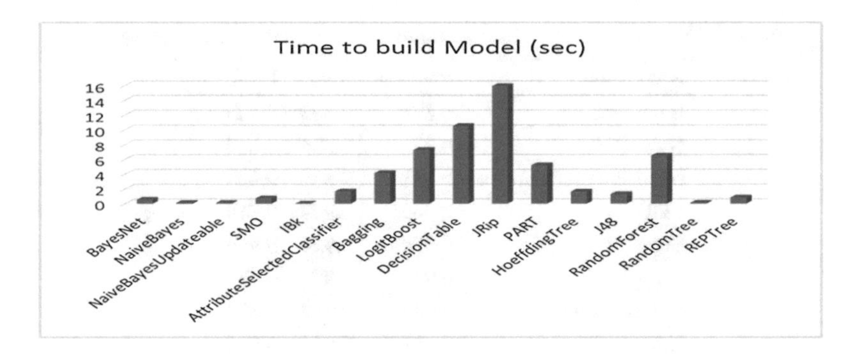

Fig. 3. Time (in seconds) take by each model

The above graph shows that Random Tree is taking very less time comparing to meta Bagging. The Random tree is more efficient and building the model for the data set in less than a second.

The graph given in Fig. 4 reveals that meta bagging is playing a vital role than the Random tree and it is giving more than 95% accuracy for the concerned data set. Random forest performs better than Random Tree.

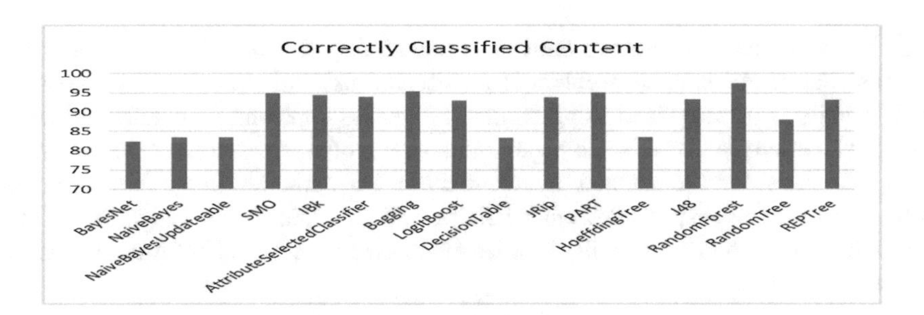

Fig. 4. Model with their accuracy

Table 1 shows results with respect to precision, recall and F-Measure. These parameters capture the performance by considering truly classified or falsely classified instances as well. Several models show high performance e.g. BayesNet, SMO, Random Forest etc. However, in terms of the quality solutions, RBCM outperforms other algorithms.

Table 1. Precision, recall and F-measure values for algorithms performance

Algorithms	Precision	Recall	F-Measure
BayesNet	0.825	0.823	0.823
NaiveBayes	0.838	0.834	0.834
NaiveBayesUpdateable	0.838	0.834	0.834
SMO	0.949	0.949	0.949
Bagging	0.954	0.954	0.954
LogitBoost	0.93	0.93	0.93
DecisionTable	0.834	0.832	0.832
JRip	0.941	0.941	0.941
PART	0.951	0.951	0.951
HoeffdingTree	0.839	0.835	0.835
J48	0.933	0.933	0.933
RandomForest	0.976	0.976	0.976
RandomTree	0.881	0.881	0.881
RBCM	0.988	0.988	0.988

9 Conclusions

This paper narrates the human activity recognition problem and formulates it as a classification problem. The problem is represented using motion based contexts and each activity is a recognized using context-awareness approach. A detailed review of existing solutions is given in the paper and a new model is presented. The commonly used classifiers for human activity recognition are explored using the benchmark dataset HAR. The results show that RBCM and Random Forest outperforms other classifiers. In the future, we will optimize generative model of RBCM to optimize its convergence time.

References

1. Parasuraman, R., Sheridan, T.B., Wickens, C.D.: A model for types and levels of human interaction with automation. IEEE Trans. Syst. Man Cybern. **30**(3), 286–297 (2000)
2. Gandetto, M., Marchesooti, L., Sciutto, S., Negroni, D., Regazzoni, C.S.: From multi-sensor surveillance towards smart interactive spaces. In: IEEE International Conference on Multimedia and Expo, vol. 1, pp. 1–641 (2003)
3. Li, Y., Shi, D., Ding, B., Liu, D.: Unsupervised feature learning for human activity recognition using smartphone sensors. In: Prasath, R., O'Reilly, P., Kathirvalavakumar, T. (eds.) Mining Intelligence and Knowledge Exploration. Lecture Notes in Computer Science, vol. 8891. Springer (2014)
4. Ronao, C.A., Cho, S.B.: Deep convolutional neural networks for human activity recognition with smartphone sensors. In: Arik, S., Huang, T., Lai, W., Liu, Q. (eds.) Neural Information Processing. ICONIP 2015. Lecture Notes in Computer Science, vol. 9492. Springer (2015)

5. Reyes Ortiz, J.L., Oneto, L., Samà, A., Parra, X., Anguita, D.: Transition-aware human activity recognition using smartphones. Neurocomputing **171**, 754–767 (2016). ISSN: 0925-2312
6. Ann Ronao, C., Cho, S.-B.: Human activity recognition with smartphone sensors using deep learning neural networks. Expert Syst. Appl. **59**, 235–244 (2016)
7. Ronao, C.A., Cho, S.B.: Human activity recognition using smartphone sensors with two-stage continuous hidden Markov models. In: 2014 10th International Conference on Natural Computation (ICNC), Xiamen, 2014, pp. 681–686 (2014). https://doi.org/10.1109/icnc.2014.6975918
8. Reyes-Ortiz, J.L., Alessandro, G., Xavier, P., Davide, A., Joan, C., Andreu, C.: Human activity and motion disorder recognition: towards smarter interactive cognitive environments. In: 2013 ESANN European Symposium on Artificial Neural Networks, Computational Intelligence and Machine Learning. Bruges (Belgium), i6doc.com publ. (2013). ISBN 978-2-87419-081-0
9. Jiang, W., Yin, Z.: Human activity recognition using wearable sensors by deep convolutional neural networks. In: Proceedings of the 23rd ACM International Conference on Multimedia, Brisbane, Australia, pp. 1307–1310 (2015). ISBN: 978-1-4503-3459-4 https://doi.org/10.1145/2733373.2806333
10. Reyes-Ortiz, J.L., Oneto, L., Ghio, A., Samá, A., Anguita, D., Parra, X.: Human activity recognition on smartphones with awareness of basic activities and postural transitions. In: Wermter, S., et al. (eds.) Artificial Neural Networks and Machine Learning – ICANN 2014. ICANN 2014. Lecture Notes in Computer Science, vol. 8681. Springer (2014)
11. Mourcou, Q., Fleury, A., Franco, C., Klopcic, F., Vuillerme, N.: Performance evaluation of smartphone inertial sensors measurement for range of motion. Sensors (2015). (ISSN 1424-8220; CODEN: SENSC9)
12. Davis, K., Owusu, E., Bastani, V., Marcenaro, L., Hu, J., Regazzoni, C., Feijs, L.: Activity recognition based on inertial sensors for ambient assisted living. In: 2016 19th International Conference on Information Fusion (FUSION), Heidelberg, pp. 371–378 (2016)
13. Jiang, W., Yin, Z.: Human activity recognition using wearable sensors by deep convolutional neural networks. In: Proceedings of the 23rd ACM International Conference on Multimedia, pp. 1307–1310, Brisbane, Australia (2015). ISBN: 978-1-4503-3459-4 https://doi.org/10.1145/2733373.2806333
14. Yang, J., Lee, J., Choi, J.: Activity recognition based on RFID object usage for smart mobile devices. J. Comput. Sci. Technol. **26**(2), 239–246 (2011)
15. Chen, L., Wei, H., Ferryman, J.: A survey of human motion analysis using depth imagery. Pattern Recognit. Lett. **34**(15), 1995–2006 (2013)
16. Ong, W., Palafox, L., Koseki, T.: Investigation of feature extraction for unsupervised learning in human activity detection. Bull. Networking Comput. Syst. Softw. **2**(1), 30–35 (2013)
17. Lara, O.D., Labrador, M.A.: A survey on human activity recognition using wearable sensors. IEEE Commun. Surv. Tutorials **15**(3), 1192–1209 (2013)
18. Chaaraoui, A.A., Padilla-López, J.R., Climent-Pérez, P., Flórez-Revuelta, F.: Evolutionary joint selection to improve human action recognition with RGB-D devices. Expert Syst. Appl. **41**(3), 786–794 (2014)
19. Ryoo, M.S.: Human activity prediction: early recognition of ongoing activities from streaming videos. In: 2011 International Conference on Computer Vision, no. Iccv, pp. 1036–1043 (2011)
20. Iglesias, J., Cano, J., Bernardos, A.M., Casar, J.R.: A ubiquitous activity-monitor to prevent sedentariness. In: Proceedings of IEEE Conference on Pervasive Computing and Communications (2011)

21. Choujaa, D., Dulay, N.: TRAcME: Temporal activity recognition using mobile phone data. In: IEEE/IFIP International Conference on Embedded and Ubiquitous Computing, vol. 1, pp. 119–126 (2008)
22. Parkka, J., Ermes, M., Korpipaa, P., Mantyjarvi, J., Peltola, J., Korhonen, I.: Activity classification using realistic data from wearable sensors. IEEE Trans. Inf Technol. Biomed. **10**(1), 119–128 (2006)
23. Jatoba, L.C., Grossmann, U., Kunze, C., Ottenbacher, J., Stork, W.: Context-aware mobile health monitoring: evaluation of different pattern recognition methods for classification of physical activity. In: 30th Annual International Conference of the IEEE Engineering in Medicine and Biology Society, pp. 5250–5253 (2008)
24. Olszewski, R.T., Faloutsos, C., Dot, D.B.: Generalized Feature Extraction for Structural Pattern Recognition in Time-Series Data (2001)
25. Naveed, M., Kitchin, D., Crampton, A.: Monte-Carlo planning for pathfinding in real-time strategy games. In: Proceedings of PlanSIG 2010. 28th Workshop of the UK Special Interest Group on Planning and Scheduling Joint Meeting with the 4th Italian Workshop on Planning and Scheduling, Brescia, Italy, pp. 125–132. PlanSIG (2010)
26. Naveed, M., Crampton, A., Kitchin, D., McCluskey, T.L.: Real-time path planning using a simulation-based Markov decision process. In: Research and Development in Intelligent Systems, vol. 28, pp. 35–48. Springer, London (2011)

Perceiving Intellectual Style to Solve Privacy Problem in Collaborative Systems

Ossama Embarak(✉), Kholoud Saeed, and Manal Ali

Department of Computer Sciences, Higher Colleges of Technology, Fujairah, UAE
{oembarak,H00307864,H00307076}@hct.ac.ae

Abstract. Privacy problem is a big challenge in collaborative systems. Such systems depend on users collected data to generate recommendations in their future visits. Site visitors give falsify information to avoid privacy disclosure; this leads to inefficient recommendations. In this paper, we address the privacy problem in collaborative systems; we suggested a new perceiving intellectual style to generate recommendations and avoiding users' privacy issues. According to the suggested approach, we were able to provide two types of recommendations, the Intellectual Node Recommendation or the Intellectual Batch Recommendation. We evaluated both recommendation types by calculating levels of coverage and precision. We found that Intellectual Batch Recommendation achieved better performance comparing to the Intellectual Node Recommendation.

1 Previous Solutions

Privacy, one of the most current challenges in web recommendation systems. All web collaboration approaches collect data explicitly or implicitly about visitors to being able to generate recommendations. Creating and maintain users' profiles represent one of the main stages in web recommendation. However, users become more concerned about their privacy because of the computers' predictions and misuse of their collected data. Inadvertently reveal personal information to other users who use the user computer [1]. When cookies are used for authentication or accessing users' profiles, anyone who uses a particular computer may have access to the information in a user's profile [2]. Any user who gains access to a user computer and hence to his profile might gain unauthorized access to a user's accounts [3].

Some users who are working together share the user's computer may gain unauthorized access to each other account on a personalized website by guessing a password. Thieves, for example, may find profile information too useful [4]. Several systems tried to handle privacy issue by using pseudonymous profiles [5], client-side profiles [6], task-based personalization [7], or by putting users in control [8], but still privacy represents one of the big issues in web recommendation and personalization systems.

Several solutions provided to tackle the privacy problem, some solutions were focusing only on the privacy problem ignoring cold start problem, scalability problem, coverage problem, etc. Some other researches tried to solve the cold start for both items

L. Barolli et al. (Eds.): EIDWT 2019, LNDECT 29, pp. 401–410, 2019.
https://doi.org/10.1007/978-3-030-12839-5_37

and user cold start problems, in addition to the privacy problem [9]. Some systems use demographical data to find similarities between site visitors. Demographic data refers to specific user characteristics such as age, gender, income, religion, marital status, language, ownership (home, car, etc.), and social position, etc. [10]. Demographic data can be used as initial characteristics for creating recommendations and solve not only the privacy problem but also the cold start problem related to the user i.e. providing recommendations when the system does not yet have any information about the user ratings. However, this method provides nonnovel recommendations as it creates generalizations about a bunch of people, demographic information represent aggregate information about groups, not about specific individuals. Site visitors should fill a form or the system force them to provide demographic information, which causes annoying for users, furthermore, these profiles will be in a static manner and need to be updated every period to reflect the reality which makes users boring.

Other systems depend on stereotypes, i.e., use standardized conception or image with specific meaning about a group of people to generate a recommendation based on the user's stereotype category [11]. A stereotype may be a conventional and oversimplified conception, opinion, or image based on the assumption that there are common attributes holds by members of the specific group. It may be a positive or negative, also it typically formed by limited knowledge about a group which the person is doing stereotyping, or false association between two variables. Stereotypes based systems represent people entirely in narrow assumptions; it may refuse to recognize a distinction between an individual and the group to which he or she belongs. Attribute-based recommendation systems collect data about both users and items' attributes. Thus, when a new item added to the website or new user visits the website, the system collects information about new item specifications and attributes, as well as for the new user asks him/her to fill a form to create his profile attributes [12]. Attribute-based technique solves both cold start problems (item and user), but main disadvantages of this technique are that the profiles are static and data collection is done by forcing users' to fill forms or give their interests by selecting from specific pre-prepared categories. Such systems repeatedly recommend the same items to the same users because it generates a static profile.

2 Suggested Method

The suggested method depend on **P**erceiving **I**ntellectual **S**tyle **P**rofile (PISP), created *PISP* profile is used to generate recommendations. PISP use integrated minds interests' roadmap for all abstract visitors regardless of their identification data. Whatever, these identification data are demographic, social, IP, or any other identification data. This approach deals with users based on their online thinking which expressed by their online selections and paths that they will follow. We assume that "*if two abstract users are similar hence they think the same way*" hence users who go through the same path have similar thoughts and tastes and should inherit benefits of that path (Fig. 2).

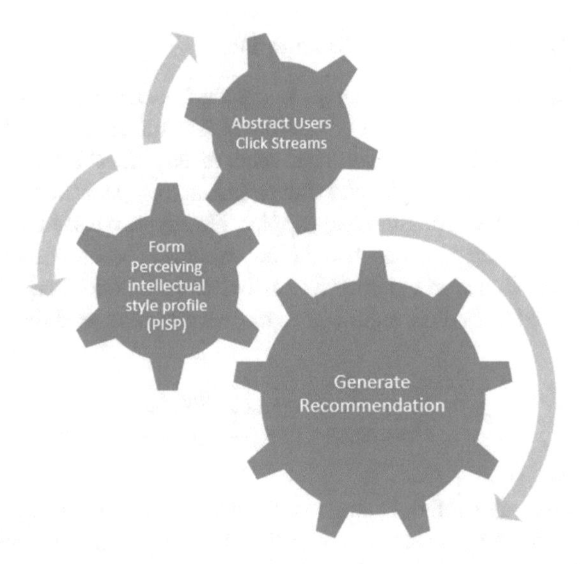

Fig. 1. Recommendation based on Perceiving Intellectual Style (PIS).

We collected users' abstract click-streams, which reflect the users' power of thinking of a specific item(s). The collected abstract click-streams are used to create maximal abstract sessions (loop-less sessions) that show the abstract users' preferences. We evaluated the collected maximal sessions to remove extremely low and extremely high sessions (to avoid the robustness problem), and then all sub-sessions absorbed into its super-sessions (to reduce the storage space without losing data quality). As a result, we got integrated routes (Elastic routes); which represent the largest abstract loop-less routes visited by abstract users through their click-streams on a specific website, which in turn used for generating the delivered recommendation sets to site visitors regardless of their data (Fig. 2).

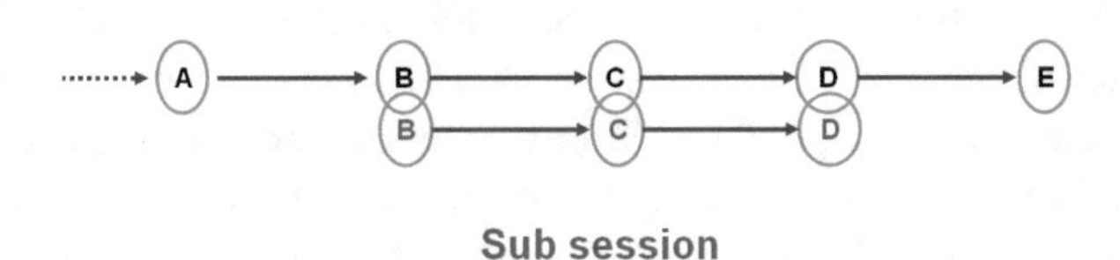

Fig. 2. Create users abstract Perceiving Intellectual Style Profile (PISP).

Provide a recommendation for users; using the presented concept, is valid not only for recurring users but also for new users, where any new user enters the website, he will navigate through a specific path, which expresses his/her thinking. Therefore, the system will be able to provide recommendations based on his maximal online preferences and based on his followed path(s). As indicated earlier, users are identified by their online Perceiving Intellectual Style. Therefore, the privacy problem using this concept vanishes.

2.1 Method Illustration and Implementation

Once the users' perceiving intellectual style session is detected, a session significance value is calculated, where a session is significant if it makes a clear difference to the stored integrated routes. Session significant value is also an estimation of how much it reflects the users' intellectual style, and it depends on the spent time by a specific user during his session on the node (item/page/topic). However, we see that extremely high and extremely low time sessions should be excluded because they involve non-important browsing. Equation (1) is used to calculate each node (item) threshold value, which in turn is used to find the significance of every session as shown by the Eq. (2).

$$Threshold(x_j) = \frac{\sum_{i=1}^{k} time(x_i)}{k} \tag{1}$$

Where the numerator refers to the total time spent on an item x_i by site users in *k* sessions, while the denominator refers to the number of sessions which contains an item x_i as a visited node. Threshold value should update with any changes in the created maximal forward sessions which include the item x_i as a node.

$$Sig(s_j) = \frac{\sum_{i=1}^{n} time(x_i) - Min_{TH}}{Max_{TH} - Min_{TH}} \tag{2}$$

Where,
$Sig(s_j)$, is the significance value of the session s_j,
$\sum_{i=1}^{n} time(x_i)$, the total time spent in all visited nodes during the session s_j,
Max_{TH}, the maximum threshold value from the session s_j items,
Min_{TH}, the minimum threshold value from the session s_j items.

All significant sessions proceed to the absorption process, if $P(s_k)$ path represents an ordered set of pages visited in session s_k, then when we get $P(s_k) \subseteq P(s_j)$, the perceiving intellectual style profile (PISP) only store s_j and update the s_j route nodes weight. Therefore, as soon as an absorption case detected, the system should update the larger session and remove the sub sessions, this leads to having only the maximal integrated routes which combines all users visits and its associated weight (Fig. 3).

We can generate integrated routes if there is an intersection between the beginning and the end of any two absorbed sessions. The main aim of generating integrated routes is to generate maximal stretchable paths of interests. Create such integrated routes reduce the number of the stored session as well as it provides more flexibility to generate recommendations using the active node technique.

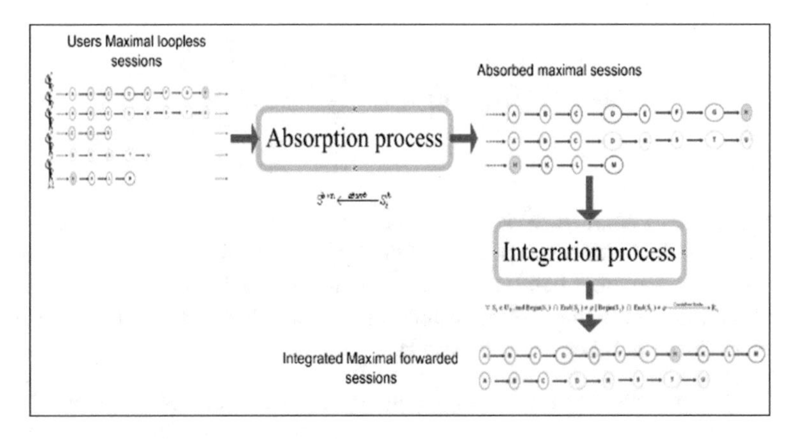

Fig. 3. Integrated routes (IR) in Perceiving Intellectual Style Profile (PISP).

2.2 Generate Recommendations

This approach enables us to generate two types of recommendations, *Intellectual Node Recommendation* or *Intellectual Batch Recommendation*. In the Intellectual Node Recommendation, the system creates a recommendation set based on directly linked nodes to the active node (page); node recommendation is not an item-to-item recommendation, because it does not depend on items attributes, but it depends on the *Intellectual* visited nodes in a specific path(s) with high relative weight (high power of desires weight). The system recommends items of high weight from the stored integrated routes IR in PISP which have a sequential association to the online new user path. The used rule to find a candidate for node recommendation is, $Find(x_i \mid A \xrightarrow{e_j} x_i \subset \mathrm{IR}_j)$ where A refers to the user selected active node and refers x_i to all items in association with the active node A and stored in the integrated route IR_j. All x_i are involved as candidates for recommendations, and only top n items selected for recommendation. While in the *Intellectual Batch Recommendation*, the recommendation set is generated using top N highly weighted nodes in his expected paths based on his/her maximal online path. The used rule to find a candidate for batch recommendation is $Find(x_i \mid CMP \subset IR_j)$, where CMP refers to the user current online maximal path, and IR_j refers to the stored integrated routes in which CMP is a subset of.

New items are also involved in the recommendation set, as indicated before; we consider newly added items as well as old items that were never visited before as new items. These new items threshold weights are equal to zero. All selected candidates for a recommendation in node and batch recommendation should be checked if they have a direct link to new items (nodes with a threshold equal to zero), and then by implementing the virtual weight Eq. (3), we selected the top n items for recommendation.

$$W_v(A \rightarrow N \mid A \rightarrow x_i) = \frac{e.w_r(A, x_i).k}{n} \cdot \frac{\sum_{i=1}^{n} R(x_i)}{\sum_{j=1}^{k} R(A_j)} \quad (3)$$

Where $W_v(A \rightarrow N \mid A \rightarrow x_i)$ is the virtual weight between the active node A and the new added item N via item x_i, and e = 1 if the hyperlink is found between items x_i and N, else e = 0. On the other hand, $w_r(A, x_i)$ refers to the number of times items A and x_i appears (purchased) together. While k refers to the number of users who rate item A, and n refers to the number of users who rate an item x_i.

$\sum_{i=1}^{n} R(x_i)$ represents the total ratings done by all users for item xi, while $\sum_{j=1}^{k} R(A_j)$ represents the total ratings done by all users for item A_j.

3 Description of Experiments

We collected users clickstreams in a website, then we calculated the threshold for each node before generating abstract routes, and finally make absorption of all sub-sessions to create abstract integrated routes (IR). This process is a continuous process which keeps updating intellectual profile (PISP). The developed pilot system generates two types of recommendations as mentioned earlier Node recommendation and Batch recommendations. We calculated precision and coverage of the generated recommendations in different visits for both cases Batch and Node using the suggested method; we select these visits far from each other to show the progress in the created recommendation sets for the wide intervals. We notice that with the increase in the coverage level, there was also an increase in the accuracy level. Coverage measures the participation level of candidate items in each recommendation set to its target set. While the precision measure percentage of recommended items appears in the target set to the total number of recommended items. Levels of coverage and precision in batch recommendations depend on to which level users trust provided recommendation sets. If users trust recommended items, then he/she will select some items that will affect the accuracy and coverage level of generated recommendations. Equation 4 is used to calculated the levels of coverage, it represents the percentage of items provided as recommended candidates and appear in the target sets to the total number of items in the target set (selected items by user during training phase is used as target set).

$$Coverage = \frac{\sum_{i=1}^{n} |R_i \cap TS_i|}{\sum_{j=1}^{k} |TS_j|} \quad (4)$$

Where numerator represents number of items found in both recommendation set and target set. While denominator represents the total number of items in the target set. The coverage level of batch recommendation was better than the coverage level of node recommendation, the reason behind is that in batch recommendation the recommendation sets generated based on the user online maximal path which reflect their intellectual style and gives high novelty as well, while in node recommendation the candidates are scattered with low level of novelty (Fig. 4).

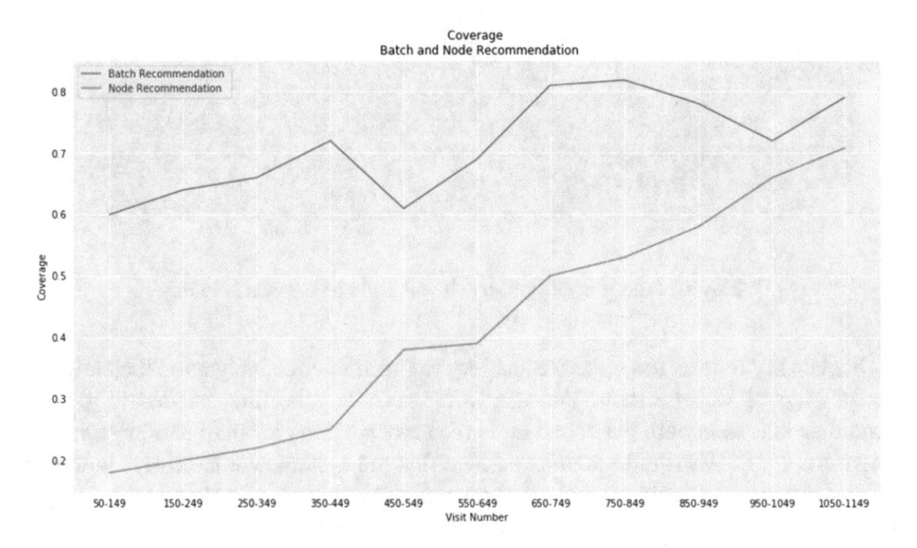

Fig. 4. Coverage of both batch and node recommendations.

Equation 5 is used to calculate precision for both batch and node recommendations. Where numerator represents number of items found in both recommendation sets and target sets. While denominator represents the total number of candidates in the recommendation sets (Fig. 5).

$$Precision = \frac{\sum_{i=1}^{n} |R_i \cap TS_i|}{\sum_{j=1}^{k} |R_j|} \tag{5}$$

Batch recommendation achieved better precision level compared with the node recommendation, which means that most users selected items from recommendation candidates. While in node recommendation, users selected items outside the recommended candidates because the same item can has semantic link with many other items and hence could appears in many recommendation sets.

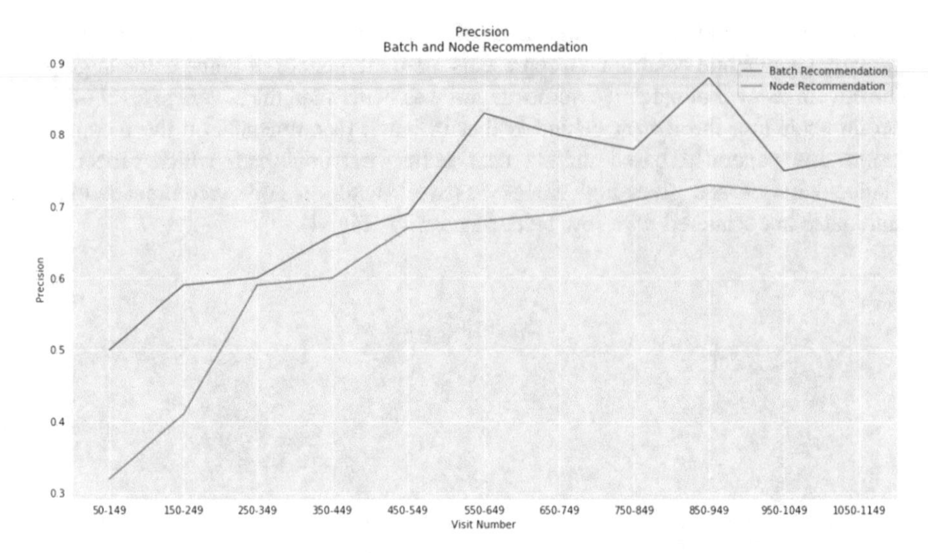

Fig. 5. Precision of both batch and node recommendations.

Scalability reflects the system capability to function effectively and efficiently with the growing of website items and users. We recorded the time needed to generate recommendations in both batch and node recommendations in small and large scale as shown in Fig. 6. We should mentioned here that the system should always bubble up the top significant candidates, where the users clickstreams collected online but profiling is processed offline, its inefficient to make profiling with minor clickstreams.

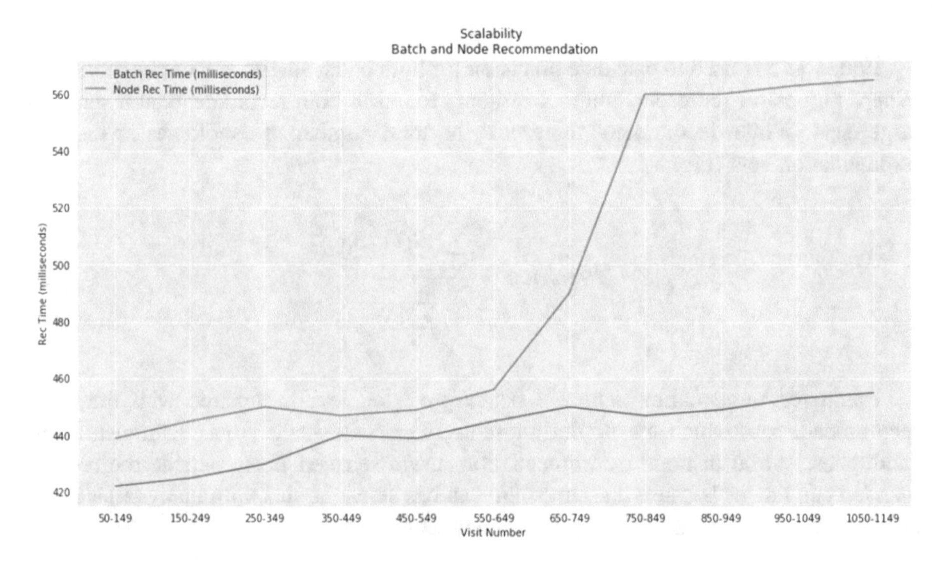

Fig. 6. Scalability of both batch and node recommendations.

Figure 6 shows that batch recommendations achieved better scalability performance compared with node recommendation. There is slightly change in the time (milliseconds) needed to generate recommendations with the increase in number of visits because the system use the top bubbled significant candidates as mentioned earlier.

4 Conclusion and Future Work

The suggested method consider users by their perceiving intellectual style on the website. We collected users' preferences without their demographic data, as well as all sessions are evaluated before maintaining it in an integrated routes. The suggested approach provides two different recommendations. Intellectual Node Recommendation (INR) that generate recommendations based on the active node; using the active node intellectual relationship with other maintained items in PISP. Intellectual Batch Recommendation (IBR) that generate a recommendation based on the user online maximal session. The experiments showed that batch recommendation achieved better performance in terms of coverage, precision and scalability compared with the node recommendations. In the future, we want to measure the impact of using semantics besides the users' intellectual style on the system performance.

References

1. Hafshejani, Z.Y., Kaedi, M., Fatemi, A.: Improving sparsity and new user problems in collaborative filtering by clustering the personality factors. Electron. Commer. Res. **18**(4), 813–836 (2018)
2. Elmisery, A.M., Botvich, D.: An enhanced middleware for collaborative privacy in IPTV recommender services (2017). arXiv preprint arXiv:1711.07593
3. Sofos, J.T., Chow, L.M., Piepenbrink, D.J.: US Patent No. 9,172,482. US Patent and Trademark Office, Washington, DC (2015)
4. Friedman, A., Knijnenburg, B.P., Vanhecke, K., Martens, L., Berkovsky, S.: Privacy aspects of recommender systems. In: Recommender Systems Handbook, pp. 649–688. Springer, Boston (2015)
5. David, S., Pinch, T.J.: Six degrees of reputation: the use and abuse of online review and recommendation systems (2005)
6. Hosea, D.F., Zimmerman, R.S., Rascon, A.P., Oddo, A.S., Thurston, N.: US Patent No. 7,979,880. US Patent and Trademark Office, Washington, DC (2011)
7. Sarwar, S., Hall, L.: Task based segmentation in personalising E-government services. In: Proceedings of the 31st British Computer Society Human Computer Interaction Conference, p. 9. BCS Learning & Development Ltd., July 2017
8. Harper, F.M., Xu, F., Kaur, H., Condiff, K., Chang, S., Terveen, L.: Putting users in control of their recommendations. In: Proceedings of the 9th ACM Conference on Recommender Systems, pp. 3–10. ACM, September 2015
9. Embarak, O.H.: A method for solving the cold start problem in recommendation systems. In: International Conference on Innovations in Information Technology, pp. 238–243 (2011)

10. Zhao, X.W., Guo, Y., He, Y., Jiang, H., Wu, Y., Li, X.: We know what you want to buy: a demographic-based system for product recommendation on microblogs. In: Proceedings of the 20th ACM SIGKDD International Conference on Knowledge Discovery and Data Mining, pp. 1935–1944. ACM, August 2014
11. Beel, J., Gipp, B., Langer, S., Breitinger, C.: Paper recommender systems: a literature survey. Int. J. Digit. Libr. **17**(4), 305–338 (2016)
12. Chen, L., Chen, G., Wang, F.: Recommender systems based on user reviews: the state of the art. User Model. User Adap. Inter. **25**(2), 99–154 (2015)

An Enhanced Knowledge Integration of Association Rules in the Privacy Preserved Distributed Environment to Identify the Exact Interesting Pattern

Sujni Paul(✉)

Department of Computer Information Science, Dubai Men's College,
Higher Colleges of Technology, Dubai, United Arab Emirates
spaul@hct.ac.ae

Abstract. Numerous research works are carried out in the field of data mining, especially in the areas of association rule mining, knowledge integration in the distributed data mining and privacy intense data mining. In the distributed data mining environment, the local data mining systems distributed across the environment. The way these local mining systems distributed in the environment, plays a major role in the process of knowledge integration. If all the local data mining systems are deployed in an organization, there will not be any impact. If the local data mining systems distributed across multiple organizations, that would cause a major impact in the process of knowledge integration. The problems are caused due to the privacy related issues and the agreement between those organizations. Though there are existing generic approaches to integrate the knowledge in the distributed mining, focus of this paper is to propose an enhanced algorithm specific to integration of association rules in the privacy protected distributed data mining environment and to find the interesting rules which are sub sets of an actual rule.

1 Introduction

Advancements in the computing and communications resulted in different kind of distributed computing environments. These environments composed of distributed data sources, computing nodes and the user community. The field of data mining deals with analyzing data sources and deriving the knowledge which can be used for different purposes. The advancements in the technology and the computing infrastructure forced the data mining technology to adapt to the distributed environment. The Distributed Data Mining (DMM) deals with mining of distributed data sources by paying careful attention to the distributed resources. In the distributed data mining infrastructure, there are distributed local data mining systems deployed. The knowledge gathered from these local systems are integrated and mined to infer the global knowledge. There are exiting approaches for integrating knowledge from the local data mining systems. Focus of this paper is, to propose an algorithm for knowledge integration in the privacy protected distributed data mining environment. The scope of this algorithm is limited to integration of association rules.

L. Barolli et al. (Eds.): EIDWT 2019, LNDECT 29, pp. 411–424, 2019.
https://doi.org/10.1007/978-3-030-12839-5_38

2 Related Works

Association rule mining is a well-researched method which is used to discover the relation between the variables. Given the distinct set of variables X and Y, X is associated to Y, if the existence of X implies the existence of Y. Apriori algorithm is a well-known algorithm based on association rule mining [1]. As the data to be mined grown tremendously, fast algorithms were developed to mine the association rules in the large data bases [3, 4]. In order to avoid the costlier process of candidate key generation in the Apriori, an extended prefix tree based FP growth tree algorithm was introduced [4]. [13] talks about the different types of association rules to be mined and how different privacy preserving techniques and algorithms for different levels of mining can be applied on them to protect the privacy of data with less information loss and high accuracy. Three parallel algorithms for mining association rules [5] have been briefed in this paper. Knowledge integration in parallel and distributed environment with association rule mining using XML data has been described in the paper [6]. Categories of patterns gained from the multi data base mining and the exceptional pattern identification from multi database mining has been explained in the paper [7]. Techniques for Mining Multiple Large Data Sources have been discussed in the paper [8]. Wu and Zhang have proposed Rule Synthesizing (RS) algorithm for synthesizing high-frequent association rules in multiple databases [9]. Discovering valid rules from the different sized data sources have been explained in [10]. Multi-Level Synthesis of Frequent Rules from Different Data-Sources has been explained in [11]. Modified algorithm for synthesizing high-frequency rules from different data sources has been discussed in [12].

3 Distributed Data Mining and Its Challenges

In Distributed Data Mining, multiple local data mining systems are deployed based on the location or volume of the data resources. Local data mining systems process and mine the local data and generate a local pattern. All the local patterns from the Local data mining systems are integrated together and mined to generate a unified global pattern. With respect to the association rule mining, local data mining systems generates local association rules based on the frequent items. The global association rule pattern is identified by integrating and mining the local patterns. The traditional algorithms designed for non-distributed data mining systems had been enhanced to adopt the distributed data mining environment. Researches carried out to find the optimal solution to integrate the knowledge from the local data mining systems and to find the useful patterns without distorting the knowledge. The renowned algorithms used in the data mining to find the frequent items are Apriori algorithm and the FP growth tree algorithm.

4 Knowledge Integration in Distributed Data Mining

Wu and Zhang [9] have proposed Rule Synthesizing (RS) algorithm for synthesizing high-frequent association rules from the multiple databases. Using this technique, every local database is mined separately at Random Order (RO) using a Single Data Mining Techniques (SDMT) to find the high-frequent association rules. Based on the association rules in the different databases, the authors have estimated weights of different databases. If $D_1, D_2 \ldots D_m$ are *m* different data sources from the branches of a large company of similar size2, and Si is set of association rules from $D_i (i = 1, 2, \ldots m)$. For given rule $X \rightarrow Y$, suppose $w_1, w_2 \ldots w_m$ are the weights of $D_1, D_2 \ldots D_m$, respectively, then the synthesizing model is defined as follows

$$supp_w(X\,U\,Y) = .w_1 * supp_1(X\,U\,Y) + w_2 * supp_2(X\,U\,Y) \ldots w_m * supp_m(X\,U\,Y) \quad (1)$$

$$conf_w(X \rightarrow Y) = .w_1 * conf_1(X \rightarrow Y) + w_2 * conf_2(X \rightarrow Y) \ldots w_m * conf_m(X \rightarrow Y) \quad (2)$$

Where $supp_w$ (Rule) is the support of rule after synthesizing, $conf_w$ (rule) is the confidence of rule after synthesizing; $supp_i$ (rule) is the support of rule in data source Di. The weight of the rule is calculated as

$$w\,Ri = \frac{Num(Ri)}{\sum_{j=1}^{n} Num(Rj)} \quad (3)$$

Where i = 1, 2 …n. *Num (R)* is the number of data sources that contain rule R, or the frequency of R in S. Then the weight of the data source is calculated as follows

$$w\,Di = \frac{\sum_{Rk \in Si} Num(Rk) * wRk}{\sum_{j=1}^{m} Num(Rh) * wRh} \quad (4)$$

This approach is designed considering all data sources in an organization have equal power to vote for a pattern. *Ramkumar and Srinivasan* proposed “Modified Algorithm for synthesizing high frequency rules from different data sources” [12]. In the pure business case, all the branches are not equal; the branches with the higher volume of transaction should have higher power in determining the global pattern. So they proposed an approach based on the transaction population.

Example: A company having four branches at metropolises and 16 branches located at urban areas is considered. Let the volume of transactions in the metropolitan branches be ten times the volume of transactions in the urban branches. Suppose the metropolitan branches support rule R1, while the urban branches support rule R2.

$$\begin{aligned} &\text{Weight of rule R1} = 4 \times 10 = 40 \\ &\text{Weight of rule R2} = 16 \times 1 = 16 \\ &\text{Normalized weight of R1} = 40/56 = 0.7143 \\ &\text{Normalized weight of R2} = 16/56 = 0.2857 \end{aligned}$$

Nedunchezhian and Anbumani proposed "Post Mining- Discovering Valid Rules from Different Sized Data Sources" [10]. They came up with another weight based modal considering the approach proposed in [9, 12], in this approach the net weight of the data source is calculated as follows. The first weight of the data source is calculated as:

$$w\,Di = \frac{\sum_{Rk \in Si} Num(Rk) * wRk}{\sum_{j=1}^{m} Num(Rh) * wRh} \tag{5}$$

The second weight of the data source is calculated as:

$$WDi = \frac{Size(Di)}{\sum_{i=1}^{n} Size(Di)} \tag{6}$$

The Net weight of the data source is calculated as:

$$Net\,weight = \frac{(wDi + WDi)}{2} \tag{7}$$

Adhikari, Ramachandrarao, Prasad, and Adhikari proposed a modal for synthesizing global patterns from the local patterns in different databases and an approach using pipelined *feedback technique* for mining multiple large data sources [8]. The framework has layers and interfaces with defined operations. There are four interfaces; interface 2/1 performs operations to produce a processed data base. Interface 3/2 applies filters, so that interested data is separated from the outlier data. Interface 4/3 mines the local data base and identifies local patterns and suggested patterns. Suggested patterns are very close to interesting patterns but failed to qualify as an interesting pattern. Interface 4/5 synthesis the global patterns (Fig. 1).

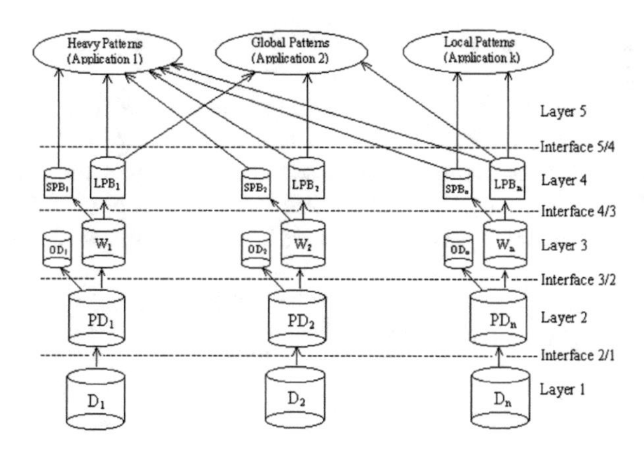

Fig. 1. Synthesizing global patterns from local patterns in different data bases

Chengqi Zhang□, *Meiling Liu_, Wenlong Nie_, and Shichao Zhang* proposed approaches to find the Identifying Global Exceptional Patterns in Multi-database Mining system. As part of this, the approach for finding the high voting pattern and global pattern have been proposed [7].

5 Problem Definition

In the distributed or multi database mining, mining happens in the local data mining systems using the Single Data Mining Techniques (SDMT). The local patterns gathered from the local systems are integrated and the unified patterns (high-voting pattern, global exception pattern, suggesting pattern) are identified using one of the approach explained in the earlier Sect. 4 “Knowledge Integration in Distributed Data Mining”. Though those approaches are being used prevalently, there are few aspects with respect to association rule mining make them little inefficient. For example, Let us assume, the following patterns are from two different local data mining systems DM1 and DM2. Assume both the DMs have equal weightage and the rule from the DM2 has more support value.

$$\text{DM1} : \text{Bread}^\wedge\text{Milk} - > \text{Butter}[\text{support} = 55\%] \quad (8)$$

$$\text{DM2} : \text{Bread}^\wedge\text{Milk} - > \text{Butter}^\wedge\text{Cheese}[\text{support} = 60\%] \quad (9)$$

Most of the existing approaches work on the pattern as a whole. So in this example, the outcome would be, the pattern from the DM2 considered to be a high voting pattern. But in the logical view, the pattern from the DM1 should have been the high-voting pattern, as it has 60% support from pattern (9) and 55% support from pattern (8).

For understanding, this is explained with an example using the approach proposed by Zhang et al. [7].

Given n databases $D_1, D_2 \ldots\ldots D_n$, they represent the databases from n branches of a large company. Let $LP_1, LP_2 \ldots\ldots LP_n$ be the corresponding local patterns which are mined from every database; And $minsupp_I$ be the user specified minimal support in the database $D_i (i = 1, 2, \ldots\ldots n)$; For each pattern P, its support in D_i is denoted by $Supp_i(P)$. We define the global support of a pattern in the databases as follows

$$Supp_g(P) = \frac{\sum_{i=1}^{Num(p)} \frac{Supp_{i(P)} - minsupi}{1 - minsupi}}{Num(P)} \tag{10}$$

6 Problem Statement

Most of the existing approaches on finding the global patterns in the distributed data mining environment, considers each local pattern as an entity and the global pattern is calculated based on the support count or the confidence of that pattern. If the global high voting pattern happens to be the subset of other patterns, then they are not considered and ignored. And few other approaches demands the data set to be available throughout the mining process, which is not possible in the case of privacy protected environment. And none of the approaches gives a structured way of storing the patterns retrieved from the local data mining systems. Sorting them in a structured way will reveal more interesting patterns. Considering these constraints, this new algorithm is proposed.

7 Proposed Methodology to Find the Exact Frequent Pattern

This algorithm considers each elements of an association rule, and stores the association rule in the association rule form itself. So that lot of knowledge can be inferred. Also, it provides an approach for calculating high voting pattern. The high voting pattern is calculated by keeping the antecedent of the rule intact, and for that fixed antecedent, it analyses its consequents to see, if any subset of the consequent can qualify as a frequent pattern, than the super set of the consequent.

7.1 Identifying Common Data Structure

The intention of this approach is to find the global frequent pattern which can also be a subset of the association rule. So, the data structure should be able to maintain the relationship between the elements in the association rule. Also, the data structure should also be flexible enough to make modification when required. And the data structure should be traversable to find different patterns. It should provide efficient ways to serve the purpose of merging or integration of the rules. Considering all the requirements and the observing the flexibilities provided by the tree data structure in FP growth tree algorithm, it is decided to use the tree structure for this purpose.

Like all the traditional tree structure, this tree starts with the root node, which is the parent of the tree. In a common association rule, there are two types of elements, the elements which identifying the actual data and the elements which are identifying the relationship between the data. Example of the later type of elements are '^' (AND), '→' (Implies). So, the nodes in a tree can contain two types of elements; they are Data Element (DE) or Operator Element (OE). For example, in the association rule, "*Bread ^ Milk - > Butter*", Bread, Milk and Butter are qualified as DE and the operators '^' and "→" are qualified as OE. Each node, along with one of the DE or PE, it also stores some attributes, for now, there are two attribute have been identified, that is End of Record (EOR) attribute which marks the end of the each association rule in the common tree structure. And the '*Support'* attributes which keeps track of the support of each node in the tree. The *support* used might not be a support value directly received from the local DM; this could be a value which is arrived after applying any of the existing syntheses algorithms also.

7.2 Plotting Association Rules into the Tree Structure

In order to plot an association rule into the tree structure, a given rule should be preprocessed to make the rule plot able and good for integration. The elements in the both antecedent and consequent of the rule should be separately sorted in a specific order, so that, all the rules which are being plotted will have a uniform order, which would make the integration of rules more efficient. Let us look at an example, how following rules can be plotted on the tree structure.

$$\text{Milk}^\wedge\text{Bread} - > \text{ Butter } [\text{support} = 55\%] \tag{11}$$

$$\text{Milk}^\wedge\text{Bread} - > \text{ Cheese}^\wedge\text{Butter } [\text{support} = 60\%] \tag{12}$$

Based on the nature of the element, it is determined that, the elements in the rules to be sorted on the alphabetical order. On this, the rules are preprocessed and converted as shown below:

$$\text{Bread}^\wedge\text{Milk} - > \text{ Butter } [\text{support} = 55\%] \tag{13}$$

$$\text{Bread}^\wedge\text{Milk} - > \text{ Butter}^\wedge\text{Cheese } [\text{support} = 60\%] \tag{14}$$

After plotting the first rule, the second rule is taken; the first DE of the second rule is 'Bread'. As it is the beginning of the second rule, the active node is a root node. The active node is examined to see, if any of its children has the DE 'Bread', as the node is already present, only the support of that node is added with the support of the rule (14). The support of the node becomes 115, which is addition of the support of the rule (14) with the existing support. Exiting support is 55 and the support of rule (14) is 60, hence it becomes 115. Likewise, all the elements of the rule (14) are plotted into the tree, which is shown in the Fig. 2.

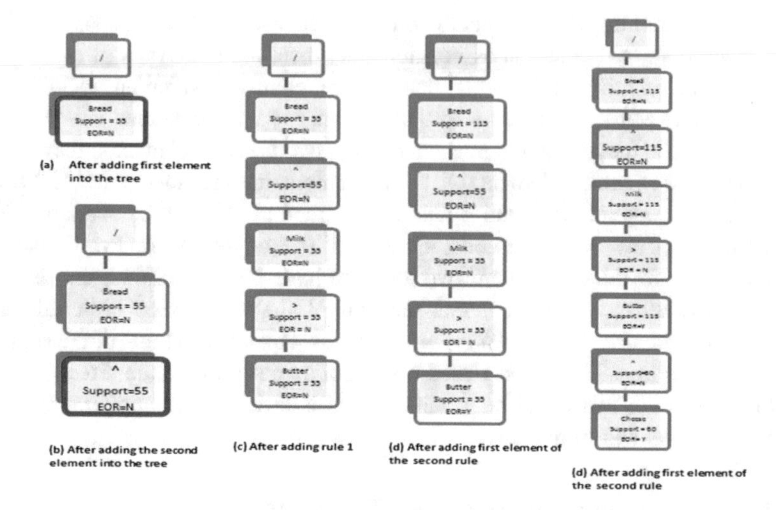

Fig. 2. Tree construction

7.3 Identifying the Global High Voting Pattern

After plotting all the rules, onto the common data structure, the high voting pattern should be identified using the tree structure. The new data structure introduced here is 'Frequency Table' which is a simple table with two columns, first column is the pattern which is inferred from the tree and the second column is the 'global Support' of that pattern. The following example will help to get the better understanding of this approach. The association rules retrieved from the different local DMs have been given in the following Table 1.

Table 1. Association rules from local D

Rule	Support
A^B -> C^D	0.73
A^B -> C	0.375
C^D -> F	0.355
A^B -> G	0.269
B^C -> E	0.236
A^B -> E^F	0.234
B^C -> F	0.222

Table 2. Support count after applying the algorithm

Rule	Global support
A^B > C	1.105
A^B > C^D	0.73
C^D > F	0.355
A^B > G	0.269
B^C > E	0.236
A^B > E^F	0.234
B^C > F	0.222

After applying the newly proposed algorithm, the tree structure will look like as shown in the Fig. 3.

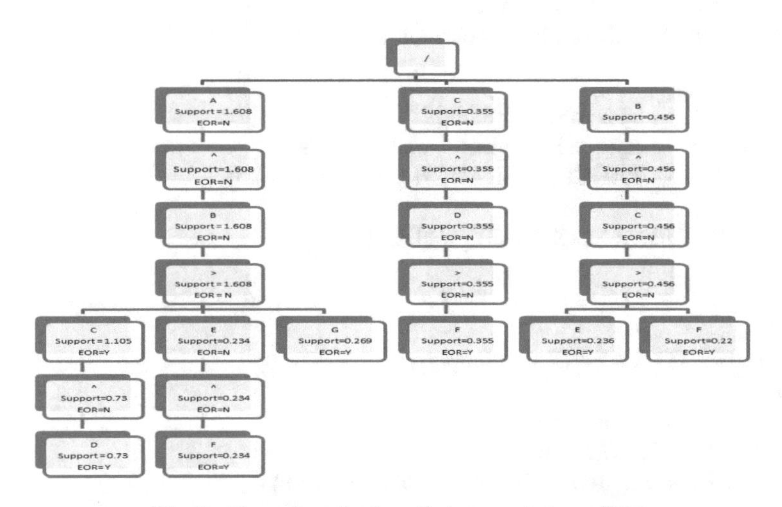

Fig. 3. Tree after plotting all the association rules

Based on the constructed tree, the frequency table obtained will look like given in the Table 2. Which identifies the pattern ‘A^B > C’ as a frequent pattern. As this pattern is frequent in one of the data base, and also subset of other frequent patterns from different local different data mining systems; the complete algorithm for the same has been shown below.

The proposed novel algorithm for identifying the exact interesting pattern is:

```
INPUT: Association rules from Local Databases.
Rule Base RB ← {LR1,LR2,….LRn}
      LRx ←   {AR1, AR2, AR3...ARn};
         ARn ←association Rule with corresponding support count.

OUTPUT: List of association rule with highest voting to least.

BEGIN
          // Tree construction
          FOR EACH (Rule in    RB )
         {
                  ADD-RULE-TO-TREE (Rule);
         }
         // get the support count for each of the branch

         current_node = root_node
          FOR EACH (child    of current_node)

         {
                   new_rule    ← child ;
                   new_rule.support ←child.support;
                   PROCESS-BRANCH (new_rule, child);
          }
          RETURN FREQUENCY_TABLE;

END

ADD-RULE-TO-TREE (Rule ,Support)
BEGIN
      current_tree_Node ← root node;
       FOR EACH (item in Rule)
         IF ( current_tree_node.child    contains item )
                  current_tree_node ← child_node;
                  child_node.support ← child_node.support + Support
         ELSE
                  INSTANTIATE node new_node ;
                  new_node.itemID ← item;
                  new_node.support ← Support;
                  current_tree_node ← new_node;

         IF   last_element_of_rule
                  set EOR ←True
END


PROCESS-BRANCH(rule, child)
BEGIN
```

```
    FOREACH (child of child)
    {
            new_rule U = child ;
            new_rule.support = child.support;
            processBranch(new_rule,child)
    }

    IF child. EOR = Y
    {
         Add the rule with the support count into the frequency table.
    }
END
```

8 Experimental Results and Evaluation

As this algorithm is a new form of this kind, there are no comparative algorithms to evaluate the current algorithm. So the current algorithm has been evaluated by giving different number of inputs. The evaluation chart is given in the following charts.

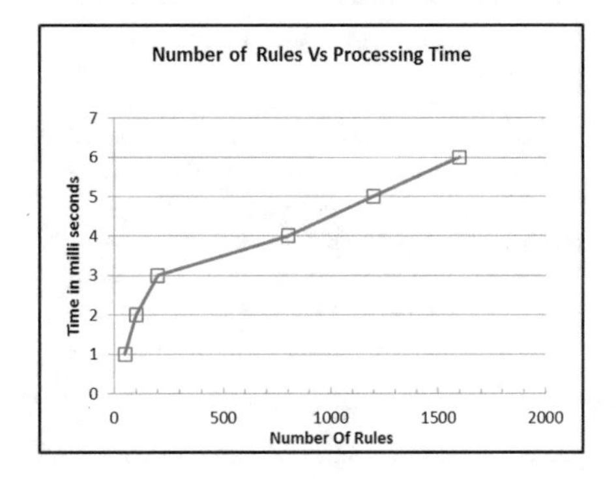

Fig. 4. Rules with different combination of elements

In the Fig. 4, the graph has been plotted for number of association rules against the time taken by the proposed algorithm.

The test is performed by having only fewer range of elements in the association rule, and it is observed that when there are fewer range of elements in the association rule, the time taken for processing is less, as the creation of new tree nodes are less. This has been plotted in the graph shown in the Fig. 5.

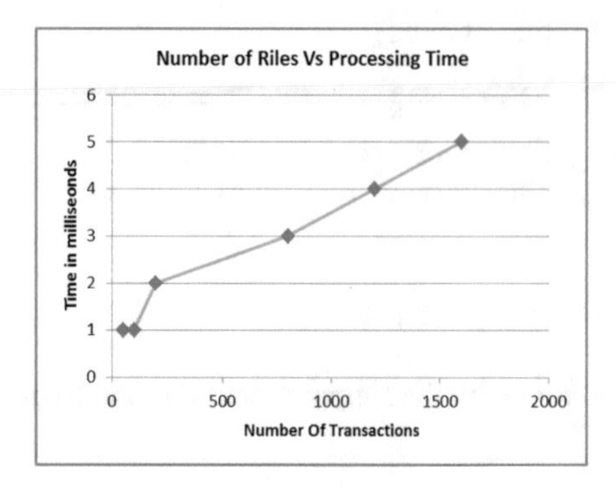

Fig. 5. Rules with same combination of elements

The tests are performed by having different number of elements in the antecedent and consequent of the rule. It is observed that when the number of elements in the rules increases, the processing time increases. This has been plotted in the graph shown in the Figs. 6 and 7.

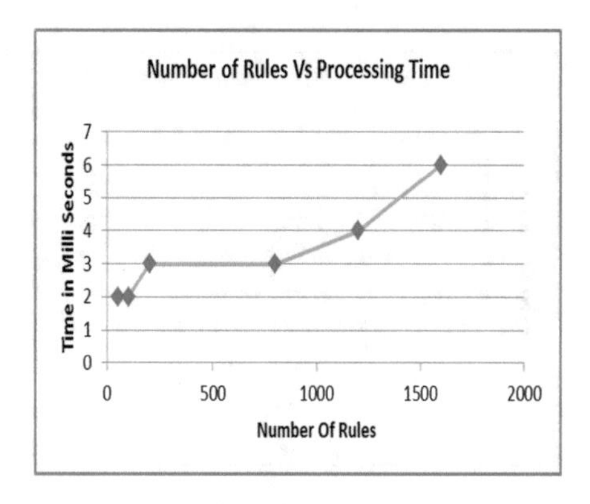

Fig. 6. Rules with 3 items in both antecedent and consequent

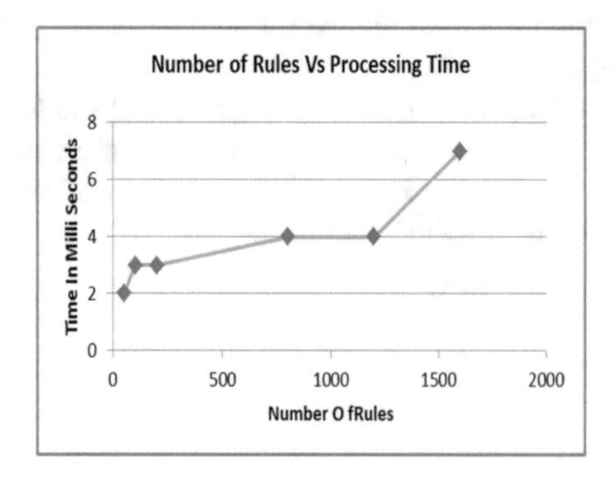

Fig. 7. Rules with 4 items in both antecedent and consequent

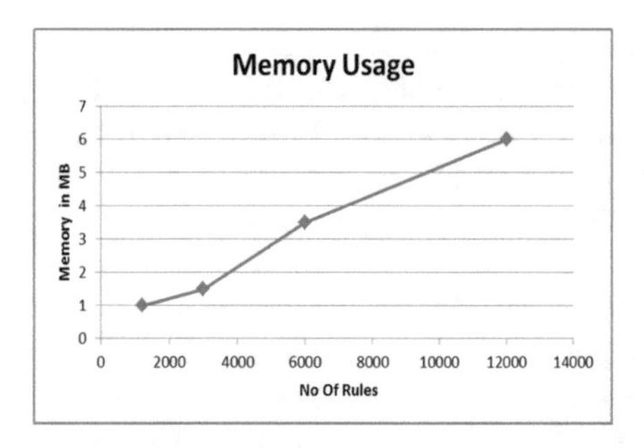

Fig. 8. Number of rules vs. memory conception

From this, it is observed that, when the range of elements in the antecedent and consequent of the association rules are fewer, then the time taken by the algorithm is less and when large variations of elements are part of the association rule, then the algorithm takes little more time (Fig. 8).

On the memory management part, it is observed that memory conception increases when the range of items in the association rules are more. But if the same items are repetitively used, then the memory conception is very less as the new nodes are not introduced, instead, same nodes support is updated accordingly.

9 Conclusion

The overall process of mining the association rules in the distributed environment using the proposed algorithm is shown in the Fig. 9.

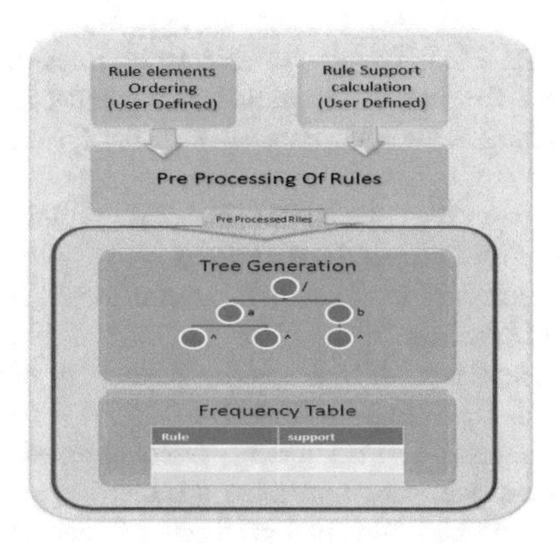

Fig. 9. Overall mining-processes using the proposed algorithm

The portion highlighted in red border is the place where the proposed algorithm fits in. This algorithm can work in combination with exiting rule synthesizing algorithms. Also it provides options for the end user to use their own ordering mechanism and support count calculation during the pre-processing stage. As the association rules are structured and stored, these stored rules can be reused to infer more knowledge out of this.

This algorithm is a first step towards the mining association rules in the association rule form itself. This algorithm will help to identify the exact high voting pattern in the association rules gathered from the different local data mining systems in the DDM. By keeping the association rules in the common structure, more knowledge can be inferred from this. This algorithm can be used directly to integrate the association rules, or it can be applied after the high voting patter identified using the existing algorithms. More investigations will be carried out to gather more knowledge using the common tree structure. The tree nodes are designed to accommodate more attributes, so they can be used to identify more knowledge out of the tree structure.

References

1. Agrawal, R., Imielinski, T., Swami, A.N.: Mining association rules between sets of items in large databases. In: Proceedings of the 1993 ACM SIGMOD International Conference on Management of Data, pp. 207–216 (1993)
2. Agrawal, R., Srikant, R.: Fast algorithms for mining association rules. In: Proceedings of 20th International Conference on Very Large Databases (VLDB 1994, Santiago de Chile), pp. 487–499. Morgan Kaufmann, San Mateo (1994)
3. Margahny, M.H., Mitwaly, A.A.: Fast algorithms for mining association rules. In: AIML 05 Conference, CICC, Cairo, Egypt, 19–21 December 2005 (2005)
4. An implementation of the FP-growth Algorithm Christian Borgelt Workshop Open Source Data Mining Software (OSDM 2005, Chicago, IL), pp. 1–5. ACM Press, New York (2005)
5. Agrawal, R., Shafer, J.C.: Parallel mining of association rules. Distrib. Syst. Online (2004)
6. Paul, S., Saravanan, V.: Knowledge integration in a parallel and distributed environment with association rule mining using XML data. IJCSNS Int. J. Comput. Sci. Netw. Secur. **8**(5) (2008)
7. Zhang, C., Liu, M., Nie, W., Zhang, S.: Identifying global exceptional patterns in multidatabase mining. IEEE Comput. Intell. (2010)
8. Adhikari, A., Ramachandrarao, P., Prasad, B., Adhikari, J.: Mining multiple large data sources. Int. Arab J. Inf. Technol. **7**(3) (2010)
9. Wu, X., Zhang, S.: Synthesizing high-frequency rules from different data sources. IEEE Trans. Knowl. Data Eng. **15**(2), 353–367 (2003)
10. Nedunchezhian, R., Anbumani, K.: Post mining - discovering valid rules from different sized data sources. World Acad. Sci. Eng. Technol. **7** (2007)
11. Ramkumar, T., Srinivasan, R.: Multi-level synthesis of frequent rules from different data-sources. Int. J. Comput. Theory Eng. **2**(2) (2010)
12. Ramkumar, T., Srinivasan, R.: Modified algorithm for synthesizing high frequency rules from different data sources. Knowl. Inf. Syst. **17**(3), 313–334 (2008)
13. Panchal, M.C., Scholar, P.G.: Privacy preserving of association rule mining: a review. Int. J. Innov. Adv. Comput. Sci. IJIACS **4**(Special Issue), September 2015. ISSN 2347 – 8616

Application of Fog and Cloud Computing for Patient's Data in the Internet of Things

Soulat Waheed[1(✉)] and Peer A. Shah[2]

[1] Department of Computer Science, COMSATS University Islamabad, Attock, Pakistan
soulatadmob@gmail.com

[2] Computer and Information Science Division, Higher Colleges of Technology, Fujairah Women's College, Fujairah, United Arab Emirates
pshah@hct.ac.ae

Abstract. The last few years have brought a sudden boost in the Internet of Things (IoT). It is considered to be the next big thing in the evolution of the Internet and an integral part of the future Internet. IoT devices that have their own storage and processing capabilities can process and store data at their end. However, the devices which don't have storage and processing resources, like sensors attached to the patient's body, collect data from the physical environment and send to some sink for processing and storage. Such sensors generate a huge amount of data, so there is a need to process and store the data efficiently. However, the cloud computing which is used as a platform for IoT has an inherent problem of latency which can cause bad monitoring and patients which need an immediate treatment can be affected. This problem can be considered in every latency sensitive application which requires real-time monitoring and processing. To solve such problems, we need a new platform for IoT related data which offers the same services as a cloud but do not have problems like a cloud. This study proposes a new solution for IoT patient's data which utilizes an intermediate layer, fog computing with cloud computing, and accelerates the awareness and response to events by removing a round trip delay to the cloud for analysis. It also offloads the gigabytes of network traffic from the core network to the local edge fog network. This work also proposes how energy efficient sensing will be done. Implementation based analysis is performed to demonstrate the performance of the proposed solution with existing solutions. Results show reduction of the delay and energy efficient sensing.

1 Introduction

The next generation of computing is not like traditional computing, because with the Internet of Things (IoT) every object is connected with one another and also with the network [1]. Radio Frequency IDentification (RFID) and sensor networks have to upgrade their communication technologies so that they will be able to compete with the challenges that IoT has and a huge amount of data traffic that it generates. Processing and storing that enormous data so that it can be presented in a seamless and efficient manner requires a platform which is known as cloud computing [2]. Cloud computing provides the resources of the storage and processing and one can easily access that data

L. Barolli et al. (Eds.): EIDWT 2019, LNDECT 29, pp. 425–436, 2019.
https://doi.org/10.1007/978-3-030-12839-5_39

and can use computing and processing skill from on demand and from anywhere. It is a technology which promises next-generation data centers and it also acts as a receiver for so many devices. It collects data from devices, analyzes the data and interprets it accordingly [3]. It also hides detail from the user and also runs in the background, so user doesn't have knowledge of how thing have been done.

In IoT, everything has to be connected with the Internet. It is possible nowadays with 4G network and Wi-Fi devices are available everywhere but to make IoT really an integral part of future Internet not only the devices which have storage and processing capabilities should be connected to the Internet but also devices and the objects which don't have storage and processing should be a part of the future Internet and IoT [4].

In the future of the Internet, in which evolution of the Internet will be seen, different objects will sense the information with the help of the sensors and then they will really interact with the physical environment and provide a different type of services. In 2011, numbers of connected devices have exceeded the number of people living on the earth, so the numbers of the connected devices are increasing day by day. Now, there are almost 9 billion devices which are connected and are expected to rise about to 24 billion in 2024 [5].

Data management and information processing is a significant topic in IoT related applications. Cloud computing allows to send all the data to the cloud for computing and storage which results in high traffic, an enormous amount of data allows the cloud to hide a lot of specification detail from the user. This bliss becomes a problem for latency-sensitive applications which require nodes in the vicinity to meet their delay requirements. Applications which require high efficiency with respect to delay, the cloud cannot play an active role in that. It also does not support a large number of nodes which is also very important due to the geographical distribution of sensors.

Applications like health care which require very low and predictable latency, should not be using cloud service because cloud hides a lot of the detail from the user and such application requires very low latency become victim of cloud computing. Sensitive applications which require real-time monitoring, simply cannot use cloud computing due to the Inherent problem of cloud which is latency. In the scenario of health care applications, as it is stated, a lot of data is generated by a body sensor and by sending all the data for computing and storage to the cloud will be expensive and also have a reasonable security issue. Patients health is very important and sometimes patient needs really an immediate treatment but by sending that data to cloud for processing creates an extra loop which results in bad monitoring of the patient and real-time processing becomes almost impossible. As delay involved in transmitting the patient data to cloud for processing and receiving the response can be high which will result in bad monitoring of patients. This delay can be reduced by using some local processing. As in the case of body area network, the sensors have very low processing capabilities, hence creating a load of local processing at sensors can result in poor performance. The best way to resolve this problem is to use a new layer between sensors and cloud that should perform local processing for critical sensor data and global processing should be performed by the cloud for noncritical data. There is also a need for a platform for building blocks of IoT, smart communication with real-time, low latency, location awareness and support of a large number of nodes which is fog computing.

Fog computing is a distributed approach in which some application servers are handled at the network edge and some of them are handled at the data center which is cloud. The purpose of fog computing is to improve efficiency and lesser use of the cloud services which are storage and processing and these services can be provided on the network edge and we will be able to overcome the delay which is inherent in the cloud models. Figure 1 shows the overall architecture of IoT with cloud computing and fog computing.

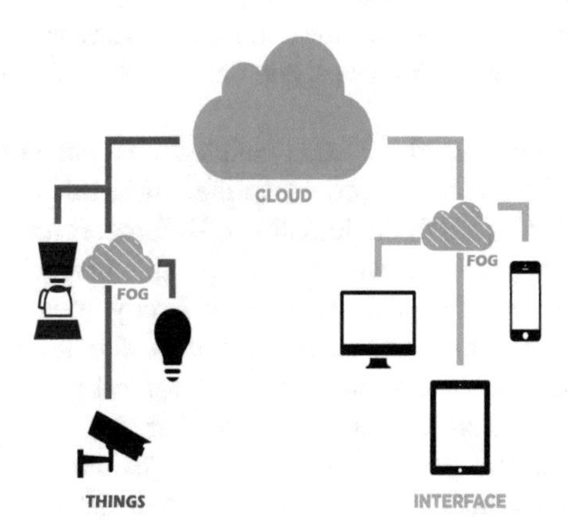

Fig. 1. IoT basic architecture with cloud and fog

The key contributions of this study are as follows:

(i) The development of a technique for IoT patient's data and its processing and management using fog and cloud.
(ii) This research helps in reducing the delay involved in cloud models.
(iii) Efficient sensing with minimized energy consumption.

2 Related Work

Capitalizing on the IoT requires a new kind of infrastructure. Today's models are not designed for the volume, variety, and velocity of data that the IoT generates. Billions of previously unconnected devices are generating more than two Exabytes of data each day [6]. An estimated 50 billion "things" will be connected to the Internet by 2020.

Work has been done in literature to solve the problems in which most of them deal the data management on the devices e.g. (Storage cards) or utilizing immediate nodes i.e. mobile phone and also storing data on computer nodes. Some of the well-known techniques from the literature are discussed below.

Cloud-based system that manages sensors data [7]. Wearable sensors collect bio signals from the user (like heart rate, ECG, oxygen, and temperature), and contextual data (like location, Activity status, etc.). Depending on the wireless technology used, this data can be forwarded directly to the Cloud infrastructure utilizing established techniques for IoT Communication. In this system wearable and mobile sensors that acquire patient bio signals, motion, and contextual information. The sensor gateway that collects all the signal from the sensor and forwards them to the cloud platform.

Cloud burst [8] is an algorithm which uses parallel mapping technique for analysis of different biological data it allows the user to connect different computer nodes with each other and run those nodes. It uses open source Hadoop implementation.

Cloud4SNP [9] is a technique for bio informatics and it is completely cloud-based. It evaluates the statistical results of microarray data.

CloudMan [10] enables separate bioinformatics researchers to easily deploy, customize, and share their entire cloud analysis environment, including data, tools, and configurations.

Doukas et al. [11] introduced a system which stores patient's data. It is also been used for processing of the data and provides different health services upon that.

Microsoft health vault [12] is the initiative of Microsoft and its purpose is to save and manage the data regarding a patient's health. This program also helps people to collect and share their medical data with their family members and doctors also. Microsoft health vault also deals and ensure with the safety and privacy. One can have access to this system through social media and can control it himself by logging in and can select the people to whom the data could be shared.

Fusion [13] is also a type of cloud system, which maintains the patient's health in a very easy and appropriate method. It also enables you to share your problems with your doctor and is very cheap also. This also provides safety and confidentiality to medical data. Fusion has held over the whole cloud which was encrypted and not has the ability to decrypt. In this way, it controls the data safety. Fusions also help you so that you can combine and share the different systems.

Bombino Gesu [14] is an Italian hospital located in Rome, which has a good reputation in research and treatment in the field of pediatrics. This hospital uses cloud system which gives fruitful results. One of the benefits of using this system is the mutual understanding amongst the staff and good relation and understanding with the patients.

Panday et al. [15] proposed a new cloud environment named Autonomic cloud environment which collects patient's data and deposit and save it in a cloud-based repository. Under this scheme, patient's can be monitored by joining mobile phone and cloud technology. The user connects the sensor with their body and runs an application on the mobile, the data is transferred to the cloud, which gives the results graphically.

Hsieh [16] developed an application which is a telemedicine application based on the cloud. It uploads in 12-lead ECG reading cloud and with its help, we can see the results of ECG on mobile. This application is very good for heart patients.

Fortino [17] introduces a body cloud which monitors the body sensor data in the cloud. Body sensor network and data collected from it managed very efficiently by this body cloud and provide different services accordingly.

Voice over IP VOIP [18] is a service which has been started for the diabetes patients. With the help of this service a patient can send a pre-recorded message to the cloud by using VOIP and in result, different health reminders are send to the patients.

Cloud-based health care service [19]. It monitors very important parameters such as blood sugar glucose and ECG. It collects data from the body sensors and with the help of Bluetooth that data is sent to the mobile device and with a mobile device that data is forwarded to the cloud for more processing.

Kim et al. proposes [20] a cloud-based system in which patients vital signals are monitored. It has 4 modules.

In these systems, authors have focused on the cloud computing platform in very sensitive applications which requires real-time monitoring but there are issues unsolved due to the Inherent problem of cloud computing which is latency. In the scenario of health care applications as it is stated a lot of data is generated by a body sensor and by sending all the data for computing and storage to the cloud will be Expensive and also have a reasonable security issue.

The above research has shown that there is a need for a platform for building blocks of IoT with real-time, low latency, location awareness and support of a large number of nodes.

3 System Working

In this section, we present the architecture of the application of fog and cloud computing with body area network. Different components of the system and their functionalities are also discussed in detail. Data gathering using body sensors, fog local processing, and cloud global processing are presented in detail.

3.1 Design Challenges

Patient's health is very important and body sensors are producing very critical data that needed to be processed efficiently so that the patient is provided with immediate treatment without any delay. Following are the design challenges for the new system:

a. To make sure only critical and real-time data is sent to the fog and non-critical to be stored in the cloud.
b. Effective use of body sensors.
c. Data sharing among fog nodes and their data integrity.

3.2 System Model

To address design challenges discussed in the previous section, this section proposes a scheme of fog and cloud-based architecture which is a layered architecture in which system has been divided into two layers: fog layer and the cloud layer. Figure 2 shows the component diagram which is the basic architecture of the proposed system. Local processing will be done on fog (Intelligent Controller) and cloud (Data Center) will be used for global processing.

3.3 System Working

In 2011 numbers of connected devices have exceeded the number of the people living on the earth, and the numbers of the connected devices are increasing day by day now

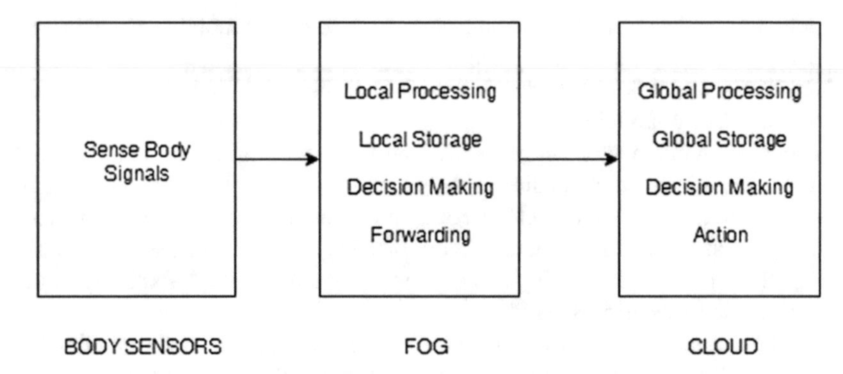

Fig. 2. Component diagram for the proposed architecture

there are almost 9 billion devices which are connected and are expected to rise about 24 billion in 2024 [5].

These are so many devices which generate a huge amount of data and storing that data temporarily is not possible. For that, there is a need for rental storage and we have to process that data in such a way that it becomes known for everyone and user. For that, we need a lot of processing which is not possible at the IoT end. Hence, we need rental processing which can be provided with the help of fog computing and cloud computing.

Analyzing patient's IoT data and reacting on delay sensitive and emergency related data also become more effective. Patients body sensors collect data and send to the fog network by means of any access network i.e. 4G, Wi-Fi or GPRS and these are processed locally, and results are sent to the cloud and any one can access the result in the form of the application which in our case is smart health care center on their respective devices. Figure 3 shows the overall communication pattern of the proposed system.

Cloud computing hides the specification of many details from the user and the user do not know about how storage, processing, and computation have been done. This bliss becomes a problem with latency sensitive application which needs very low delay and wants a fast response from the server due to the nature of their service which they are providing.

Fog computing extends the cloud computing services to the edge of the network. Also, fog and cloud share the same resources (storage, processing, computing) and they also have almost the same functions and attributes. Fog computing is the best model because of its efficiency and tailor-made for the latency sensitive applications and also provide the temporary storage and processing also filter the data which should be sent to the cloud.

Keeping in view this entire thing, now we present more detailed, layered architecture of our system. It is a 6 - layer architecture. These layers are described in Fig. 4.

In the physical and virtualization layer, physical nodes, Wireless sensor networks, virtual nodes, and sensor networks are managed and maintained according to the needs of the system. A network of body sensors is managed which have been placed at the bodies of the patient and constantly sensing the patient condition.

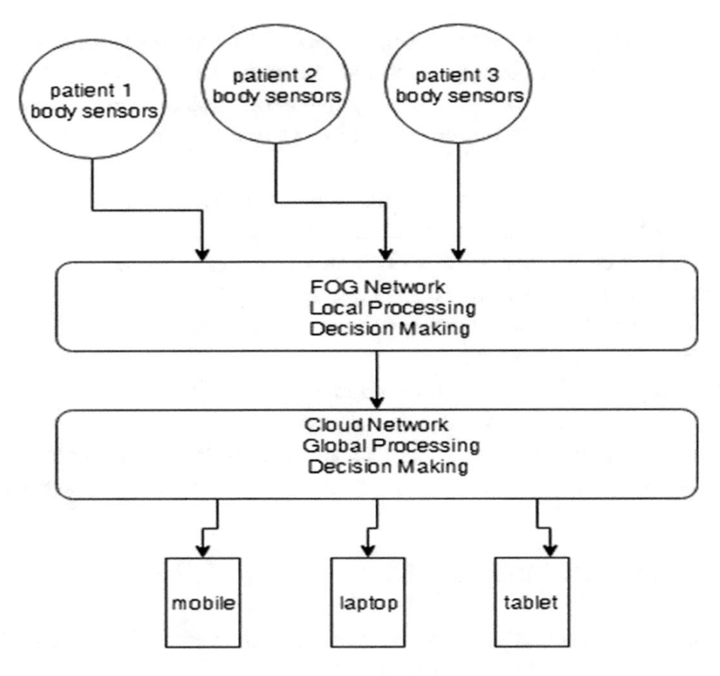

Fig. 3. Overall communication pattern

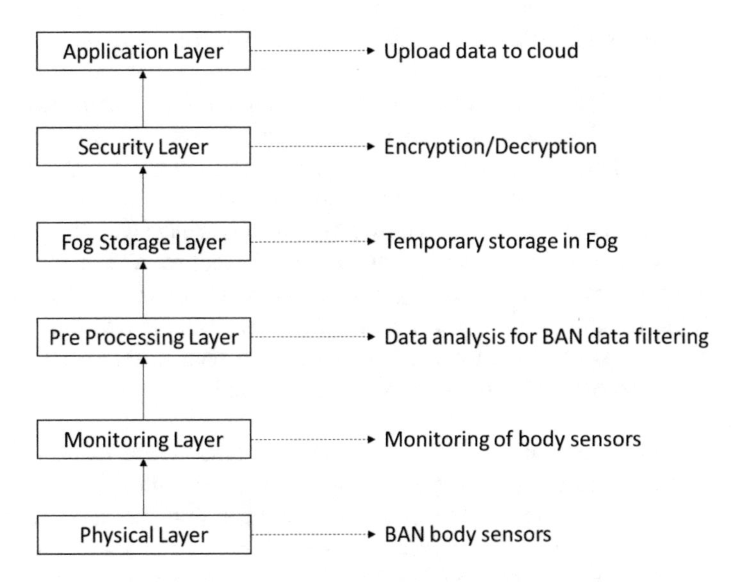

Fig. 4. The detailed layered architecture of the system

Monitoring layer is a second layer in which monitoring of the different component of the system is done. It monitors the activities of the underlying nodes and networks. Monitoring of the body sensors, power monitoring which node is performing what task at what time, and what is required from it next is monitored here. So that node is monitored based on their energy level and constrained of energy consumption and effective measures can be taken. It is very important because sensors have very low power and bad monitoring of sensor results in loss of power which result is bad monitoring of patients also.

Pre-processing layer performs data management related tasks. It analyzes the collected data, perform data filtering and in the end, more meaningful and necessary data is generated. Fog computing work is done on pre processing layer.

Temporary storage layer stores this data temporarily on fog resources. Once the data is uploaded to the cloud there is no more need to store that locally, that data is removed from the local resource.

Security layer provides the security of the patient data because of the privacy of the patient. This is why the security layer is so important.

Now the transport layer sends this relevant data to the cloud, burdening the core network to a minimum and allowing the cloud to create more useful services.

4 Implementation and Results

To evaluate the proposed system, testbed implementation is performed on HP Machines with core i3 processor installed RAM of 4 GB having hard disk of 500 GB and 64-bit windows 8 operating system.

Many machines are configured in the test bed that composes a complete network. Machines are performing the duty of Fog and Cloud respectively and some of the machines are connected with sensors. An application is developed at the sensing machine that sends data after a periodic time interval to the fog machine where data is processed and stored temporarily. After processing of data, the decision is taken whether this data should be sent to the cloud or it is processed back to sensors. Data sent to the cloud are stored permanently for future use and notifications.

For the implementation of our system, we used Java language with Eclipse as a tool for the development. Client-Server programs are developed. Client programs are installed on sensor machines and on fog machine both client and server programs are installed. Server program is used to get data and processed data sent from the sensors, and a client program on Fog machine is used to send data to the cloud for further processing and storage of data for permanent basis and only server program is installed on cloud machine.

Figure 5 shows the results of delay computed for critical data of patients in two different scenarios: one with the proposed system and other with the existing system.

The program was executed 10 times and for each iteration, the delay is computed for the proposed idea and existing idea and the results are plotted in Fig. 5. These results clearly show at each iteration that the delay for the proposed idea when sensing is done with fog is low as compared to the existing solution when sensing is done without fog computing. The variation in delay at each iteration is due to the different size of data being sent by the sensors.

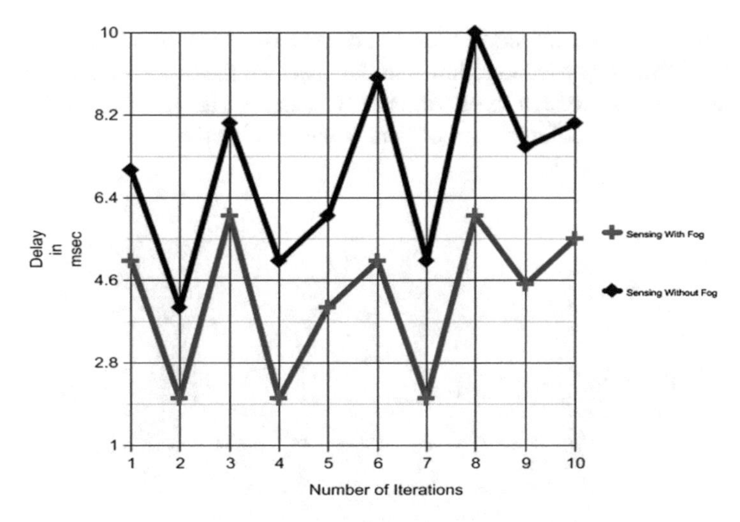

Fig. 5. Delay for critical data

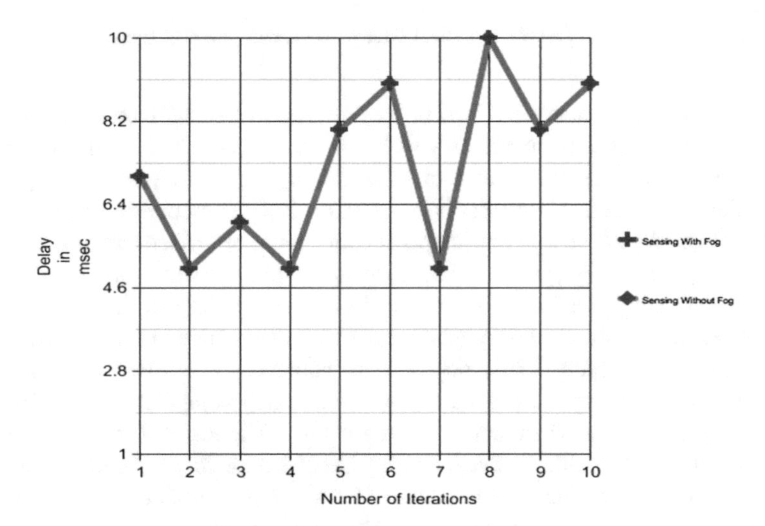

Fig. 6. Delay for noncritical data

Figure 6 shows the delay involved in the noncritical data as a function of a number of iterations. In this case, the application was executed again 10 times and for each iteration, delay is computed for the proposed idea and the existing idea. The results are plotted in Fig. 6. Delay obtained for both of the solutions i.e. the proposed idea and existing idea is the same. It is because, both solutions handle noncritical data in the same manner as it was handled by the existing solutions of cloud computing when data is sent for processing and storage to cloud, so no processing is involved at intermediate layer i.e. fog. All the data is noncritical, so is being sent to cloud for processing and storage. This cloud's computation on noncritical data can be used in near future. Results clearly show at each iteration, that the delay for the proposed idea and existing idea when sensing is done with or without fog is same. The variation in delay at each iteration is due to the different size of data being sent by the sensors.

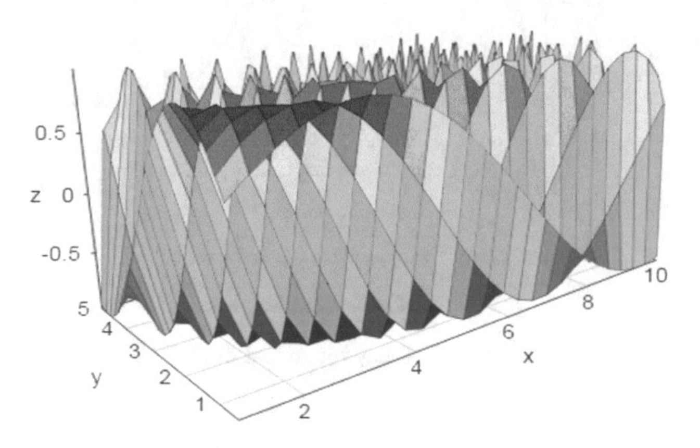

Fig. 7. Delay for critical data with varying sizes of data

Figure 7 shows the delay involved in the proposed system with three variables. The first variable is number of iterations which has been represented by the value x, the second variable is different size of data being sent by the sensors and it is represented by the value y and the third variable is delay which has been represented by variable z.

Figure 8 shows the energy consumption comparison of the sensors of the proposed system and the existing system. The graph is plotted with the passage of time in seconds when the application on the test-bed was executed to test the functionality of the proposed idea. It shows the cumulative energy consumption at different time stages of a sensing node. Result clearly depicts that the energy consumed when sensing is done with fog computing is low at each cumulated energy computation bar. The energy consumed by the sensor when there is no fog is high. The reason for this difference in energy consumption is due to the proposed idea because in proposed idea no computation is done by sensing nodes because they are dumb terminals which sense data and sent to the fog for the processing and storage and actually storage and processing is done by fog and cloud so the energy consumed by sensor is low. In case, when there is no fog computing environment is involved sensor sent data directly to the cloud and in that particular case dumb terminal cannot be used, rather sensors have to process some data before sending it to the cloud. This local processing is actually done to minimize the delay which causes an increase in energy consumption. In case of proposed idea, no processing is involved at the sensor end so the energy consumption is low.

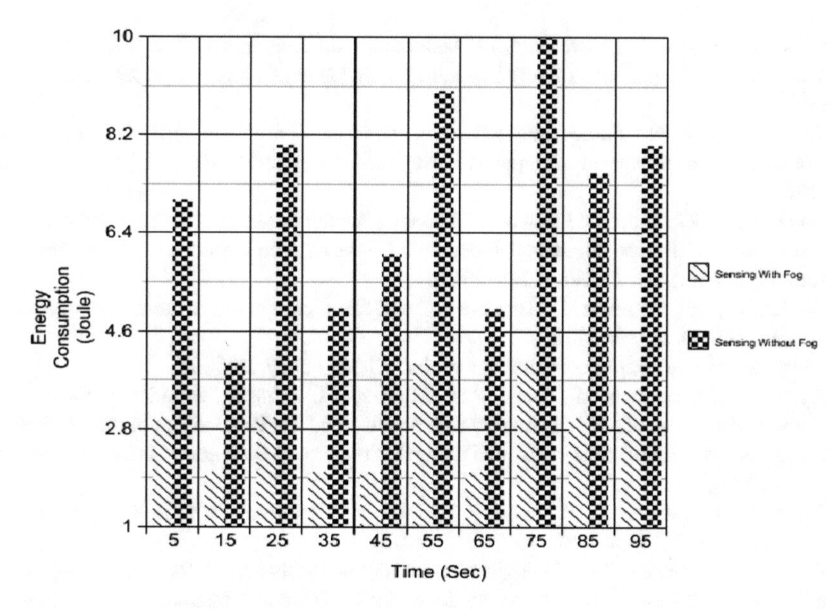

Fig. 8. Energy consumption comparison of sensors

5 Conclusion

This work proposed a new solution for the Internet of Things in general and IoT patient's data in specific. Though existing solutions used cloud computing as a platform for the patient's data but delay and the latency which is inherent in the cloud models result in poor performance and the bad monitoring of the patients which leads to poor treatment of the patients. In our proposed solution, we use a new platform of fog computing which offers same services as that of a cloud on the edge of the network but with more efficiency and with very less delay which is very critical in sensitive applications like health care applications in which patients health is our first priority. As the proposed scheme suggests that patient's critical data processing will be done in the fog and patients which need immediate treatment can be provided services by overcoming the round trip delay to cloud for analysis, which results in effective monitoring of patients.

References

1. Panda, S.S., Veeranjaneyulu, M., Nayak, R.: Internet of Things (IoT), gives life to non living. IJRCCT **5**(2), 043–046 (2016)
2. Osborne, J.: Internet of Things and cloud computing. In: Internet of Things and Data Analytics Handbook, pp. 683–698 (2017)
3. Hosseinpoor, M., Dehghani, S., Roshan, S.: Internet of Things in cloud computing. Int. J. Adv. Res. Comput. Sci. Electron. Eng. (IJARCSEE) **6**(3), 19 (2017)

4. Gubbi, J., Buyya, R., Marusic, S., Palaniswami, M.: Internet of Things (IoT): a vision, architectural elements, and future directions. Future Gener. Comput. Syst. **29**(7), 1645–1660 (2013)
5. Tan, L., Wang, N.: Future Internet: the Internet of Things. In: 2010 3rd International Conference on Advanced Computer Theory and Engineering (ICACTE), vol. 5. IEEE (2010)
6. Botta, A., De Donato, W., Persico, V., Pescapé, A.: On the integration of cloud computing and Internet of Things. In: 2014 International Conference on Future Internet of Things and Cloud (FiCloud), pp. 23–30. IEEE (2014)
7. Doukas, C., Maglogiannis, I.: Bringing IoT and cloud computing towards pervasive health care, pp. 922–926, July 2012
8. http://sourceforge.net/projects/cloudburst-bio/. 15 May 2017 1:40 pm
9. Agapito, G., Cannataro, M., Guzzi, P.H., Marozzo, F., Talia, D., Trunfio, P.: Cloud4SNP: distributed analysis of SNP microarray data on the cloud. In: Proceedings of the International Conference on Bioinformatics, Computational Biology and Biomedical Informatics, p. 468. ACM (2013)
10. Afgan, E., Chapman, B., Taylor, J.: CloudMan as a platform for tool, data, and analysis distribution. BMC Bioinform. **13**(1), 315 (2012)
11. Doukas, C., Pliakas, T., Maglogiannis, I.: Mobile healthcare information management utilizing Cloud Computing and Android OS. In: 2010 Annual International Conference of the IEEE Engineering in Medicine and Biology Society (EMBC). IEEE (2010)
12. https://www.healthvault.com/. 12 May 2017 4:45 pm
13. Basu, S., Karp, A.H., Li, J., Pruyne, J., Rolia, J., Singhal, S., Suermondt, J., Swaminathan, R.: Fusion: managing healthcare records at cloud scale. Computer **45**(11), 42–49 (2012)
14. Bosch-Andersen, L.L.: Hospital uses cloud computing to improve patient care and reduce costs (2011)
15. Pandey, S., Voorsluys, W., Niu, S., Khandoker, A., Buyya, R.: An autonomic cloud environment for hosting ECG data analysis services. Future Gener. Comput. Syst. **28**(1), 147–154 (2012)
16. Hsieh, J.-c., Hsu, M.-W.: A cloud computing based 12-lead ECG telemedicine service. BMC Med. Inform. Decis. Mak. **12**(1), 77 (2012)
17. Fortino, G., Parisi, D., Pirrone, V., Di Fatta, G.: BodyCloud: a SaaS approach for community body sensor networks. Future Gener. Comput. Syst. **35**, 62–79 (2014)
18. Piette, J.D., Mendoza-Avelares, M.O., Ganser, M., Mohamed, M., Marinec, N., Krishnan, S.: A preliminary study of a cloud-computing model for chronic illness self-care support in an underdeveloped country. Am. J. Prev. Med. **40**(6), 629–632 (2011)
19. Kaur, P.D., Chana, I.: Cloud based intelligent system for delivering health care as a service. Comput. Methods Programs Biomed. **113**(1), 346–359 (2014)
20. Kim, T.-W., Kim, H.-C.: A healthcare system as a service in the context of vital signs: proposing a framework for realizing a model. Comput. Math Appl. **64**(5), 1324–1332 (2012)

Adoption of Emerging Technologies in Supply Chain Operations for Cost Reduction and Enhancement of Shareholder Wealth: A Case Study of UAE Organization

Benjamin S. Bvepfepfe(✉), Amjad khan Suri, and Ali El Asad

Higher Colleges of Technology, Fujairah Women College, Fujairah, UAE
{bbvepfepfe, asuri, aasad}@hct.ac.ae

Abstract. Technology and innovation are considered to be the primary driving forces of competitive advantage for most organisations, with evidence of developments in new technologies, e.g., AI IoT, Block chains and most recently FINTECH platforms for financial and accounting operations. These technologies provide businesses and managers immense opportunities for adoption towards cost reduction, improved customer service and enhanced shareholder wealth. However if not checked properly, the same can be a 'disruptive' technology that will threaten the performance and competitiveness for businesses. This paper explored which digital technologies have been adopted for supply chain management and how these technologies have brought reduction in operational costs and enhanced shareholder value. The research paper highlights the key technologies that are adopted within UAE and demonstrate how these technologies contribute to organizational performance and enhanced shareholder wealth. It is evident that organisations are investing in technologies like, RFID and ERP for cost reduction and shareholder wealth enhancement.

1 Introduction

For a number of years now the internet has become an essential part of human and business interactions. In fact, Lowe et al. suggested then, that the internet had evolved from being a mere research resource to become a global information hub and network offering a wide range and variety of service opportunities [17]. The internet has readily been accepted by many, transforming from previous fears about its adoption to become a source of opportunities for all aspects of human interactions. It is also an acceptable fact that the internet has become inexpensive, easy to use with various platforms and data presentations. Businesses have been on the forefront to maximize the potential of the internet and associated web technologies. On the same note, the technology industry has progress further and there is an increasingly perceived vision that computing will one day be the 5th utility after water, electricity, gas, and telephony, [5] and this can be evidenced by the penetration of such services like WIFI and Cloud computing platforms [5].

To be precise, technology and innovation are considered to be the primary driving forces of competitive advantage for most organisations, with evidence of developments in new technologies, e.g., WIFI, Cloud Computing, artificial intelligence (AI), internet

L. Barolli et al. (Eds.): EIDWT 2019, LNDECT 29, pp. 437–446, 2019.
https://doi.org/10.1007/978-3-030-12839-5_40

of things (IoT), Block chains and most recently FINTECH platforms for financial and accounting operations [4]; Leong and Sung [16]. These technologies provide businesses and managers immense opportunities for adoption towards cost reduction, improved customer service and enhanced shareholder wealth. However if not checked properly, the same can be a 'disruptive' technology that will threaten the performance and competitiveness for businesses. The forth revolution, or Industry 4.0 has been of interest to both academics and practitioners as they attempt to predict future trends within the IoT arena. Industry 4.0 here encompasses numerous technologies and associated paradigms, including Radio Frequency Identification (RFID), Enterprise Resource Planning (ERP), Internet of Things (IoT), cloud-based manufacturing, and social product development [18].

This calls for adequate knowledge regarding these available technologies, new process technologies and new products on offer on the technological environment. There is evidence from both small to medium enterprises (SMEs) and large firms showing that successful innovation and sustainable competitiveness is not just the result of technological innovation, but is also heavily dependent on what has been called 'management of technological innovation' [13]. It is argued that organizations need to complement technological innovation with non-technological innovation concerning processes, people and the organization for sustainable performance [14].

2 Supply Chain Operations and Technological Environment

Supply chains comprise of activities and networks of organizations that work together and independently to produce and deliver goods and services from source right up to consumption. The key flows that are essential for effective performance of supply chain operations are physical flows of goods, money flows and information flows [7]. The effectiveness of supply chains is determined on how efficient these key flows are to ensure the right product/service of the right quantity and quality is delivered to the right place, at the right time and cost/price [18, 19]. This is where supply chain management derived its strategic position in boardrooms as a source of competitive edge for business, offering opportunities for enhancement of shareholder value and customer service. Christopher also posited that much of product or service costs are incurred outside the organization's boundaries, which means, supply chain activities upstream and downstream have a large bearing on the final product costs and subsequently, price changed to the customer [7]. The arguments therefore are that effective management of logistics and supply chain operations can enhance productivity and reduce costs. On the other hand is value added service offering for businesses. Most products have direct or indirect substitutes to an extent that they may be considered 'commodities' in the market place. The only differentiators outside price come from the value adding provided in the key service deliveries as provided by supply chain operations.

Christopher stated that supply chain operations can be tailor-made to suite distinct market value segments for different groups of customers [7]. Customer service has gained popularity on the market as a key differentiator in those areas like delivery, after sales or technical support. This is another dimension that has elevated supply chain management to a strategic role because of its value added to both the customer and the other stakeholders of the business. It is therefore an acceptable conclusion that logistics and supply chain management offers businesses with immense opportunities for both cost reduction and value creation. For example, cost reduction through improvements in asset utilization, productivity and waste reduction/removal; value creation through superior service and product offering for different market segments. Challenge is whether businesses have seen this source of competitive edge and value creation and how they are taking advantage of supply chain management for sustainable performance. Porter [22] argued that a firm in itself does not offer competitive advantages, but rather through the value chain that comprise of various activities and processes performed by the different partners of the supply chain [23].

Supply chains are operating in an environment of rapid changes, brought about by market dictates and technological changes. Evidence abound from research that efforts are being undertaken to develop the digital technologies for managing the interface between supply chain networks and other key activities of supply chain management. Managing these supply chains go beyond basics of processes automation and transaction between supply chain partners. Whilst these are important facets in managing supply chains, real value is created when managers can see the bigger picture in their supply chain operations so that enabling technologies become essential for this aspect of managing the entire supply chain networks. The role played by these technologies in sustaining the strategic role supply chain management has enabled the much-needed integration of operations towards common goals [26, 27]. Opportunities for e-supply chain management from through ERP networks are immense for both B2B and B2C market sectors.

Advances in adoption of technologies in supply chain operations through information networking and exploitation of data generated from emerging data technologies such as ERPs; Data Centers, Cloud technologies and Block chains is gaining momentum [28, 29]. Surprisingly, a good number of managers appear oblivious of the competitive advantages that these technologies can drive within their market segments. The Middle East and UAE in particular, would do well by embracing a culture that not only talk about technologies, but also take calculate risks in adoption of these enabling technologies. For example, Fintech has long been a question of great interest in a wide range of fields. Fintech as defined by FSB [9] as "technologically enabled financial innovation that could result in new business models, applications, processes, or products with an associated material effect on financial markets and institutions and the provision of financial services" [11]. Fintech applications ranges from enabling technologies (e.g. mobile banking) to disrupted financial services (e.g. transaction banking) [26].

In the traditional system, technological presence in banking and financial system is graded by ATM usage, online banking or credit cards whereas, fintech tend have a very definite role in driving new business models to address issues that traditional system would not able to answer. Introduction of innovative technological platforms prepare individuals, companies and the government to operationalize process in an easier, efficient and low-cost manner. Moreover, Fintech will help financial sector by managing risk efficiently and share more information among stakeholders to reduce information asymmetry. Similarly, many startups in ASEAN countries use Fintech services in the form of mobile banking services to the unbanked and under-banked clients like wing money in Cambodia, wavemoney in Myanmar, match move in Singapore and Momo, an e-wallet and a payment app in Vietnam.

For gulf countries, in United Arab Emirates alone, 60 to 65% of the population remains either unbanked or barely banked, according to research conducted by mobile wallet start-up Tripple while World Bank project 80% of the UAE population is outside the current financial system who have no access to insurance policies, credit cards, bank accounts and loan options. Suri and Purohit conducted a study on expatriates financial literacy level in UAE and found that expatriates financial literacy is low and suggested measures to increase their literacy level [29]. In a study conducted by Master card in Saudi Arabia revealed that majority of its population is unbanked.

This paper aims to explore which digital technologies have been adopted for supply chain management and how these technologies have brought reduction in operational costs and enhanced shareholder value.

The key objectives will be to:

- Investigate which technologies are adopted for information flows for the selected organization(s)
- Collect data on adoption of technologies in sample organization within the region
- Assess the impact of such technologies in reduction of costs and value-added
- Recommend key technologies that can enhance organizational performance and shareholder wealth

3 Methodology

The research started with secondary research on the concepts of emerging technologies and their adoption within supply chain operations. Secondary data was collected and analyzed from selected organizations within the UAE. A qualitative approach was considered more appropriate and provides more opportunities for revelation on goings on in supply chain operations of the chosen organisation. The organization was selected purposively in order to gain deeper insights into the key technologies and their impact on overall performance from a supply chain perspective. Internet based data collection on Emerging Technologies for supply chain operations. This was then scaled down to technologies for specific supply chain roles, within the UAE e.g., procurement, production, distribution logistics. Follow-up interviews were conducted with managers within the selected organisation to clarify on adoption and specific functional application of the technologies.

4 Findings

The findings highlight the key technologies that are adopted within UAE, demonstrate how these technologies have contributed to organizational performance, and enhanced shareholder. A numbers of technologies are mentioned in most supply chain literature. These are shown in Fig. 1 below.

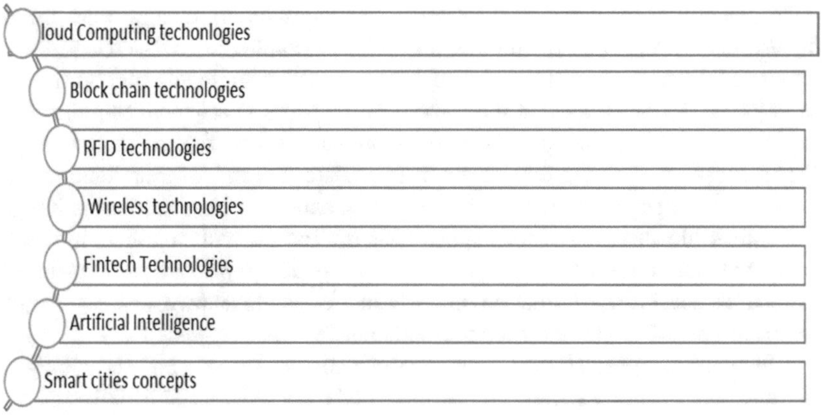

loud Computing techonlogies
Block chain technologies
RFID technologies
Wireless technologies
Fintech Technologies
Artificial Intelligence
Smart cities concepts

Fig. 1. Enabling technologies for supply chain operations (Source: Authors)

The above technologies appear to form the back bone of emerging technologies that are considered to provide value-adding and potentially sustainable competitive advantages. The uptake and adoption of same is dependent on an organisation's strategy, scope of operations and capacity for technological application. In addition, vehicle tracking systems capable of monitoring engine performance during trips, vehicle location provide greater flexibility in operations, improving asset utilization and productivity and thereby reducing operational costs. Further analysis of data revealed the following applications for 'upstream' the supply chain operations as shown in Fig. 2.

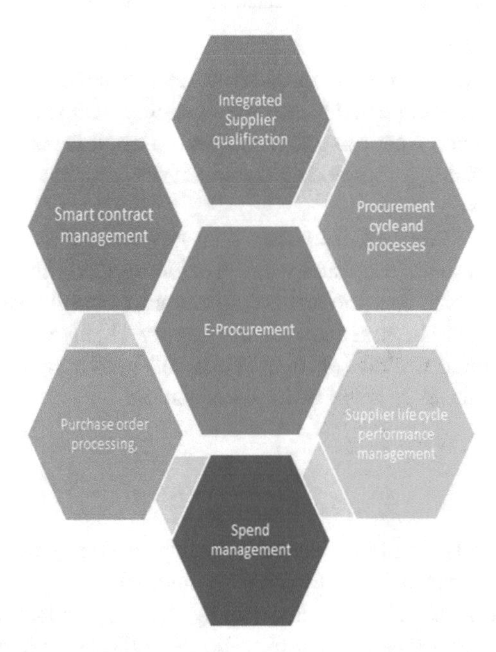

Fig. 2. Enabling technologies for upstream supply chain operations

The objectives of any supply chain operations is to ensure that the right product/service in the right quantity and quality is supplied or delivered to the right customer and the right place, time and cost. The above technologies support the supply chain objective by ensuring that the right supply source is selected. Supplier qualification is key to this procurement process and will reduce risk of contracting with unreliable suppliers. This requires access to up to date data and information on supplier. The case study organization has used a single procurement platform that is integrated with suppliers' downstream. This enables B2B transaction with suppliers in real time, reducing manual interventions, expediting and reducing procurement cycle times. Opportunities for further reduction in ordering costs, inventory holding results in reduction of cost of goods and therefore enhancement of the gross margin.

Further analysis revealed the following applications 'downstream' the supply chain within logistics operations, i.e., before the goods are shipped and delivered to the customer (Fig. 3).

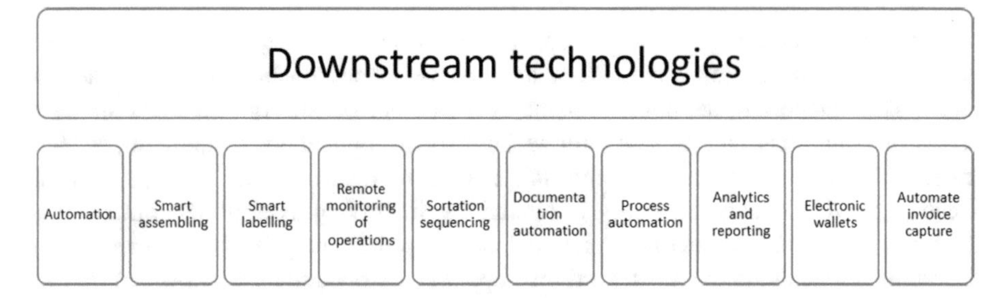

Fig. 3. Enabling technologies for downstream supply chain operations, (Source: Authors)

Sample organization recently adopted block chain technologies, although it is still in its infancy stages. It is expected to provide visibility of operations from production wells to the end customer. According to the organization, this technology provide some single platform for tracking, monitoring the quantities and financial values across the various business units.

Fintech adoption in the UAE is evident with government and regulated policies in place to support the system. The framework provide opportunities for traction services like bill payments, for example, the technology supports mobile wallets, a concept that has become popular within a section of the supply chain community. The sample organization has adopted wallets for its business and individual customers. Again this technology is still in its early stage of implementation.

5 Discussion

It is clear is that there is a wide variety of emerging technologies that can provide the desired integration between an organizations and its supply chain partners, with resultant differentiation in service, value adding and cost reductions [23]. Emerging technologies like, radio frequency identification (RFID); global positioning systems

(GPS) and low-power wide area network (LPWAN) are applied in logistics operations for asset tracking and material flows monitoring systems in warehouses, production, security during transportation [2]. These technologies also provide real time information on status; availability and location for both planning and execution. RFID capabilities for real time information transmission coupled with interface connectivity with enterprise resource planning systems, (ERP) like SAP, is a significant contributor to value adding within supply chain operations. Assets including current assets can be tracked and traced from source right through production until they are transformed into finished or semi-finished goods for delivery to the customers. For example, the same technologies have been successfully applied for time sensitive and perishable products that may require special handling and temperature control during transit, thereby reducing damages and deterioration to products in supply chain pipelines.

E-procurement platforms enhances the value adding role for strategic and tactical sourcing for organizations and their supply chain partners [10, 24]. Whether centralized or decentralized procurement structure, e-procurement has provided organizations with opportunities to minimize supply base and maximize cost reduction shareholder value. Sample organization reported that by integrating together multiple suppler catalogues into a single supply data base, organizations have been able to reduce the purchasing order cycles.

Adapting Christopher's framework on ultimate competitive advantage for organization, this paper categorized these technologies and plotted same along the quantum, [7] (see Fig. 4). The ideal technologies would provide the organization with both value-added customer service at reduced cost, the ultimate technologies fit for purpose.

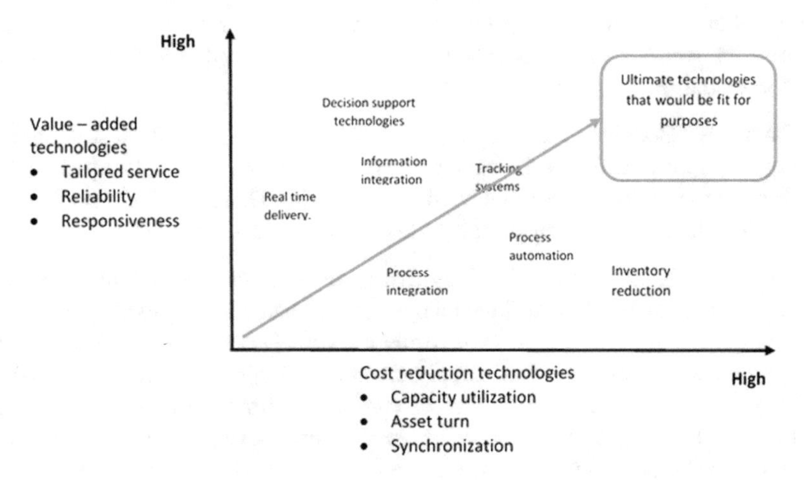

Fig. 4. Enabling technologies value/cost matrix (Adapted from [7])

We categorize value added emerging technologies as those that provide easy access and real time information across the network. Technologies supporting management decision and integrating business operations are also categorized under this group as depicted in Fig. 4 above. On the other hand, availability of information in electronic form can lead to cost savings through reduction in the processing, printing, and distribution of paper documents. These are grouped together with technologies that improve productivity in the group for cost reduction technologies. Therefore, successful selection and management of emerging technologies is critical in coping with the changes that may be necessary [12]. This is even more pressing for supply chain operations in developing and emerging markets, where there is inadequate capacity and under developed infrastructure provisions [6]. However, as note by Demirgüç-Kunt et al., the payment industry claim lion's share in the Fintech sector in developing countries in 2017, with reports that digital payments rose from 32% to 42% in 3 years' time period [8]. In Kenya alone, 73% of the population use mobile money, Zimbabwe is around 97% [20] and other developing countries also showing rapid progress [20].

Fintech has a promising potential in consumer finance applications such as payments, financing and digital currency [25]. Global online e-commerce giant Amazon purchase of Dubai's Souq.com has given new dimension to online shopping. Nearly 80% of all online transactions in MENA region are settled on a cash on delivery basis with most of the payment are made after goods are delivered to the customers. This complicates logistics and sometimes customers decline to purchase goods. To address the payment woes, an e-commerce platform "cash basha" in Jordan allows unbanked customers more flexibility for online purchases and minimizing the risk of failed orders. There are various applications where Fintech is making supply chain process easier on two applications of Fintech, i.e., payments and financing. For purposes of this research paper focus will be made on payments as this is considered key to supply chain management. One of the main flows in a supply chain is money flows such that improvements in this flow is likely to compliment the other key flows especially materials. Banks can also be expected to contribute to the money flows through collaborative innovation as far as payment systems are concerned. The cash-to-cash cycle time is essential in supporting material flows

(1) Payments: Companies are trying to incorporate easy payment solutions and develop apps to simplify the payment services. The payment industry claim lion's share in the Fintech sector in developing countries. Demirgüç-Kunt *et al.,* provide ample evidence of digital payments in developing countries [8].

Block chain is one other technology that is gaining attention within supply chain operations. Block chain is a distributed ledger that is shared among participating parties to record transaction and creates trust in the network [1, 21]. Block chain is helping industry in better tracking the food supply chain network, reduce costs across logistics and financing [3]. It has applications in cutting the time in delivering product to customers and brings accountability at every step—several pioneer initiatives in reviewing block chain principles for example, hawk to protect personal data [15]. This will help in securing the privacy of information and reduce leakage of information to the outsiders. Halalchain app in Dubai enable consumers to track and trace halal products and ease operations on checking counterfeit product from non-halal sources.

6 Conclusion

The two key flows in supply chains are material and information we investigate the integrations of both information and material flows between supply chain partners and their effect on operational performance. Information integration consists of information technology and information sharing, such that effective information integration will have a positive effect on material integration. When there is effective flows through material integration, this will have a positive effect on operational performance and sustainable competitive advantage. However, this variety of technologies are likely to present challenges to organizations and their managers if they are not carefully selected and implemented successfully. The argument posited herein is that only technologies that are fit for purpose will ensure sustainable survival of business. This is based on the notion that adopting enabling technologies for inappropriate process or adopting inappropriate technologies operations will not provide the desired output. In fact, there are inherent risks associated with these technologies that may end up disrupting the level of performance and expectations. A wide range of enabling technologies are available for supply chain operations and to date several of these have been successfully adopted. The scope for adoption is wide and future trend is dependent on adaptability and adoption of these technologies. The paper argues for careful selection of the technologies to consider both value-added and cost reduction advantages attributes for the ultimate goals of sustainable competitiveness and shareholder value maximization.

References

1. Abeyratne, S.A., Monfared, R.P.: Blockchain ready manufacturing supply chain using distributed ledger (2016)
2. Angeles, R.: RFID technologies: supply-chain applications and implementation issues. Inf. Syst. Manag. **22**(1), 51–65 (2005)
3. Apte, S., Petrovsky, N.: Will blockchain technology revolutionize excipient supply chain management? J. Excipients Food Chem. **7**(3), 910 (2016)
4. Barclay, I., Preece, A., Taylor, I.: Defining the collective intelligence supply chain. arXiv preprint arXiv:1809.09444 (2018)
5. Buyya, R., Yeo, C.S., Venugopal, S., Broberg, J., Brandic, I.: Cloud computing and emerging IT platforms: vision, hype, and reality for delivering computing as the 5th utility. Future Gener. Comput. Syst. **25**(6), 599–616 (2009)
6. Bvepfepfe, B.S.: Supply chain network and logistics management. In: Adewole, A., Struthers, J. (eds.) Logistics and Global Value Chains in Africa, pp. 45–89. Palgrave Macmillan, Cham (2019)
7. Christopher, M.: Logistics & Supply Chain Management. Pearson, Harlow (2016)
8. Demirgüç-Kunt, A., Klapper, L., Singer, D., Ansar, S., Hess, J.: The Global Findex Database 2017: Measuring Financial Inclusion and the Fintech Revolution. World Bank. Washington, DC (2018)
9. Financial Stability Board (FSB): Financial stability implications from FinTech supervisory and regulatory issues that merit authorities' attention (2017)

10. Gabhart, K., Bhattacharya, B.: Service Oriented Architecture Field Guide for Executives. Wiley, New York (2008)
11. Global Findex Database: Measuring Financial Inclusion and the FinTech Revolution. World Bank, Washington, D.C. (2017)
12. Gunasekaran, A., Ngai, E.W.: Expert systems and artificial intelligence in the 21st century logistics and supply chain management. Expert Syst. Appl. **1**(41), 1–4 (2014)
13. Gunasekaran, A., Dubey, R., Singh, S.P.: Flexible sustainable supply chain network design: current trends, opportunities and future (2016)
14. Johnston, M.W., Marshall, G.W.: Sales Force Management: Leadership, Innovation. Technology. Routledge, London (2013)
15. Kosba, A., Miller, A., Shi, E., Wen, Z., Papamanthou, C.: Hawk: the blockchain model of cryptography and privacy-preserving smart contracts. In: Proceedings of the 2016 IEEE Symposium on Security and Privacy (SP) (2016)
16. Leong, K., Sung, A.: FinTech (Financial Technology): what is it and how to use technologies to create business value in FinTech way? Int. J. Innov. Manag. Technol. **9**(2), 74–78 (2018)
17. Lowe, H.J., Lomax, E.C., Polonkey, S.E.: The World Wide Web: a review of an emerging Internet-based technology for the distribution of biomedical information. J. Am. Med. Inform. Assoc. **3**(1), 1–14 (1996)
18. Lu, Y.: Industry 4.0: a survey on technologies, applications and open research issues. J. Ind. Inf. Integr. **6**, 1–10 (2017)
19. Monczka, R.M., Handfield, R.B., Giunipero, L.C., Patterson, J.L.: Purchasing and Supply Chain Management. Cengage Learning, Boston (2015)
20. Marumbwa, J.: Exploring the moderating effects of socio-demographic variables on consumer acceptance and use of mobile money transfer services (MMTs) in Southern Zimbabwe. Am. J. Ind. Bus. Manag. **4**(2), 71 (2014)
21. Peck, M.E.: Blockchains: how they work and why they'll change the world. IEEE Spectr. **54**, 26–35 (2017)
22. Porter, M.E.: Competitive Advantage: Creating and Sustaining Superior Performance. Simon and Schuster, New York (2008)
23. Prajogo, D., Olhager, J.: Supply chain integration and performance: the effects of long-term relationships, information technology and sharing, and logistics integration. Int. J. Prod. Econ. **135**(1), 514–522 (2012)
24. Rothaermel, F.T.: Strategic Management. McGraw-Hill Education, New York (2015)
25. Scholz, N.: The Relevance of Crowdfunding. Springer Fachmedien Wiesbaden, Wiesbaden (2015)
26. Schueffel, P.: Taming the beast: a scientific definition of Fintech. J. Innov. Manag. **4**, 32–46 (2016)
27. Stevens, G.C., Johnson, M.: Integrating the supply chain 25 years on. Int. J. Phys. Distrib. Logistics Manag. **46**(1), 19–42 (2016)
28. Suri, A.K., Sonal, P.: An analysis of personal financial literacy among Expatriates in United Arab Emirates. Contemp. Rev. Middle East **4**(3), 1–19 (2017)
29. Underwood, S.: Blockchain beyond bitcoin. Commun. ACM **59**(11), 15–17 (2016)

Hybrid Approach for Heart Disease Prediction Using Data Mining Techniques

Monther Tarawneh(✉) and Ossama Embarak(✉)

HCT, ALfujairah, UAE
{mtarawneh, oembarak}@hct.ac.ae

Abstract. Heart disease is one of the significant reason of death and disability. The shortage of Doctors, experts and ignoring patient symptoms lead to big challenge that may cause death, disability to the patient. Therefore, we need expert system that serve as an analysis tool to discover hidden information and patterns in hear disease medical data. Data mining is a cognitive procedure of discovering the hidden approach patterns from large data set. The available massive data can used to extract useful information and relate all attributes to make a decision. Various techniques listed and tested here to understand the accuracy level of each. In previous studies, researchers expressed their effort on finding best prediction model. This paper proposes new heart disease prediction system that combine all techniques into one single algorithm, it called hybridization. The result confirm that accurate diagnose can be taken by using a combined model from all techniques.

1 Introduction

Heart disease is considered the main reason for death in the world. The heart disease diagnosis is the process of detecting or predicting heart disease from patient's records. Doctors may not able to diagnose patient properly in a short time, especially when the patients suffer from more than one disease. Therefore, heart disease diagnose is a complex task that require experience and knowledge. Improper diagnose may cause death, or disability to the patient. Disease prediction Model can support medical professionals and practitioners in predicting heart disease. The huge amount of data that can be collected using digital devices (by the patient itself of in hospital) can used with data mining techniques to diagnose patient and predict diseases. This paper analyses various types of classification and prediction techniques used in heart disease prediction. Also, propose a hybrid approach that combine all techniques into a single one to combine all functions and produce accurate diagnose.

2 Related Work

Huge amount of medical data generated by healthcare devices is large and complex to be analyzed by traditional methods. Data mining is used to improve the process by discovering patterns and features in large complex data. Several techniques have been presented to give accurate medical diagnose for various diseases. Data mining

L. Barolli et al. (Eds.): EIDWT 2019, LNDECT 29, pp. 447–454, 2019.
https://doi.org/10.1007/978-3-030-12839-5_41

techniques are uses to eliminate interrelated and redundant data and to find hidden pattern from data since the data are high dimensional. Most known datamining tasks are association rules, feature selection, classification, clustering, prediction and sequential patterns.

Classification techniques are widely used in healthcare, since they are capable of processing large set of data. The common used techniques in healthcare are Naïve Bayesian, support vector machine, Nearest neighbor, decision tree, Fuzzy logic, Fuzzy based neural network, Artificial neural network, and genetic algorithms [1]. Machine learning with classification can be efficiently applied in medical applications for complex measurements. Modern classification techniques provide more intelligent and effective prediction techniques for heart disease [2]. Many studies have been stationing on prediction of heart diseases; by applying different data mining techniques based on some feature selection techniques in order to get accurate classification using the optimal features set.

ANFIS is a heart disease prediction model based on coactive neuro-fuzzy inference system [3]. The model diagnosed disease by using various techniques include neural network adaptive capabilities, fuzzy logic qualitative approach and genetic algorithm. ANFIS was evaluated in term of classification accuracy and training data, the result showed a great potential in heart disease prediction with a very little mean square error. ANFIS is an adaptive neuro fuzzy inference system to train the neural network in order to predict heart diseases and cancer in diabetic patients based on some factors like age, obesity and some other factors related to life style [4]. The input nodes in neural network are constructed based on the input attributes and the hidden nodes are used to classify given input based on training dataset. GAFL use genetic algorithm and fuzzy logic [5]. The role of genetic algorithm was to provide the optimal solution for the features selection problem to help the diagnosing system. Fuzzy logic role was to develop a classification model using fuzzy inference system. The accuracy of this model was 86% using stratified k-fold technique with specificity and sensitivity 0.9 and 0.8 respectively. With adaptive group-based-k-nearest-neighbor algorithm (AGKNN) help along with feature subset selection using PSO. This model reduced the cost of various medical tests and helps the patients to take a preventative measures well in advanced and showed very good prediction accuracy compared to traditional methods. According to this experiment 0.4% of people are under the risk of both cancer and heart disease if they are having diabetes.

A simple and reliable features selection method proposed to determine heartbeat case using weighted principal component analysis (WPCA) method [6]. In pre-processing stage they enlarged the ECG signal's amplitude and eliminated the noises. The total accuracy for this model was 93.19%. A hybrid classification technique by [7] which is a correlation based filter selection algorithm and support vector machine classifier, were features ordered according to their absolute correlation value with respect to the class attribute. The top k-features have been selected from an ordered features list. Classification accuracy was measured using SVM classifier with and without extended features of the reductive dataset. This model observed high accuracy in the case of three of five high dimensional used datasets with very less number of features up to 100% using the proposed hybrid technique with square extended feature.

A prototype intelligent heart disease prediction system (IHDPS) developed [8] using decision tree, Naïve Bayes and neural network data mining techniques and utilizing 909 records with 15 features from Cleveland heart diseases dataset. IHDPS can predict the likelihood of patients getting a heart disease and enables a significant knowledge such as relationship between related heart disease medical factors in addition to its ability to answer a complex "what if" queries. The result showed that each of used data mining techniques has its own strength side in grasping the defined mining goals objectives. Naive bayes was the best classifier to predict heart disease with accuracy 95% and to identify the significant influence and relationship in the medical input associated with predictable heart disease. Decision tree showed the best accuracy 99.61% to identifying the impact and relationship between the medical attributes in relation to the heart disease predictable state. for the last two goals Naïve Bayes was the best to identify heart disease patients characteristics and to determine the attributes values that differentiate nodes favoring and disfavoring the predictable state for patients with and without heart disease. IHDPS was exploited DMX (data mining extension) query language for model creation, training, prediction and content access, and evaluated using lift chart and classification matrix method. The most effective model to predict heart disease patients was Naïve Bayes with 86.53% accuracy followed by neural network with less than 1% difference and decision tree get the best accuracy for predicting patients with no heart disease since the model was evaluated on data set contain patients with and without heart disease.

Two phases experiment has been done to understand how machine learning techniques can help in comprehending the level of risk associated with heart disease using information gain and gain ratio feature selection techniques [9]. The 1st phase is to applying feature selection techniques on commonly used 13 attributes. The 2nd phase is to use 75 attributes related to heart anatomic structure such as heart blood vessels from four heart disease datasets and to test the used feature selection techniques accuracy using Naïve Bayes, decision tree, support vector machine, logistic regression, and multi-layer perception and adaboost classifiers. The use of 13 attribute was not enough to understand the risk level of heart disease unlike the use of 75 attribute with feature selection techniques which improve the classification accuracy. Decision tree with adaboost get the best accuracy 98% on three datasets.

A dataset with 303 instances and 54 features exploited to study the role of feature selection and data mining techniques in diagnosis of coronary artery disease [10]. Information gain and confidence were used to determine the features effectiveness and select the optimal set of them. SVM, Naïve Bayes, bagging algorithm and neural network data mining classification techniques was used to evaluate the proposed feature selection techniques in addition to confusion matrix to detect the sensitivity, specificity and accuracy. Researchers were used LAD (left anterior descending), LCD (left circumflex) and RCA (right coronary artery) recognizers to create three new features in order to recognize if the three major coronary arteries are blocked. The achieved accuracy of this experiment was 98.08%.

A novel approach for normal and coronary artery disease conditions automated detection using heart rate signal Introduced by [11]. Discrete wavelet transformation (DWT) was used to decompose the heart rate signals into frequency sub-hand, in total the 3rd level of detail coefficient will be 76 in number. Principle component analysis, linear discriminate analysis and independent component analysis dimensionality reduction techniques were applied on DWT extracted coefficient and reduced the dimensionality from 67 to 10 features (coefficient). The remaining features applied to SVM, Gaussian Mixture Model, Probabilistic Neural Network and KNN classifiers. The accuracy of this approach was 96.8%.

Back Propagation algorithm methods [12] compares classification techniques. The author efficiency and deliver high accuracy from the heart disease prediction. A comparative analysis of accuracy on heart disease prediction [13] used Naïve Bayes and SVM, logistic regression. The highest accuracy (80%) of heart disease prediction is SVM. Various types of clustering techniques compared on heart disease prediction [14], it shows that the best performance accuracy of cluster algorithm make density based cluster. Mai and her partners [15] compare 3 classification algorithm c4.5, j4.8 and nagging algorithm and conclude that the best performance algorithm in heart disease prediction with 81.41% accuracy is bagging algorithm. The study was based on UCL repository heart disease data set.

Combining techniques in a hybrid model may result high accuracy. A hybrid approach that combine genetic and Naïve Bayes [16] produce high accuracy against others techniques, the authors investigate most used techniques and chose the best two. Another hybrid machine learning model comprising of genetic algorithm, SVM and regression analysis [17]. As we can see that each algorithm used by each technique has some functions that help in heart disease prediction. To get the best result, we can combine the output of each algorithm and compare, it is called hybridization [18].

All researchers made a major effort in present an optimal heart disease diagnosis systems based various techniques. The achievement of high classification accuracy remains the common objective of all those different models which are expanded from simple to complex features types from supervised to unsupervised and semi-supervised features selection and from simple to more advanced techniques.

3 Proposed Method

The proposed Method contains three phases as depicted in Fig. 1. Starts with preprocessing phase where data filtered and classified before any processing. The output of this phase goes into number of classification techniques where these techniques evaluated tom eliminate low performance one. Then we combine the result and look at the patient history to give a decision (negative/positive) of heart attack. The steps of each phase described in details below.

Phase 1: Preprocessing

1. Pass the data set through a filtering which replace all missing values with a value inferred from the column values using the concept of mutation in evolutionary algorithm. This a very important step especially for real dataset.

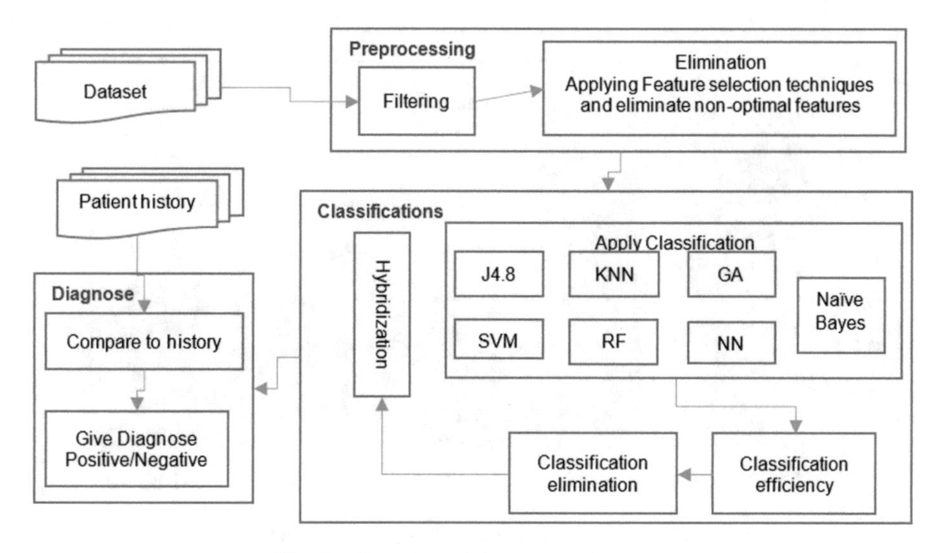

Fig. 1. Overview of the proposed system

2. Apply feature selection techniques (Information gain, gain ratio, reliefF, symmetrical uncertainty, and oneR feature selection) on the data set to eliminate non-optimal features. Features with zero rank or clearly low in comparison to others should be eliminated, only 2 features eliminated.

Phase 2: Classification

1. Apply number of classifications techniques on the output of the first phase.
2. Classification accuracy, precision, recall and f-measure will be used to evaluate the efficiency of the used techniques, Fig. 2 shows the classification results of the original data.
3. Eliminate low efficiency algorithms based on the evaluations from previous step. This process done by comparing the values of accuracy, precision, recall and f-measure for each feature to determine the consistency of the classification on the data set. We notice that Naïve Bayes and SVM always perform better than others and never been eliminate, tree decision eliminated a couple of times. Where KNN is most of the time get eliminated.
4. Apply Hybridization, where we combine the results from the chosen Classification.

Phase 3: Diagnose

1. If the patient history is available, compare the result with the patient history. Patient profile can available in the hospitals or clouded by the health services in the country.
2. Give a diagnose result with either positive or negative heart attack diagnose.

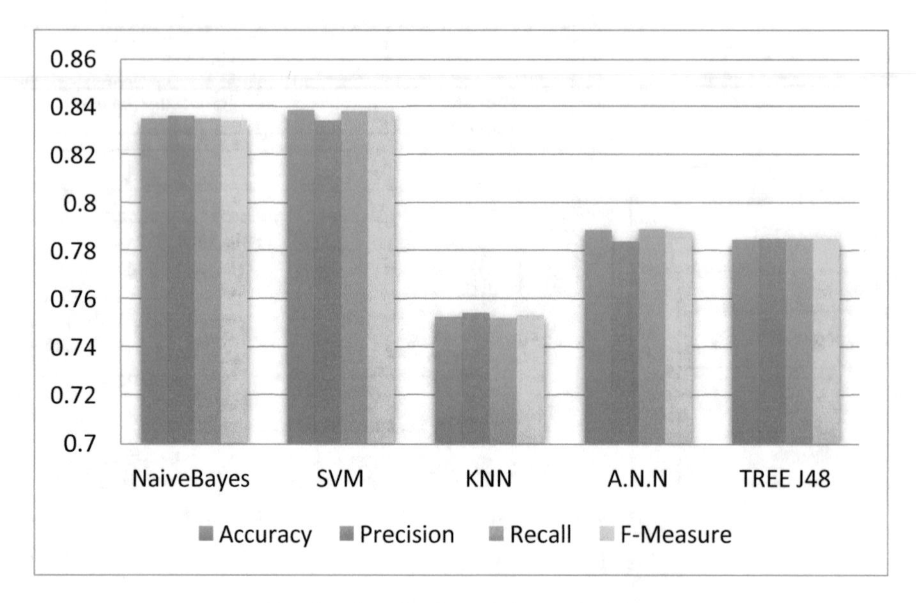

Fig. 2. Classification of the original data

4 Results

Our experiment applied on publicly available heart disease data set from UCI machine learning repository. Cleveland heart disease dataset contain fourteen numerical features and 303 instances [19]. This data set usually uses for researches about heart disease, we used this data set due its reasonable number of features and instances as well as its free from noisy data and contain a very little missing data. We achieved accuracy of 89.2%, we were able to reduce the number of features from 14 to 12 without loss in the accuracy as computed by Naïve Bayes, SVM, KNN, NN, J4.8, RF, and GA on the whole data set.

We notice that Naïve Bayes and SVM are always give better accuracy than other classification techniques, where the last one is KNN. However, the hybridization approach is recommended since each technique has its own functions that help in heart disease prediction, and combining these techniques will combine all functions on all features. The performance study of different data mining algorithms in prediction of Heart disease as given below in Fig. 3.

We have done some analysis on the number of classifications techniques used on the second phase. The time will increase by increasing the number of classification techniques, it might be a drawback for the proposed system. However, the accuracy increases by adding more classification techniques.

The proposed system can used in hospitals to help doctors make a quick diagnose or test new ones on some cases. Students in medical colleges may wish to use this system to learn and test their learning.

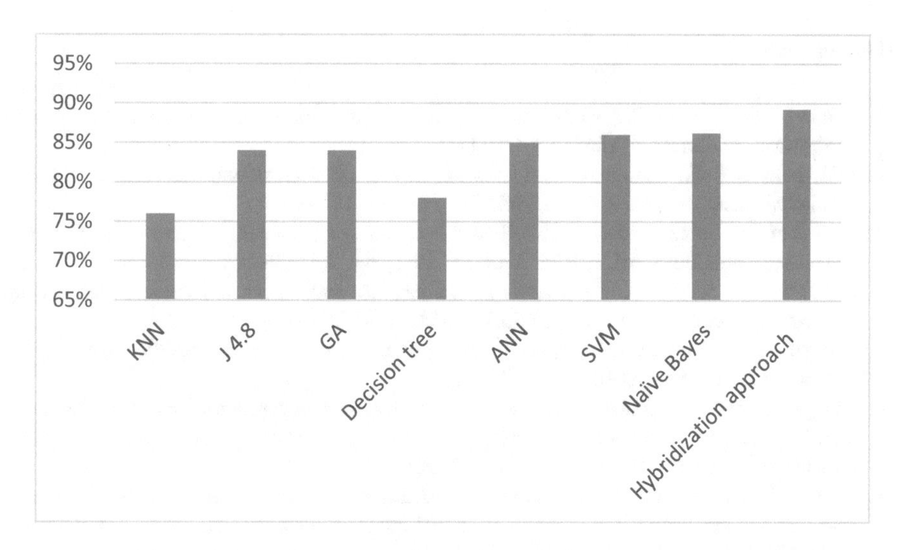

Fig. 3. The performance study of different data mining algorithms

5 Conclusion

Patient's electronic medical records usually contain relevant, irrelevant and redundant features. Doctors face a challenge to predict and diagnose heart diseases quickly and correctly due to this inefficient number of features. Different studies investigates feature selection techniques role in extracting the optimal set of features to being used in predicting and diagnoses heart disease with various methodologies and different classification accuracy.

Here we have investigated most of the used classification techniques in data mining and apply them on Cleveland data set. Some of the techniques perform better always such as Naïve Bayes and SVM, while others depend on the selected features. The main goal in this paper is to investigate available data mining techniques to predict heart disease and compare them, then combine the result from all of them to get most accurate result. The focus was on the classification and prediction methods. The accuracy of the algorithms can improved by hybridization or combining algorithm into single powerful algorithm. The new algorithm can be used as expert system in hospitals to help doctors in diagnose heart disease quickly and save life. Also, can used for education purpose in medical schools.

6 Future Work

In future, we are planning to introduce an efficient Remote heart disease prediction system to monitor and predict the heart disease based on the patient data collected from remote devices. Better accuracy id the main goal.

References

1. Kumari, M., Godara, S.: Comparative study of data mining classification methods in cardiovascular disease prediction 1 (2011)
2. Purusothaman, G., Krishnakumari, P.: A survey of data mining techniques on risk prediction: heart disease. Indian J. Sci. Technol. **8**(12) (2015)
3. Parthiban, L., Subramanian, R.: Intelligent heart disease prediction system using CANFIS and genetic algorithm. Int. J. Biol. Biomed. Med. Sci. **3**(3) (2008)
4. Kalaiselvi, C., Nasira, G.: Prediction of heart diseases and cancer in diabetic patients using data mining techniques. Indian J. Sci. Technol. **8**(14) (2015)
5. Santhanam, T., Ephzibah, E.: Heart disease prediction using hybrid genetic fuzzy model. Indian J. Sci. Technol. **8**(9), 797–803 (2015)
6. Yeh, Y.-C., et al.: A reliable feature selection algorithm for determining heartbeat case using weighted principal component analysis. In: 2016 International Conference on System Science and Engineering (ICSSE). IEEE (2016)
7. Dubey, V.K., Saxena, A.K.: Hybrid classification model of correlation-based feature selection and support vector machine. In: IEEE International Conference on Current Trends in Advanced Computing (ICCTAC). IEEE (2016)
8. Krishnaiah, V., Narsimha, G., Chandra, N.S.: Heart disease prediction system using data mining technique by fuzzy K-NN approach. In: Satapathy, S., Govardhan, A., Raju, K., Mandal, J. (eds.) Emerging ICT for Bridging the Future - Proceedings of the 49th Annual Convention of the Computer Society of India (CSI) Volume 1, pp. 371–384. Springer, Cham (2015)
9. Dominic, V., Gupta, D., Khare, S.: An effective performance analysis of machine learning techniques for cardiovascular disease. Appl. Med. Inform. **36**(1), 23–32 (2015)
10. Alizadehsani, R., et al.: A data mining approach for diagnosis of coronary artery disease. Comput. Methods Programs Biomed. **111**(1), 52–61 (2013)
11. Giri, D., et al.: Automated diagnosis of coronary artery disease affected patients using LDA, PCA, ICA and discrete wavelet transform. Knowl.-Based Syst. **37**, 274–282 (2013)
12. Al-Milli, N.: Backpropagation neural network for prediction of heart disease. J. Theor. Appl. Inf. Technol. **56**(1), 131–135 (2013)
13. Dbritto, R., Srinivasaraghavan, A., Joseph, V.: Comparative analysis of accuracy on heart disease prediction using classification methods. Int. J. Appl. Inf. Syst. (IJAIS) (2016). ISSN 2249-0868
14. Pandey, A.K., et al.: Datamining clustering techniques in the prediction of heart disease using attribute selection method. Heart Dis. **14**, 16–17 (2013)
15. Shouman, M., Turner, T., Stocker, R.: Using decision tree for diagnosing heart disease patients. In: Proceedings of the Ninth Australasian Data Mining Conference-Volume 121. Australian Computer Society, Inc. (2011)
16. Singh, N., Firozpur, P., Jindal, S.: Heart disease prediction system using hybrid technique of data mining algorithms (2018)
17. Agrawal, A., et al.: Disease prediction using machine learning (2018)
18. Shirwalkar, N., et al.: Human heart disease prediction system using data mining techniques (2018)
19. Cleveland Database. https://archive.ics.uci.edu/ml/datasets/heart+Disease

A Hybrid Optimization GNA Algorithm for the Quadratic Assignment Problem Solving

Kefaya Qaddoum(✉) and Azmi Al Azzam

Higher Colleges of Technology, Abu Dhabi, UAE
{kqaddoum, aalzaam}@hct.ac.ae

Abstract. The quadratic assignment problem (QAP) was considered one of the most significant combinatorial optimization problems due to its variant and substantial applications in real life such as scheduling, production, computer manufacture, chemistry, facility location, communication, and other fields. QAP is an NP-hard problem that is impossible to be solved in polynomial time when the problem size increases, hence heuristic and metaheuristic approaches are utilized for solving the problem instead of exact methods. Optimization plays a significant role in easing this problem. In this paper, we will provide a solution to optimize QAP. In the QAP problem, there is a total of facilities (departments, company's,…etc.) that must be located to minimize the flow (amount of material to be exchanged). Thus, the objective function is composed by multiplying both distances between the locations and the flow among these facilities. Global Neighborhood (GNA) Algorithm will be used to optimize the QAP problem, and the solution will also be compared to the well-known Genetic Algorithm (GA).

1 Introduction

Optimization serves many different fields of research. In general, most processes in research contains different parameters which need to be optimized to better solve problems despite all given constraints. The optimization problem mainly consists of an objective function and a set of constraints. The objective function can be in a mathematical form or combinatorial form. In the mathematical structure, the decision variables consist of one or more than one dimension. The other type of optimization problems includes of combinatorial problems where the decision variables are discrete or combination of numbers like the traveling salesman problem (TSP) and the quadratic assignment problem (QAP). Once the objective function of the optimization problem is formulated, and all the constraints are defined, then the main issue is to solve this problem. The optimal solution is usually the best values of the decision variables or the best scenarios. This optimal solution should give us the best performance or best fitness for the objective function. Different deterministic and heuristic algorithms have been proposed to solve optimization problems, although until now there is no evidence that one optimization algorithm can be best suited to address all optimization problem as stated by the No Free Lunch theorem (NFL) [1].

In the last few decades, a new smart kind of heuristic algorithms has emerged. This kind of algorithms was named meta-heuristics and was first introduced by Glover [2]. The meta-heuristic algorithms are just like the basic heuristic with additional features

L. Barolli et al. (Eds.): EIDWT 2019, LNDECT 29, pp. 455–464, 2019.
https://doi.org/10.1007/978-3-030-12839-5_42

that allow them to explore the entire search space. The term meta-heuristic was derived from the composition of two Greek words. A heuristic is derived from the verb heuriskein which means "to find," while the suffix Meta means "beyond, in an upper level." Before this term was widely used, meta-heuristic algorithms were often called modern heuristics [3].

The computational drawbacks of mathematical techniques and methods (i.e., complex derivatives, sensitivity to initial values, and a large amount of enumeration memory Required) have forced researchers to rely on meta-heuristic algorithms based on simulations and some degree of randomness to solve optimization problems [4]. Although, these meta-heuristic approaches are not exact or deterministic and they do not always give the optimal solution, but in most cases, they give a near optimal solution with less effort and time than the mathematical methods [5].

The meta-heuristic algorithms are general purpose stochastic search methods simulating natural selection and biological or natural evolution [6]. Different meta-heuristic algorithms have been developed in the last few decades imitating and emulating different processes. The biological evolutionary processes inspired some of these meta-heuristic algorithms; such as the evolutionary strategy (ES) [7], the evolutionary programming introduced by Fogel [8], De Jong [9] and Koza [10] and the genetic algorithm (GA) proposed by Goldberg [11] and Holland [12]. Some meta-heuristic algorithms were emulating different animal behaviors; like the tabu search (TS) introduced by Glover [13], the ant colony algorithm (ACA) by Dorigo et al. [14], Particle Swarm Optimization (PSO) by Kennedy and Eberhart [15], Harmony Search (HS) algorithm [16], Bee Colony Optimization (BCO) by Nakrani and Tovey [17]. Other meta-heuristic algorithms were inspired by different physical and natural phenomena like the simulated annealing (SA) proposed by Kirkpatrick et al. [18] and Gravitational Search Algorithm (GSA) by Rashedi et al. [19].

Swarm intelligence (SI) [20] is a particular type of meta-heuristic algorithms used to solve the optimization problems. SI algorithms are inspired by the natural swarms or species such as fish schools, bird swarms, bees, ants, insects' colonies and animal herds. Most of the SI algorithms study and imitate the behavior of swarm's members and there to locate the food sources. The food source usually represents the optimal solution. SI algorithms include many algorithms such as ant colony optimization (ACO) [21], particle swarm optimization (PSO) [22]. GNA is one of the new algorithms that can be used to optimize combinatorial problems. GNA was introduced and used in previous work to optimize different issues [23].

2 GNA Algorithm

The algorithm proposed in this paper is used to optimize combinatorial problems. The combinatorial problems could have more than one local and global optimal value within the search space values. The proposed methodology will work to find the optimal value among these optimal local switching between exploration and exploitation. Research allows for exploring the whole search space. Exploitation allows focusing the search next to the neighborhood of the best solution of generated solutions.

To explain the methodology of the GNA algorithm, assume we have a discrete function that we need to optimize and let us say that we need to minimize this function (without loss of generality).

So the objective function we have is:

$$g = f(x_1, x_2, \ldots, x_n) \tag{1}$$

Where,

$x_1, x_1 \ldots x_n$, are the different combinations of the solution sequence; we can think of these combinations as the city sequence in the TSP problem.

We need to find the optimal combination or solution $x_1, x_1 \ldots x_n$ that will give the optimal (minimum) value for the objective function (g). In general, if each of the variables $x_1, x_1 \ldots x_n$ can be chosen in $n_1, n_2, \ldots,\ n_n$ ways respectively, then if we want to enumerate all the possible solutions this will yield (n_1 × n_2 × … × n_n) Solutions. However, this process could take several hours or days depending on the size of the problem. Thus, using a meta-heuristic approach is better even if it does not always give the optimal solution, but in most cases, it will provide an answer that is close to the optimal solution with less computational power.

According to the GNA algorithm, a set of (m) random solutions are first randomly generated from the set of all possible solution, where: $(x\_1,\ x\_2, \ldots, x\_n)$ can be chosen in $(n\_1,\ n\_2, \ldots, n\_n)$ ways.

The generated solutions will then look like:

$$(x\_1^\wedge q, x\_2^\wedge q, \ldots, x\_n^\wedge q)$$

where

The fitness for the above solution will be evaluated, and this can be done by substituting them in the objective function (g).

The solutions are then sorted according to their fitness obtained from the objective function:

$$f(s_1) < f(s_2) < f(s_3) < \ldots < f(s_m);\ s_1 = (x'_1, x'_2, \ldots, x'_n)$$

is the solution sequence with best fitness.

The best combination (s1) is then used as a good measure for the optimal local solution, and it is also initially set as the best-known solution.

In the next iteration, 50% of the (m) generated solutions will be produced near the best solution neighborhood by using a suitable move operator.

The other 50% of the (m) generated solutions will be still made from the whole search space, and the reason for that is to allow for the exploration of the search space, because if we just choose the solutions close to the best solution we will only be able to find the local solution around this point, and since the function that need to be optimized could have more than one local optima, which might lead us to get stuck at one of these local optima.

Next, the best solutions from the above (m) solutions (50%, 50%) are calculated. The new value for the best solution is compared to the best-known solution, and if it were found to be better, it would replace it.

The procedure is then repeated until a specified stop criterion is met. This stops rules can be a pre-specified number of iterations (t), or when there is no further improvement on the final value of the optimal solution we obtained.

The flowchart for the proposed GNA is shown in Fig. 1.

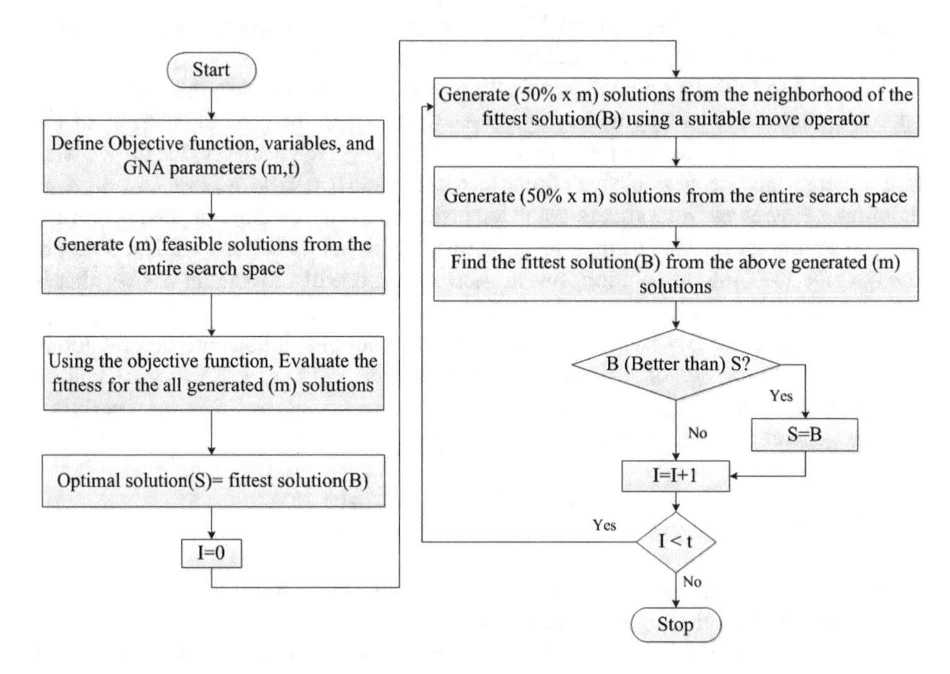

Fig. 1. Flowchart for Global Neighborhood Algorithm (GNA).

3 Quadratic Assignment Problem

Quadratic Assignment Problem (QAP) was mathematically modeled by Koopsman and Beckman in 1957 and applied in the field of economic activities [1]. QAP is one of the NP-hard combinatorial optimization problems. The problem aims at assigning n number of facilities to n number of locations with minimum cost. When the problem solving is achieved, each facility will be awarded to a location, and any facility should not be left empty. As a permutation problem, QAP can be mathematically formulated as follows:

$$\mathrm{cost}(\pi) = \sum_{i=1}^{n} \sum_{j=1}^{n} d_{i,j} \cdot f_{\pi(i),\pi(j)}$$

Where d is an *nxn* dimensional matrix at which the distances between the locations are kept, and *f* is *the annxn* dimensional matrix in which the flow cost between the facilities are maintained. The aim is to find the permutation array that may obtain the minimum value of cost function.

In the QAP problem, there is a total of facilities (departments, company's,...etc.) that must be located to minimize the flow (amount of material to be exchanged). Thus, the objective function is composed by multiplying both distances between the locations and the flow among these facilities Fig. 2.

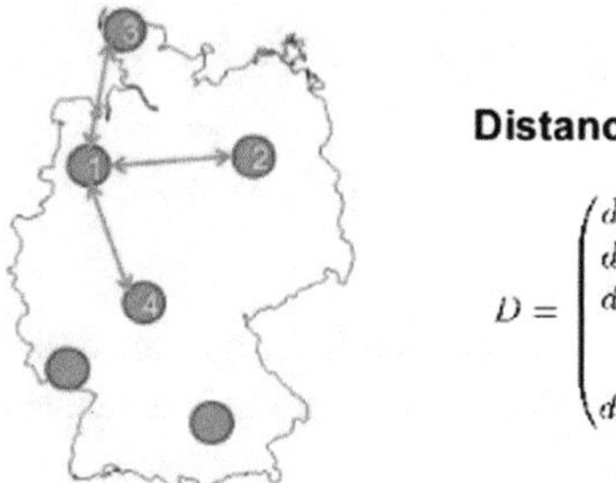

$$D = \begin{pmatrix} d_{1,1} & d_{1,2} & d_{1,3} \cdots & d_{1,n} \\ d_{2,1} & d_{2,2} & d_{1,3} \cdots & d_{2,n} \\ d_{3,1} & d_{3,2} & d_{3,3} \cdots & d_{3,n} \\ \vdots & \vdots & \ddots & \vdots \\ d_{n,1} & d_{n,2} & d_{n,3} \cdots & d_{m,n} \end{pmatrix}$$

Fig. 2. QAP architecture

This is the fifth of the QAP test problems of [24]. Twelve departments are to be placed in twelve locations with four in the top row, four in the middle row, and four in the bottom row. The objective is to minimize flow costs between the placed departments. The flow cost is (flow × distance), where both flow and distance are symmetric between any given pair of departments.

Below is the flow and distance matrix where rectilinear distance is the upper half. The optimal flow costs are 289 Table 1.

Table 1. Flow-distance matrix

Flow\dist.	1	2	3	4	5	6	7	8	9	10	11	12
1	–	1	2	3	1	2	3	4	2	3	4	5
2	5	–	1	2	2	1	2	3	3	2	3	4
3	2	3	–	1	3	2	1	2	4	3	2	3
4	4	0	0	–	4	3	2	1	5	4	3	2
5	1	2	0	5	–	1	2	3	1	2	3	4
6	0	2	0	2	10	–	1	2	2	1	2	3
7	0	2	0	2	0	5	–	1	3	2	1	2
8	6	0	5	10	0	1	10	–	4	3	2	1
9	2	4	5	0	0	1	5	0	–	1	2	3
10	1	5	2	0	5	5	2	0	0	–	1	2
11	1	0	2	5	1	4	3	5	10	5	–	1
12	1	0	2	5	1	0	3	0	10	0	2	–

4 Experimental Results and Analysis

The GNA algorithm was used to solve the Quadratic Assignment Problem (QAP). The QAP problem consists of some 12 departments that need to communicate with each other based on the flow given in the matrix above. To optimize this QAP problem, we want to find the optimal location of these different departments that gives the minimum total cost (distance × flow). Thus, the objective (value) function for the QAP is provided by:

$$C(s) = \sum_{i=1}^{12} \sum_{n=j}^{12} d(P_i, P_j) \times f(P_i, P_j)$$

Where:

C(S): The total cost for a given sequence
d(P_i, P_i): The distance between the assigned locations for department i and j
f(P_i, P_j): The cost (flow) between department i and j The sequence of a possible candidate solution will look like:

1	5	7	8
11	2	3	12
6	10	4	9

Note that in this example we assigned department 1 to location one and department 2 to location 6. Therefore, to calculate the total cost, we should calculate the value from each department to the other.

To solve the QAP problem, we have to find the optimal sequence (S) that will give the minimum cost. If all the possible solutions are to be checked, then a total number of the combinations will be (N!) for asymmetric QAP or (N!/2) for the symmetric QAP. If the amount of the department (N) is small, then all the combination can be tried, and a deterministic optimal solution can be found. However, if the amount of the department is large, then checking all the possible solution will take a very long time, and the complexity of the QAP problem will grow exponentially. For this reason, different meta-heuristic algorithms have been widely used to solve QAP problems.

In this paper, the GNA algorithm is used to solve the QAP problem. The optimal solution for this problem is known and documented (289). The GNA algorithm was implemented using MATLAB software, and the total number of solutions (m) generated at each iteration was 50. At, each iteration 25 feasible solutions were created from the whole search space, and the other 25 solutions were designed from the neighborhood of the fittest solution. The neighborhood move operator that was used in our case is the two-opt swap; where two departments were randomly chosen and swapped. The code was run for different times, and at each time the obtained optimal solution and the runtime were recorded. The stopping criteria used was 100 iterations. The results for the GNA algorithm from the MATLAB code are shown in Table 2.

Table 2. MATLAB output for using GNA to solve the QAP problem

Run	The fitness of optimal solution	No. of iterations	Run time (Sec)
1	293	100	15.32
2	293	100	14.89
3	296	100	15.02
4	301	100	16.03
5	291	100	17.03
6	289	100	16.04
7	291	100	15.34
8	306	100	14.89
9	291	100	15.09
10	293	100	16.22

As can be seen from Table 2, in the ten different run times, we obtained a near-optimal solution that is very close to the known optimal solution (289). Furthermore, a single statistical analysis using MINTAB was also conducted to test the optimal Solution as follow:

H0: Mean = 289
Ha: Mean $\neq$ 289

The one sample t-test using Megastat showed the following results:
One-Sample T: Solution

```
Test of mu = 289 is not = 289
Variable     N       Mean       StDev       SE Mean            95% CI
T      P
Solution   10     294.1       5.27      1.67      (290.63, 298.17)
3.24    0.0102
```

This means that we fail to reject the null hypothesis (Mean = 289) since P-value = 0 (< 0.05)

Also, the confidence interval for the mean is (290.63, 298.17) which means that on average we are 95% confident that the way is within this interval.

The results of the GNA algorithm were also compared to the Genetic algorithm (GA).

The parameters for the genetic algorithm were as the following:

Generation size: 50
Crossover probability: 95%
Mutation probability: 0.05%
Number of iterations: 10000

Table 3 shows that the runtime for the GA is more than twice the run time for the GNA, and the solution obtained by the GA is not always close to the known optimal solution. Megastat software was used to conduct statistical analysis between the means of the two optimal solutions obtained by both GA and GNA.

Table 3. MATLAB output for using GA to solve the QAP problem

Run	The fitness of optimal solution	No. of iterations	Run time (Sec)
1	293	100	25.90
2	312	100	27.47
3	306	100	28.23
4	296	100	28.37
5	312	100	27.13
6	316	100	29.91
7	296	100	24.59
8	302	100	26.42
9	293	100	29.33
10	296	100	27.82

The 2 sample t-test using mega stat showed the following results:

Two-Sample T-Test and CI: GNA, GA

```
Two-sample T for GNA vs. GA
     N    Mean  StDev  SE Mean
GNA  10  294.4   5.27     1.67
GA   10   302.2  8.7      2.75

Difference = mu (GNA) - mu (GA)
Estimate for difference:  -7.8
95% CI for difference:   (-14.702, -0.898)
T-Test of difference = 0 (vs not =): T-Value  = -2.42  P-Value
= 0.0295  DF = 9
```

A 2-Sample t-test showed that there is a significant difference between the optimal solution obtained from both GNA and GA, P-value = 0.295.

And also 95% CI for difference: (–14.702, –0.898).

Since the difference is always negative as indicated by the CI. This shows us that on average the solution is always greater for the GA than it is for the GNA. Therefore, the fittest solution provided by the GNA is better and closer to the optimal solution.

5 Conclusion

In this paper, The GNA meta-heuristic optimization method was used to solve the QAP problem. This optimization method is a population-based algorithm; since it starts with generating a set of random solutions from the search space for the optimization problem. The proposed algorithm can be used to solve combinatorial optimization problems. These combinatorial problems are usually more difficult to answer than other continuous optimization problems. The methodology of this algorithm was elaborated and 12 department QAP optimization problem was solved using the GNA. The QAP optimization problem was also solved using a genetic algorithm (GA), and the result was compared to the GNA. Statistical analysis was conducted using mega stat software, and it was found that the GNA showed better performance, and the results obtained were very close to the known optimal solution. Future studies can include different variants for the basic GNA algorithm as well as various instances of QAP.

References

1. Wolpert, D.H., Macready, W.G.: No free lunch theorems for optimization. IEEE Trans. Evol. Comput. **1**, 67–82 (1997)
2. Glover, F.: The future paths for integer programming and links to artificial intelligence. Comput. Oper. Res. **13**(5), 533–549 (1986)
3. Reeves, C.R.: Modern Heuristic Techniques for Combinatorial Optimization Problems. Blackwell Scientific, Oxford (1993)
4. Lee, K.S., Geem, Z.W.: A new structural optimization method based on the harmony search algorithm. Comput. Struct. **82**, 781–798 (2004)
5. Lee, K.S., Geem, Z.W.: A new meta-heuristic algorithm for continuous engineering optimization. Comput. Methods Appl. Mech. Eng. **194**(2005), 3902–3933 (2005)
6. Omran, M.G., Mahdavi, M.: Global-best harmony search. Appl. Math. Comput. **198**, 643–656 (2008)
7. Rechenberg, I.: Cybernetic solution path of an experimental problem, Royal Aircraft Establishment, Library Translation no. 1122 (1965)
8. Fogel, L.J., Owens, A.J., Walsh, M.J.: Artificial Intelligence Through Simulated Evolution. Wiley, Chichester (1996)
9. De Jong, K.: Analysis of the behavior of a class adaptive systems, Ph.D. Thesis. University of Michigan, Ann Arbor, MI (1975)
10. Koza, J.R.: Genetic programming: a paradigm for genetically breeding populations of computer programs to solve problems, Report No. STAN-CS-90- 1314. Stanford University, Stanford, CA (1990)
11. Goldberg, D.E.: Genetic Algorithms in Search Optimization and Machine Learning. Addison-Wesley, Boston (1989)
12. Holland, J.H.: Adaptation in Natural and Artificial Systems. University of Michigan Press, Ann Arbor (1975)
13. Glover, F.: Heuristic for integer programming using surrogate constraints. Decis. Sci. **8**(1), 156–166 (1977)
14. Dorigo, M., Maniezzo, V., Colorni, A.: The ant system: optimization by a colony of cooperating agents. IEEE Trans. Syst. Man Cybernet. **26**(1), 29–41 (1996)
15. Kennedy, J., Eberhart, R.: Particle swarm optimization. In: IEEE International Conference on Neural Networks Perth, Australia, pp. 1942–1948 (1995)
16. Geem, Z.W., Kim, J.H., Loganathan, G.: A new heuristic optimization algorithm: harmony search. Simulation **76**(2), 60 (2001)
17. Nakrani, S., Tovey, C.: On honey bees and dynamic server allocation in internet hosting centers. Adapt. Behav. **12**(3–4), 223 (2004)
18. Kirkpatrick, S., Gelatt, C., Vecchi, M.: Optimization by simulated annealing. Science **220** (4598), 671–680 (1983)
19. Rashedi, E., Nezamabadi, H., Saryazdi, S.: GSA: a gravitational search algorithm. Inform. Sci. **179**(13), 2232–2248 (2009)

20. Engelbrecht, A.P.: Fundamentals of Computational Swarm Intelligence. Wiley, Hoboken (2006)
21. Dorigo, M., Gambardella, L.M.: Ant colony system: a cooperative learning approach to the traveling salesman problem. IEEE Trans. Evol. Comput. **1**(1), 53–66 (1997)
22. Eberhart, R., Kennedy, J.: A new optimizer using particle swarm theory. In: Proceedings of the Sixth International Symposium on Micro Machine and Human Science MHS 1995, pp. 39–43. IEEE (1995)
23. Alazzam, A., Lewis III, H.: A new optimization algorithm for combinatorial problems. (IJARAI) Int. J. Adv. Res. Artif. Intell. **2**(5) (2013)
24. Nugent, C.E., Vollman, T.E., Ruml, J.: An experimental comparison of techniques for the assignment of facilities to locations. Operations (1968)

An Enhanced Model for Abusive Behavior Detection in Social Network

Kefaya Qaddoum(✉), Israr Ahmad, Yasir Javed, and Ali Rodan

Higher College of Technology, Al Ain Women's College, Abu Dhabi, UAE
{kqaddoum, iahmad, yjaved, arodan}@hct.ac.ae

Abstract. Due to the growing use of social media, incidents of online abuse are also on rise. Online abusive behavior is defined as the use of electronic devices connected through internet for offensive activities. It is mostly in the form of comments containing abusive words about others, which affect the target users' psychology and depresses them. This paper is aimed at devising method for detecting abusive behavior using supervised learning techniques. Two hypotheses are presented to extract features for detection of offensive comments. The initial experiments show that using features using our proposed method has better accuracy than the traditional feature extraction techniques like TF-IDF.

1 Introduction

Pickering [1] examined discussions in social media on the STEM (Science, Technology, Engineering, and Mathematics) through the span of multi-year to discover who are the most critical individuals that may impact STEM instruction talks and the patterns in the themes of discussions. The discoveries uncover that the most compelling clients speak to associations even though the people affect such discussions. They talked and explored the education framework throughout the last two decades. It has been endeavoring to reevaluate itself from just intellectual arrangement for academics or employable works (know-what and know-why) to a broadened stage centering on the advancement of students' 21st-century capabilities (know-how and know-who). They utilized a social network analysis (SNA) approach to investigate the idea of the STEM training network on twitter analyzing 146,976 tweets, posted by 69,459 users, both in terms of the people as well as the content. They examined multiple assigned tags like #science (tweets = 7,349) followed by #education (tweets = 5,301) and #steam (tweets = 4,300). The results of this study explored the key players and the sorts of data that are available on the Twitter social network. The authors claimed that these discoveries would guide educators throughout the STEM education pipeline.

An approach used by Li [2] for identifying anomalous behaviors in communication networks for detecting user behavior. They focused directly on user's behavior rather than examining the patterns shaped by multiple users. An email dataset gotten from the email log records of a college email server was used. Because of security issues, the email addresses are anonymized, and the substance of the messages are not open, they answer on non-printed data, for example, the timestamp, sender email locations and beneficiary email addresses with 81,244 messages and 1,238 sender emails. Based on

L. Barolli et al. (Eds.): EIDWT 2019, LNDECT 29, pp. 465–471, 2019.
https://doi.org/10.1007/978-3-030-12839-5_43

user historical behaviors, they constructed the benchmark of behaviors and then measure the turns of behavior in each snapshot. They introduced a transform process to create a comparable normalized score as the indicator of user behavior in each snapshot. The experiment on email dataset has demonstrated that their approach was easily interpretable for the user behaviors. Exception scores succession instinctively showed the difference in client conduct. They encouraged to spot critical occasions from vast masses of system depictions. If a few depictions with numerous irregular clients, at that point they deduced the event of unforeseen occasions in that time event. Future work could center on enhancing the approach utilizing more highlights, or a superior benchmark determination.

Chen et al. [3] explored deep neural network learning to extract knowledge from the social media tools. They used neural networks learning to mine text data gathered from social media tools. The convolutional neural networked are designed to classify test data into pre-defined categories. The text is represented using a bag of words. The neural network performance is measured using precision, recall, and f-score. The neural networks are compared with Support vector machines and feedforward neural networks. The results show that convolutional neural networks perform better than its rivals do.

Twitter data set is evaluated by using deep learning neural networks by [4]. The authors applied neural networks to predict geographical location by extracting knowledge from the text data. This work used three datasets: GeoText, TwUS, and TwWORLD for the evaluation of the proposed neural network architecture. The architecture of the neural network includes loss function as activation function for each neuron and connection weights are randomly initialized. The neural network performance is compared with a different setting, i.e., sigmoid, linear and RELU activations. The accuracy rate is used to measure the performance. The results show that RELU performs better than other sets.

Ivaschenko et al. [5] explored an agent-based approach to model the user behaviors for social media applications. The model is set up by integrating several agents together to perform behavior modeling. Each agent is autonomous in terms of its learning processes and performs a sub-task in user behavior modeling. For example, one agent is only to do user profile, and another agent does context-awareness. The design of such a model needs a multi-agent approach. The model gets social media data as input and produces posts and comments depending on the user profile and the current context of the information. The multi-agent approach is implemented using an asynchronous communication-based architecture. The experiments are performed on a user profile where the user profile of three years is investigated.

Gupta et al. [6] used particle swarm based neural network training algorithm to detect if a URL is infectious for a phishing attack or not. The neural network design is based on a feed-forward architecture and weights are initialized using the random initialization and then tuned it using particle swarm optimization. The authors used the MATLAB implementation of the neural networks. For the evaluation of the algorithms, the Phish tank dataset is used. The neural network training and testing are validated using Holdout. The performance of the neural networks is measured using mean squared error (mse). The results show that the hybrid of neural networks and particle swarm optimization (PSO) is better than backpropagation based neural networks and PSO.

Li [7] investigated centrality and degree of graphs for social networks. The social network graphs were analyzed to profile individual behavior in a social network from a communication point of view. The network is searched using four parameters: indegree centrality, out-degree centrality, degree centrality, between centrality. These parameters depend on many links in a social network, e.g., number of friends in a Facebook account. The authors measure these parameters for two individuals. The results show that such measures can be used to study an individual's characteristics regarding its social interaction and profiling in a given social network.

2 Proposed Framework

This section proposes the methodology and framework for classification of comments as shown diagrammatically in Fig. 1. The steps involved are Normalization, standard Feature extraction, additional feature extraction, feature selection and finally classification.

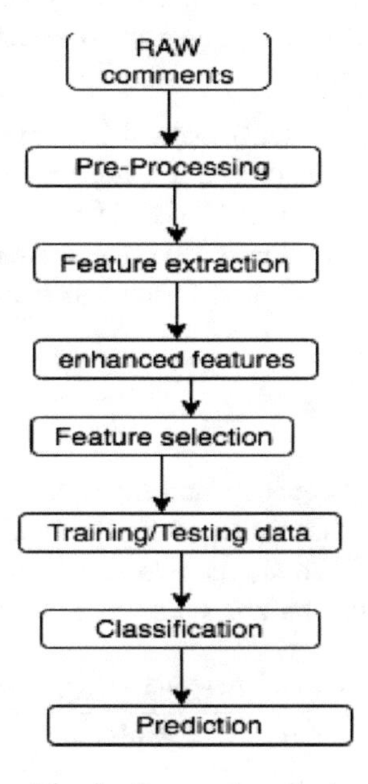

Fig. 1. Proposed method

(a) Pre-processing

The Dataset used contains a list of comments and respective labels. These are to be converted into the feature list, which is used by the machine-learning algorithm. For this, different Natural language processing techniques are used to obtain an accurate representation of the comments in feature list form. Various techniques were used based on our observations.

Data is taken from repository of Thomas Davidson, Dana Warmsley, Michael Macy, and Ingmar Weber. 2017. "Automated Hate Speech Detection and the Problem of Offensive Language. Removing unwanted strings: For the comments to be used by machine-learning algorithms should be in a standard form. Their raw comments present in the dataset should be fixed before feeding the dataset to the classifier; an automatic pre-processing procedure assembles the comments for each user and chunks them into sentences. For each sentence in the sample dataset, an automatic spelling and grammar correction process precedes the introduction of the sample dataset to the classifier. With the help of WordNet, correction of spelling and grammar mistakes in the raw sentences occurs by tasks such as deleting repeated letters in words, deleting meaningless symbols, splitting long words, transposing substituted letters, and replacing the incorrect and missing letters in words. As a result, words missing letters, such as "speling," are corrected to "spelling"; misspelled words, such as "korrect," changes to "correct".

(b) Feature Selection

If all the features are considered, they will be order of some hundred thousand and therefore, the machine learning algorithms cannot handle them altogether. Therefore, selection of best features was required. A statistical hypothesis method known as "Chi-Squared test" was used on our feature matrix to select k best features where k is parameter roughly equal to 5000.

Chi-Square Method: chi-square (X2) method is commonly used for selecting best features. This metric calculates the cost of a feature using the value of the chi-squared statistics for class. Initially, a hypothesis H0 is assumed that the two features are unrelated, and it is. The initial hypothesis H0 is the assumption that the two features are unrelated, and it is tested by chi-squared formula as is shown in Eq. (1)

$$\chi^2 = \sum \frac{(oij - eij)^2}{eij} \tag{1}$$

Where *Oij* is the observed frequency, and *Eij* is the expected frequency, asserted by the null hypothesis. Higher the value of (X2), greater the evidence against the hypothesis H0, hence more related are the two variables. Lesser the value of (X2), the hypotheses tends to be true, the variables are independent.

$$X^2 = \sum \frac{(\text{observed} - \text{expected})^2}{\text{expected}} \tag{2}$$

Naïve Bayes (NB) is a statistical classifier that predicts class membership probability and classifies input data set upon them, which means it calculates probability whether input dataset belongs to a certain class or not and get the whole dataset classified in this way. As [6] and [7] indicated it's based on the Bayes rule which is defined as follows:

$$\mathrm{P(C|A) = ((A|C).\, P(C))/(P(A))}$$

(a) Classification

Once the features are built, we extract the best features using chi-squared test and apply the machine learning algorithms to train models on it. We have used Support Vector Machine (SVM) on our feature data. SVM algorithm maps the training data into feature space using kernel functions and then separates the dataset using large hyperplane. We have used a linear kernel function.

$$K(x_i, x_j) = x_j x_i^T \tag{3}$$

Where K (xi, xj) represents the dot product of input data points xi mapped into sizeable dimensional feature space xj by transformation function.

Machine learning techniques—Naïve Bayes (NB) and SVM—are used to perform the classification using Weka software, where 10-fold cross-validation was conducted in this experiment. To fully evaluate the effectiveness of users' sentence offensiveness value we fed them sequentially into the classifiers, and It uses offensive words as the base feature to detect the offensive user. Offensive words and user language features are not compensating to each other to improve the detection rate, which means they are not independent. In contrast, incorporating with user language features, the classifiers have better detection. While all three types of features are used to improve the classification rate, style features and content features are more valuable than structure features in user offensiveness classification. One possible reason is that once the number of actively offensive words beyond a certain amount, the user who posts the comments is considered being offensive.

(b) Performance Measures

Accuracy: Depends on the ability of the classifier to rank patterns, but also on its ability to select a threshold in the ranking used to assign patterns to the positive class if above the threshold and to the negative class if below. [8] Mentioned the equation to calculate it by the following equation, noting that an abbreviation of TP indicates true positive predictions, TN True Negative predictions, FP False positive predictions and FN False Negative predictions:

$$\text{Accuracy} = (\text{TP} + \text{TN})/(\text{TP} + \text{FP} + \text{TN} + \text{FN})$$

Precision: Measures proportion of truly predicted positive results. It was found to be dealing with substitution and insertion errors which can be calculated through the following equation as mentioned by [14]:

$$\text{Precision} = \text{TP}/(\text{TP} + \text{FP})$$

Recall: Measures the truly predicted positive and were correctly identified. It deals with substitution and deletion errors, [14] had stated that the following formula could calculate it:

$$\text{Recall} = \text{TP}/(\text{TP} + \text{FN})$$

Area Under Curve (AUC): Measures the classifiers skill in ranking a set of patterns according to the degree to which they belong to the positive class, but without actually assigning patterns to classes.

At first, we built feature list containing standard feature extraction. Then we trained our algorithms based on these feature lists and the best accuracy achieved was of 79%. Then, we included occurrence of pronouns and skip-gram as features, which enhanced the accuracy and outperformed in this too with 80%. The test datasets used for our experiment contained nearly 4320 unlabeled comments. Also, we tried to train the system with all the features using SVM.

An experimental result suggests that comments targeted towards peers help in detecting abusive behavior more efficiently. Table 1 shows the accuracy, precision and recall values where Fig. 2 demonstrates the increase in accuracy by adding extra features.

Table 1. Performance

Iterations	AUC score	Recall	Precision
1200	82.84	0.72	0.753
1630	83.84	0.71	59.64
1600	83.92	0.71	0.791

We further examined the classification rates for different feature sets using Naïve Bayes and SVM classifiers. Since the rates vary from time to time, we run each instance 15 times and take the average. The classification rate is independent of the number of users and the number of sentences. Generally, SVM produces more accurate classification results.

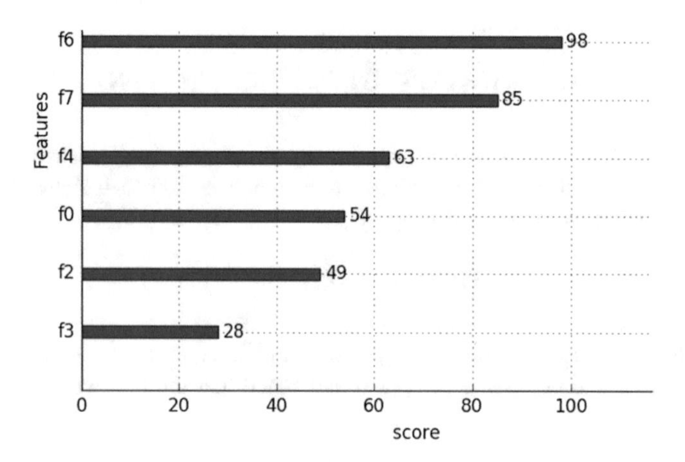

Fig. 2. AUC score with features increase

3 Conclusion

In this study, we investigate existing machine learning methods in detecting offensive contents. We presented a new hypothesis for feature extraction which can be helpful in detecting abusive contents. The result is the probability of comment being offensive to participants. Results show that our hypothesis increases the accuracy by 4% and can be used to detect the comments that are targeted towards peers. We improved the traditional style features, structure features, and context specific features to predict better a user's potential to send out offensive content in social media. The experimental result shows that user's sentence offensiveness prediction and user's offensiveness estimate algorithms outperform traditional learning-based approaches regarding precision and recall. We believe that such a language-processing model can help online offensive language monitoring, and ultimately building a finer online environment.

References

1. Pickering, T.A., Yuen, T.T., Wang, T.: STEM conversations in social media: implications on STEM education. In: 2016 IEEE International Conference on Teaching, Assessment, and Learning for Engineering (TALE), Bangkok, pp. 296–302 (2016). https://doi.org/10.1109/TALE.2016.7851810
2. Li, Q., Liu, P.: Detecting user behavior anomalies in communication networks. In: 2017 IEEE 2nd International Conference on Cloud Computing and Big Data Analysis (ICCCBDA), Chengdu, pp. 384–388 (2017). https://doi.org/10.1109/icccbda.2017.7951943
3. Chen, Y., Lv, Y., Wang, X., Wang, F.Y.: A convolutional neural network for traffic information sensing from social media text. In: 2017 IEEE 20th International Conference on Intelligent Transportation Systems (ITSC), Yokohama, pp. 1–6 (2017). https://doi.org/10.1109/itsc.2017.8317650
4. Lourentzou, I., Morales, A., Zhai, C.: Text-based geolocation prediction of social media users with neural networks. In: 2017 IEEE International Conference on Big Data (Big Data), Boston, MA, pp. 696–705 (2017). https://doi.org/10.1109/bigdata.2017.8257985
5. Ivaschenko, A., Khorina, A., Isayko, V., Krupin, D., Bolotsky, V., Sitnikov, P.: Modeling of user behavior for social media analysis. In: 2018 Moscow Workshop on Electronic and Networking Technologies (MWENT), Moscow, pp. 1–4 (2018). https://doi.org/10.1109/MWENT.2018.8337258
6. Gupta, S., Singhal, A.: Phishing URL detection by using artificial neural network with PSO. In: 2nd International Conference on Telecommunication and Networks (TEL-NET), Noida, pp. 1–6 (2017). https://doi.org/10.1109/tel-net.2017.8343553
7. Li, H.: Centrality analysis of online social network big data. In: 2018 IEEE 3rd International Conference on Big Data Analysis (ICBDA), Shanghai, pp. 38–42 (2018). https://doi.org/10.1109/ICBDA.2018.8367648
8. Whittaker, E., Kowalski, R.M.: Abusive behavior via social media. J. Sch. Violence **14**(1), 11–29 (2015)

Proposal of a Regional Culture Inheritance System by AR Technology

Hengyi Li[1] and Tomoyuki Ishida[2](✉)

[1] Ibaraki University, Hitachi, Ibaraki 316-8511, Japan
17nm725r@vc.ibaraki.ac.jp
[2] Fukuoka Institute of Technology, Fukuoka, Fukuoka 811-0295, Japan
t-ishida@fit.ac.jp

Abstract. In this research, we implement a regional culture inheritance system by digital archiving of regional cultural assets. This system realizes inheritance of regional culture and dissemination of information on cultural activities. This system is a smartphone application utilizing Web-GIS and AR (Augmented Reality) technology. By utilizing Web-GIS technology, this system provides users with multi-dimensional information visualization and location information reference. Also, by utilizing AR technology, this system provides users with browsing of regional culture contents superimposed on real space.

1 Introduction

Each region of Japan has various regional cultures such as unique regional history and tradition. However, in recent years the inheritance of regional culture is critical due to various factors such as depopulation and aging due to the sharp decline of population, changes in lifestyle and value accompanying the economic growth of Japan, aging of historic buildings and environmental destruction. In such a critical situation, a mechanism for preserving and inheriting the regional culture is strongly required [1–3].

On the other hand, due to the spread of mobile terminals such as smartphones in recent years, AR (Augmented Reality) application attracts attention. By using this AR technology, it became possible to reproduce the regional culture that existed in the past on the real space.

2 Research Objective

In this research, we focus on knowledge of regional culture and construct the regional culture inheritance system. This system enables sharing via the Internet by digital archiving of regional cultural assets. This system creates opportunities to learn about regional culture and knowledge, and disseminates these information to the world. In addition, this system visualizes the location and tradition related to regional culture using electronic map. This research promotes understanding of regional culture by visualizing on the map by linking regional tradition and location information. Furthermore, this system guides users to the location where regional culture exists through location information presented on the map by using AR technology. This system

L. Barolli et al. (Eds.): EIDWT 2019, LNDECT 29, pp. 472–481, 2019.
https://doi.org/10.1007/978-3-030-12839-5_44

provides the user with media information of regional culture (photo, outline of text-based regional culture) as an AR. In this research, we provide rediscovery of regional culture of local residents and promotion of interest in regional culture by the regional culture inheritance system.

3 System Configuration

The system configuration of this system is shown in Fig. 1. This system has Web-GIS function and AR function. This system manages the location of regional cultural contents by the Web-GIS function, and transmits regional culture information by AR function.

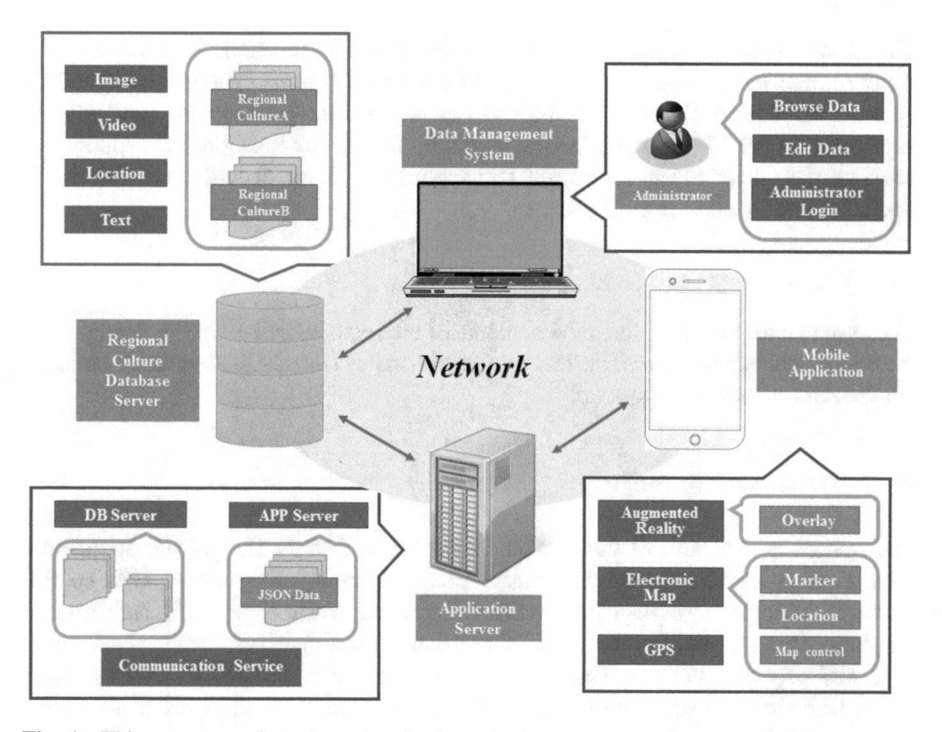

Fig. 1. This system consists of a regional culture database server, an application server, a mobile application, and a data management system.

3.1 Regional Culture Information Database Server

The regional culture information database server stores contents of regional culture. This database server transmits regional culture information (moving images, images, photographs, etc.) according to a request received from the application server.

This system deals with the traditional culture of Hitachi City, Ibaraki Prefecture, “Hitachi Furyumono” and “Hitachi Sasara” as content of regional culture. “Hitachi Furyumono” designated an important intangible folk cultural property. Also, “Hitachi Sasara” designated a cultural property by Ibaraki Prefecture.

3.2 Application Server

The application server manipulates the database in response to a request received from the mobile application. The information received from the database is transmitted to the mobile application. In addition, this server updates the data in the database in response to the request received from the data management system.

3.3 Mobile Application

The mobile application has Web-GIS function and AR function. Smartphones and tablet terminals have camera function and GPS function to use AR technology. This application acquires the location information in real time by the GPS function and superimposes the AR object on the real world by the camera function. This application acquires data from the database via the application server according to the user’s operation and displays it on the screen.

3.4 Data Management System

The system administrator stores the contents of the regional culture in the database via the data management system. Further, the administrator can edit the registered contents via the data management system.

4 System Architecture

The system architecture of this system is shown in Fig. 2. The mobile application consists of user interface, activity manager, asynchronous task manager, GPS control manager, camera control manager, media object manager, street map API, and network interface. The application server consists of data output manager, database edit manager, and network interface. The data management system consists of user interface, data edit manager, data show manager, and network manager. The Database Server stores contents of regional culture (latitude and longitude, images, texts, moving images, etc.).

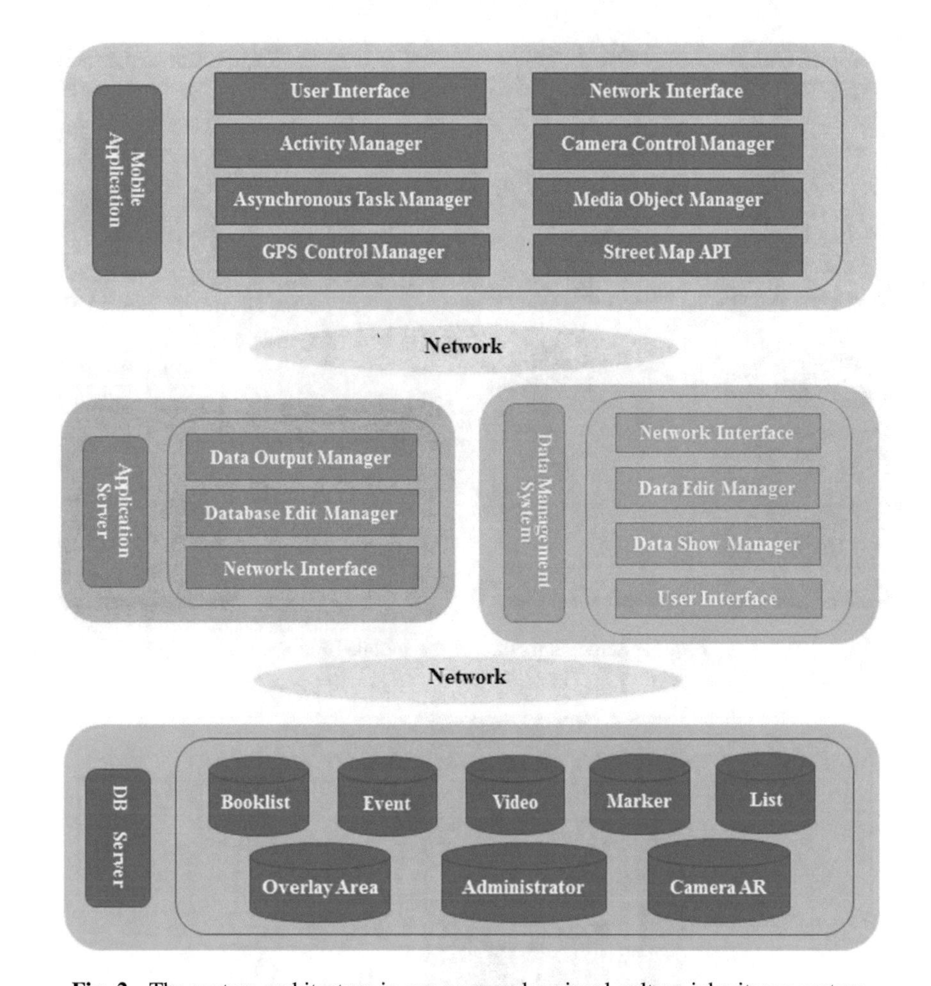

Fig. 2. The system architecture in our proposed regional culture inheritance system.

5 Prototype System

This chapter described the prototype of mobile application and data management system.

5.1 Mobile Application

Mobile terminals have portability and processing performance to realize AR. Therefore, we implemented the mobile application. Through this mobile application, users can experience regional culture anytime and anywhere. The screen structure of the mobile application is shown in Fig. 3.

On the digital content screen, the user can browse digitized regional culture contents. In addition, the user can control (move, enlarge, reduce) regional culture content by touch operation. As a result, the user can browse the details of the digital contents.

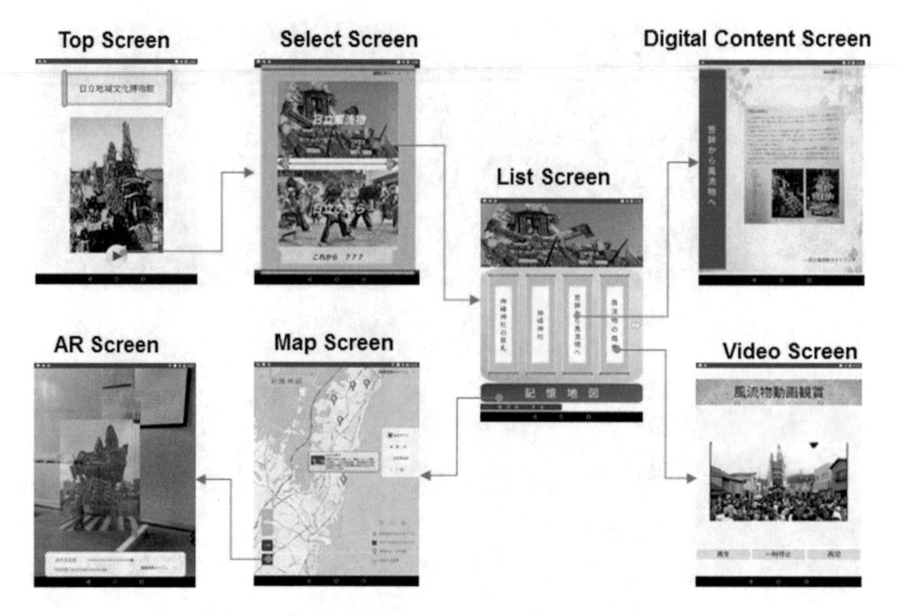

Fig. 3. Screen structure of the mobile application.

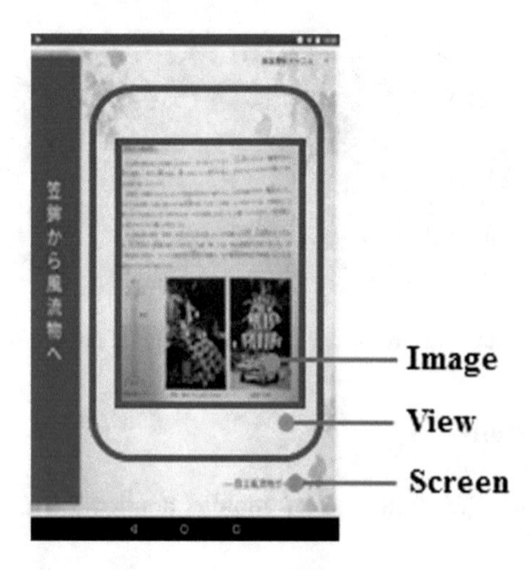

Fig. 4. Structure of the digital content screen.

The structure of the digital content screen is shown in Fig. 4. In this screen, a layer structure is constructed based on the screen and an image view is generated. Digital contents are drawn on the generated image view.

When the user selects the video button on the list screen, the screen changes to the video screen (Fig. 5). The video screen provides the user with videos related to regional culture. When the user selects the "Play" button at the bottom of the video screen, the user can view various video contents.

Fig. 5. The video screen is composed of “Play” button, “Pause” button, and “Restart” button.

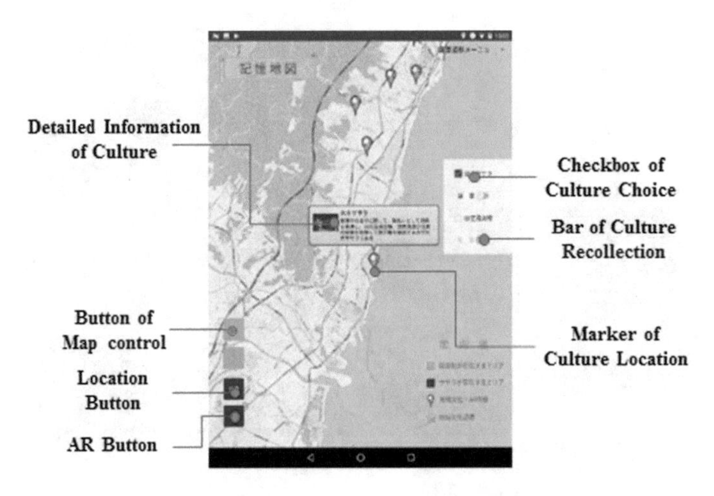

Fig. 6. The map screen is composed of markers, location button, AR button, and the like.

When the user selects the map button on the list screen, the screen changes to the map screen (Fig. 6). The map screen is an electronic map that visualizes regional cultural contents on a map. On the map screen, the registered regional culture contents are displayed as markers. When the user selects a marker, detailed information on the selected regional culture is displayed. As a result, the user can learn the regional cultural contents on the map screen.

When the user selects the AR button on the map screen, the screen changes to the AR screen (Fig. 7). This application has the function of location-based AR. The location-based AR is a technique for displaying the AR object on the camera view according to the location information. In this research, we registered the latitude and

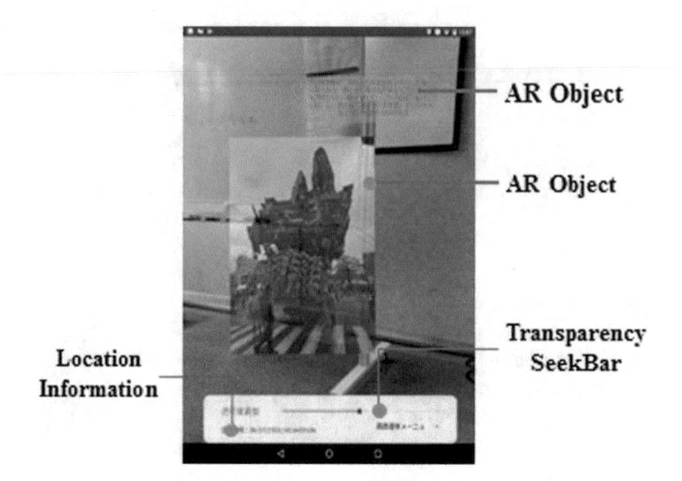

Fig. 7. Users can experience regional cultural contents as AR on the AR screen.

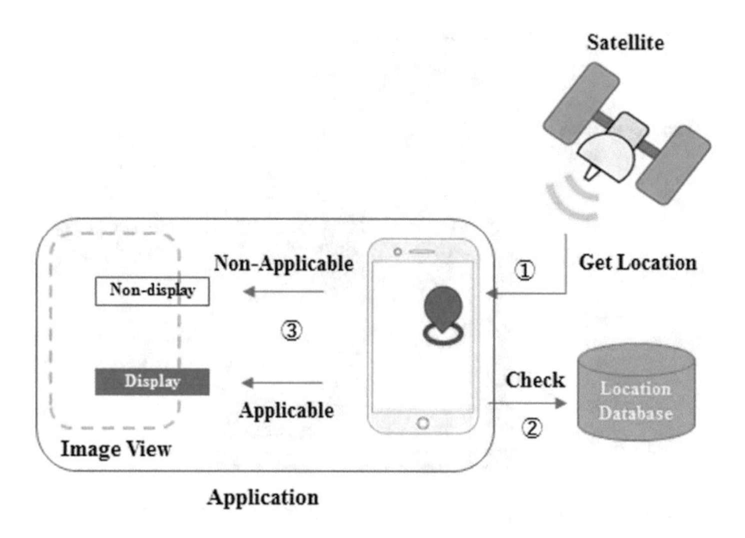

Fig. 8. Specifications for realizing location-based AR.

longitude of the place where regional culture exists in the system. When the user activates the AR screen at the location where the regional culture contents are registered, detailed information on the regional culture contents is displayed as AR.

The specifications for realizing location-based AR is shown in Fig. 8.

The application receives location information from the satellite. The received location information is collated with the location information stored in the Location Database. Display/Non-display of Image View (AR Object) is managed according to the collation result.

5.2 Data Management System

In this research, we have constructed the data management system that enables administrators to edit databases storing regional cultural contents from the Web. By managing regional culture contents via the Web system, the administrator can easily

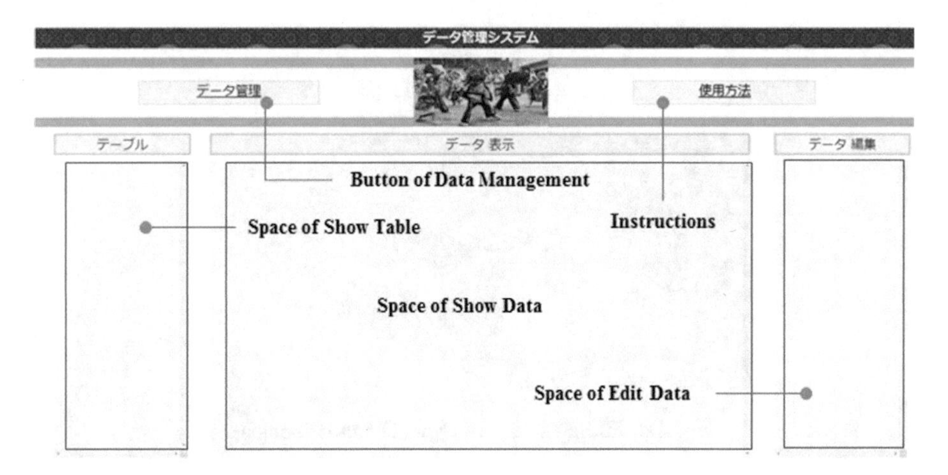

Fig. 9. The data management system consists of login function, data list display function, and data editing function.

Fig. 10. Regional culture contents registered in the regional culture database server.

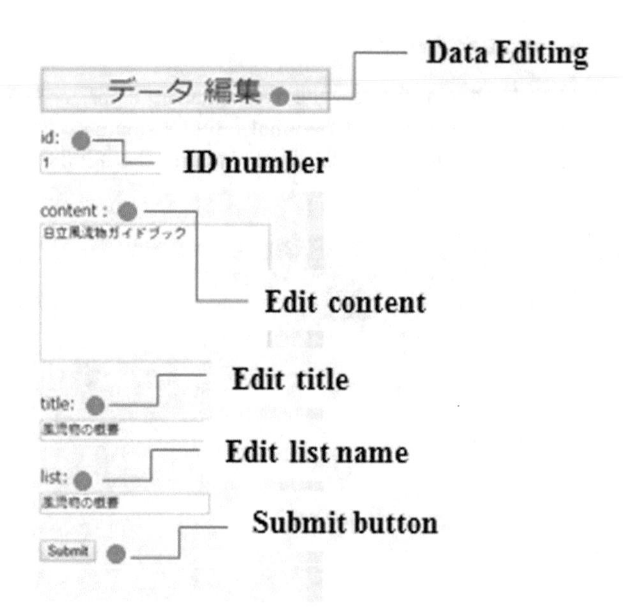

Fig. 11. Editing screen of regional culture content.

manage the content data. As shown in Fig. 9, the data management system has a login function of the administrator, a data list display function, and a data editing function.

The content data of the registered regional culture is displayed in the center of the data management system as shown in Fig. 10. The administrator can fix registered images and information by selecting a correction button. When the administrator selects the correction button, a data editing screen is displayed as shown in Fig. 11. The administrator can modify the ID, detailed information, titles, etc. of the registered contents on this screen.

6 Conclusion

In this research, we have constructed the regional culture inheritance system aiming at dissemination and succession of regional culture. The regional culture inheritance system visualizes the location where the regional culture exists on the electronic map, and provides the user with regional tradition as AR information. By providing regional tradition as AR information, media information of regional culture is displayed superimposed on the real world photographed by the camera of the mobile terminal. In addition, this application transmits information on regional culture to the world via the Internet. This system consists of the regional culture database server, the application server, the mobile application, and the data management system. We stored various regional culture information on the regional culture database server. In addition, users can experience various regional cultures on the mobile application via the application server. The mobile application has functions of browsing and manipulating digital contents, Web-GIS function for confirming location information, and AR function. The data management system manages the regional culture contents provided by the mobile application to the user. The administrator can edit the data stored in the database via the data management system.

References

1. The Agency for Cultural Affairs, Government of Japan, "Historical and Cultural Basic Plan" Establishment Technical Guide. http://www.bunka.go.jp/seisaku/bunkazai/rekishibunka/pdf/guideline.pdf. Accessed Oct 2018
2. Kaga City, Kaga City Historical and Cultural Basic Plan. https://www.city.kaga.ishikawa.jp/data/open/cnt/3/191/1/11290.pdf. Accessed Oct 2018
3. Fukui City, The Fifth Fukui City Comprehensive Plan. https://www.city.fukui.lg.jp/sisei/plan/plan/p000612_d/fil/007.pdf. Accessed Oct 2018

Clinical Trial on Computer-Aided Design in Pre-operative Alveolar Bone Grafting in Cleft Lip - Cleft Palate Patient

Krit Khwanngern[1,2(✉)], Vivatchai Kaveeta[1], Suriya Sitthikham[1,2], Watcharaporn Sitthikamtiub[1], Sineenuch Changkai[1], and Tanyakorn Namwong[1]

[1] CMU Craniofacial Center, Faculty of Medicine, Chiang Mai University, Chiang Mai, Thailand
{krit.khwanngern,suriya.sitthikham,watcharaporn.sit, sineenuch.c}@cmu.ac.th, vivatchai.kaveeta@gmail.com, warapan.wn@gmail.com
[2] Center of Data Analytics and Knowledge Synthesis for Healthcare, Chiang Mai University, Chiang Mai, Thailand

Abstract. Cleft lip and cleft palate are facial anomalies. One of the treatment for the patients is alveolar bone grafting. In Thailand, spiral CT scan has been extensively utilized in the diagnosis in order to examine the lesions on facial bone. However, spiral CT scans are expensive, relatively large and immobile. In this study, we evaluated the application of moveable CT scan. The important pre-operative information such as the shape of the alveolar cleft, the volume of the alveolar cleft, teeth, and tip of dental roots around alveolar cleft are required for treatment planning. Normally, plastic surgeons measured and evaluated these quantities in the operating room. Differently, the cleft lip and cleft palate patients underwent mobile CT for the pre-operative period. And with computer-aided design software, plastic surgeons worked collaboratively with engineers to process output and perform measurement pre-operatively. Our study shows that the pre-operative information from computer-aided design can increase understanding for alveolar cleft anatomy. The most important advantage of mobile CT and computer-aided design software is avoidance of dental root injury during alveolar bone graft operation.

1 Introduction

Cleft lip and cleft palate are facial anomalies. Cleft lip is physical separation of the upper lip to gum and the front portion of the roof of the mouth that can occur one side or both sides of the face. Cleft palate is a split of the soft palate and the hard palate. The symptoms of cleft palate are more severe than cleft lip. Figure 1 shows a few examples of patients with cleft lip and cleft palate.

One of the treatment procedures for the patients is alveolar bone grafting. The patients who received a bone graft before permanent teeth are fully formed usually has a better chance of success. X-ray scan is performed to see the growth

L. Barolli et al. (Eds.): EIDWT 2019, LNDECT 29, pp. 482–491, 2019.
https://doi.org/10.1007/978-3-030-12839-5_45

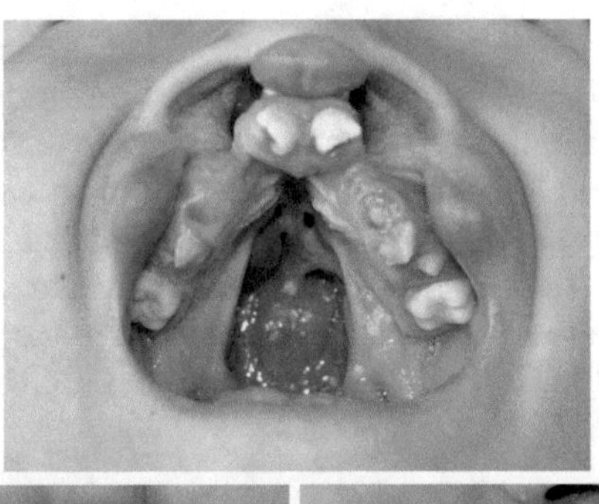

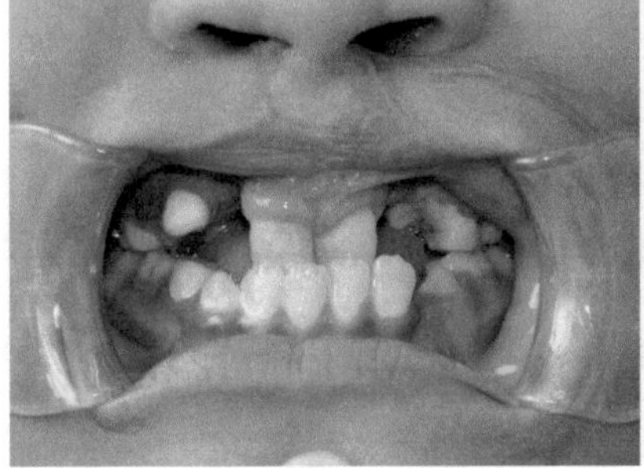

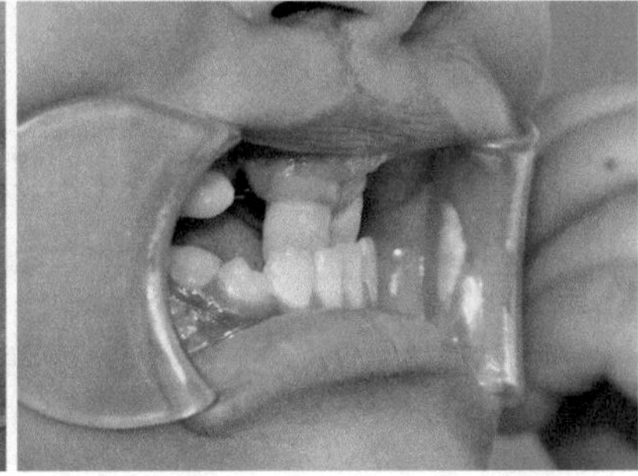

Fig. 1. Patients with cleft lip and cleft palate.

of the teeth in the alveolar cleft region to evaluate the appropriate time period for performing alveolar bone graft. If the lateral incisor is located within the alveolar cleft, the alveolar bone graft will be performed between the ages of 8 and 10. If it is located in canine, the graft will be performed between the ages of 9 and 11 [1]. In these cases, the more pre-operative information surgeons have about alveolar cleft and teeth around alveolar cleft, the better design of the operation which can benefit patients safety. Some important pre-operative alveolar cleft information including position, shape and volume of the teeth and the tip of dental roots. These information enable surgeons to correctly calculate the required bone volume. With precise value, they can identify the best position for drilling and performing bone grafting to substitute the missing part.

Presently in Thailand, 3D X-ray Computed Tomography (CT), Spiral CT scans have been utilized in the treatment to examine the condition of injury which occurs in the facial bone. The spiral CT scan is a relatively large machine that stationary and expensive. Small and local hospitals with a limited budget need to refer patients to get the treatment at the bigger hospitals which result in the delayed diagnoses. On the other hand, the movable CT scan is more convenient and operates faster. It is suitable to be used in multiple situations such as during surgery, in the emergency room, and with the patients who are unable to move. If the patients receive an X-ray examination immediately, the patients can receive the treatment on time which extremely reduces patients loss rate. Furthermore, patients receive a lower dose of radiation when compared with regular medical X-ray CT.

There are many applications of computer-aided design in surgical operation such as dental implant and prostheses examination [2,3]. The computer-aided software included with mobile CT has many functionalities. Beside its ability to visualize the volumetric data output from the scan, it can provide an accurate measurement of the underlying organ which is difficult or impossible to examine without an operation. Bone graft operation was selected because the CT scans were already performed as standard procedure of the treatment process. As the result, applying mobile CT scan did not cause any additional treatment cost or risk to the patient.

2 Method

2.1 General Aspects of Method Development

Moveable 3D X-ray computed tomography had been developed by the National Electronics and Computer Technology Center (NECTEC) and the National Metal and Materials Technology Center (MTEC) under the National Science and Technology Development Agency (NSTDA) in collaboration with Faculty of Medicine, Prince of Songkla University with the working principles described in [4].

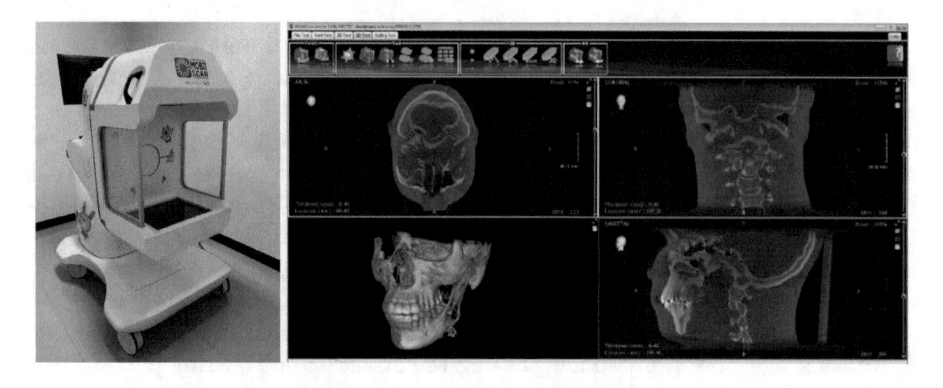

Fig. 2. (Left) Moveable 3D X-ray computed tomography (Mobile CT). (Right) 3D visualization and computer-aided design software.

The mobile CT used in this study is shown in Fig. 2 (left). The benefit of Mobile CT is the size of the machine which is much smaller than multi-slice CT. Also, The patients receive a lower dose of radiation compared with a traditional CT scan. The gathered output data go through an image reconstruction algorithm to create 3D models of the mouth, jaw, and face of the patients. The outputs of the reconstruction are shown in 2D and 3D through the visualization software as shown in Fig. 2 (right).

2.2 Design and Testing

Study Population and Inclusion Criteria

1. Nine patients with cleft lip and cleft palate who receive normal services from the hospital, or referred to get this particular treatment. Seven of which were scanned on one side of the mouths. Two others were scan on both left and right side.
2. Age between 8–20 years old and not pregnant.
3. Conscious and able to sign consent for enrollment in the study.
4. Diagnosed by the physician who provides treatment that the X-ray CT was needed for diagnosis and following up treatment.

3 Data Measurement

3.1 Statistical Analysis

The following characteristics were retrieved and included in the analysis:

1. Height of the center of the left teeth (HA).
2. Height of the center of the right teeth (HB).
3. Volume of alveolar cleft

Measurement of each variable had been conducted by two methods, measuring by mobile CT with computer-aided software, and actual measurement in the operating room. Afterwar we evaluated the differences in the measured values between both methods and standard error of the mean. All analyses were performed using STATA software version 12.0 (STATA Corp, College Station, Texas, USA) and SPSS version 17.0 (SPSS Inc., Chicago, USA).

3.2 Ethical Considerations

All participants were informed and signed consent prior to participation in this study. This study was approved by the Faculty of Medicine, Chiang Mai University Ethics Committee.

4 Result

Figure 3 shows the differences between measuring by using computer-aided design and actual measurement in the operating room by the physician who performed the surgery. Figure 3a, b and c are the patient with abnormality on the left side of the face. The results of measuring by using computer-aided design were displayed in Table 1. While the results from measuring in the operating room were shown in Table 2.

Table 1. Computer-aided design measurement for the patient from Fig. 3 (left).

Height of the center of left side teeth (HA)	13.26 mm
Height of the center of right side teeth (HB)	11.21 mm
Volume of alveolar cleft	675 mm^3

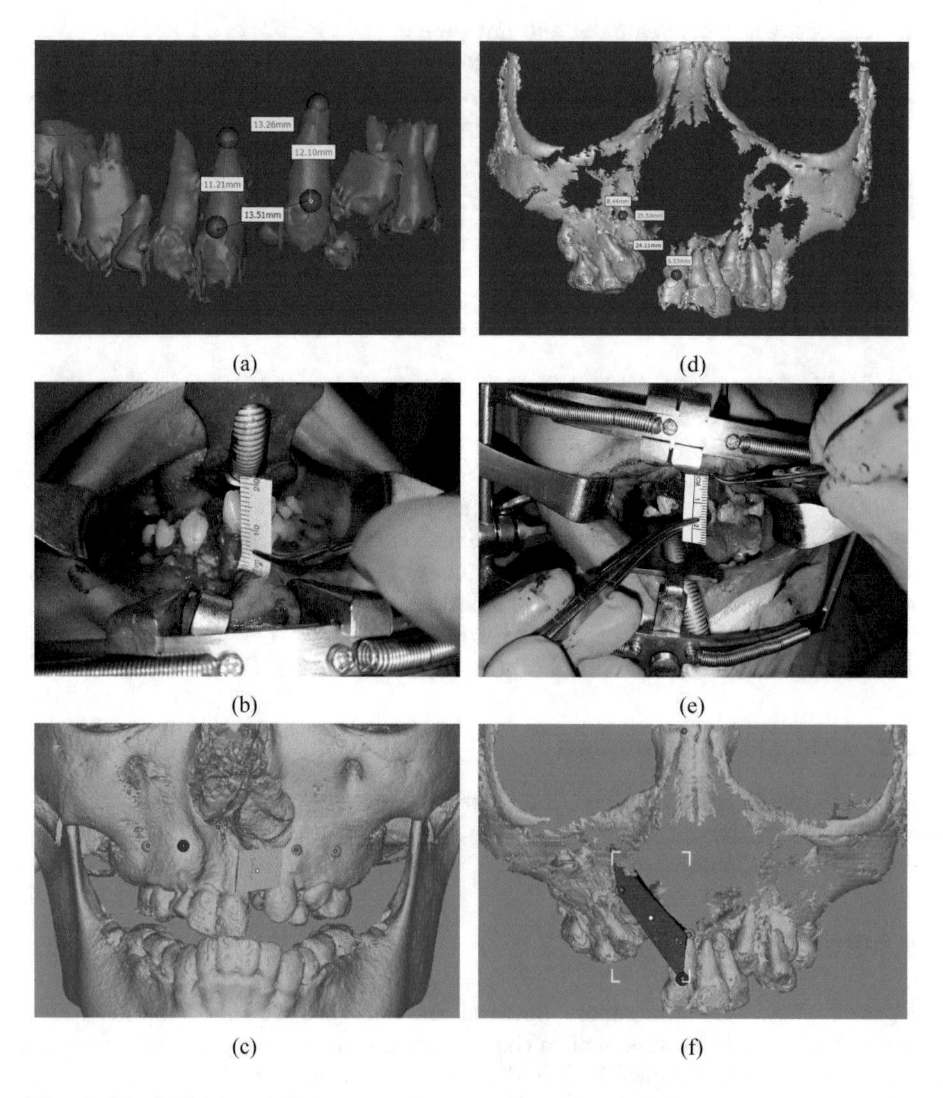

Fig. 3. (Top) Height of the center of the teeth measuring by computer-aided design software, (Center) Height of the center of the teeth measuring by the actual measurements in the operating room, (Bottom) Volume of the alveolar cleft measurement by computer-aided design

Table 2. Actual measurement in operating room for the patient from Fig. 3 (left).

Height of the center of left side teeth (HA)	6 mm
Height of the center of right side teeth (HB)	8 mm
Volume of alveolar cleft	Could not be measured

Table 3. Computer-aided design measurement for the patient from Fig. 3 (right).

Height of the center of left side teeth (HA)	8.53 mm
Height of the center of right side teeth (HB)	8.44 mm
Volume of alveolar cleft	722 mm^3

Table 4. Actual measurement in operating room for the patient from Fig. 3 (right).

Height of the center of left side teeth (HA)	9 mm
Height of the center of right side teeth (HB)	Could not be measured
Volume of Alveolar cleft	Could not be measured

Figures 3d, e and f are the pictures of a patient with the abnormality on the right side of the face. Results of measuring by using computer-aided design were displayed in Table 3. While the results from measuring in the operating room were shown in Table 4.

The results of all samples are shown in Table 5. Note that the empty cells in the tables indicate the immeasurable cases with visual obstruction prevented the actual measurement in the operating room.

5 Discussion

By comparing the measurement results of the holes (alveolar cleft) in the mouths of 11 samples between computer-aided design on mobile CT output and actual measurements in the operating room, the conclusions are as follows:

Measurement Possibility. As seen from Table 5, from the total of 11 samples with facial anomalies which were measured by computer-aided design and actual measurement in the operating room, we found that computer-aided design could measure 100% of the cases for the height of the center of the left and right side teeth and hole volume. While the actual measurements in the operating room often could not be performed, shown as empty cells in Table 5. Measuring the height of the center of left side teeth measurement could be conducted on 9 samples, 81.8%, the height of the center of right side teeth could be measured from only 6 samples, 54.6%. The volume of the holes could not be measured at all.

Table 5. The measuring results classified by measurement methods, computer-aided design and actual measurements in the operating room.

ID	Height of the center of the left teeth (mm)		Height of the center of the right teeth (mm)		Volume of alveolar cleft (mm^3)	
	Computer-aided design	Actual measurement in the operating room	Computer-aided design	Actual measurement in the operating room	Computer-aided design	Actual measurement in the operating room
1	10.07	10.00	9.74	—	72031	—
2	13.56	13.00	11.66	10.00	1770	—
3	13.08	11.00	13.80	13.00	2541	—
4	8.53	8.72	10.39	10.00	485	—
5	12.10	6.00	11.21	8.00	675	—
6	8.53	9.00	8.44	—	722	—
7	5.04	5.00	11.38	—	610	—
8	9.29	5.00	11.10	—	1510	—
9	11.10	—	8.83	8.00	1084	—
10	6.44	13.00	12.89	—	84396	—
11	9.33	—	11.28	7.00	723	—

Note that missing cells indicate the immeasurable cases.

This immeasurability was cause by the gum tissue visually obstructs the view of the teeth root. In some cases in order to make an actual measurement in the operating room, the surgeon needs to cut open more gum tissue which is excessively invasive operation and cut put more risk on very serious complications such as oronasal fistula and infection. The measurement by computer-aided design software which is noninvasive and usually obtainable clearly have an advantage over the operating room measurement.

Measurement Difference. Table 6 and Fig. 4 show the differences of the height of the center of the left and right side teeth between measuring by computer-aided design and actual measurement in the operating room. The difference on the left and right side teeth in average were 0.66 mm, and 1.86 mm respectively.

Height measurement by computer-aided design and actual measurements showed small differences. In some patients, this is due to the position that needs to be measured or tooth roots locate deeper than the reachable areas. As the result, the surgeon needs to perform an operation to open the wound wider and deeper to allow the measurement which can ll bad effect on the treatment.

Sometimes tooth root is obstructed under bone so the results of measurement rely on the experience of estimating the positions by operating surgeons instead. When the measured values in the operating room are less than real value, it may cause serious consequences due to root damage.

For the volume of the alveolar cleft measurement, the depth could not be measure in an actual measurement in operating room because there is no device

Table 6. Differences between measuring by computer-aided design and actual measurements in the operating room

ID	HA differences[a]	HB differences[b]
1	0.07	—
2	0.56	1.66
3	2.08	0.80
4	−0.19	0.39
5	6.10	3.21
6	−0.47	—
7	0.04	—
8	4.29	—
9	—	0.83
10	−6.56	—
11	—	4.28
Mean ± SD	0.66 ± 3.53	1.86 ± 1.55

[a]**HA** Height of the center of the left teeth.
[b]**HB** Height of the center of the right teeth.
Note that missing cells indicate the immeasurable cases.

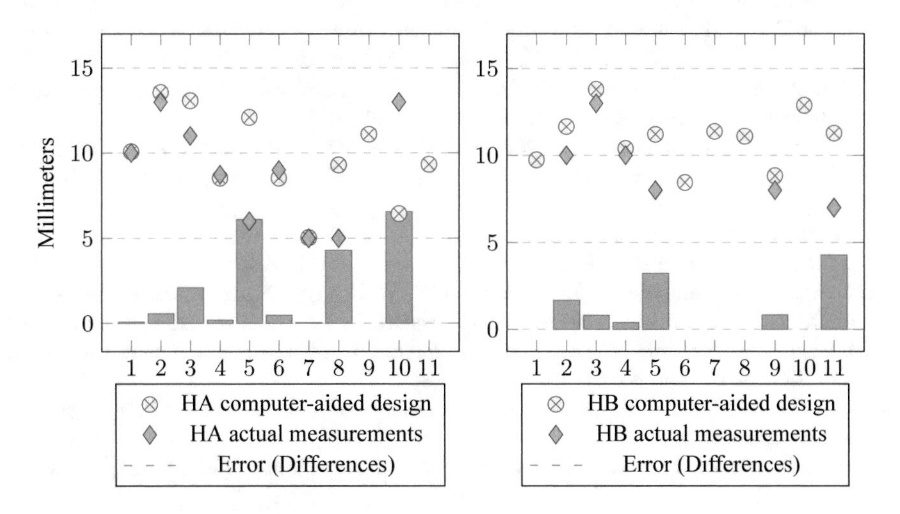

Fig. 4. Differences between measuring results by using computer-aided design and actual measurements in the operating room (HA Height of the center of left side teeth, HB Height of the center of right side teeth. Note that missing data indicate the immeasurable cases)

that can reach the spot that needs to be measured. Instead in an actual practice, these number were estimated based on surgeon experience. On the other hand, by using computer-aided design, the measurement can be conducted even prior to the surgery.

As the result, measuring by computer-aided design is able to provide precise values with explicit advantage in term of eliminating the need for estimation. Also in case of root spot that cannot be seen by eyes, computer-aided design in pre-operative can get the correct measurement noninvasively instead of open gum operation. In conclusion, computer-aided design can be more convenient, more precise and have the potential to save valuable time of surgeons who perform the operation and lead to improvement of the patient safety.

Acknowledgements. This work was supported by the Faculty of Medicine Research, Chiang Mai University, Chiang Mai, Thailand. We thank National Electronics and Computer Technology Center, National Metal and Materials Technology Center, Chiang Mai for providing mobile CT along with computer-aided design software. We thank Center of Data Analytics and Knowledge Synthesis for Healthcare, Chiang Mai University, Chiang Mai, Thailand for their assistance with the experiment and experimental result analysis. The authors would also like to extend gratitude to all volunteers and personnel who participate in this study.

References

1. Rudick, R.A., Miller, D., Bethoux, F., Rao, S.M., Lee, J.C., Stough, D., Reece, C., Schindler, D., Mamone, B., Alberts, J.: The multiple sclerosis performance test (MSPT): an iPad-based disability assessment tool. J. Vis. Exp. JoVE **88** (2014)
2. Naitoh, M., Nabeshima, H., Hayashi, H., Nakayama, T., Kurita, K., Ariji, E.: Postoperative assessment of incisor dental implants using cone-beam computed tomography. J. Oral Implantol. **36**(5), 377–384 (2010)
3. Wouters, V., Mollemans, W., Schutyser, F.: Calibrated segmentation of CBCT and CT images for digitization of dental prostheses. Int. J. Comput. Assist. Radiol. Surg. **6**(5), 609–616 (2011)
4. Whitaker, J., McFarland, H., Rudge, P., Reingold, S.: Outcomes assessment in multiple sclerosis clinical trials: a critical analysis. Mult. Scler. J. **1**(1), 37–47 (1995)
5. Cohen, J.A., Reingold, S.C., Polman, C.H., Wolinsky, J.S., International Advisory Committee on Clinical Trials in Multiple Sclerosis, I.A.C., et al.: Disability outcome measures in multiple sclerosis clinical trials: current status and future prospects. The Lancet Neurol. **11**(5), 467–476 (2012)
6. NECTEC: National electronics and computer technology center (2018). https://www.nectec.or.th
7. MTEC: National metal and materials technology center (2018). https://www.mtec.or.th
8. NSTDA: National science and technology development agency (2018). https://www.nstda.or.th

9. Pauwels, R., Jacobs, R., Bosmans, H., Pittayapat, P., Kosalagood, P., Silkosessak, O., Panmekiate, S.: Automated implant segmentation in cone-beam CT using edge detection and particle counting. Int. J. Comput. Assist. Radiol. Surg. **9**(4), 733–743 (2014)
10. Scarfe, W., Li, Z., Aboelmaaty, W., Scott, S., Farman, A.: Maxillofacial cone beam computed tomography: essence, elements and steps to interpretation. Aust. Dent. J. **57**, 46–60 (2012)

Large-Scale Simulation of Site-Specific Propagation Model: Defining Reference Scenarios and Performance Evaluation

Zeshan Hameed Mir(✉)

Faculty of Computer Information Science, Higher Colleges of Technology (HCT), PO Box 4114, Fujairah, United Arab Emirates
zhameed@hct.ac.ae

Abstract. The vehicular network research community continue to strive for improvements in the accuracy and validity of simulation-based studies. While there have been several new enhancements, most of the network simulators lack comprehensive support to represent real wireless channel characteristics, especially at the physical (PHY) layer. The test-bed has been a widely-accepted alternative which is capable of bringing realism to the performance evaluation. However, higher implementation cost and scalability issues prevent test-bed from being a viable solution for large-scale performance evaluation studies. Therefore, realistic simulation frameworks are highly sought-after to reduce the implementation cost and complexity for city-wide testing of novel networking protocols and algorithms. In this paper, a set of common reference scenarios are introduced in the city of Fujairah, UAE. These reference scenarios exhibit different wireless channel characteristics specific to the requirements in vehicular communication. Next, a generic simulation framework is described that combines a suite of simulation tools. The framework utilizes a site-specific, geometry-based vehicular propagation approach which models different characteristics of a wireless link more accurately. Finally, through extensive simulation, each reference scenario is evaluated in terms of received signal strength, packet delivery ratio, and reliable communication range.

1 Introduction

A widely acceptable simulation model is a founding pillar for the implementation and the evaluation of the protocols and algorithms in vehicular ad hoc networks (VANETs). Varying assumptions and different level of details through which simulation models are developed make it difficult to compare and reproduce research results [1,2]. These limitations necessitate the need for defining a standard set of reference scenarios to improve the overall efficiency and quality of research in the area of wireless communication in general and vehicular communications in particular [3–5].

Vehicular communications enable a wide variety of applications and use cases in the domain of road safety, traffic efficiency, and infotainment [6]. The

L. Barolli et al. (Eds.): EIDWT 2019, LNDECT 29, pp. 492–503, 2019.
https://doi.org/10.1007/978-3-030-12839-5_46

effectiveness of these applications and the supporting protocols and algorithms are measured either by developing them in real-world test-beds or through simulations [7]. While test-bed implementations [8] provide site-specific and accurate results, they incur substantial cost and are less scalable. On the other hand, simulation-based approaches are cost-effective and scalable, i.e., capable of conducting large-scale evaluations covering city-wide geographical areas which involves thousands of vehicles. Despite several recent enhancements and integration of network simulators with mobility generator tools such as [7,9], simulations are less accurate and inflexible. Another limiting issue is the lack of realistic wireless channel characteristic at the Physical (PHY) layer. Most of the network simulators rely on gross assumptions regarding the surrounding environments and implement a simplified version of propagation models to achieve lower implementation and computational complexity.

This paper aims at underlining the importance of three factors. (1) Definition of a well-defined set of reference scenarios which allow realistic emulation of vehicular networking situations in different types of environments such as urban, suburban, open streets and highways. The scenario definitions often include representations of building and vehicle geometry outlines, vehicular mobility traces, and application-specific parameters settings. (2) Integration of measurement-based, site-specific channel propagation models for Vehicle-to-Vehicle (V2V) communications. (3) Development of generic simulation framework which can easily combine reference scenario definitions and propagation models to evaluate the use cases and application-specific vehicular communication performance by means of computer simulations.

Since the selection of parameters that are used to characterize reference scenarios and configure applications have a profound impact on radio wave propagation, a parameterized simulation framework is highly sought-after. For example, if an application is tested for a simple scenario with a single isolated corner of a building located on a single street, the simulation results obtained through such situation wouldn't be applicable to dense urban intersection surrounded by multiple buildings of varying heights. Similarly, road network (street or highways) and time-varying traffic densities play a critical role in terms of wireless communication performance [10]. The presence of different types of obstructions such as buildings, foliage, and nearby vehicles severely adds to the path loss of vehicular wireless communication. Thus, the radio propagation model must be correctly implemented in sufficient details and configured to simulate the given reference scenarios realistically.

In this paper, the first attempt is to define and characterize a set of common reference scenarios by considering different simulation environments in the city of Fujairah, United Arab Emirates (UAE). Next, a two-tiered [11] simulation framework proposed in [2,12,13] is utilized to perform large-scale simulations of the vehicular channel model. The simulation framework is composed of three major components. (1) Extracting Building geometry details. (2) Generating realistic traffic and vehicular mobility patterns. (3) Geometry-based [14–17] efficient propagation model for V2V communication (GEMV2) [18]. An extensive

Table 1. Characteristics of reference scenarios

Property/Space	Open	Wide urban	Narrow urban	Downtown
Lanes	2–6 (High)	2+1 (Low)	1–2 (Low)	2–4 (Medium)
Speed	25–40 m/s (High)	20–30 m/s (Medium)	15 m/s (Low)	10–30 m/s
Density	High	Low	Low	High
Surrounding	Low	Low	Medium	High
Building height	None or Short	Low-Medium	Low	Medium-High

set of simulations were performed by implementing a number of typical reference scenarios using the simulation framework. The V2V communication performance is analyzed in terms of key metrics such as received signal power, packet delivery ratio and reliable communication range. The results indicate that the site-specific environmental details have a significant impact on vehicular networking performance.

This paper is organized as follows. Section 2 defines the characteristics of common reference scenarios in an urban environment for VANET performance evaluation. Section 3, gives an overview of the simulation framework and explains the data sources required to implement a reference scenario followed by the details of the Geometry-based V2V propagation model. Section 4 describes the performance evaluation of the representative reference scenarios for the city of Fujairah in UAE. Section 5 completes the paper with a brief conclusion.

2 Characterizing Reference Scenarios in an Urban Environment

In order to highlight the importance of site-specific information and its impact on the V2V propagation path loss prediction, we considered four different types of reference scenarios. The main difference lies in the availability of a detailed and accurate database of location and density of obstructing objects, i.e., building, foliage, and vehicles. Table 1, summarizes the several characteristics of different reference scenarios.

- Open Space: Such scenarios are often characterized by either complete absence of building and other types of obstructions or very sparsely distributed shorter building structures (single-story) with heights ranging from 4 m to 6 m. The road layout consists of multilane (typically 2 to 3 lanes), bidirectional main roads with multiple large intersections and roundabouts. The intersections are usually medium to densely crowded with vehicles traveling at the speeds from 60 km/h up to 80 km/h. The general environment is characterized as relatively flat with only a few buildings in direct proximity to the relevant road segments.
- Wide Urban Space: A series four-way intersections with multiple stories building structures, typically 3 to 6 stories (i.e., 15 m to 30 m) in height.

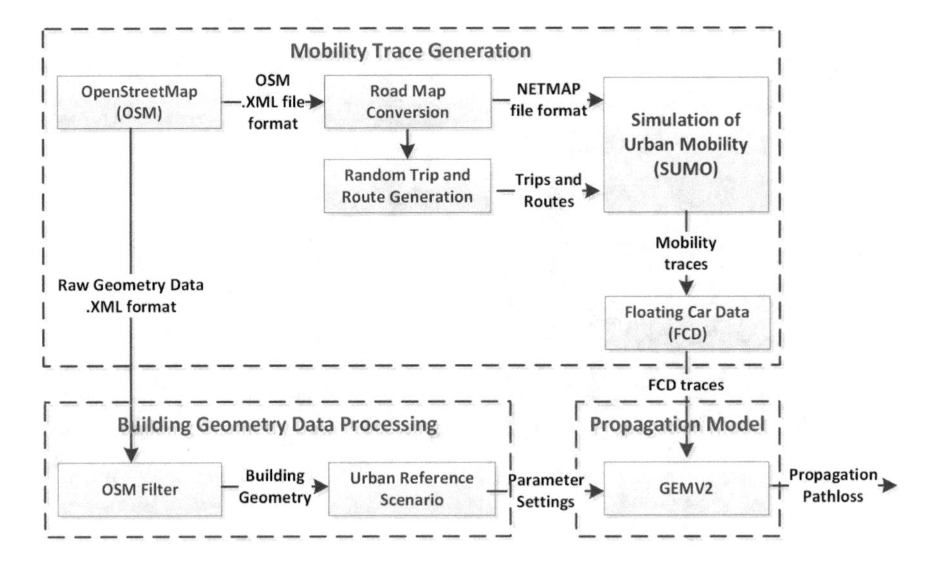

Fig. 1. Simulation framework for V2V communication performance evaluation.

The intersections can be classified as major crossroad with up to 3–4 lanes (with additional turn lanes) in each direction. The distance from the road center and buildings are wider with each quadrant of the intersection featuring a mix of the residential and commercial structure of different heights.

- Narrow Urban Space: This reference scenario is generally characterized by homogenous building heights between 4 to 12 stories (i.e., 15 m to 50 m). Buildings are of a mixed-use type, utilized for both office and residential purposes and located close to the roundabouts. The intersection topologies are almost rectangular in shape with a diameter of the roundabouts are approximately up to 30 m with one lane roads of width between 8 m to 10 m.
- Downtown Space: Usually in downtown areas, both sides of the roads are flanked by high-rise and tall building structure. We considered the total height of the buildings 50 m or more that makes the downtown area with no significant height differences.

3 Defining Scenarios and Simulation Model

This section describes the implementation details of the reference scenarios and simulation model. For this purpose, data from multiple sources has to be extracted and processed in necessary details to create a realistic representation of the site-specific environment surrounding the communicating vehicles. Figure 1 shows the schematic diagram of the simulation framework with three main components.

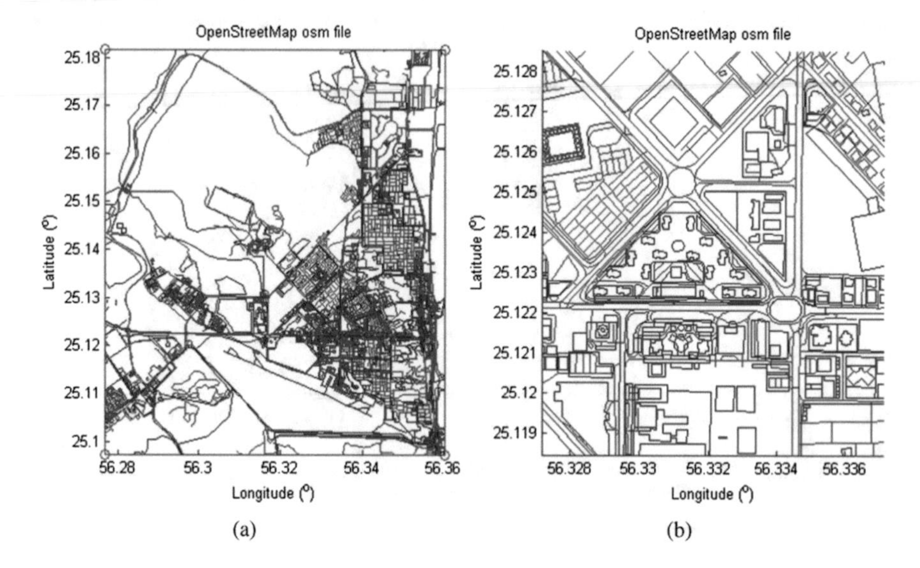

Fig. 2. An urban V2V reference scenario at the city-wide scale. (a) Building data of Fujairah, UAE obtained from OpenStreetMap (OSM) [19]. (b) A detailed (zoomed-in) version of (a).

3.1 Building Geometry Data

In order to implement the reference scenario, a case study in the city of Fujairah, UAE is undertaken. The building geometry details are obtained from the OpenStreetMap (OSM) [19] for the selected region. The outlines of the buildings are given as closed, two-dimensional polygons, along with other relevant information. For example, in [18] the building outlines are used to measure the diffraction off the building walls using multiple Knife Edge model. The GEMV2 propagation model takes into account single-interaction reflected and diffracted rays while measuring the received signal strength. The raw building data goes through a filtering step to keep the relevant information only. The filtering step employs a suite of utility software like OSMOSIS [20] and JOSM [21]. Figure 2(a) illustrates the road network and building outlines extracted from the OSM database, with red lines representing the road network and blue lines depicting the building outlines. Figure 2(b) shows the zoomed-in or detailed version of the selected region.

3.2 Vehicular Mobility Trace Data

Generating vehicular mobility traces at city-wide is another critical aspect of realistic reference scenarios. For this purpose, road network layout is extracted from the OpenStreetMap (OSM) [19] for the simulation area. To actually produce the vehicular mobility traces Simulation of Urban Mobility (SUMO) [22] is utilized. The obtained XML file format is converted to the NETMAP file

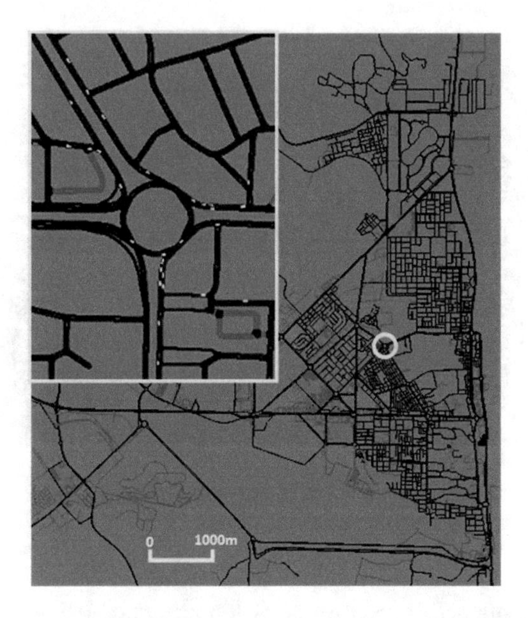

Fig. 3. Road network layout, obtained from OpenStreetMap (OSM) [19] and illustrated in SUMO-GUI [22]. Inset image shows vehicular traffic near a roundabout.

format as expected by the SUMO and also used to generate random trips and routes. The route selection step decides on the traffic flow during the simulations. The SUMO combines the road network, trip and route selection data to generate the vehicular mobility traces in the Floating Car Data (FCD) format. The FCD format contains movement information about each individual vehicle such as position, current velocity, and direction. Figure 3 shows the traffic flow and vehicles using SUMO-GUI, where each vehicle is represented by a yellow color rectangular shape.

3.3 GEMV2 Propagation Model

The Geometry-based efficient propagation model for V2V communication (GEMV2) [18], combines the outlines of buildings, vehicles movement trace and measurement-based site-specific propagation models to calculate several different propagation mechanisms and characterizes them into three link types i.e., line-of-sight (LOS), non-line-of-sight (obstructed by buildings) or NLOSb and non-line-of-sight (blocked by surrounding vehicles) or NLOSv. In this paper, GEMV2 is used to study the impact of different environments especially the presence or absence of buildings and their densities as the source of obstructions (i.e., NLOSb type links only) on important networking and communication performance metrics such as received signal power, packet delivery ratio and reliable communication range.

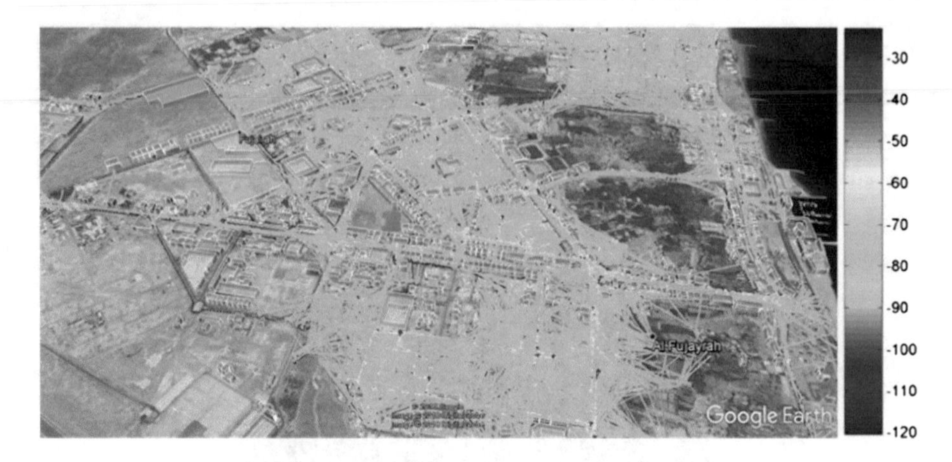

Fig. 4. Google Earth [23] visualization of the simulation scenario using GEMV2 [18]. The colorbar represents received power in dBm.

GEMV2 is a geometry-based propagation model which combines three propagation mechanisms (1) path loss, (2) small-scale fading and (3) large-scale fading. In GEMV2, these three propagation mechanisms are measured differently for each link types, i.e., LOS, NLOSb, and NLOSv. However, GEMV2 achieve lower computational complexity and higher scalability without sacrificing accuracy by calculating large-scale variations and small-scale variation of each link types. The large-scale variations are calculated deterministically and models path-loss and shadowing whereas small-scale variations are estimated stochastically, and models multi-path as a result of diffractions, reflections, and Doppler spread by taking into account number and size of the object in the close proximity of the communicating vehicles.

Depending on the link type, GEMV3 applies different propagation model to calculate the large-scale variations. For example, Two-ray ground reflection, Log-distance and multiple Knife Edge diffraction models are used to estimate large-scale variations for LOS, NLOSb, and NLOSv link types, respectively. The small-scale variations are measured by using a zero-mean normal distribution with minimum and maximum standard deviation (σ_{min}, σ_{max}). The values for σ_{min} and σ_{max} are obtained for each link type via extensive measurement-based studies. These studies were site-specific and conducted under different environmental settings such as open and urban spaces with lower and higher traffic densities. Further details on the parameters with corresponding values and the algorithm working can be found in [18].

Figure 4 illustrates the Google Earth visualization of the simulation instance using GEMV2. The building geometry is depicted in white color, and the links between the communicating vehicles are shown as colored lines. The line color represents the received signal power in dBm.

Table 2. Simulation parameters and values

Simulation parameters	Values
Number of Vehicles	250 to 1400
Simulation Duration	100 s
Data Rate	3 Mbps
Minimum Receiver Sensitivity	−85 dBm
Maximum Transmission Power	18 dBm
Maximum Transmission Range	300 m
Antenna Height	0.2 m
Path Loss Exponent	2 to 4
Vehicle Speed	20 km/h to 60 km/h
Simulation Area	8 km^2

4 Performance Evaluation

This section evaluates the performance of vehicular networks by implementing a few of the representative reference scenarios. Since, an urban environment can vary drastically with open, narrow, wider spaces and downtown; the simulations were conducted over a vast region which covered approximately 8.0 km^2 of area. The selected simulation is centered around (25.132°, 56.336°) coordinates [23].

The main focus is to perform site-specific simulations at different areas within the selected region each with diverse characteristics and measure their impact on vehicular communication performance in terms of main metrics, such as Received Signal Power (dBm), Packet Delivery Ratio (PDR), and Reliable Communication Range (m).

4.1 Simulation Environment

The Geometry-based propagation model for V2V communication (GEMV2) [18] version 1.0 is utilized. During the simulation the number of vehicles was varied between 250 to 1400, traveling between randomly chosen origins and destination. The average speed is selected within a range between 20 km/h to 60 km/h. During the simulations, the maximum transmission power and transmission range are set to 18 dBm and 300 m, respectively. Table 2, gives the summary of simulation parameters and their values.

4.2 Simulation Results

Figure 5 shows the received power (dBm) for different scenarios as the distance between the communicating vehicles increases. Generally, the received signal strength decreases as the distance increases. The obtained received power depends significantly on the surrounding environment. The open space scenario

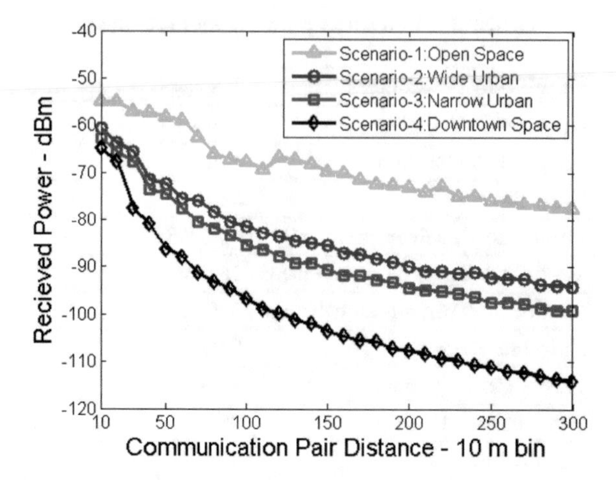

Fig. 5. Received signal strength (dBm) vs. distance (m) between the communicating vehicles.

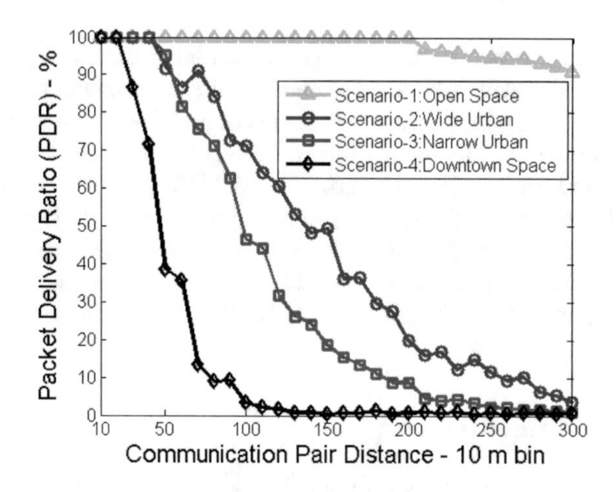

Fig. 6. Packet Delivery Ratio (%) vs. distance (m) between the communicating vehicles.

is often characterized as relatively flat terrain. Therefore the absence of building and other sources of obstructions resulted in lower degradation of received signal strength. At the smaller distances, clear line-of-sight paths are available, however, as the distance increases, the presence of obstructions causes difficult channel conditions. As for the wider and narrower urban spaces, the received power is comparable for the closer vehicle distance (up to 50 m). However, with broader urban areas the vehicles experience line-of-sight conditions for a longer duration as compared with narrower urban areas, where direct links between two vehicles approaching the center of the intersection will almost certainly be blocked by buildings until each vehicle is traveling relatively close to each other. Finally, in downtown space, difficult channel conditions are observed due to the absence of line-of-sight.

Figure 6 shows the impact of different scenarios on the Packet Delivery Ratio (PDR). The threshold on minimum receiver sensitivity as described in the Dedicated Short-Range Communications (DSRC) standard [24] is used to measure the PDR. The PDR depends on the minimum receiver sensitivity threshold value, which is −85 dBm for the data rate of 3 Mbps. In the open space scenarios, a significant number of packets were delivered to the farthest possible neighboring vehicles. The PDR remained well above 90% even at the distance of 300 m between the communicating vehicles. For the broader and narrower space scenario, most of the packets were delivered at shorter distances, i.e., up to 50 m. However, a sharp decline in PDR is observed for a distance of 60 m and above. As for the downtown area, a significant number of packets were lost for 30 m and above distances.

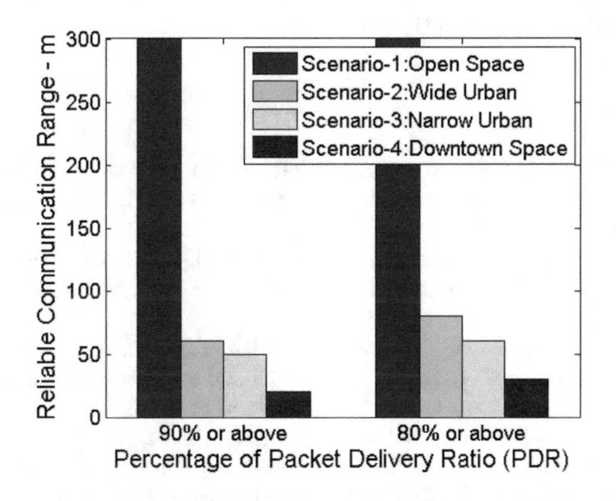

Fig. 7. Reliable communication range (m) with PDR 90% and 80% or above.

Figure 7 shows the reliable communication range (m), which is defined as the maximum distance over which the PDR value is 90% and 80% or above. In the open space scenarios, the reliable communication range is found to be 100% of the maximum communication range, i.e., 300 m. In wide urban space scenarios, the reliable communication range is 20% and 27% of the maximum communication range for 90% and 80% PDR, respectively. As for the narrower space scenarios, the reliable communication range is to some extent less than the wider space scenarios. Finally, in the downtown scenarios, the communication range over which packets can be transmitted reliably is considerably shorter, i.e., 20 m to 30 m which accounts for 7% and 10% of the maximum communication range, respectively.

5 Conclusion

In this paper, a standard set of reference scenarios are defined for conducting large-scale vehicular propagation model simulations. The reference scenarios include details that are necessary to create realistic simulation environments surrounding the communicating vehicles. These environments exhibit varying characteristics that are specific to the requirements in vehicular networking and communication performance evaluation. Details are also included regarding a generic simulation framework which implements the typical simulation scenarios by employing a suite of simulators and toolchain. The simulation framework mainly consists of three components, i.e., extracting building geometry data, generating vehicular mobility traces and geometry-based V2V propagation model. By using a case study for the city of Fujairah in UAE, extensive simulations are conducted. Simulation result shows significant variability among different simulation environments namely open space, wide space, narrow space and downtown in terms of performance metrics such as received signal power, packet delivery ratio and reliable communication range.

References

1. Al-Bado, M., Sengul, C., Merz, R.: What details are needed for wireless simulations? - A study of a site-specific indoor wireless model. In: 2012 IEEE INFOCOM, pp. 289–297 (2012)
2. Mir, Z.H., Filali, F.: Large-scale simulations and performance evaluation of connected cars - a V2V communication perspective. Simul. Model. Pract. Theory **73**, 55–71 (2017)
3. Dennis, M.R., Thomas, J., Thomas, W., Ulrich, T., Kürner, T.: The IC 1004 urban Hannover scenario - 3D pathloss predictions and realistic traffic and mobility patterns. In: European Cooperation in the Field of Scientific and Technical Research, COST, 8th IC1004 MC and Scientific Meeting, Ghent, TD(13)08054 (2013)
4. Möller, A., Baumgarten, J., Mir, Z.H., Filali, F., Kürner, T.: Realistic simulation scenario for hybrid LTE/IEEE 802.11p vehicular communication. In: IEEE 9th European Conference on Antennas and Propagation (EuCAP) (2015)
5. Dreyer, N., Möller, A., Baumgarten, J., Mir, Z.H., Kürner, T., Filali, F.: On building realistic reference scenarios for IEEE 802.11p/LTE-based vehicular network evaluations. In: 2018 IEEE 87th Vehicular Technology Conference (VTC Spring), pp. 1–7 (2018)
6. Mir, Z.H., Filali, F.: Applications, requirements, and design guidelines for multi-tiered vehicular network architecture. In: 2018 Wireless Days (WD), pp. 15–20 (2018)
7. Rondinone, M., Maneros, J., Krajzewicz, D., Bauza, R., Cataldi, P., Hrizi, F., Gozalvez, J., Kumar, V., Röckl, M., Lin, L., Lazaro, O., Leguay, J., Härri, J., Vaz, S., Lopez, Y., Sepulcre, M., Wetterwald, M., Blokpoel, R., Cartolano, F.: iTETRIS: a modular simulation platform for the large scale evaluation of cooperative its applications. Simul. Model. Pract. Theory **34**, 99–125 (2013)
8. Brahim, M.B., Mir, Z.H., Znaidi, W., Filali, F., Hamdi, N.: Qos-aware video transmission over hybrid wireless network for connected vehicles. IEEE Access **5**, 8313–8323 (2017)

9. Sommer, C., German, R., Dressler, F.: Bidirectionally coupled network and road traffic simulation for improved IVC analysis. IEEE Trans. Mob. Comput. **10**(1), 3–15 (2011)
10. Bylykbashi, K., Spaho, E., Barolli, L., Xhafa, F.: Impact of node density and TTL in vehicular delay tolerant networks: performance comparison of different routing protocols. Int. J. Space Based Situated Comput. **7**(3), 136–144 (2017)
11. Zafar, B., Mir, Z.H., Shams, S.M.S., Ikram, M., Baig, W.A., Kim, K., Yoo, S.: On improved relay nodes placement in two-tiered wireless sensor networks. In: 2009 IEEE Military Communications Conference (MILCOM), pp. 1–7 (2009)
12. Mir, Z.H., Filali, F.: Simulation and performance evaluation of vehicle-to-vehicle (V2V) propagation model in urban environment. In: 2016 7th International Conference on Intelligent Systems, Modelling and Simulation (ISMS), pp. 394–399 (2016)
13. Mir, Z.H., Filali, F.: A simulation-based study on the environment-specific propagation model for vehicular communications. Int. J. Veh. Telemat. Infotain. Syst. (IJVTIS) **1**(1), 15–32 (2017)
14. Sommer, C., Eckhoff, D., German, R., Dressler, F.: A computationally inexpensive empirical model of IEEE 802.11p radio shadowing in urban environments. In: 2011 8th International Conference on Wireless On-Demand Network Systems and Services (WONS), pp. 84–90 (2011)
15. He, D., Liang, G., Portilla, J., Riesgo, T.: A novel method for radio propagation simulation based on automatic 3D environment reconstruction. In: 2012 6th European Conference on Antennas and Propagation (EuCAP), pp. 1445–1449 (2012)
16. Nuckelt, J., Rose, D., Jansen, T., Kürner, T.: On the use of openstreetmap data for V2X channel modeling in urban scenarios. In: 2013 7th European Conference on Antennas and Propagation (EuCAP), pp. 3984–3988 (2013)
17. Wang, X., Anderson, E., Steenkiste, P., Bai, F.: Improving the accuracy of environment-specific channel modeling. IEEE Trans. Mob. Comput. **15**(4), 868–882 (2016)
18. Boban, M., Barros, J., Tonguz, O.: Geometry-based vehicle-to-vehicle channel modeling for large-Scale simulation. IEEE Trans. Veh. Technol. **63**, 4146–4164 (2014)
19. OpenStreetMap Foundation (OSM), 5 November 2018. http://www.openstreetmap.org
20. Osmosis, 5 November 2018. http://wiki.openstreetmap.org/wiki/Osmosis
21. Java OpenStreetMap (JOSM) editor, 5 November 2018. http://josm.openstreetmap.de/
22. DLR - Institute of Transportation Systems: SUMO - Simulation of Urban MObility, 5 November 2018. http://sumo-sim.org
23. Google Earth Pro V7.1.8.3036, Fujariah, United Arab Emirates (UAE) - Imaginary Date 5/11/2018, Accessed 5/11/2018, 25°07′54.8330″N, 056°20′08.9270″E, 5 November 2018. https://www.google.com/earth/
24. ASTM E2213-03: Standard Specification for Telecommunications and Information Exchange Between Roadside and Vehicle Systems - 5-GHz Band Dedicated Short-Range Communications (DSRC), Medium Access Control (MAC), and Physical Layer (PHY) Specifications (2018)

Security Approach for In-Vehicle Networking Using Blockchain Technology

Maher Salem(✉), Moayyad Mohammed, and Ali Rodan

Department of Computer Information and Sciences,
Higher Colleges of Technology, Fujairah, United Arab Emirates
{msalem1, mmohammed, arodan}@hct.ac.ae

Abstract. Security is nonnegotiable key point for in-vehicle networking. However, all communication between Electrical Control Unites (ECU) still suffer from security drawbacks like highly processing time or preserving confidentiality, integrity and authenticity. In this paper, we propose an approach to assess the feasibility of a private Blockchain technology to overcome the aforementioned drawbacks. In this approach, we consider in-vehicle networking contains two parts, namely, central (or connected) gateway (cGW) and switches. cGW and switches are Blockchain nodes, wherein Blockchain consensus protocols are what keep all the nodes on a network synchronized with each other. The approach considers any communication type between ECUs as an individual event, which can be a transaction, data entry or application execution. A use case of secure communication between two ECUs is presented as an evaluation mechanism for securing in-vehicle networking using the proposed Blockchain approach.

1 Introduction

Recent In-vehicle networking architecture includes ECUs, which are intercommunicating with each other. All ECUs are attached to various buses like CAN, CAN-FD, LIN, MOST, FlexRay [1]. The vehicle industry is expanding rapidly and hence security becomes one of the major key point in vehicle communications.

Most of in-vehicle networking topology are now domain based architecture and each domain contains all related ECUs attached to it. In addition, most of modern vehicles have now the Ethernet as a new bus system. Ethernet is intended to connect inside the vehicle high-speed communication requiring sub-systems like Advanced Driver Assistant Systems (ADAS), navigation and positioning, multimedia, and connectivity systems. For hybrid (HEVs) or electric vehicles (EVs), Ethernet will be a powerful part of the communication architecture layer that enables the link between the vehicle electronics and the Internet where the vehicle is a part of a typical Internet of Things (IoT) application. Figure 1 shows the recent and most likely future in-vehicle networking architecture [2].

As seen in Fig. 1, all subnetworks with related ECUs are connected to the same domain, which are connected to a cGW. All ECUs are communicating with each other through the domain controllers and the cGW is managing all the communication processes.

L. Barolli et al. (Eds.): EIDWT 2019, LNDECT 29, pp. 504–515, 2019.
https://doi.org/10.1007/978-3-030-12839-5_47

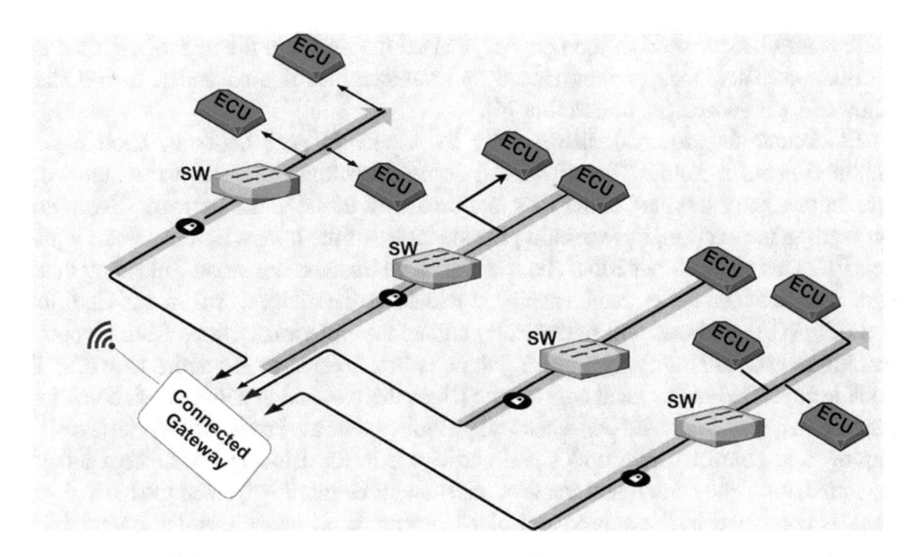

Fig. 1. Overview of modern and future in-vehicle networking architecture

With the considered architecture in Fig. 1 and the communication to the external resources using WiFi, Bluetooth or LTE, the vehicle becomes one of the complex and complicated networks. This architecture allows more vulnerabilities to be emerged and new attack vectors from the external networks to threaten the in-vehicle communication. Moreover, the process of authentication and preserving the integrity inside the vehicle network between all ECUs now needs more time for identity verification from the controller and the cGW as well, which in turn is considered time consuming. On the other hand, used cryptographic HW or SW solutions such as the Trusted Platform Module (TPM), Hardware Secure Module (HSM), Software Guard Extension (SGX), or Trusted Zone (TZ) is adding extra processing time, particularly with asymmetric methods and each one is installed in different OEM vehicle manufacturer, e.g. AUDI is implementing NXP HW solution from TPM or HSM while BMW mostly SGX or TZ. This non de-facto security solution is also considered in time processing issue [3].

There are currently several security approaches and architectures in the in-vehicle networking to secure the data, prevent any unauthorized access, or to preserve the integrity. All of these approaches are securing the vehicle but they still suffer from the processing time and the lack to immutability. In the next section, some of the recent security architecture will be discussed. In this article, we adopt the Blockchain technology and propose an approach to overcome the previous shortages and strengthen security and privacy inside the vehicle networks.

The Blockchain is defined as an open ledger that offers decentralization to the parties. In addition, it also offers transparency, immutability, and security. It has many features including being open, distributed, ledger, P2P and permanent. The function of a Blockchain is straightforward. As it is a peer-to-peer network, a user needs to start a transaction. Once done, a block is allocated to the said transaction. The transaction block is also broadcasted to the network, and all the nodes in the network get the said information. The block is then mined and validated. It is also added to the chain, followed by a successful transaction [4].

Blockchain is managed distributedly by a peer to peer network. Each node is identified using a Public Key (PK). All communications between nodes, known as transactions, are encrypted using PKs and broadcast to the entire network. Every node can verify a transaction, by validating the signature of the transaction generator against their PK. This ensures that Blockchain can achieve trustless consensus, meaning that an agreement between nodes can be achieved without a central trust broker, e.g. Certificate Authority (CA). A node will periodically collect multiple transactions from its pool of pending transactions to form a block, which is broadcasted to the entire network. The block is appended to the local copy of the Blockchain stored at a node if all constituent transactions are valid. A consensus algorithm such as Proof of Work (PoW) is employed to control which nodes can participate in the Blockchain. Once a block is appended, it (or the constituent transactions) cannot be modified, since the hash of each block is contained in the subsequent block in the chain, which ensures immutability. A node can change its PK (i.e. identity) after each transaction to ensure anonymity and privacy [5].

Consensus algorithms are a decision-making process for a group, where individuals of the group construct and support the decision that works best for the rest of them. It's a form of resolution where individuals need to support the majority decision, whether they liked it or not [4]. List of All Consensus Algorithms

- Proof-of-Work
- Proof-of-Stake
- Delegated Proof-of-Stake
- Leased Proof-Of-Stake
- Proof of Elapsed Time
- Practical Byzantine Fault Tolerance
- Simplified Byzantine Fault Tolerance
- Delegated Byzantine Fault Tolerance
- Directed Acyclic Graphs
- Proof-of-Activity
- Proof-of-Importance
- Proof-of-Capacity
- Proof-of-Burn
- Proof-of-Weight

Regarding the network between nodes, there are three types of network, namely decentralized, centralized and distributed. Since in-vehicle networking ECUs are related to one domain and all are controlled by the cGW, we adopted the centralized network approach where all the nodes come under a single authority.

The rest of this article is divided as follows, Sect. 2 discusses some related work regarding in-vehicle security and Blockchain. Section 3 presents our approach by deploying Blockchain technology. Secure communication mechanism using the proposed Blockchain will be presented in Sect. 4. Finally, Discussion and conclusion are presented in Sect. 5.

2 Related Work

Cybersecurity in vehicle communication attracts researcher to propose and implement a security solution to protect vehicle internal and external communication. However, till now security still a major issue in vehicle communications. Rajbahadur et al. conducted a survey about anomaly detection techniques in vehicle communication using 3 dimensions with several subcategories. The main result was that most of prior research evaluated their methods from simulations. Proposed techniques ignored safety of the vehicles while focusing on cybersecurity [6]. IOActive published a white paper about the vulnerabilities in automotive industry and they concluded that vulnerabilities have decreased in both impact and likelihood. In addition, they showed that most common attack vectors are internal software components and network-connected applications [7]. Li et al. presented also some explanation and declaration about considering attacks and improving security in the connected vehicle cloud computing. They have discussed and investigated all articles in the journal and give a professional insight regarding cybersecurity and the attacks against vehicle networks [8].

Singh and Kim have discussed the challenges of automotive security in hardware and software, and propose a security architecture for automotive security and also mention future research challenges in automotive cyber security [9]. They have defined possible future security issues related to intelligent vehicles such as secure communication and secure routing. This article supports our approach by emphasizing on the need for a novel security approach to protect vehicle communications.

Even pioneers in the automotive industry like NXP reviewed todays' ECUs, especially from a semiconductor technology perspective. After that they reviewed it regarding the potential of future vehicle networks, it has described future ECUs along with the limitations and opportunities. They have concluded that the domain based architectures will be introduced on the short- to mid-term while for the central computing platform items like safety, reliability and cost still need to be answered especially for the central Module [10]. On the other hand, Onuma et al. investigated the case of updating the ECU with less processing time and to avoid any attacker exploiting exposed vulnerabilities. This article emphasized on the weakness of in-vehicle networking specially on ECUs update process [11].

For the last decade a lot of proposed solutions have been presented to improve security of vehicle communications. In this regard, Zeng et al. have presented a comprehensive survey discussing all in-vehicle networks based on three factors, system cost, data transmission capacity, and fault-tolerance capability. Then they have assured the importance of the gateway in connected vehicle, and finally presented some security threats issues on the in-vehicle networks [12]. Their contribution was very clear about the importance of having connected gateway and emphasizing on the importance of security on in-vehicle communication.

Wang et al. have also proposed a distributed anomaly detection system using hierarchical temporal memory (HTM) to enhance the security of a vehicular controller area network bus [13]. The HTM model can predict the flow data in real time, which depends on the state of the previous learning. This technique is also oriented to detect and minimize the abnormal behavior inside the vehicle network. In addition, Woo et al.

showed that even wireless attack is physically possible using a real vehicle and malicious smartphone application in a connected car environment. They proposed a novel security protocol for CAN networks and used CANoe for the evaluation and experiment. The result delivered promising security protocol better than existed one in regards to authentication delay and communication load [14]. Many other security proposals for securing communications have been demonstrated and achieved good result which again spot on the importance of security in vehicle, such as [15–18]. Moreover, some significant topics describe secure communication with the cloud and enhancing it with a secure storage concept, [28, 29].

All of the above proposed methods and many more others have proven feasible security improvement. However, the main issues of the authentication and processing time are still existed. Therefore, Blockchain technology recently is the newest solution to avoid the aforementioned issues. As a start, a good survey about involving Blockchain in several applications to improve the security can be found in [19]. The authors have provided an overview of the application domains of Blockchain technologies in IoT, e.g. Internet of Vehicles, Internet of Energy, Internet of Cloud, Fog computing. One of the professional solution that utilizes Blockchain is presented by Alam in his thesis about securing in-vehicle communication [20]. The author proposes the use of symmetric key cryptography and elliptic curve-based Public Key Encryption (PKE) for ensuring confidentiality and the use of digital signature for ensuring integrity and authenticity. He introduces Blockchain in vehicles to protect the stored data of ECUs. The experiment study was conducted on Docker and ARM processor based Raspberry Pei. In our proposed article, we have used the concept in [20] and improve it not only to protect the data but also to control the communications between all ECUs. According to other applications of Blockchain in vehicle, Dorri et al. [5] proposed an optimized Blockchain instantiation for the Internet of Things (IoT) called Lightweight Scalable Blockchain (LSB). It is a decentralized approach that secure and preserve the privacy of all automotive ecosystem. They proposed the LSB approach that solved the problem of high processing time of the consensus algorithm. Moreover, they discussed some attack scenarios like DDoS and how the LSB method protect against it. In our article, we will utilize also some concepts from the LSB in reducing the processing time. However, we still believe that a centralized and private Blockchain is suitable for in-vehicle networking. On the other side, Cebe et al. proposed a permissioned Blockchain to manage all collected data by the vehicle [21]. They integrated Vehicular Public Key Management (VPKI) to the proposed Blockchain to provide membership establishment and privacy. Next, they designed a fragmented ledger that will store detailed data related to vehicle such as maintenance information/history, car diagnosis reports, etc. Yang et al. have proposed a decentralized trust management system in vehicular networks based on Blockchain technique where vehicles can validate the received messages from neighboring vehicles using Bayesian Inference Model [22]. In the before mentioned article, the proposed method gather data from vehicles and rank it. Then they generate a block in the Blockchain.

Finally, and unfortunately, we cannot cover all proposed work due to limitation and space issue. However, further details and resources about using Blockchain in vehicle security can be found in [23–26].

3 Proposed Blockchain Approach

In the domain-based architecture, ECUs are grouped by their functionalities and placed in the same communication bus (called a domain). Every domain is controlled by a controller, which is called a switch (domain controller). In this architecture, every ECU collects data from its sensors, processes the data, takes a decision, and works on that decision or sends the processed data to other ECUs. Nodes are connected through the connected/central gateway (MasterNode). A node can send data to other nodes through the MasterNode.

Since the number of switches or domain controllers in the current or future in-vehicle communication architecture is limited and connected to one cGW (or may be many cGWs), a centralized Blockchain with a single authority, i.e. permissioned, is feasible and suites the internal structure of the vehicle. However, if we consider the vehicle external communications with the infrastructure such V2X then a centralized approach may not be suitable. From this point of view, the general overview of our proposed Blockchain approach is demonstrated in Fig. 2.

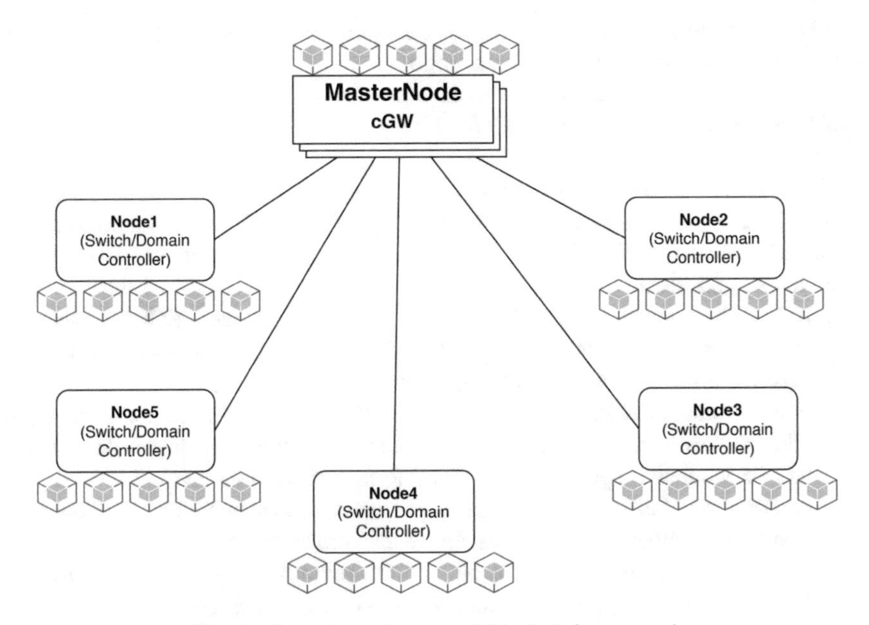

Fig. 2. Overview of proposed Blockchain approach

In the previous figure, MasterNode is permitting the authority for each node to get involved in the network or not. In addition, each node gets updated by the recent Blockchain after a block is validated, approved and created then added to the Blockchain. The MasterNode shares blocks with the nodes where the integrity is preserved by the hashing mechanism applied. The internal structure of a single block can be shown in Fig. 3.

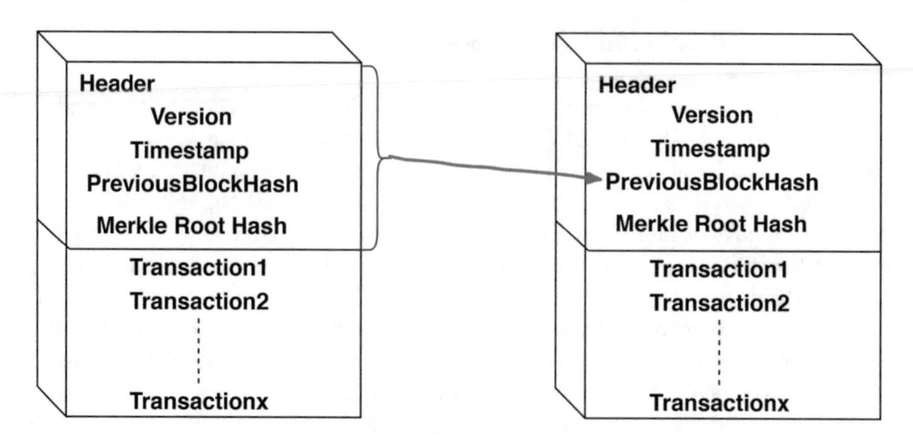

Fig. 3. Internal structure of a single block in the chain

The procedure of how the approach works is described in the following points:

1. Every ECU sends encrypted and signed data to the corresponding Node to preserve the confidentiality, integrity and authenticity
2. All ECUs data are stored in the Blockchain of each Node
3. Inter-communication between ECUs is only allowed when the Node grants permission for its ECU
4. The MasterNode monitors and verifies all Nodes, say, all Blockchain data with each Node is encrypted to avoid any impact between Nodes if one is compromised.
5. MasterNode stores all public keys for all Nodes and ECUs to keep verify their signatures and identity.
   a. If ECU1 attached to Node1 needs to communicate with ECU3 attached to Node3, Node1 verifies ECU1 and ECU3 identity from the MasterNode. Once verified, a permission is granted to ECU1.
   b. ECU1 sends a transaction to Node1. Node1 in turn shares the transaction with other Nodes to vote based on a consensus algorithm, then based on voting, it will be validated, approved and added to the transactions list.
   c. After approval of all Nodes on group of transactions then a block is added to the front of the Blockchain with the following information:
      i. Hash value for the current block and previous block. In this regard, each node has a copy of the blockchain, Merkle tree is inside each block for integrity
      ii. Block version and header added to the MasterNode for monitoring and history issues
   d. The communication is granted and executed in a way that ECU3 decrypt the transaction using its private key.

The following message sequence diagram in Fig. 4 shows the processes of securing and validating communications between two ECUs.

The previous figure shows the main and simple process to secure the communication between two nodes and how the MasterNode is monitoring the process and granting accesses for each ECU.

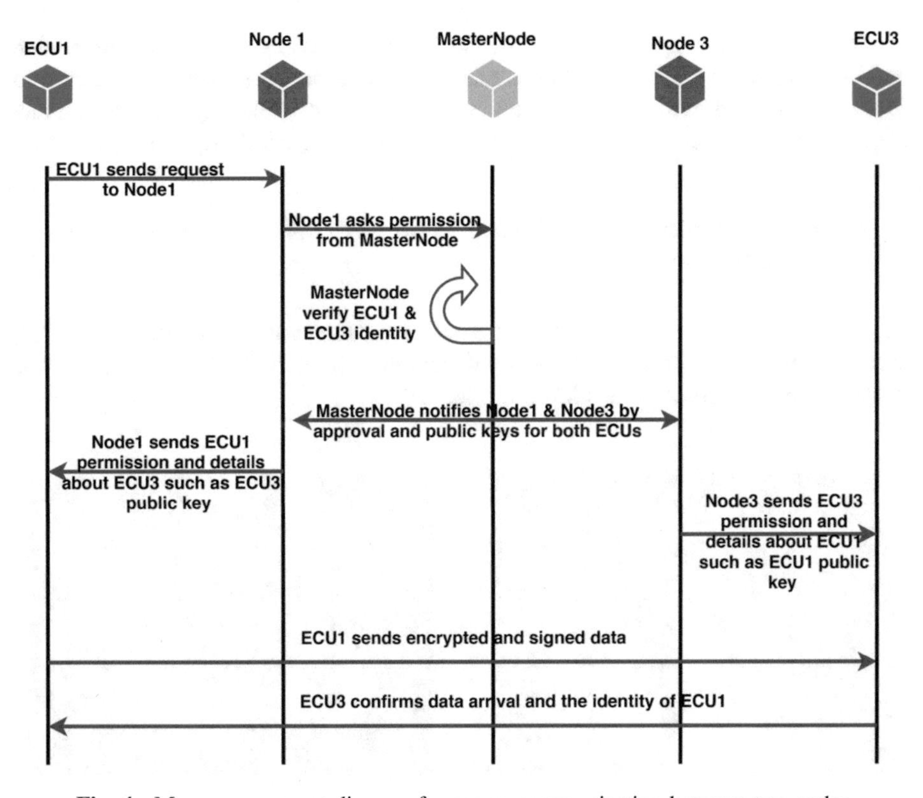

Fig. 4. Message sequence diagram for secure communication between two nodes

- ECU1 and ECU3 need to communicate.
- ECU1 sends signed request to Node1 asking for permission
- Node1 asks the MasterNode if this communication is allowed
- MasterNode contains all public keys for all components. It verifies the signature of ECU1 and ECU3 to preserve the authenticity and integrity.
- Once verified, MasterNode notifies Node1 and Node3 that both ECUs can communicate. Moreover, it sends the public keys of the two ECUs for the nodes.
- Each node shares the public key with the ECU and grants it a permission to communicate
- ECU1 can now encrypt the data using the public key of ECU3 and signs it using its private key
- ECUs can now securely communicate
- All of these transactions will be validated and verified by the MasterNode and once all nodes vote for validation, a new block is added to the Blockchain in the MasterNode
- The hash value will be then calculated for the current block and the MasterNode shares the updates with the Nodes.
- All communication between nodes and MasterNode is signed for identity verification.

In the previous diagram, all communication is secured by asymmetric encryption and the identity is verified by the signatures. And all transactions have been considered to add a new block in the chain. Therefore, Confidentiality, Integrity and Authenticity are all preserved and the MasterNode is monitoring and controlling all the process.

4 Secure Communication Using Proposed Blockchain Approach

For this case study, we adopted the secure communication process between two ECUs. Both ECUs contains an MPC5646C microcontroller from NXP [27]. The process is proposed by F. Juergen as displayed in Fig. 5.

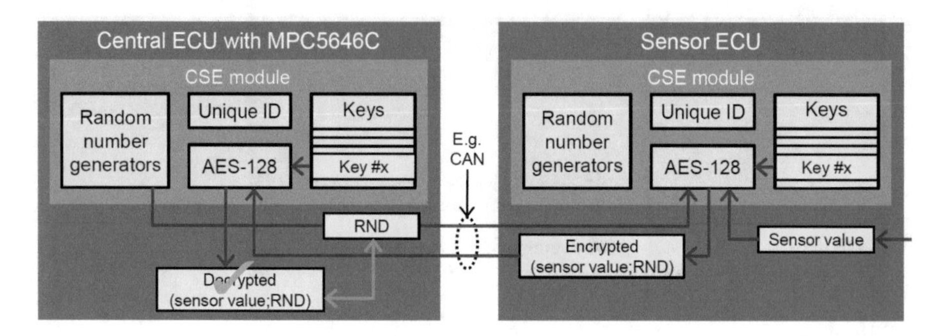

Fig. 5. Secure communication between two ECUs

The main idea of this method in Fig. 5 is to preserve integrity, authenticity and confidentiality. The Cryptographic Service Engine (CSE) contains already from the OEM all keys between all ECUs for communication. The symmetric encryption shortages like using single-key for encryption and key distribution is known and have been solved by asymmetric methods. Therefore, using public key method for encryption and private key for signing in our approach is suitable and more feasible for in-vehicle communication.

Applying our Blockchain approach in the previous process in Fig. 5 can be interpreted as the following:

1. Central ECU sends a signed request together with a random number to the intended Node
2. The intended Node verifies it and forwards it to the MasterNode
3. The MasterNode will validate the identity of Central ECU and encrypt the random number with the public keys of central ECU and sensor ECU
4. MasterNode sends permission approval and the public key of central ECU to the Node of sensor ECU. The same action for central ECU.
5. Each Node communicates with the related ECU with the following information: permission granted, public key of other node, encrypted random number.

6. Central ECU will decrypt the random number to verify MasterNode identity
7. Sensor ECU will decrypt the random number to verify MasterNode identity, then it encrypts the random number again using central ECU public key.
8. Sensor ECU sends this to the central ECU, which in turn decrypt it to check Sensor ECU identity
9. When everything is validated and verified, both ECU can now securely communicate
10. All of the above transactions will be stored in one block and all Nodes vote for approval.
11. If they are approved, the MasterNode will calculate a new hash value for all transactions, then add a new block to the Blockchain and share it with all Nodes

After applying our proposed approach, all components have been identified (Authenticity), data is encrypted between ECUs (Confidentiality) and hash values have been generated (Integrity).

5 Discussion and Conclusion

All communication between ECUs still suffer from security drawbacks like highly processing time or preserving confidentiality, integrity and authenticity. We propose an approach to assess the feasibility of a private Blockchain technology to overcome the aforementioned drawbacks. Blockchain nodes in this approach are the cGW as MasterNode and switches as Nodes. According to the nature of in-vehicle networking, the best Blockchain structure is centralized where the permission is granted by the MasterNode. We present a use case of how two ECUs can communicate together in a secure communication channel and how the Blockchain interact in this regard. To demonstrate the secure communication between ECUs, we present a message sequence diagram to show all internal processes between the MasterNode and intended nodes. Then, we apply the proposed approach on a real example from NXP microcontrollers.

All communications between ECUs are secure and valid, in which a block of all these transactions has been created by the MasterNode and then added to the beginning of the chain. Finally, MasterNode updates all Nodes with the latest Blockchain. So, the processing time is minimized, and Confidentiality, Integrity and Authenticity are preserved. We believe, the proposed Blockchain is feasible and can be applied on the in-vehicle communication. As a future outlook, more improvements on the proposed approach will be performed and a real experimental and comparison study will take place to provide a practical evidence on the success of the approach.

References

1. Wolf, M., Weimerskirch, A., Paar, C.: Secure in-vehicle communication. In: Lemke, K., Paar, C., Wolf, M. (eds.) Embedded Security in Cars. Springer, Berlin (2006)
2. Hank, P., Müller, S., Vermesan, O., Van Den Keybus, J.: Automotive Ethernet: in-vehicle networking and smart mobility. In: DATE 2013 Proceedings of the Conference on Design, Automation and Test in Europe, Grenoble, France, pp. 1735–1739. IEEE (2013)

3. Wu, Z., Zhao, J., Zhu, Y., Li, Q.: Research on vehicle cybersecurity based on dedicated security hardware and ECDH algorithm. In: Intelligent and Connected Vehicles Symposium, SAE Technical Paper 2017-01-2005 (2017)
4. 101 Blockchains Homepage. https://101blockchains.com. Accessed 12 Dec 2018
5. Dorri, A., Steger, M., Kanhere, S.S., Jurdak, R.: BlockChain: a distributed solution to automotive security and privacy. IEEE Commun. Mag. **55**(12), 119–125 (2017)
6. Rajbahadur, G.K., Malton, A.J., Walenstein, A., Hassan, A.E.: A survey of anomaly detection for connected vehicle cybersecurity and safety. In: IEEE Intelligent Vehicles Symposium (IV), Changshu, China, pp. 421–426 (2018)
7. Thuen, C.: Commonalities in Vehicle Vulnerabilities. IOActive, Seattle (2018)
8. Li, H., Lu, R., Mahmoud, M.: Security and privacy of connected vehicular cloud computing. IEEE Netw. **32**(3), 4–6 (2018)
9. Singh, M., Kim, S.: Security analysis of intelligent vehicles: challenges and scope. In: International SoC Design Conference (ISOCC), Seoul, South Korea, pp. 13–14 (2017)
10. NXP Homepage. https://www.nxp.com/docs/en/white-paper/FVNECUA4WP.pdf. Accessed 12 Dec 2018
11. Onuma, Y., Terashima, Y., Kiyohara, R.: ECU software updating in future vehicle networks. In: International Conference on Advanced Information Networking and Applications Workshops (WAINA), Taipei, Taiwan, pp. 35–40 (2017)
12. Zeng, W., Khalid, M., Chowdhury, S.: In-vehicle networks outlook: achievements and challenegs. IEEE Commun. Surv. Tutorials **18**(3), 1552–1571 (2018)
13. Wang, C., Zhao, Z., Gong, L., Zhu, L., Liu, Z., Cheng, X.: A distributed anomaly detection system for in-vehicle network using HTM. IEEE Access **6**, 9091–9098 (2017)
14. Woo, S., Jo, H.J., Lee, D.H.: A practical wireless attack on the connected car and security protocol for in-vehicle CAN. IEEE Trans. Intell. Transp. Syst. **16**(2), 993–1006 (2015)
15. Petit, J., Shladover, S.E.: Potential cyberattacks on automated vehicles. IEEE Trans. Intell. Transp. Syst. **16**(2), 546–556 (2015)
16. Dardanelli, A., Maggi, F., Tanelli, M., Zanero, S., Savaresi, S., Kochanek, R., Holz, T.: A security layer for smartphone-to-vehicle communication over bluetooth. IEEE Embed. Syst. Lett. **5**(3), 34–37 (2013)
17. Fernández, P.J., Santa, J., Bernal, F., Skarmeta, A.: Securing vehicular IPv6 communications. IEEE Trans. Dependable Secure Comput. **13**(1), 46–58 (2016)
18. Salem, M., Buehler, U.: Transforming voluminous data flow into continuous connection vectors for IDS. Int. J. Internet Technol. Secured Trans. **5**(4), 307–326 (2014)
19. Ferrag, M.A., Derdour, M., Mukherjee, M., Derhab, A., Maglaras, M., Janicke, H.: Blockchain technologies for the Internet of Things: research issues and challenges. In: CoRR - Computing Research Repository - arXiv, pp. 1–14 (2018)
20. Alam, S.: Securing vehicle Electronic Control Unit (ECU) communications and stored data. School of Computing, Queen's University, Kingston, Ontario, Canada (2018)
21. Cebe, M., Erdin, E., Akkaya, K., Aksu, H., Uluagac, S.: Block4Forensic: an integrated lightweight blockchain framework for forensics applications of connected vehicles. IEEE Commun. Mag. **56**, 50–57 (2018)
22. Yang, Z., Yang, K., Lei, L., Zheng, K., Leung, V.C.M.: Blockchain-based decentralized trust management in vehicular networks. IEEE Internet Things (2018)
23. Jiang, T., Fang, H., Wang, H.: Blockchain-based Internet of vehicles: distributed network architecture and performance analysis. IEEE Internet Things (2018)
24. CryptoTec Homepage. https://www.cryptotec.com/wp-content/uploads/2018/01/Blockchain_for_Automotive_CryptoTec_EN.pdf. Accessed 12 Dec 2018
25. CUBE Homepage. https://cryptorating.eu/whitepapers/CUBE/CUBEWhite_Paper-V1.3.pdf. Accessed 12 Dec 2018

26. Sharma, P.K., Moon, S.Y., Park, J.H.: Block-VN: a distributed blockchain based vehicular network architecture in Smart City. J. Inf. Process. Syst. **13**(1), 184–195 (2017)
27. NXP Homepage. https://www.nxp.com/docs/en/supporting-information/DWF13_AMF_AUT_T0112_Detroit.pdf. Accessed 12 Dec 2018
28. Wang, X.A., Yang, X., Li, C., Liu, Y., Ding, Y.: Improved functional proxy re-encryption schemes for secure cloud data sharing. Comput. Sci. Inf. Syst. **15**(3), 585–614 (2018)
29. Wang, X.A., Liu, Y., Zhang, Z., Yang, X.: Improved group-oriented proofs of cloud storage in IoT setting. Concurrency Comput. Pract. Exp. **30**(21), e4781 (2018)

Towards Optimizing Energy Efficiency and Alleviating Void Holes in UWSN

Abdul Mateen[1], Nadeem Javaid[1(✉)], Muhammad Bilal[2], Muhammad Arslan Farooq[3], Zahoor Ali Khan[4], and Fareena Riaz[5]

[1] COMSATS University Islamabad, Islamabad 44000, Pakistan
nadeemjavaidqau@gmail.com
[2] School of Computing and IT, Centre for Data Science and Analytics, Taylor's University, Subang Jaya, Malaysia
[3] Government College University, Lahore, Pakistan
[4] Computer Information Science, Higher Colleges of Technology, Fujairah 4114, UAE
[5] University of Kotli, AJK Campus, Azad Kashmir 1100, Pakistan
http://www.njavaid.com

Abstract. Underwater Wireless Sensor Networks (UWSNs) are promising and emerging framework having a wide range of applications. The underwater sensor deployment is beneficial; however, some factors limit the performance of the network, i.e., less reliability, high end-to-end delay and maximum energy dissipation. The provisioning of aforementioned factors have become challenging task for the research community. In UWSNs, battery consumption is inevitable and has a direct impact on the performance of the network. Most of the time energy dissipates due to the creation of void holes and imbalanced network deployment. In this work, a routing protocol is proposed to avoid the void holes problem and extra energy dissipation, due to which lifespan of the network increases. To show the efficacy of our proposed routing scheme, it is compared with state of the art protocols. Simulations result show that the proposed scheme outperforms the counterparts.

Keywords: GEDPAR · Void holes · Energy efficiency · Underwater Wireless Sensor Network (UWSNs) · Depth adjustment · Transmission range

1 Introduction

The planet Earth, on which we live our lives, consists of 70% water. Whereas, the oceans hold more than 90% of total water. This much quantity shows the importance of the water medium. To explore the underwater medium for getting and sharing the important information, a network is deployed in a specific region. Information transmission using Underwater Wireless Sensor Networks (UWSNs) is one of the emerging technologies and is used for the betterment of ocean observation systems. Applications of UWSNs range from aquaculture

L. Barolli et al. (Eds.): EIDWT 2019, LNDECT 29, pp. 516–527, 2019.
https://doi.org/10.1007/978-3-030-12839-5_48

to oil industry; instrument monitoring to climate recording; pollution control to predictions on natural disasters; search and survey purposes to submarine purposes.

The UWSN consists of several sensor nodes and these nodes acquire information and transmit it towards the next node closer to the sink[1] [18]. This sink may be the onshore data center or simple sensor node over the water surface. In data forwarding procedure, the source[2] node generates data packets and communicate with its neighbors to find the potential node. Afterwards, potential neighbor node finds the next potential node from its neighbors and transmit data packet towards that potential node. To find the potential neighbor from the forwarder node, some criteria and routing procedure are defined. This criterion may base upon efficient energy utilization or alleviation of void holes.

The design of routing protocols has paramount importance in UWSN. These protocols indicate the routing path for data from the source node at bottom towards the sinks node at the surface of ocean. Expressly, these protocols face the different challenges which are associated with the underwater medium, e.g., limited battery resources, interference, noise, reliable Packet Delivery Ratio (PDR), high propagation delay, movements of sensors and void holes.

Efficient energy usage is the most important while designing a routing protocol. As sensor nodes in water have limited resources (already discussed). The batteries are non-removable and have limited energy storage. This issue provides a strong base for the efficient utilization of batteries. Mostly, energy dissipates during the processes of data packet transmission and reception. The efficient energy usage depends upon various factors. For instance, the initial position and number of anchor nodes; sensor nodes and the way in which nodes are deployed. The deployment of a network must be one of the two types (1) sparse deployment and (2) dense deployment. The sparse deployment leads toward the creation of a void hole and dense deployment results in an excessive amount of sensors failure.

The energy and network stability have a direct relation. As, more will be the energy of sensors, longer will be the stability of the network and vice versa. Void holes are areas within the transmission range of a network where a node cannot find its next neighbor or forwarder. The void holes creation has following reasons (1) node becomes dead due to a lot of energy usage and (2) no forwarder node.

Localization of sensor network in underwater is indispensable. The gathered data is useless until it is not correlated with the specified position of the sensor node. Localization in UWSNs is very important as it has many useful applications, e.g., target tracking, underwater environment monitoring, pollution control and geographic routing protocols. Nevertheless, UWSNs cannot use Global Positioning System (GPS) due to high energy dissipation and high attenuation of RF signals [7] and [19].

In this work, we proposed GEographic and opportunistic routing with Depth and Power Adjustment Routing (GEDPAR) as a routing technique. GEDPAR is

[1] Sink: This word is alternatively used for sink node, sonobuoy, destined node and destination node.

[2] Source: The words source node and initial node are alternatively used for source.

compared with GEographic and opportunistic routing with Depth Adjustment Routing (GEDAR) and Layered Multi-path Power Control (LMPC). Simulations are performed in order to check the effectiveness of our proposed scheme.

The remainder of the paper is organized as follows: Sect. 3 provides the brief overview of state of the art work. Problem statement elaborates in Sect. 3. Section 4 represents the proposed system model. Discussion on the simulations is given in Sect. 5. Finally, Sect. 6 summarizes the paper.

2 Related Work

In this section, we review and compare some recent works on the base of covering a specific area of UWSNs. The papers which cover the energy efficiency and void holes are compared in Sect. 2.1. Additionally, the papers that cover the concept of localization or geographic routing are compared in Sect. 2.2. Moreover, Sect. 2.3 presents the comparison of topological control based schemes. In the end, the concept of a void hole is presented in Sect. 2.4.

2.1 Energy Efficiency Based

The papers [1–4] propose different schemes and protocols to enhance the energy-efficiency. The papers [1] and [2] both are using multi-hops techniques. The paper [1] is focusing on network reliability, mobility management, PDR and energy efficiency. On the other hand, the paper [2] is only focusing on energy efficiency. Both the papers [1] and [2] achieve their objectives; however, end-to-end delay is compromised. The authors in papers [3] and [4] mainly focus on reliability by covering one-hop from the forwarder node. The proposed scheme EBLE from the paper [3] aims to minimize the energy dissipation with packet size management. The objective is successfully achieved on the cost of delay. The cooperative routing is used in paper [4] for data reliability and mobility management, while PDR and efficient energy usage are main aims. The objectives are achieved successfully; however, the network performs poorly in sparse network deployment.

The works [8–11] are also using energy efficiency techniques. The works [8] and [10] provide the reliability. Both of works discuss the concept of multi-hoping. The proposed scheme in the work [8] is beneficial for a large amount of data packets; however, this proposed technique does not performed well in sparse network deployment. The MLPR from [10] looks toward the efficient path for routing by utilizing minimum energy. For the implementation of MLPR, more memory is required for the extra operations at each node. The energy dissipation schemes; SDVF and EBULC are proposed in works [9] and [11], respectively. Both schemes consider mobility management for decreasing the energy consumption in UWSNs. Results show that end-to-end delay in the works [9] and [11] is enhanced.

The energy efficiency is focused in the works [12–15]. In [12], some data collection methods are discussed which used minimum energy for data transmission from source to the destination. In both [12] and [13], mobility management is considered, while in the [12], reliability and packet size management is not considered. Nevertheless, the works [13–15] focus on the reliability of the network. Additionally, [13] considers both types of forwarding strategies; single-hop and multi-hop. While [12,14] and [15] only focus on single-hop from the current node. Moreover, the work in [13] considers the security issues of UWSNs. While in [12], the authors discusses the problems of getting route information. In [14], the complexity of the network is a major challenge. Additionally, paper [15] works for energy efficiency by managing the size of data packet.

2.2 Localization Based

The authors in [7,16–18] and [21] discuss the geographic or localization-based routing. The work in [17] and [18] review the works in which the concept of localization based routing is used. Both of these above, discuss reliability and none of them work on mobility management or packet size management. Moreover, in [17] and [18], the concept of single-hop and multi-hop is devised. The challenges which are discussed in these works are: high interference, limited batteries of sensor nodes, low bandwidth and malicious attacks. The work in [16] achieves the higher PDR by finding the locations of alive nodes. Afterwards, the data packets are sent to these alive nodes, accordingly. The challenges discussed in [21] are: localization, feasible hardware, relevant simulation tools and low power gliders.

2.3 Topology Control Based

In [2,5,6] and [19] proposed topology control based solutions. TCEB and GARM schemes are proposed for controlling the topology of UWSNs in [2] and [6], respectively. In addition, the [5] classifies different topological protocols. From [5], reliability and mobility is discussed. The work [5] focuses on single-hop and multi-hop while the work [2] only focus on next forwarder node. The challenges that discussed in [2,5,6] and [19] are: high attenuation, mobility of sensor nodes, energy efficiency, low bandwidth, connectivity loss, high bit rate error, high deployment cost, complexities and optimal location of glider. Using dynamic topological strategy, work in [2] achieves energy efficiency and the work in [6] enhances both PDR and energy efficiency. In [19], mobility management is a major consideration using EEL and the concept of multi-hoping. In addition, the work [19] achieves better simulation results from compared ones.

2.4 Void Hole Based

The concept of a void hole is presented in [20]. Void holes are the regions within the network range from where further data delivery is not possible. In other words, if a forwarder node does not have any further node for data packet transmission then this node is called void node and the area where transmission is not possible in called void holes. TORA is presented in [20] in order to avoid the void holes. The proposed scheme uses the concept of multi-hoping to avoid void holes and to improve energy efficiency. Nevertheless, reliability and complexity of this scheme are not discussed.

3 Problem Statement

In UWSN, each sensor has limited resources and requires effective utilization of these resources. Efficient energy consumption has a major contribution to stabilize the network for long term communication. In UWSNs, the packet is sent from the source node to the sink node using different relay nodes. If a node cannot find a forwarder node in its transmission range, it causes hindrance in the network during communication.

In order to avoid the void holes in UWSNs, a routing protocol namely GEDAR presented in [23]. GEDAR addresses the issue by adjusting the depth of nodes; however, the process of depth adjustment consumes lots of energy. In [24], LMPC routing technique addresses the efficient data transmission by making the binary tree from route node. However, binary tree generation consumes high energy and lead towards the transmission overhead. To overcome the aforementioned problems, a routing protocol namely GEDPAR is proposed for avoiding the void holes and eliminating the extra energy consumption.

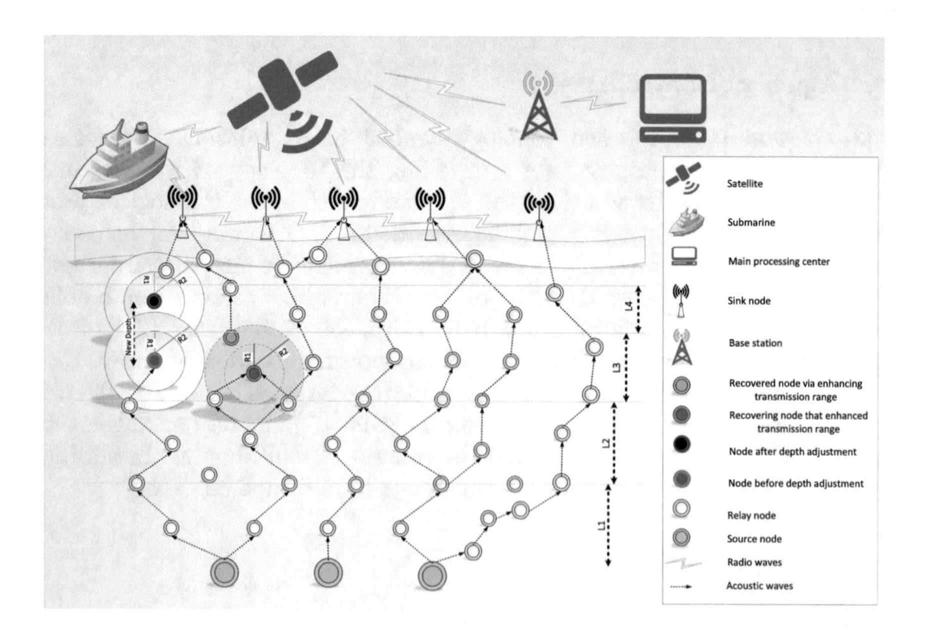

Fig. 1. Proposed system model

4 System Model

In this section, our proposed system model is presented in Fig. 1. The system model consists of source nodes, relay nodes and sonobuoys. Source node forwards data packets toward the destined sonobuoys during transmission. The proposed protocol follows multi-hoping feature for packets transmission. Source and relay nodes only use acoustic signals while radio waves are used for communication among sink node, submarine, satellite, base station and the main processing unit.

In the proposed system model, sensor nodes are randomly deployed in underwater medium. Nevertheless, sink nodes are deployed at the sea surface. The same transmission range and energy are assigned to each sensor node. Moreover, each sensor node has also the ability to adjust their depth from the lower layer to the upper layer. During depth adjustment, nodes only move in vertical direction. The process of depth adjustment occurs in the case when a node cannot find its next forwarder even by increasing the transmission range. There are three different cases that are elaborated through the proposed system model.

4.1 GEDAR

GEDAR is an opportunistic and depth adjustment-based routing protocol. In GEDAR, each packet is sent to the forwarding set which consists of several neighbors. Algorithm 1 shows the procedure of periodic beaconing in GEDAR. This procedure requires S and D. Where, κ represents beacon messages. Lines (4–16) elaborate on the overall procedure for distance and neighbor calculations. Lines (8–11) add neighbors to the neighbor list. Line 6 shows that this procedure repeats for each and every source node.

4.2 LMPC

In LMPC, multiple layers are made vertically by dividing the network for efficient transmission. As, working of LMPC is totally depending on the layers and we have already mentioned that noise in deep water is less than the shallow water. So, the size of a layer in deep water is high and vice versa for shallow water. This size of a layer has an inverse relation with noise, greater the attenuation of noise lower will be the layer size and vice versa.

4.3 GEDPAR

GEDPAR is our proposed routing protocol. For this protocol, GEDAR and LMPC are taken as benchmark schemes. In GEDPAR, layering concept is taken from the LMPC and depth adjustment is taken from the GEDAR. GEDPAR takes transmission enhancement step on the appearance of void holes. Transmission enhancement takes some extra energy; however, most of the void holes are removed in this process. If a node cannot cover the void hole even by increasing the transmission range then depth adjustment takes place for that node. The procedure for periodic beaconing is same as in GEDAR (see Sect. 4.1).

Algorithm 1. Periodic beaconing

1: node (S, D)
2: network deployment
3: κ: beacon message
4: **if** beacon is timed out **then**
5: κ.coordinates = distance (node)
6: **if** node ϵ N **then**
7: **for** s ϵ S **do**
8: **if** $\lambda_s = 0$ **then**
9: add in κ neighbor list (s.id, x-coordinates, y-coordinates)
10: $\lambda = 1$
11: **end if**
12: **end for**
13: **end if**
14: broadcast λ
15: set new timeout
16: **end if**

Algorithm 2. Void hole recovery

1: LMPC(node)
2: **if** current node is void = 1 **then**
3: stop beacon messages
4: **end if**
5: $\nu = \emptyset$: no neighbor node
6: ν: set of next forwarder nodes
7: Δ: set of void nodes
8: n_v: is current void node
9: **if** $|\nu| > 0$ **then**
10: enhance transmission radius
11: $dist = \sqrt{(x_v - x_u)^2 + (y_v - y_u)^2}$
12: **if** $dist \leq r_c$ **then**
13: goto (23)
14: **else**
15: **for** $n_u \ \epsilon \ \nu$ **do**
16: $dist = \sqrt{(x_v - x_u)^2 + (y_v - y_u)^2}$
17: **if** $dist \leq r_c$ **then**
18: $(x_v - x_u)^2 + (y_v - y_u)^2 + (z_v - z_u)^2$
19: $\nu = \nu \ \mathrm{U} \ z_v$
20: **end if**
21: **end for**
22: n_v moves to new calculated depth
23: **end if**
24: current node is void = 0
25: **end if**

Algorithm 2 involves the steps for the recovery of the void hole. First of all, value for the current node is set to "1" for its identification and stop beacon messages. The symbol $\emptyset$ shows that current node has no neighbor. In other words, it is a void node. ν is set that contains the record of next forwarder nodes. Δ and n_v are the symbolic representations of void nodes set and current void node, respectively. The distance for each forwarder near the current void is calculated. Afterwards, this distance is compared with the transmission range. If the distance is less then the transmission range, it means that the next forwarder node is within the range of the current forwarder node and vice versa. In case, if no forwarder node exists within transmission range then depth adjustment takes place and the status for the void node is set "0" from "1".

5 Simulation and Discussion

Simulations are performed in order to check the effectiveness of the proposed scheme. The results of our proposed technique are compared with GEDAR and LMPC. GEDPAR is greedy opportunistic routing protocol in which next forwarder node is selected on the criteria of minimum distance from the current node. In the proposed protocol, firstly, current node enhance transmission range when it finds no neighbor in its transmission range. After that, if current forwarder still not able to find any node in its range then it executes depth adjustment. During depth adjustment, the node moves from deeper layer to the shallow one.

5.1 Network Parameters Setting

The network is deployed over the area of 1500 m × 1500 m × 1500 m. The number of nodes and sinks are 100 and 45, respectively. Initially, nodes are deployed randomly. The initial transmission range of each node is 245 m and nodes can transmit up to 270 m using some extra energy. This happens only when current forwarder cannot find the next node in its transmission area. The initial energy of each node is 100 J. The velocity of acoustic waves and bandwidth for the network is considered 1500 m/s and 3000 kHz, respectively. Transmission energy, reception energy and idle time energy is considered as 2 W, 0.1 W and 10×10^{-3} W, receptively. Size of hello packet is 100 bytes while the size of all other packets is 150 bytes.

5.2 Simulation Results

Figure 2 depicts the depth adjustment of nodes. We can see from the Fig. 2 that most of the depth adjustment is done during the start of network deployment. Once the network is deployed and initial depth adjustments are done then there exist only a few occasions on which depth adjustment is required. A large amount of energy is dissipated during the process of depth adjustment. So, we make sure that the depth adjustment only occurs when it is necessary. Otherwise, try to

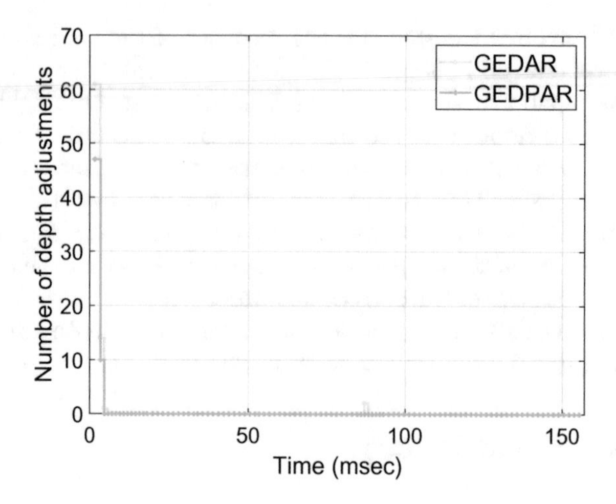

Fig. 2. Depth adjustment

avoid the nodes by enhancing the transmission range. It is clear from the Fig. 2 that in GEDPAR routing protocol nodes require fewer depth adjustments as compare to GEDAR. This step further involved in less energy dissipation.

The throughput of proposed routing protocol is compared with GEDAR and LMPC. Figure 3 shows this comparison and assure the efficiency of proposed scheme. According to simulation results, LMPC performs better than GEDAR while GEDPAR outperforms both GEDAR and LMPC. The efficiency of the proposed scheme is better than LMPC and GEDAR by the percentage of 13% and 37%, respectively.

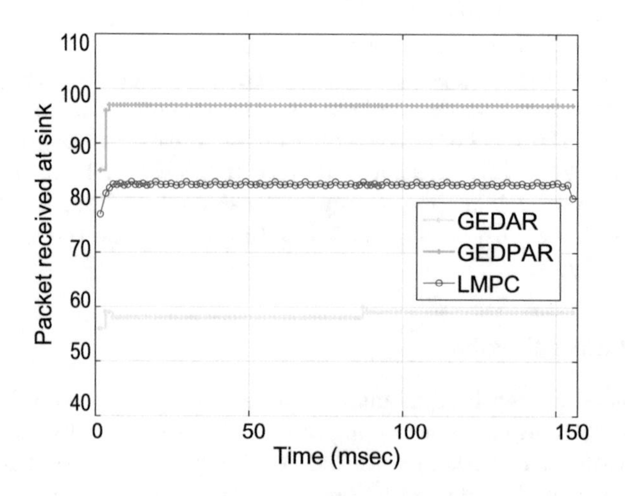

Fig. 3. Throughput

Figure 4 depicts the total energy consumption of network when different routing protocols are implemented. GEDPAR consumes less energy as compared to the GEDAR and LMPC. GEDAR consumes more energy because it focuses on depth adjustment during the void hole avoidance. Depth adjustment takes 15 J energy for one meter while transmission range enhancement takes less energy than depth adjustment. LMPC uses multiple transmissions for one packet which becomes a major cause in energy dissipation. Our proposed routing protocol consumes less energy because it tries to cover the void hole by increasing the transmission range. GEDPAR only change its depth when no forwarded node is found even by increasing transmission area. According to Fig. 4, GEDPAR outperforms GEDAR and LMPC.

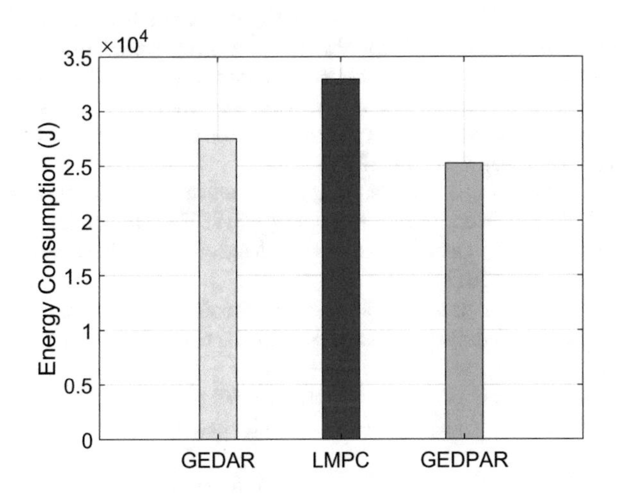

Fig. 4. Total energy consumption

6 Conclusion

In current work, imbalance and unnecessary energy dissipation is avoided by covering the void hole in an efficient way. We propose a routing protocol namely GEDPAR for void hole recovery. In order to show the productiveness of the proposed protocol, comparative analysis is performed with the existing state of the art protocols: GEDAR and LMPC. Simulation results show that GEDPAR outperforms GEDAR and LMPC in terms of throughput by the percentage of 13% and 37%. However, the proposed protocol minimizing the energy consumption at the cost of delay.

References

1. Khasawneh, A., Latiff, M.S.B.A., Kaiwartya, O., Chizari, H.: A reliable energy-efficient pressure-based routing protocol for the underwater wireless sensor network. Wireless Netw. **24**(6), 2061–2075 (2018)
2. Hong, Z., Pan, X., Chen, P., Su, X., Wang, N., Lu, W.: A topology control with energy balance in underwater wireless sensor networks for IoT-based application. Sensors **18**(7), 2306 (2018)
3. Wang, H., Wang, S., Zhang, E., Lu, L.: An energy balanced and lifetime extended routing protocol for underwater sensor networks. Sensors **18**(5), 1596 (2018)
4. Khan, A., Ali, I., Rahman, A.U., Imran, M., Mahmood, H.: Co-EEORS: cooperative energy efficient optimal relay selection protocol for underwater wireless sensor networks. IEEE Access (2018)
5. Ahmed, F., Wadud, Z., Javaid, N., Alrajeh, N., Alabed, M.S., Qasim, U.: Mobile sinks assisted geographic and opportunistic routing based interference avoidance for underwater wireless sensor network. Sensors **18**(4), 1062 (2018)
6. Sher, A., Khan, A., Javaid, N., Ahmed, S., Aalsalem, M., Khan, W.: Void hole avoidance for reliable data delivery in IoT enabled underwater wireless sensor networks. Sensors **18**(10), 3271 (2018)
7. Nayyar, A., Puri, V., Le, D.-N.: Comprehensive analysis of routing protocols surrounding Underwater Sensor Networks (UWSNs). In: Balas, V., Sharma, N., Chakrabarti, A. (eds.) Data Management, Analytics and Innovation, pp. 435–450. Springer, Singapore (2019)
8. Wu, F.-Y., Yang, K., Duan, R.: Compressed sensing of underwater acoustic signals via structured approximation l_0 norm. IEEE Trans. Veh. Technol. **67**(9), 8504–8513 (2018)
9. Khosravi, M.R., Basri, H., Rostami, H.: Efficient routing for dense UWSNs with high-speed mobile nodes using spherical divisions. J. Supercomputing **74**(2), 696–716 (2018)
10. Gomathi, R.M., Manickam, J.M.L.: Energy efficient shortest path routing protocol for underwater acoustic wireless sensor network. Wireless Pers. Commun. **98**(1), 843–856 (2018)
11. Hou, R., He, L., Hu, S., Luo, J.: Energy-balanced unequal layering clustering in underwater acoustic sensor networks. IEEE Access **6**, 39685–39691 (2018)
12. Iwata, M., Tang, S., Obana, S.: Energy-efficient data collection method for sensor networks by integrating asymmetric communication and wake-up radio. Sensors **18**(4), 1121 (2018)
13. Muhammed, D., Anisi, M.H., Zareei, M., Vargas-Rosales, C., Khan, A.: Game theory-based cooperation for underwater acoustic sensor networks: taxonomy, review, research challenges and directions. Sensors **18**(2), 425 (2018)
14. Jan, M.A., Tan, Z., He, X., Ni, W.: Moving towards highly reliable and effective sensor networks (2018)
15. Yildiz, H.U., Gungor, V.C., Tavli, B.: Packet size optimization for lifetime maximization in underwater acoustic sensor networks. IEEE Trans. Industr. Inf. (2018)
16. Khalid, M., Cao, Y., Ahmad, N., Khalid, W., Dhawankar, P.: Radius-based multipath courier node routing protocol for acoustic communications. IET Wireless Sens. Syst. (2018)
17. Latif, K., Javaid, N., Ahmad, A., Khan, Z.A., Alrajeh, N., Khan, M.I.: On energy hole and coverage hole avoidance in underwater wireless sensor networks. IEEE Sens. J. **16**(11), 4431–4442 (2016)

18. Wang, H., Wen, Y., Lu, Y., Zhao, D., Ji, C.: Secure localization algorithms in wireless sensor networks: a review. In: Bhatia, S., Tiwari, S., Mishra, K., Trivedi, M. (eds.) Advances in Computer Communication and Computational Sciences, pp. 543–553. Springer, Singapore (2019)
19. Yuan, Y., Liang, C., Kaneko, M., Chen, X., Hogrefe, D.: Topology control for energy-efficient localization in mobile underwater sensor networks using Stackelberg game. arXiv preprint arXiv:1805.12361 (2018)
20. Rahman, Z., Hashim, F., Rasid, M.F.A., Othman, M.: Totally Opportunistic Routing Algorithm (TORA) for underwater wireless sensor network. PloS ONE **13**(6), e0197087 (2018)
21. Heidemann, J., Stojanovic, M., Zorzi, M.: Underwater sensor networks: applications, advances and challenges. Phil. Trans. R. Soc. A **370**(1958), 158–175 (2018)
22. Javaid, N., Majid, A., Sher, A., Khan, W., Aalsalem, M.: Avoiding void holes and collisions with reliable and interference-aware routing in underwater WSNs. Sensors **18**(9), 3038 (2018)
23. Coutinho, R.W.L., Boukerche, A., Vieira, L.F.M., Loureiro, A.A.F.: Geographic and opportunistic routing for underwater sensor networks. IEEE Trans. Comput. **65**(2), 548–561 (2016)
24. Xu, J., Li, K., Min, G., Lin, K., Qu, W.: Energy-efficient tree-based multipath power control for underwater sensor networks. IEEE Trans. Parallel Distrib. Syst. **23**(11), 2107–2116 (2012)

Cluster-Based Routing Protocols with Adaptive Transmission Range Adjustment in UWSNs

Muhammad Awais[1], Zahoor Ali Khan[2], Nadeem Javaid[1(✉)], Abdul Mateen[1], Aymen Rasul[1], and Farooq Hassan[3]

[1] COMSATS University Islamabad, Islamabad 44000, Pakistan
nadeemjavaidqau@gmail.com
[2] Computer Information Science, Higher Colleges of Technology, Fujairah 4114, UAE
[3] University of Lahore, Islamabad Campus, Islamabad 44000, Pakistan
www.njavaid.com

Abstract. Nowadays, limited battery lifespan in Underwater Wireless Sensor Networks (UWSNs) is one of the key concerns for reliable data delivery. Traditional transmission approaches increase the transmission overhead, i.e., packet collision and congestion, which affects the reliable data delivery. Additionally, replacement of the sensors battery in the harsh aquatic environment is a challenging task. To save the network from sudden failure and to prolong the lifespan of the network, efficient routing protocols are needed to control the excessive energy dissipation. Therefore, this paper proposes two cluster-based routing protocols. The proposed protocols adaptively adjust their transmission range to keep maximum neighbors in their transmission range. This transmission range adjustment helps the routing protocols to retain their transmission process continuous by removing void holes from the network. Clusters formation in both proposed protocols makes the data transmission successful, which enhances the Packet Delivery Ratio (PDR). A comparative analysis is also performed with two state-of-the-art protocols named: Weighting Depth Forwarding Area Division, Depth Based Routing (WDFAD-DBR) and Cluster-Based WDFAD-DB (CB-WDFAD-DBR). Simulation results show the effectiveness of the proposed protocols in terms of PDR, Energy Consumption (EC) and End to End (E2E) delay.

Keywords: Energy efficient · Void hole · Shortest path approach · Reliable data delivery · Clustering

1 Introduction

Nowadays, recent advances in Underwater Wireless Sensor Network (UWSN) motivated many researchers for the development of various applications in the scientific and environmental era, i.e., data collection, monitoring of underwater

L. Barolli et al. (Eds.): EIDWT 2019, LNDECT 29, pp. 528–539, 2019.
https://doi.org/10.1007/978-3-030-12839-5_49

equipment and disasters prevention [1,2]. In UWSNs, during long-term communication acoustic waves are preferred instead of radio waves because of their low absorption rate and scattering. Moreover, dynamic environmental changes, limited lifespan of the battery and high End to End (E2E) delay with high Energy Consumption (EC) are the adverse characteristics of UWSN [3,4]. To enhance the lifetime of the network, some of the researchers focus on multi-hopping techniques. To enhance the stability of the network, multi-hop techniques are preferred. These promising techniques minimize the EC with affordable E2E delay and help many other UWSN applications, i.e., temperature collection and environmental data collection. Some of the researchers have focused on transmission range adjustment [5] for successful packet transmission. However, limited battery lifespan in an underwater environment is a challenging task. In UWSN, it is really difficult for the sensors to use solar power chargers because sunlight is unable to reach in the depth of the sea. These techniques still have to be enhanced because sensors are vulnerable to the sea water corrosion and marine animals activities.

The best routing path selection is also important to route the data packets from the source node towards the destined sink. Therefore, energy limitations play an important role in the designing of a routing protocol in UWSNs. In [17], only two metrics are considered to efficiently utilize the nodes battery: the depth of the nodes and next forwarder node selection on the basis of 2-hop neighbors information. Although the probability of void holes occurrence and inefficient EC is reduced during the nodes communication, the probability of void hole occurrence still exists in their work. Therefore, to minimize the aforementioned problem, cluster-based transmission range adjustment strategy is adopted in [5]. The void hole occurrence problem is minimized in this work. However, the extra power is needed for the transmission range adjustment. In this paper, to tackle the problem of extra energy dissipation and void hole avoidance, we have performed clusters formation using shortest and efficient path selection and also proposed two routing protocols named: Shortest Path-based Weighting Depth Forwarding Area Division and Depth Based Routing (SPB-WDFAD-DBR) and Breadth First Shortest Path-based WDFAD-DBR (BFSPB-WDFAD-DBR). The proposed clustering protocols make clusters and improve PDR. Residual energy of nodes is also taken into account for efficient data forwarding and transmission range adjustment. Proposed protocols perform reliable data delivery with affordable E2E delay and enhance the lifespan of node's battery in UWSNs. The main contributions, features, and achievements of this paper are summarized as follows:

- In order to avoid the void hole problem two efficient routing protocols named: SPB-WDFAD-DBR and BFSPB-WDFAD-DBR are proposed.
- The concept of adaptive transmission range adjustment is implemented to avoid the void holes.
- The proposed protocols are compared with state-of-the-art protocols named: WDFAD-DBR and CB-WDFAD-DBR.

The rest of the paper is organized as follows: Sect. 2 presents the literature review of the state of the art protocols. Section 3 describes the system model of the proposed protocols. Simulation and results are discussed in Sect. 4. Finally, Sect. 5 ends with conclusion and future work of the paper.

2 Literature Review

In this section, state of the art routing protocols are discussed for depth understanding of data load balancing, E2E delay and EC to maximize the throughput of UWSN. Few of these protocols are presented as follows:

Energy-aware and energy efficient schemes are proposed in [7–12]. DAE also balances the load per node to prolong the network lifetime, similarly, as in [8] and [10]. Chaotic compressive sensing for secure data transmission is used to reduce the number of retransmissions [8]. In the same way, based on random access, the shortest retransmission based strategy is also proposed for complex environments in UWSN.

Furthermore, a Cooperative Energy Efficient Optimal Relay Selection Protocol (Co-EEORS) is proposed in UWSN [9]. Combine information of location and depth is used to select the destination node [7,9]. In [9], after the successful receiving, destination node acknowledges the source node about the successful reception of data packets at the destination. A novel routing scheme is proposed in [10] with two mobile sinks for efficient data collection. Moreover, a new metric 'mobility sink utility ratio' is introduced to check the usage of mobile sinks during the collection of data. Afterward, in [11], the author proposed a Cuckoo Optimization Algorithm (COA) which is basically the combination of three techniques, i.e., geo-routing, duty-cyclic routing, and multi-path routing. The COA protocol transmits data hop by hop and selects the route using power consumption and energy content of the current node. However, the authors in [7] have not taken the dynamic environmental changes into account [9,11]. The authors in [8] has not considered the EC on compression and during the selection of next forwarder node. Delay in packet transmission at the sink and their recombining cost is not considered in [10].

Khan *et al.* and Wang *et al.* proposed Energy Efficient Routing and Void Hole Avoidance (E2RV) scheme, Energy-aware and Void Hole Avoidable Routing protocol (EAVARP) in [12] and [13]. The 2-hops information is used to cover the void hole region in [12]. E2RV and EAVARP maintain the energy depletion of the network to prolong the network lifetime. Furthermore, opportunistic directional forwarding strategy is used for aforementioned parameters in EAVARP. In addition, cyclic transmission is avoided in EAVARP. E2RV performed well in terms of PDR and minimum energy depletion [12,13]. Additionally, avoiding the void holes, looping, and energy efficacy EAVARP outperformed [13]. However, distance dependency on energy and communication overhead is not focused [12,13]. Spare regions affect the processing time of network is not considered in [13].

In [14–16] novel approach named: Clustered based Energy Efficient Routing (CBEER) protocol, Energy balanced unequal layering clustering (EULC) and

Clustered-based energy efficient routing(CBE2R) protocol are proposed to prolong the network lifetime. These are energy efficient and cluster-based schemes in UWSNs. CBEER finds the shortest path [14], EULC and CBE2R use layering concept based on cluster formation to transmit the data packets [15,16]. Furthermore, all above-mentioned schemes are evaluated using extensive simulations and it is noticed that these protocols performed out in terms of enhanced PDR with minimum EC [14–16]. However, sparse regions overhead and communication overhead is not discussed in [14] and [15].

In [17], weighting depth and forwarding area division depth based routing WDFAD-DBR is proposed. Weighting sum of depth difference of 2-hops is used for the selection of next forwarding hop in WDFAD-DBR. WDFAD-DBR not only consider the current depth but also considered the expected next hop. Proposed protocol reduces the EC and E2E delay efficiently.

Authors in [18], implemented a new strategy for data-centric communication based on the named data network in wireless sensor networks. Efficient communication is achieved by the authors. Additionally, this paper also presents the modification in named data network to maximize the performance of communication. Reverse path creation is specifically discussed in this paper for the next hop node selection. Broadcast nature of the wireless channel is introduced by the authors in this paper. Proposed protocol minimize the redundant transmissions efficiently. In [19], the authors proposed an adoptive status update method. This method minimizes the limitations of beacon messaging. A quality of service solution is presented for real-time applications. The proposed protocols divide the traffic into different types to make better understanding. EC is reduced by the authors with a fair distribution of traffic.

In [20], an adoption method to aggregate the network resources of smart devices is proposed. The resource adoption is dynamic. The trade-off between the number of smart devices and throughput is also discussed by the Toawa *et al.* In [21], an algorithm named Cooperation Forwarding Data Gathering Strategy Based on Random Walk Backup (CFDGSBRWB) is proposed to prolong the network lifetime in wireless sensor networks. CFDGSBRWB Consists of data back up based random walk method and cooperating data gathering strategy. Proposed method reduces the packet loss ratio enhance fault tolerance, significantly. Gupta *et al.* formulate a dynamic mesh routing algorithm in [22]. Proposed algorithm considers both static and dynamic traffic demand. Proposed algorithm provides optimal performance for all traffic demands.

3 System Model and Description

In this section, we discussed the system models of our proposed protocols. The system models of the proposed protocols are presented and discussed in detail.

In state-of-the-art protocols, WDFAD-DBR and CB-WDFAD-DBR are implemented. WDFAD-DBR considers two metrics for the selection of next forwarder: depth of the node and 2-hop neighbor's information to find the next forwarder [17]. However, there exits some chances of the void hole occurrence.

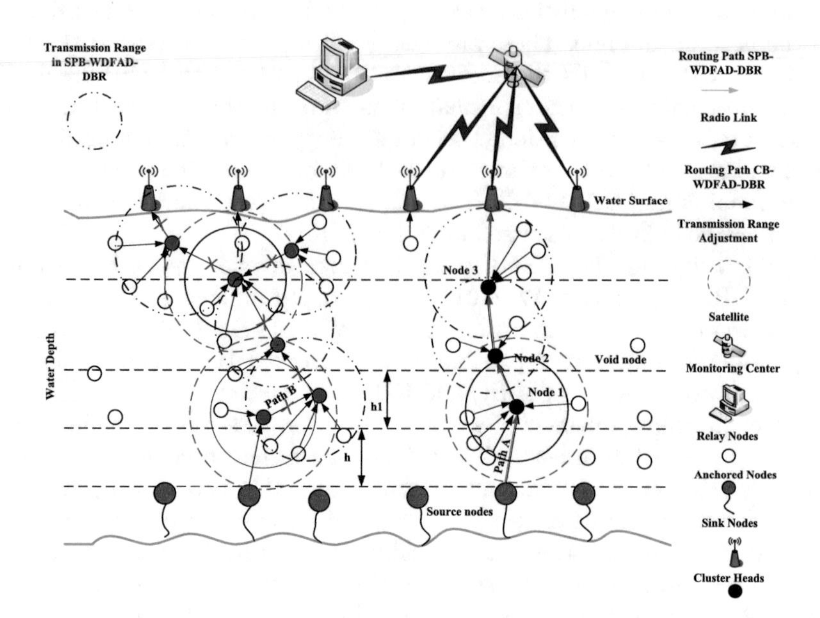

Fig. 1. System model of SPB-WDFAD-DBR

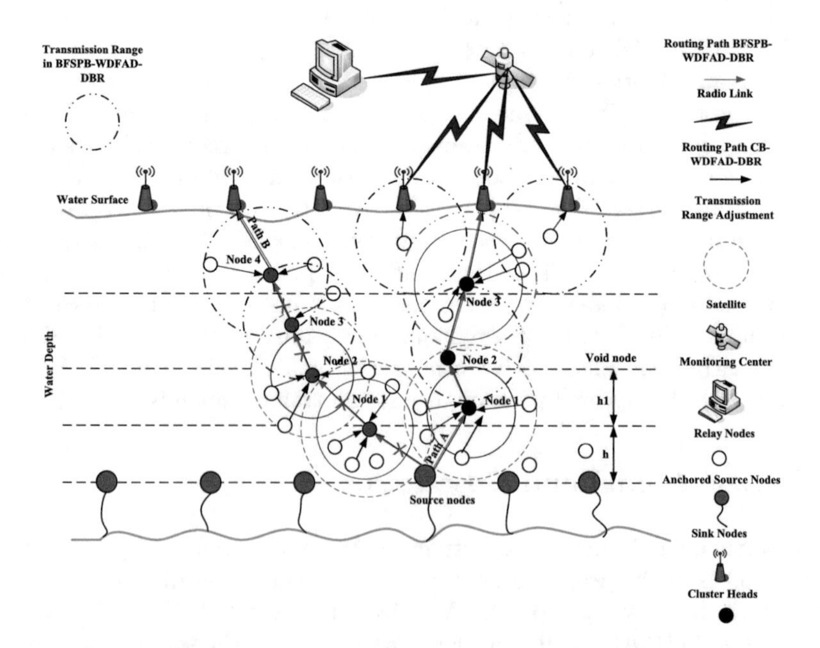

Fig. 2. System model of BFSPB-WDFAD-DBR

In CB-WDFAD-DBR [5], cluster formation reduces the APD to reduce the collisions between packets for reliable data delivery. The CB-WDFAD-DBR protocol performs transmission range adjustment which needs extra power to adjust their transmission range. The transmission range adjustment keeps maximum neighbor relay nodes in its transmission range. Therefore, two cluster-based routing protocols SPB-WDFAD-DBR and BFSPB-WDFAD-DBR as shown in Figs. 1 and 2 are proposed. Basically, both proposed protocols formulate clusters based on 'Dijkstra' and 'Breadth First Search (BFS)' algorithm to make the routing path shortest and efficient for reliable packet delivery. This strategy minimizes the possible collisions due to clusters formation and reduces the probability of void hole occurrence. The energy dissipation during transmission is minimized which will be used later in adaptive transmission range adjustment. The transmission range adjustment extends the transmission range of the current node and makes the transmission continuous without any hindrance.

In proposed protocols, we have formed clusters to restrict the wireless channels to avoid collisions. In both protocols, the 3-dimensional network architecture with multiple sinks is supposed [5], which is composed of relay, anchor and sink nodes as shown in Figs. 1 and 2. The relay nodes are placed at different depths, which receives and forward the data packets towards the sink by collecting the data packets from anchor nodes, while anchor nodes are fixed at the bottom of the sea surface. These anchor nodes are basically used to sense the data and to gather the data. Meanwhile sink nodes are at the surface of the sea and housed with both radio and acoustic modems to communicate with the nodes deployed in the water and in the terrestrial environment, respectively. Additionally, it is supposed that sink nodes have high energy and they know the location of all relay nodes. The relay nodes are deployed by considering the mobility capability of sensors and assumed the initial deployment is randomly done using planes or ships to cover the desired area. The relay nodes collect the data packets which are then forwarded to their Cluster Head (CH) and then that CHs forwards the data to the destined sink. The radio modems are used for communication between the sinks and different control stations. While, acoustic waves are used for communication between relay and anchor nodes. The transmission process is followed by different clusters having CH and relay nodes. In each cluster, there exists a CH which receives the data packets from the relay nodes and transmit the packets to next forwarder CH existing in its transmission range. The CH and simple relay nodes can directly transmit the data packets towards the respective sink if that node or CH finds a sink directly in its transmission range. Initially, their exits no CH in the network. Meanwhile, the CH is selected on the base of the maximum residual energy node. Based on that residual energy 'Dijkstra and BFS' algorithm provides the shortest possible path from source node towards the destination sink. The shortest path nodes are then selected as CH and these CHs collect the information from their neighbors using relay nodes, if anchor node is not directly in the communication range of respective CH and then forward aggregated data towards the next CH. In this way, data is transmitted to the destined sink using CH to CH transmission. The process remains continuous until all packets reaches the sink node. The sink nodes are positioned at the sea surface to destine the packets to the control centers. In the proposed protocols system model, we assumed that:

- The sinks are connected with other sinks to balance the data packets flow.
- If a packet is acknowledged at the sink, it is supposed that packet is successfully transmitted to the control station.

3.1 Proposed Protocols

In this subsection, the detail description of the proposed protocols and their algorithms is presented.

SPB-WDFAD-DBR: In SPB-WDFAD-DBR, 'Dijkstra' algorithm is used to find the shortest efficient path from source node to destined nearest sink. The selected nodes using 'Dijkstra' algorithm are selected as CH. The CHs collect the data packets from their neighbor nodes available in their transmission range. If some of the nodes are out of range then protocol performs adaptive transmission range adjustment to keep the maximum neighbors in its transmission range. In addition, CH to CH transmission begins. Finally, the data packets reach the destined sink. These sinks handover the data packets to the respective control station.

BFSPB-WDFAD-DBR: In BFSPB-WDFAD-DBR, 'BFS' algorithm is used to select the shortest routes in breadth vise. Then, opportunistic routing is performed by our routing protocol. The route with a minimum active number of nodes is selected as the final route. These nodes after checking their residual energy perform the job of CH. The CHs collect the data from their neighbor nodes and this CH communicate with their neighbor CHs and finally transmit the packets to the destined sink. Then, sinks transmit the packet to the base station to make the communication successful.

4 Simulations Results and Discussion

In this section, evaluation of the proposed protocols against state of the art protocols, i.e., WDFAD-DBR [17] and CB-WDFAD-DBR [5] is evaluated. The detailed description is given in the following subsections.

4.1 Simulations Setup

In the simulation environment, nodes are deployed randomly in UWSN environment of dimensions $10 \times 10 \times 10\,\text{km}^3$ with 9 sinks. We assume the transmission range, DR and PS of nodes up to 300 m, 16 kbps, and 72 bytes [5], respectively. The total energy of network is initialized with 100 J; where the EC during reception of the data packets and transmission rate during the transmission is kept

158 mW and 50 W. During the network execution the nodes deployment varies from 100–500, which are basically anchored with 10 relay nodes. To control the node's movement, node propagation speed in seawater is considered up to 2 m/s. Moreover, the propagation speed of the acoustic wave is kept 1500 m/s along with 4 kHz bandwidth. In the proposed protocols, relay nodes are placed at different depth, which receives and forward the data packets towards the sink by collecting the data packets from anchor nodes. Acknowledge packet size is kept 50 bits in the network along with the header size of 11 bytes in data packets. The comparative analysis is performed against the state-of-the-art protocols [5] and [17].

4.2 Performance Metrics

In this subsection, the performance of the proposed protocols, in terms of EC, PDR and delay.

4.3 Performance Comparison

For the performance comparison, we performed a comparative analysis of our proposed protocol with two states of the art protocols. Comparative analysis is performed in terms of EC, PDR and E2E delay.

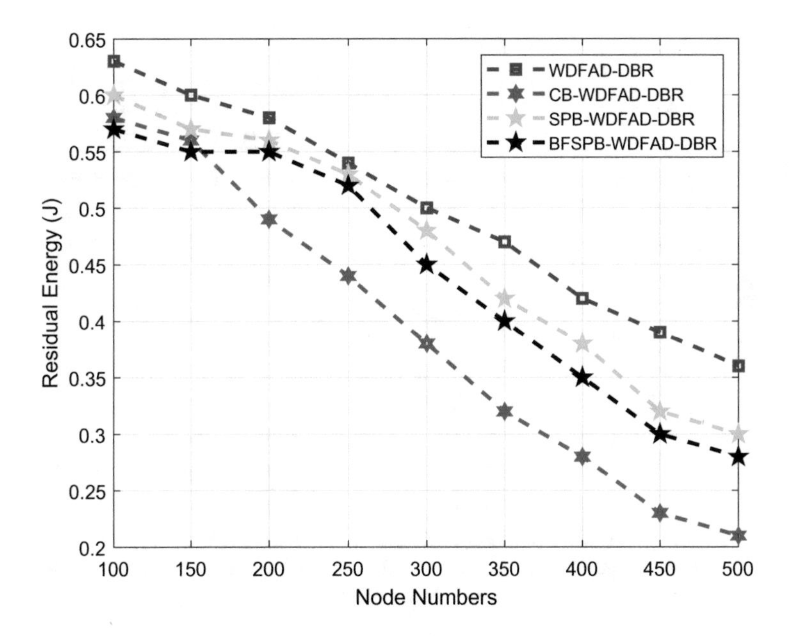

Fig. 3. Residual energy (J)

Residual Energy: The remaining energy of the nodes after the communication is defined as the residual energy of the network. The residual energy of the proposed and state of the art protocols. It is clearly demonstrated from the Fig. 3 that with the increase in a number of nodes EC also increases. Figure 3 shows that the residual energy of WDFAD-DBR is higher than CB-WDFAD-DB and proposed protocols. The reason for least EC in WDFAD-DBR is that WDFAD-DBR consumes energy during depth adjustment and forwarding area division only to get the neighbors information. Meanwhile, in CB-WDFAD-DBR, the extensive energy is consumed on adaptive transmission range adjustment and during the communication phase. Cluster formation and participation of every node in the transmission also results in high EC in CB-WDFAD-DBR. The proposed protocols adaptively adjust their transmission range in the network, when the current node finds no node in its transmission range. The power saved during the shortest path based clustering and breadth-first shortest path based clustering is used as the extra power needed for this transmission range adjustment. Then, the current node forwards the data packets to the forwarder node without any packet loss using the 2-hop neighbor's information. In both proposed protocols, clustering mechanism is adopted which increases the EC of the nodes with affordable E2E delay. It is clearly demonstrated from the figure that the EC varies directly with the increase in the number of nodes, which results in a continuous decrease in the residual energy.

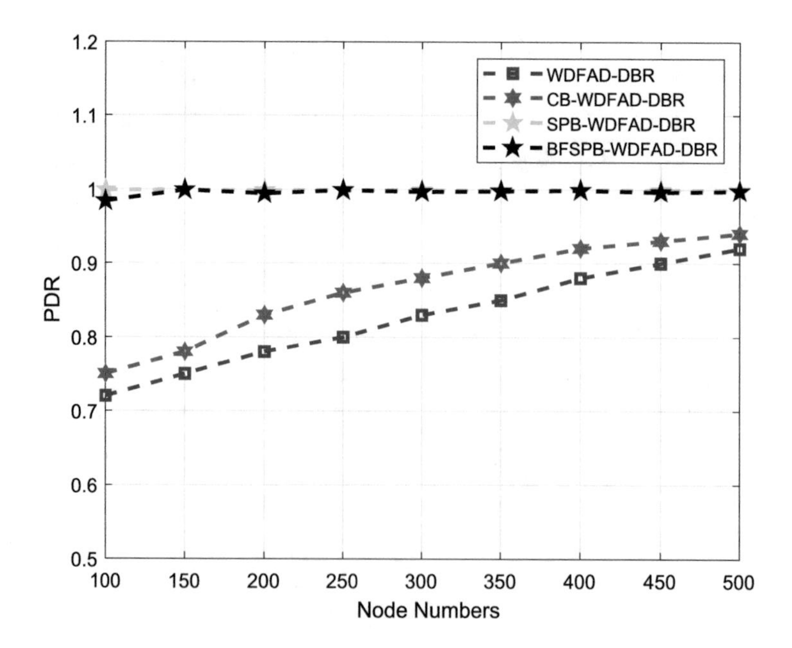

Fig. 4. Packets delivery ratio

PDR: The PDR of both existing and state of the art protocols is shown in Fig. 4. Both proposed protocols are depicting almost similar behavior. In both protocols, PDR is increasing as the number of relay nodes is increasing (high nodes density). The main reason behind this enhanced PDR is that probability of void hole occurrence decreases with the increase in nodes density. In all aforementioned protocols, the probability of void hole occurrence is trying to be minimized. The state of the art protocol named WDFAD-DBR considers only 2-hop neighbor's information for further nodes selection which does not completely eliminates the chance of void hole occurrence. This chance of void hole occurrence results in packet drop, and decrease in PDR. The PDR of CB-WDFAD-DBR is slightly higher than WDFAD-DBR because of clustering. The void hole occurrence is completely minimized in the proposed protocols which increase the lifespan of the network with high PDR.

From the Fig. 4, it is clearly demonstrated that proposed protocols outperformed in terms of PDR than WDFAD-DBR and CB-WDFAD-DBR protocols. In both proposed protocols, each CH is selected using the 'Dijkstra and BFS' algorithm on the basis of their residual energy. Afterward, CHs to CHs communication begins to transmit the data from the source towards the destined sink. Meanwhile, CH nodes adjust their transmission ranges adaptively as per requirement. In both proposed protocols, packet drop ratio is lesser than both benchmark protocols. Hence, both proposed protocols have high PDR.

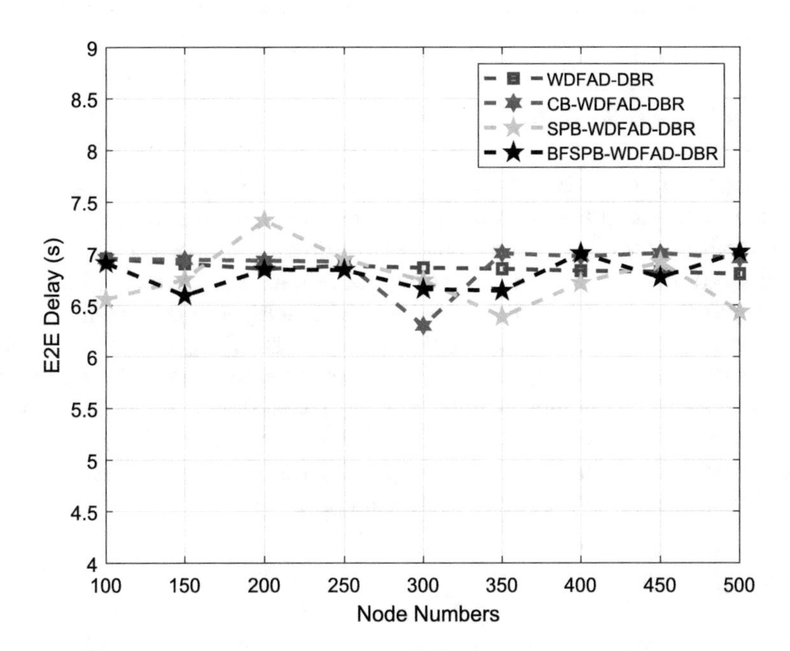

Fig. 5. E2E delay (s)

E2E Delay: The E2E delay of existing and proposed protocols is shown in Fig. 5. Delay of WDFAD-DBR is less than other clustering protocols because WDFAD-DBR finds the next forwarder on the base of the node's depth difference, which decreases E2E delay because of depth based next forwarder node selection. The CB-WDFAD-DBR and proposed protocols overcome the void hole problem that results in affordable E2E delay. In SPB-WDFAD-DBR and BFSPB-WDFAD-DBR the network forms clusters to minimize the APD which directly affects E2E delay of the network. However, with the increase in node density number of collisions between the packets increase and packets have to be transmitted again. In addition, transmission range adjustment also consumes some time during which CH holds the data packets. Therefore, it is evident from Fig. 5 that proposed protocols have high PDR by paying the cost of delay.

4.4 Performance Trade-Off

In this subsection, we review the existing and proposed protocols to discuss the existing trade-offs. The proposed protocols outperform in adaptive transmission range adjustment to keep maximum neighbors in its transmission range. The extra cost needed to fulfill this adjustment is saved by minimizing the APD in our proposed protocols. During clustering of nodes in proposed protocols, we noticed an existing trade-off between the two performance parameters: PDR and EC. The PDR is enhanced on the cost of EC which increases the reliable data delivery.

5 Conclusion and Future Work

In this paper, energy efficient routing protocols are proposed for reliable data transmission. These routing protocols enhance the network lifespan and reduce the probability of void holes generation by adjusting the transmission range of the nodes, adaptively. In addition network lifespan is increased. The proposed protocols enhance the PDR with affordable E2E delay and by paying the cost of EC. The proposed protocols use the 'Dijkstra and BFS' algorithm and fulfill the need of extra power needed for adaptive transmission range adjustment. This adjustment keeps maximum neighbors in its transmission range for maximum data collection. Simulations show the efficacy of the proposed protocols. In the future, we will minimize the packet drop ratio and E2E delay further by using artificial intelligence and data science techniques.

References

1. Akyildiz, I.F., Pompili, D., Melodia, T.: Underwater acoustic sensor networks: research challenges. Ad Hoc Netw. **3**(3), 257–279 (2005)
2. Wu, H., Chen, M., Guan, X.: A network coding based routing protocol for underwater sensor networks. Sensors **12**(4), 4559–4577 (2012)

3. Basagni, S., Petrioli, C., Petroccia, R., Spaccini, D.: CARP: a channel-aware routing protocol for underwater acoustic wireless networks. Ad Hoc Netw. **34**, 92–104 (2015)
4. Francois, R.E., Garrison, G.R.: Sound absorption based on ocean measurements. Part II: boric acid contribution and equation for total absorption. J. Acoust. Soc. Am. **72**(6), 1879–1890 (1982)
5. Sher, A., Khan, A., Javaid, N., Ahmed, S., Aalsalem, M., Khan, W.: Void hole avoidance for reliable data delivery in IoT enabled underwater wireless sensor networks. Sensors **18**(10), 3271 (2018)
6. Yildiz, H.U., Gungor, V.C., Tavli, B.: Packet size optimization for lifetime maximization in underwater acoustic sensor networks. IEEE Trans. Ind. Inf. (2018)
7. Rahim, S.S., Ahmed, S., Hadi, F.-E., Khan, A., Usman Akhtar, M., Javed, L.: Depth-based adaptive and energy-aware (DAE) routing scheme for UWSNs. EAI Endorsed Trans. Energy Web **5**(17), e6 (2018)
8. Li, X., Wang, C., Yang, Z., Yan, L., Han, S.: Energy-efficient and secure transmission scheme based on chaotic compressive sensing in underwater wireless sensor networks. Digit. Signal Process. **81**, 129–137 (2018)
9. Khan, A., Ali, I., Rahman, A.U., Imran, M., Mahmood, H.: Co-EEORS: cooperative energy efficient optimal relay selection protocol for underwater wireless sensor networks. IEEE Access **6**, 28777–28789 (2018)
10. Yahya, A., Islam, S., Akhunzada, A., Ahmed, G., Shamshirband, S., Lloret, J.: Towards efficient sink mobility in underwater wireless sensor networks. Energies **11**(6), 1471 (2018)
11. Dosaranian-Moghadam, M., Amo-Rahimi, Z.: Energy efficiency and reliability in underwater wireless sensor networks using cuckoo optimizer algorithm. AUT J. Electr. Eng. **50**(1), 93–100 (2018)
12. Khan, G., Dwivedi, R.K.: Energy efficient routing algorithm for void avoidance in UWSN using residual energy and depth variance
13. Wang, Z., Han, G., Qin, H., Zhang, S., Sui, Y.: An energy-aware and void-avoidable routing protocol for underwater sensor networks. IEEE Access **6**, 7792–7801 (2018)
14. Tuna, G.: Clustering-based energy-efficient routing approach for underwater wireless sensor networks. Int. J. Sens. Netw. **27**(1), 26–36 (2018)
15. Hou, R., He, L., Hu, S., Luo, J.: Energy-balanced unequal layering clustering in underwater acoustic sensor networks. IEEE Access **6**, 39685–39691 (2018)
16. Ahmed, M., Salleh, M., Ibrahim Channa, M.: CBE2R: clustered-based energy efficient routing protocol for underwater wireless sensor network. Int. J. Electron. **105**(11), 1916–1930 (2018)
17. Yu, H., Yao, N., Wang, T., Li, G., Gao, Z., Tan, G.: WDFAD-DBR: weighting depth and forwarding area division DBR routing protocol for UASNs. Ad Hoc Netw. **37**, 256–282 (2016)
18. Kuniyasu, T., Shigeyasu, T.: Data-centric communication strategy for wireless sensor networks. Int. J. Space-Based Situated Comput. **8**(1), 30–39 (2018)
19. Serhan, Z., Diab, W.B.: Energy efficient QoS routing and adaptive status update in WMSNS. Int. J. Space-Based Situated Comput. **6**(3), 129–146 (2016)
20. Togawa, K., Hashimoto, K.: Cooperative and priority based on dynamic resource adaptation method in wireless network. Int. J. Space-Based Situated Comput. **8**(1), 40–49 (2018)
21. Chen, L., Liu, L., Qi, X., Zheng, G.: Cooperation forwarding data gathering strategy of wireless sensor networks. Int. J. Grid Util. Comput. **8**(1), 46–52 (2017)
22. Gupta, B.K., Patnaik, S., Mallick, M.K., Nayak, A.K.: Dynamic routing algorithm in wireless mesh network. Int. J. Grid Util. Comput. **8**(1), 53–60 (2017)

A Unicast Rate-Based Protocol for Video Streaming Applications over the Internet

G.-A. Lusilao Zodi(✉), T. Ankome, J. Mateus, L. Iiyambo, and J. Silaa

Department of Computer Science, Namibia University of Science and Technology, 5 Storch Street, P. B. 13388, 9000 Windoek, Namibia
{gzodi, tankome, jmateus, liiyambo, jsilaa}@nust.na

Abstract. This paper presents a unicast rate-based transport protocol that regulates transmission rate for video streaming over best-effort networks such as the Internet. The protocol runs on top of the Real-time Transport Protocol and relies on the feedbacks reports of its sister protocol Real-time Transport Control Protocol to control congestion. The proposed protocol uses the square increase multiplicative decrease rules to alter its transmission rate and operates in a TCP-friendly way towards TCP flows. In addition, the protocol includes two control criteria: the cumulative jitter and the delay factor to allow incipient congestion detection prior to loss of video packets. With the addition of these criteria, the transmission rate is adjusted in a way that anticipates loss of video packets. Network parameters such as packet loss ratio and Round-Trip Time as well as the TCP-friendly shared throughput are computed at the receiver, and the results sent back to the sender using in the receiver packet report. Upon reception of the feedback information, the sender adjusts the sending rate to match the network's available capacity. The performance evaluation results using both network related metrics and video quality measurement show a performance improvement in terms of frame loss rate and peak signal to noise ratio, providing the receiver with an image of better visual quality than the classical TCP-Friendly Rate Control (TFRC) protocol.

1 Introduction

Since their inception in the early 1990s, video streaming applications have experienced an explosive growth and transformation, from novel applications into one of the main manner in which people experience the Internet today. The Internet however is a best effort transmission platform with no quality of service (QoS) guarantees, which creates several challenges for video streaming applications. The major one being the fluctuations in the network throughput, which do not often match the video traffic pattern and negatively affect its service-related requirements, characterized by stringent delay, jitter and high throughput. Video streaming over the Internet remains possible if the network is provisioned to accommodate the bit rate requirement of the video. However, issues arise when new requests for network's capacity applications cause an aggregate traffic load larger than the network's capacity. When such a situation occurs, the network experiences long queues at routers, which lead to excessive delay, high delay jitter and

L. Barolli et al. (Eds.): EIDWT 2019, LNDECT 29, pp. 540–551, 2019.
https://doi.org/10.1007/978-3-030-12839-5_50

loss of video packets. This problem, known as network congestion, not only compromises the video throughput performance and impairs its perceptual quality but may also interrupt the display of the transmitted video.

For streaming applications, congestion control can take the form of rate adjustment. Rate adjustment is a process whereby a receiver informs the sender of the transmission rate at which he is prepared to accept packets. In the dominant Internet's Transmission Control Protocol (TCP), this information is fed back in the advertised window field of every acknowledgement packet and transmission rate is increased by one more packet per RTT interval if a packet loss is not reported. If packet loss is reported, the current rate is decreased to one-half [1]. This congestion control mechanism is to a great extent responsible for the remarkable stability of the Internet [2, 3]. However, the halving of the transmission rate for a single packet loss causes dramatic sending rate fluctuations and is not suitable for streaming video applications. The transport protocol User Datagram Protocol (UDP) on the other hand is also not entirely adequately fashioned for real-time applications.

Streaming video commonly relies on the Real-time Transport Protocol (RTP) for data transmission. The RTP header contains fields such as sequence numbers that facilitate detection of packet loss and reordering along the end-to-end path between source and receiver. Although originally designed to work well for multicast groups on very large scales, more applications today use RTP for small-multicast groups such as video conferences and unicast Video on Demand. RTP is further enhanced by its sister protocol Real-time Transport Control Protocol (RTCP), which provides quality of service functionality through its source and receiver reports. RTP runs over User Datagram Protocol UDP rather than TCP because it only includes basic transport layer functionalities to guarantee fast delivery of data to the destination.

Given the lack of in-built congestion control mechanism within UDP and with due consideration that the majority of Internet traffic is still TCP-based, it becomes important to regulate UDP video traffic in a TCP-friendly manner so as to achieve fair bandwidth sharing towards TCP traffic. Therefore, UDP sources need to be enhanced with congestion control algorithms to preserve the stability of the Internet. The algorithms should not only reduce the transmission rate in a smooth manner in response to congestion to suit streaming video but must also operate in a TCP-friendly manner towards competing TCP flows [3, 4].

This paper proposes a TCP-friendly congestion control protocol named RTP Rate-Based Square Increase Multiplicative Decrease (RRB-SIMD) for streaming video where an interaction between the video source and the receiver is essential to monitor the network state. The protocol adjusts the video transmission rate at the encoder every time that a change in the network conditions is reported back by the receiver. RRB-SIMD runs on top of the RTP protocol, and takes advantage of its sister protocol Real-time Transport Control Protocol (RTCP) for quality of service control. The contribution in this work resides in the way RRB-SIMD's sender detects congestion and adjusts the transmission rate using the square increase multiplicative decrease rule. RRB-SIMD also includes novel congestion control criteria based on delay to detect incipient congestion prior to a loss of a packet and to act in a way that can anticipate packet loss.

Simulation results using both network's metrics and video quality measurements show that RRB-SIMD is TCP-friendly and performs better than TCP-Friendly Rate Control (TFRC) in term of frame loss ratio jitter and hence, improved user-perceived quality.

The rest of the paper is organized as follows: Sect. 2 discusses a number of unicast protocols proposed in the literature with the emphasis on TFRC and Rate-Based Square Increase Multiplicative Decrease (RB-SIMD). Section 3 describes the proposed RRB-SIMD protocol and Sect. 4 the simulation environment used to evaluate the performance of the RRB-SIMD protocol. Section 4 discusses the simulation results, and finally Sect. 5 concludes.

2 Related Work

Congestion control protocols can be classified into window-based or rate-based depending on either the sending rate is adjusted at the sender using the congestion window size or the transmission rate based on received information about congestion levels experienced along the communication path [5]. Datagram Congestion Control Protocol (DCCP) [6] is a rate based protocol proposed to provide congestion control to delay-sensitive applications with relaxed packet loss requirements. DCCP protocol uses the transport protocol UDP enhanced with basic TCP functions to make it connection-oriented like TCP. In doing this, DCCP benefits in two ways. Firstly the protocol is guaranteed traversal of network firewalls which blocked UDP packets and secondly it is provided with the possibility to negotiate some parameters during session initiation, such as the congestion control algorithms. DCCP, however, inherits the halving rate of TCP, which is critical for streaming applications.

An open source software that provides web browsers and mobile applications with real-time communication via an application programming interface is proposed in [7]. The client web-browsers are organised in the shape of a binary tree along which the video and audio data are relayed in the multi-hop fashion by the open source Web Real-Time Communication (WebRTC) protocol.

In [8], the authors proposed an online rate adaptation where the video's bit rate is changed on the fly to react to network status changes and the dynamic in the sequence of video's frames to provide better-perceived quality. The method relies on new feedback control mechanism for video streaming, which consider the interactions among video rate control, RED active queue management, and received video quality to maintain the video packet loss probability at certain level in order to achieve a consistent distortion level of the video.

A two-pass rate control algorithm that uses hierarchical structures for encoding video sequences to obtain better objective and subjective visual qualities is proposed in [9]. The proposed solution assigned to the frame to be encoded a specific rank, and when network's throughput fluctuates, uses the quantization parameter of frame of rank i-1 to derive the next quantization parameter of the corresponding frame.

TFRC [10] is another rate-based protocol for best-effort networks. TFRC emulates TCP flow steady state behavior. Its function consists of carrying streaming video applications, avoiding the abrupt halving of the sending rate when congestion occurs.

TFRC does not incorporate flow control at the receiver but includes the slow start mechanism of TCP, which ends on detection of the first packet loss, after which the protocol enters the congestion avoidance. During congestion avoidance, TFRC relies on packet loss metrics as well as the RTT to compute the acceptable rate at which to transmit video packets. Packet loss is counted at the receiver during a predefined period time, and send the current loss event rate back to the sender using a feedback report. The receiver computes the loss interval as the total number of packets drop between two consecutive loss events. TFRC presents the advantage of having a relatively steady sending rate while still being responsive to congestion. However, the protocol is quite sensitive to packet loss events and round-trip time estimates, and in some situations, its throughput can overshoot or undershoot the throughput of TCP under the same network conditions [4, 10].

Rate-Based Square Increase Multiplicative Decrease (RB-SIMD) [11] is a rate based-protocol proposed to emulate the properties of SIMD. In RB-SIMD the conventional TCP feedback mechanism is replaced with some periodic feedback from the destination. The source sends his packets as a continuous batch, with each batch identified with a control action number. Packets within a batch are sent at a constant rate, which is determined by the last control action. The destination returns a single feedback for each batch. If the feedback is a congestion notification, then the sender decrease his sending rate

$$\begin{aligned} r_{max} &= r_t \\ r_{t+\Delta} &= r_t \times (1-\beta) \\ r_0 &= r_{t+\Delta} \end{aligned} \tag{1}$$

where $r_{t+\Delta}$ is the new transmission rate, r_t the current rate, and r_{max} and r_0 the rate equivalents of w_{max} and w_0. Otherwise the returning feedback is a normal feedback, in which case the sender may increase, maintain or even decrease his current transmission rate [9]. If the rate can be increased, then the sender uses Eq. 2

$$r_{t+1} = r_t + K \times 2 \times \sqrt{\alpha(r_t - r_t)} \tag{2}$$

where r_{t+1} is the new transmission rate and $0 < K < 2$ a parameter depending on RTT. The choice of K is also discussed in [9].

3 RTP Rate-Based Square Increase Multiplicative Decrease (RRB-SIMD)

RRB-SIMD [12] is a unicast sender-based rate control protocol implemented on top of RTP and which takes advantage of RTCP reports to assess network congestion and adjusts the transmission rate. The protocol operates in a TCP-friendly way and adjusts the transmission rate in a smooth manner to preserve the user-perceived quality. The computation of the TCP-friendly shared transmission rate is performed at the receiver side. The receiver uses packet loss ratio as a control criterion for congestion detection. However, since packet loss occurs after congestion, RRB-SIMD includes both a cumulative jitter and a delay factor schemes to detect incipient congestion prior to loss of a packet. Since RTP was originally designed for large multicast sessions, RRB-SIMD increases its responsiveness through the quadratic increase of the transmission rate and uses the cumulative jitter and delay factor scheme to detect congestion prior to loss of a packet. The protocol is described in three stages, are described below.

3.1 Control Variables

This section presents the computation of control variables used in the evaluation of the network conditions and computation of the TCP-friendly transmission rate. Considering a case where a packet index i is transmitted successfully, let S_i denote the RTP timestamp of packet i, R_i the arriving time in RTP timestamp units of packet i and $D_i = R_i - S_i$ the one way delay experienced by the packet.

3.1.1 Average Cumulative Jitter

The cumulative jitter is the amounts of playback delayed provided for in order to avoid discarding delay video frames at the client side [12, 13]. The cumulative jitter is correlated with the level of congestion at the bottleneck link of the path and is obtained as follows. The receiver computes the jitter using the algorithm defined in RFC 3550 [14], that is, for two sequentially packets indexed i and $i+1$, the delay jitter J_i is expressed as follows.

$$J_i = D_{i+1} - D_i \tag{3}$$

The cumulative jitter CJ_k for a sequence of k jitter values, is then computed as,

$$CJ_k = J_1 + \cdots + J_k \tag{4}$$

The average of cumulative jitter A_k is then derived from a sequence of k cumulative jitter values as follows.

$$A_k = \frac{CJ_1 + \cdots + CJ_k}{k} \tag{5}$$

Finally, the receiver calculates two values of the average cumulative jitter, the average since the start of the session, referred to as long running cumulative average (longAvg) and the average measured between two successive RTCP reports referred to as a short running average (shortAvg).

3.1.2 Delay Factor

The delay factor d_f is a measure of the buffer occupancy at the bottleneck link of the path from the sender to the receiver. It is computed periodically on a round basis (a round being the time interval between two RTCP packets) and is obtained as follows.

For each received packet i, the receiver computes the delay D_i and updates the minimum delay D_{min} and the maximum delay D_{max}. From D_{min} and D_{max} the receiver derives the maximum queuing delay QD_{max} as follows

$$\mathrm{QD_{max}} = \mathrm{D_{max}} - \mathrm{D_{min}} \tag{6}$$

The queuing delay $\mathrm{QD_i}$ is a measure of the path congestion and is obtained as,

$$\mathrm{QD_i} = \mathrm{D_i} - \mathrm{D_{min}} \tag{7}$$

The average value of queuing delay of all received packets within a round is computed,

$$\mathrm{avgQD} = \frac{\sum_{i=1}^{n} \mathrm{QD_i}}{\mathrm{n}} \tag{8}$$

where n is the number of packet received within a round. Finally, the receiver obtains the delay factor $\mathrm{d_f}$ as the ratio of the average queuing delay to the maximum queuing delay.

$$\mathrm{d_f} = \frac{\mathrm{avgQD}}{\mathrm{QD_{max}}} \tag{9}$$

Values of the delay factor $\mathrm{d_f}$ range between [0, 1], with zero indicating an empty buffer and one a full buffer at the bottleneck link. Thus, it is possible to predefine threshold values to assess the network condition. In RRB-SIMD two threshold values are considered, the low-threshold value $\mathrm{T_1}$ and the high-threshold value $\mathrm{T_2}$ both chosen in the range $[0, 1]$. The two threshold values serve to classify the shared available bottleneck throughput into three regions, the non-congested region $[0, \mathrm{T_1}]$, the loaded region $]\mathrm{T_1}, \mathrm{T_2}[$ and the congested region $]\mathrm{T_2}, 1]$.

3.1.3 Packet Loss Measurement

It is important to notice that packet loss is unavoidable due to best-effort nature of the Internet along with other reasons, such as the difference between the time a receiver generates report and the time the sender receives the report. This difference delays the rate adjustment and may cause excessive packet loss during network congestion. In RRB-SIMD, loss event is measured at the receiver using the RTP sequence number field. The receiver then minimises the effect that single spurious loss may have on the packet loss estimation, by smoothing the values of packet loss using Eq. 10

$$\widehat{l} = \frac{\sum_{i=0}^{n} w_i l_i}{\sum_{i=0}^{n} w_i} \tag{10}$$

where $\widehat{l}$ is the smooth value of packet loss rate, l_i the measured packet loss rate and w_i the weight. Values of the weight are chosen such as a newer packet loss event has a high weight while the weights gradually decay to zero for older packet loss events. The following series $(1, 1, 1, 1, 0.8, 0.6, 0.4, 0.2)$ of weight values are used in the implementation of RRB-SIMD, for $n = 8$. More details about this smoothing technique are provided in [14].

3.1.4 Round-Trip Time Estimation

The parameter K used in Eq. 8 to compute the transmission rate r_t is inversely proportional to RTT [15]. Thus, for the receiver to compute the transmission rate when the network is not congested, he needs first to compute the RTT. RTT is computed as in [16], that is, using the RTP timestamp field and the arrival time of the packet at the receiver. Assuming that the arriving packet has the index i, first the receiver calculated the one way delay D_i to the sender and, computed the RTT using the one way delay as shown in Eq. 11

$$\mathrm{RTT} = (1 + \varphi)\mathrm{D_i} \tag{11}$$

where the value of φ is unknown. To estimate φ, the receiver uses RTT_{eff} the effective estimation of RTT provided for in the RTCP reports. RTCP protocol requires the receiver to store in its reports to the sender the timestamp of the most recent report from the sender in the t_{LSR} field, and the delay between the reception of the last sender report and the transmission of the receiver report in the t_{DLSR} field. With this information and the time A when the sender receives the receiver RTCP report, the sender computes the effective RTT_{eff} as follows.

$$\mathrm{RTT_{eff}} = \mathrm{A} - \mathrm{t_{LSR}} - \mathrm{t_{DLSR}} \tag{12}$$

The sender then computes an appropriate value for the parameter φ using Eq. 13

$$\varphi = \frac{\mathrm{RTT_{eff}}}{\mathrm{D_i}} - 1 \tag{13}$$

To avoid the effect of instant RTT high value in the computation of the transmission rate, the receiver uses the composite filter of degree-0 [17] to smooth the RTT process, and thus the oscillations in the computed transmission rate.

3.2 Decision and Adaptation Mechanism

During congestion avoidance, the receiver assesses the congestion level using a mechanism that combines the delay factor, short and long running cumulative average jitter and packet loss. The receiver first examines the packet loss rate event; if packet loss rate is greater than zero then the receiver concludes that the network is congested. Otherwise, the receiver undertakes further investigation using the short and long running average cumulative jitters, and the delay factor.

When the short running average is greater than the long running average the network workload is interpreted as increasing and the receiver concludes that there is upcoming congestion (incipient congestion). RRB-SIMD reinforces the cumulative jitter scheme with the delay factor d_f. This is done to reduce the risk of unnecessary decrease of the transmission rate each time that incipient congestion is detected through the cumulative jitter scheme. When, a measure of d_f is substantially larger than the specified high-threshold value T_2, the receiver safely concludes that the network is congested and uses Eq. 9 to compute the TCP-friendly share bandwidth. Otherwise, if the delay factor is lower than a specified low-threshold value T_1 ($T_1 < T_2$) the transmission rate is increased using the relationship in Eq. 10 or maintained for a value of d_f between the two threshold values, that is, $T_1 < d_f < T_2$.

The receiver then feeds this information into the Receiver Report (RR) and uses the extension mechanism of RTP/RTCP to communicate the transmission rate to the sender. The sender, upon reception of the receiver's RTCP reports, adapts the transmission rate.

At the beginning of a connection, the sender starts with an initial rate r_i, which is set to a very low value. The receiver, which has not experienced any packet loss, increases the rate by no more than one packet per RTT. For this reason, it calculates the new rate r_t

$$r_t = r_i + S/RTT \tag{14}$$

where S is the size of the packet. This process of increasing the rate by no more than one packet per round reduces the risk of bandwidth overshoot at the bottleneck link. Once a loss is registered the receiver uses Eq. 9 to compute the transmission rate. RRB-SIMD then enters immediately into the congestion avoidance mode. During connection start-up only packet loss is considered as an indication of congestion and that the initial values of r_{max} and r_0 used during congestion avoidance are obtained at the end of this process.

3.3 Parameters Determination

The choice of (α, β) has a direct impact on protocol responsiveness to network bandwidth availability. For example, a flow with a large β is very sensitive to bandwidth variation while a small value of α will reduce the aggressiveness of the flow, that is, its ability to probe for the available bandwidth. In [11] a value of $\beta = 0.0625$ is suggested, this value is considered in the implementation of RRB-SIMD.

4 Simulation Environment and Results

RRB-SIMD protocol is tested using the network Simulator NS-2 enhanced with Evalvid-RA framework [17, 18] that supports rate adaptive. MPEG-4 is used to compress the concat.yuv video file at 25 fps and to generate 30 different encoded video clips with quantizer step in the range of 2 to 31. The concat.yuv combines various video sequences of different complexity, which can better simulate video transmission as its temporal resolution changes over time.

Network metrics that are computed during post-processing include the throughput, end-to-end delay and the jitter. For quality evaluation of the video, we used the Peak Signal to Noise Ratio (PSNR) and the Mean Opinion Score (MOS) values. The concat. yuv video is stored in a server and then transferred from the server to the receiver using the topology in Fig. 1. We used several Pareto distributed ON/OFF sources that emulate the behavior of competing short TCP connections. The number of competing short TCP flows is set to 5, with each web's sender transmitting its packets at 200 Kbps to a web's receiver RN (N = 1, 2, …, 5). The network consists of several access links, all having a bandwidth set to 100 Mbps and the buffer size at the bottleneck to 130 packets. The delay for all access links are set to 5 ms except for the bottleneck link, which is set to 30 ms. The bottleneck link is configured to have a bandwidth of 12 Mbps. A number N of long live TCP background flows are also concurrently run to realize congestion at the bottleneck link. Packets from a long life TCP application flow from sender (SN) to receiver (RN), with N being the sender and the receiver identification number. All, the plots presented in this section are obtained by varying the number of long life TCP flows from 1 to 20 to realize different network congestion conditions.

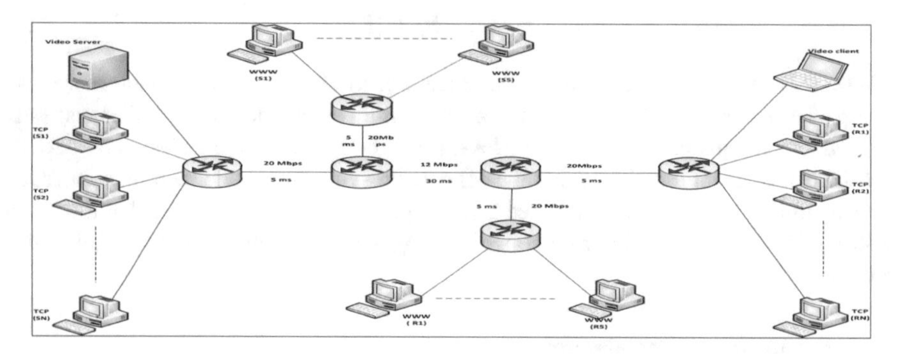

Fig. 1. Network topology

Figure 2 illustrates the average throughput results of both RBB-SIMD and TFRC protocols. From the plot, after a transient behavior at start-up, the average throughput of both TFRC and RRB-SIMD protocols show a steady behavior as the number of long life flows exceeds 11. Overall TFRC realizes the higher average throughput than RRB-SIMD. However, the throughput improvement in TFRC comes with a cost of higher frames loss. Figure 3 illustrates the percentage of frame loss as the number of TCP flows increases. An approximate 10% frame loss is reached at worst scenario for TFRC (N = 17), while the average frame loss in the case of RRB-SIMD did not exceed 7%, for the all the simulation. Sometimes spikes and dips in frame loss are observed when not expected as it was the case with throughput. For example, a spike is observed in point 3, followed by a dip in point 4, this means that the frame loss is higher in point 3 than point 4, when one would expected the inverse. This situation can be explained by the traffic pattern of short time WWW flows, which may vary from one case to another.

Figure 3 expresses the frame loss results in term of average peak-signal to noise ratio. As expected, the lower frame loss observed when streaming with RRB-SIMD is transformed to an higher average PSNR. Higher PSNR means better visual quality. When the number of competing TCP flow is less than 7, the network condition is good, and the video is streamed with an excellent visual quality. As the number of competing TCP traffic increases, the network conditions rapidly deteriorate; this leads to transmission of video content at very low rate.

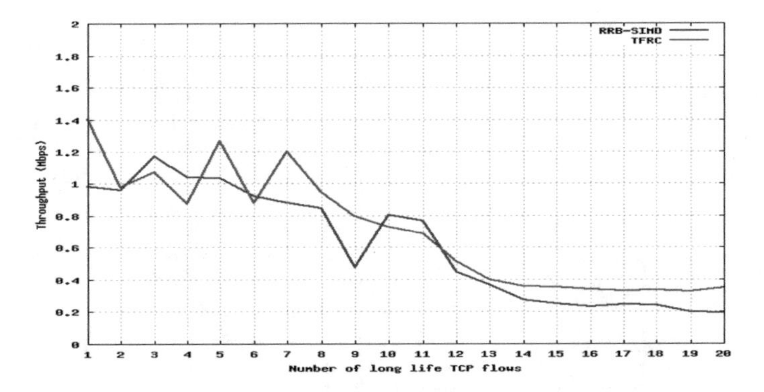

Fig. 2. Average throughput of RRB-SIMD and TFRC flows

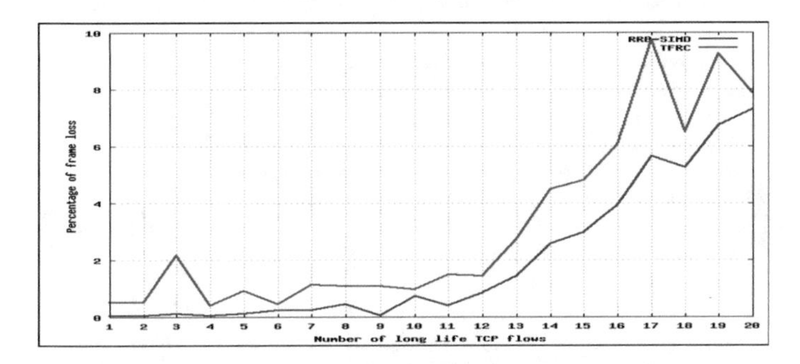

Fig. 3. Average frame lost for RRB-SIMD and TFRC

Figure 4 illustrates the behavior of the average cumulative jitter for both protocols. The result is split into two parts to better illustrated the changes on the average cumulative jitter. From the first plot, it can be observed that the cumulative jitter in RRB-SIMD is lower than in TFRC when the number of flows less or equal to 7. In this case a short buffer can be provided for at the decoder to alleviate the variation of the delay jitter. When the number of TCP flows exceeds 7, the average cumulative jitter shows an unexpected behavior in most cases and TFRC has realized the better performance. TFRC proves to have the lower average cumulative jitter than RRB-SIMD when the number of competing TCP traffic exceeds 16, i.e., in a very scarce network throughput situation.

Figure 5 illustrates the average delay behavior of both RRB-SIMD and TFRC. From the plot, both protocols realized on average a delay within the limit of conversational applications (150 ms). The average delay in both cases did not exceed 100 ms, and RRB-SIMD realizes the better average delay as the network conditions deteriorate; which deterioration is obtained by increasing the number of long life TCP flows. This performance improvement is a result of a very low conservative transmission rate and the delay factor scheme, which bound the average delay of RRB-SIMD between two threshold values.

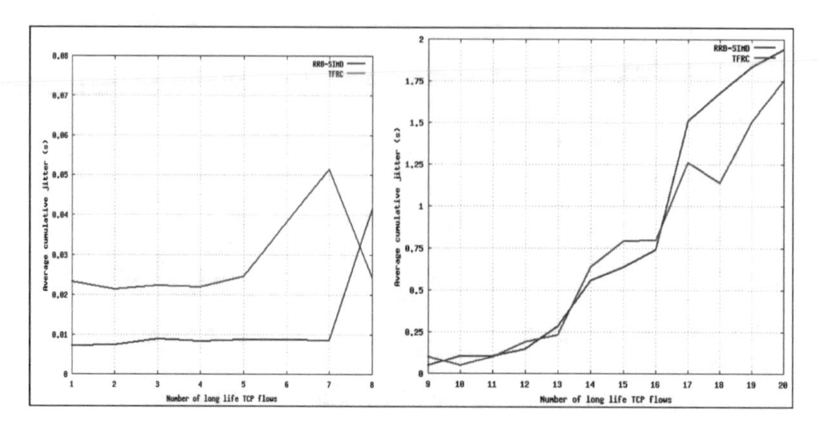

Fig. 4. Average cumulative jitters of RRB-SIMD and TFRC

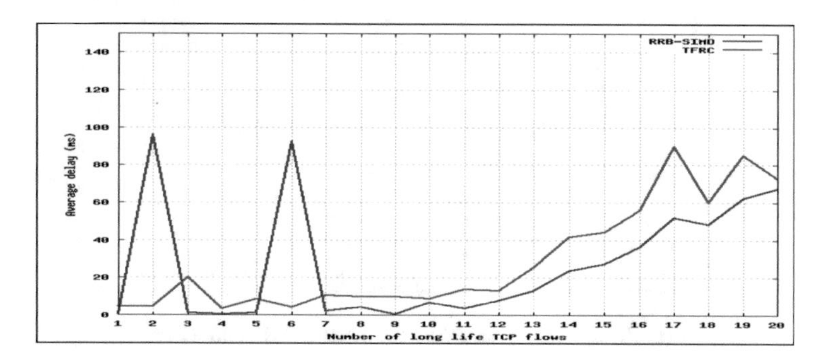

Fig. 5. Average delay of RRB-SIMD and TFRC

5 Conclusion

In this paper, a transport protocol named RRB-SIMD for video streaming over the Internet that is capable of providing better QoS support than the existing TCP-friendly rate control is presented. Here the cumulative jitter, an important QoS parameter for multimedia data is considered in addition to packet loss ratio for congestion control. The cumulative jitter scheme is reinforced with a delay factor to reduce risk of unnecessary decrease of the transmission rate each time that incipient congestion is reported through the cumulative jitter scheme.

The performance evaluation was obtained using various simulations with the network simulator NS2 and trace files derived from real video clips similar to those find on the Internet. RRB-SIMD was compared to the established standard TFRC protocol, and the following conclusions were drawn from the results: RRB-SIMD is TCP-friendly, as the shared bandwidth was almost equally distributed between RRB-SIMD and TCP flows. RRB-SIMD has realized a much better performance compared to TFRC in terms of loss frame rate, delays, PSNR and jitter results, providing the end user with an image with better quality than TFRC. With a low average jitter, a short decoder buffer at the receiver can cope with the variation in the packets arrival time.

References

1. Floyd, S., Handley, M., Padhye, J.: A comparison of equation based and AIMD congestion control. In: ACIRI, pp. 1–12, May 2002
2. Jacobson, V.: Congestion avoidance and control. In: Proceedings of ACM SIGCOMM, pp. 314–329, August 1998
3. Floyd, S., Fall, K.: Promoting the use of end-to-end congestion control in the Internet. In: Proceedings of ACM/IEEE Transactions in Networking, pp. 458–472, May 1999
4. Sterca, A., Hellwagner, H., Boian, F., Vancea, A.: Media-friendly and TCP-friendly rate control protocols for multimedia streaming. Proc. IEEE Trans. Circuits Syst. Video Technol. **26**(8), 1516–1531 (2016)
5. Talaat, M.A., Attiya, G.M., Koutb, M.A.: Enhanced TCP-friendly rate control for supporting video traffic over the Internet. Can. J. Electr. Comput. Eng. **36**(3), 135–140 (2013)
6. Balan, V., Eggert, L., Niccolini, S., Brunner, M.: An experimental evaluation of voice quality over the datagram congestion control protocol. In: Proceedings of the IEEE INFOCOM 2007, Anchorage, AK, pp. 2009–2017, May 2007 [DCCP]
7. Ito, D., Niibori, M., Kamada, M.: Real-time web-cast system by multihop WebRTC communications. Int. J. Grid Util. Comput. **9**(4), 345–356 (2018)
8. Huang, Y., Mao, S., Midkiff, S.F.: A control-theoretic approach to rate control for streaming videos. IEEE Trans. Multimedia **11**(6), 1072–1081 (2009)
9. Carlucci, G., De Cicco, L., Holmer, S., Mascolo, S.: Congestion control for web real-time communication. IEEE/ACM Trans. Netw. **25**(5), 2629–2642 (2017)
10. Nor, S.A., Hassan, S., Ghazali, O., Omar, M.H.: Enhancing DCCP congestion control mechanism for long delay link. Int. Symp. Telecommun. Technol. **2012**, 313–318 (2012)
11. Rejaie, R., Handley, M., Estrin, D.: RAP: end to end rate-based congestion control mechanism for real-time streams in the Internet. In: Proceedings of IEEE INFOCOM, pp. 1337–1345, March 1999
12. Lusilao-Zodi, G.A., Dlodlo, M.E., De Jager, G., Ferguson, K.L.: RRB-SIMD: RTP rate-based SIMD protocol for media streaming applications over the Internet. In: Annual Communication Networks and Services Research Conference, pp. 69–76 (2011)
13. Bouras, C., Gkamas, A., Kiomourtzis, G.: Adaptive smooth multicast protocol for multimedia data transmission. In: International Symposium on Performance Evaluation of Computer and Telecommunication Systems, pp. 269–276 (2008)
14. Sivarajah, J., Armitage, D.W., Allinson, N.M.: New TCP-friendly, rate-based transport protocol for media streaming applications. Proc. IEEE Commun. **151**(3), 280–286 (2004)
15. Hartanto, F., Sirisena, H.R.: Cumulative Inter-ADU Jitter concept and its application. In: Proceedings of the 20th International Conference on Computer Communications and Networks, USA, pp. 531–534 (2001)
16. Lusilao-Zodi, G.A., Dlodlo, M.E., De Jager, G., Ferguson, K.L.: Round-Trip time estimation in telecommunication networks using composite expanding and fading memory polynomials. In: 15th IEEE Mediterranean Electro-technical Conference (MELECON 2010), pp. 1581–1585, April 2010
17. Schulzrinne, H., Casner, S., Frederick, R., Jacobson, V.: RTP: a transport protocol for real-time application. RFC3550, July 2003
18. Klaue, J., Rathke, B., Wolisz, A.: Evalvid: a framework for video transmission and quality evaluation. In: Proceedings of the 13th International conference on modelling techniques and tools for computer performance evaluation, Illinois, pp. 255–272 (2003)

TCP with Network Coding Performance Under Packet Reordering

Nguyen Viet Ha(✉) and Masato Tsuru

Kyushu Institute of Technology, 680-4 Kawazu, Iizuka-shi, Fukuoka 820-8502, Japan
nguyen.viet-ha503@mail.kyutech.jp, tsuru@cse.kyutech.ac.jp

Abstract. The adverse impact of packet reordering besides packet loss is significant on the goodput performance of TCP (Transmission Control Protocol), a dominant protocol for reliable and connection-oriented transmission. With the primary purpose of improving the TCP goodput in lossy networks, the Network Coding technique was introduced. TCP/NC (TCP with Network Coding) is a promising approach which can recover lost packets without retransmission. However, the packet reordering has not been considered, and no study on that issue is found for TCP/NC. Therefore, in this paper, we investigate the goodput performance degradation due to the out-of-order reception of data or acknowledgment packets and propose a new scheme for TCP/NC to estimate and adapt to the packet reordering. The results of our simulation on ns-3 (Network Simulation 3) suggest that the proposed scheme can maintain the TCP goodput well in a wide range of packet reordering environments compared to TCP NewReno as well as TCP/NC.

1 Introduction

Transmission Control Protocol (TCP) has a long history and is still used widely in many applications as a primary connection-oriented transport protocol for reliable and in-order transmission of a byte sequence. One of the important features of TCP is the congestion control based on the congestion window (CWND). TCP will decrease the sending rate by reducing CWND when detecting a packet loss, which is necessary for fair bandwidth sharing and works almost correctly on conventional wired single-path networks. However, when being applied to complex and challenging environments, TCP exposes a weakness in maintaining the goodput performance due to lossy links as well as packet reordering.

On lossy links, packets are lost not by congestion but by physical link errors. In such conditions, since TCP cannot distinguish the type of losses and consider any packet loss as a congestion signal, TCP decreases the CWND mistakenly, resulting in a seriously lower goodput. Some TCP variants have been proposed to overcome this issue, e.g., TCP Westwood+ [1], but they are not useful in heavy loss environments. Another approach is combining Network Coding with TCP (called TCP with Network Coding - TCP/NC) [2].

L. Barolli et al. (Eds.): EIDWT 2019, LNDECT 29, pp. 552–563, 2019.
https://doi.org/10.1007/978-3-030-12839-5_51

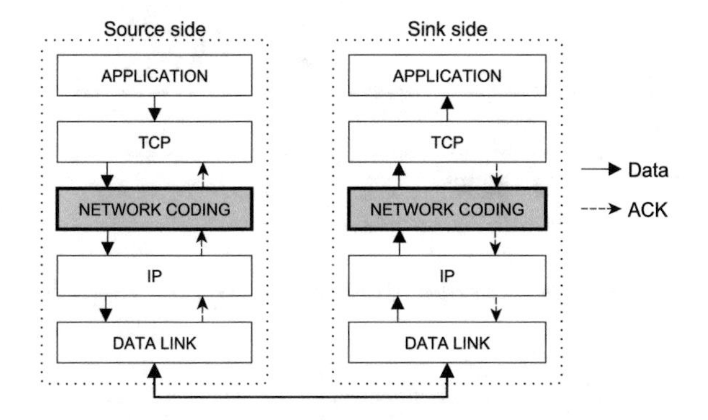

Fig. 1. NC layer in TCP/IP model

In TCP/NC, a new NC sub-layer is added between TCP and Internet layer shown as Fig. 1 to control the packet loss issue. This sub-layer become an intermediate handler after and before the TCP layer sends and receives the packet, respectively. At sending side, NC sub-layer combines n original TCP segments to m combination packets with $m > n$. At receiving side, the sink is expected to recover all n original segments if the number of lost packets is no more than $m - n$.

The principal purpose of TCP/NC is improving the goodput performance in lossy networks. Therefore, many variants of TCP/NC have been developed for this demand. Some exemplary contributions are mentioned following. TCP/NC with Enhanced Retransmission (TCP/NCwER [3]) can improve the retransmission process by sending multiple retransmission in one Round Trip Time (RTT); besides, all the retransmission are encoding to prevent the lost again which lead to TCP Timeout (TO). Self-Adaptive NC-TCP (SANC-TCP [4]), Adaptive NC (ANC [5]), and Dynamic Coding (DynCod [6]) focus on the channel condition estimation (link loss rate) and NC parameters adaptation (n and m) to work well in the practical channels frequently changed over time. Especially, TCP/NC with Loss Rate and Loss Burstiness Estimation (TCP/NCwLRLBE [7]) can estimate the channel condition of burst loss environments (both link loss rate and loss burstiness) and be flexible in adjusting the NC parameters without disrupting the current settings. Our study in [8] also solved the problem of Acknowledgment (ACK) packet loss by adding new information in the NC-ACK header of the ACK packet to convey the reception information of not only the current packet but also the previous packets.

Besides the packet loss, packet reordering also causes goodput performance degradation. There are many reasons causes the packet reordering, such as multipath routing for load balancing, route fluttering, and Link-layer retransmission. Packet reordering (i.e., out-of-order packet arrival) happens not only on data packets (referred to as forward-path reordering) but also on ACK packets (referred to as reverse-path reordering). The simple illustration of packet reordering is shown in Fig. 2. p_1, p_2, and p_3 sent in the order, but they are received out

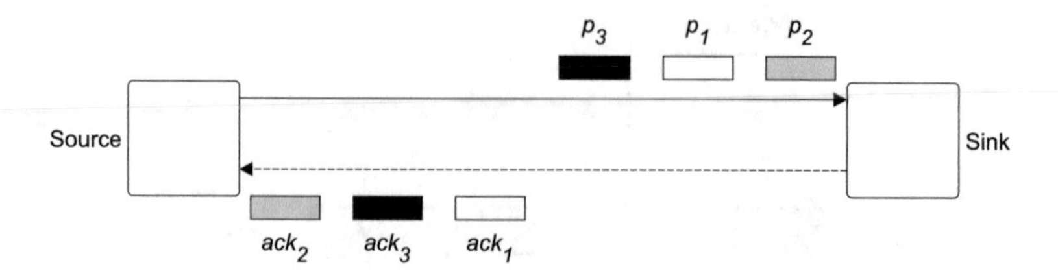

Fig. 2. Illustration of forward-path reordering and reverse-path reordering

of order at the sink. In the reverse side, ACK packets are transmitted in the order of ack_2, ack_1, and ack_3, but ack_1 arrives after ack_2 and ack_3 at the source. In the TCP operation, when the sink receives a packet out of order, the sink will put the same ACK number with the previous ACK packet into the current ACK packet. When the source receives many duplicated ACK packets, it enters to retransmission process and decreases CWND mistakenly. The goodput performance of not only the regular TCP but also the current TCP/NC variants will be affected. Another hand, reverse-path reordering will not impact the performance of the regular TCP, but it is dangerous in some TCP/NC variants. If the information conveyed in the ACK packet is used to estimate the channel conditions, e.g., calculating the loss burstiness, receiving out of order ACK packet will adversely affect the estimation process.

In this paper, we propose a new scheme to estimate the reordering conditions, such as reordering length, and the number of duplicated ACK to react the reordering affectation. This scheme can change the NC parameters based on not only loss/burstiness but also the reordering information. Besides adjusting the NC parameters, we introduce the Pause-ACK mechanism to helps the source delay some the duplicated ACK in the determined period to wait for the proper ACK packet. This mechanism can avoid the source receive not incorrect ACK packet and enter to retransmission mistakenly. Consequently, the goodput performance will maintain stable.

The remainder of this paper is organized as follows. Section 2 introduces the overview of TCP/NC. Section 3 explains the detail of the proposed scheme. Simulation evaluation is presented in Sect. 4 and conclusion is given in Sect. 5.

2 TCP/NC Overview

2.1 Network Coding in Protocol Stack

TCP/NC is proposed with the main responsibility of handling the packet loss to robustness in the lossy channel without any modifications of the TCP protocol. Therefore, new NC sub-layer is complemented and placed between TCP and network layer shown in Fig. 1. This sub-layer handles the incoming and outgoing packets from TCP and network layer, respectively. It works transparently with other layers; thus, TCP/NC can apply to any current devices. If NC sub-layer can recover all packet losses, the TCP layer is unaware of the loss events; thus, the goodput is not affected by the lossy channel. Besides, NC sub-layer will return ACK packet with ACK number determining based on the degree of freedom and the seen/unseen definition [2]. It means that the sink can return the different ACK number for every received packet without waiting for decoding all the packets. When the sink receives enough combination packet, all original packets will be decoded. Therefore, the CWND is kept increasing even though some combination packets are lost. Thus, the goodput performance is stable through lossy channels.

2.2 Coding Process

TCP/NC allows the source to send m combination packets (C) created from n original packets (p) with $m \geq n$ using Eq. (1) where α is the coefficient (encoding process). If the number of lost combinations is less than $k = m - n$, the sink can recover all the original packets using the received combinations without retransmission except for the case of the linearly dependent combinations (decoding process). TCP/NC using a sliding method to combine the original packets into a combination packet with the number of combined packets in one combination packet (referred to as the sliding window) is $k + 1$. Besides, α is selected randomly; thus, the coding algorithm is also called Random Linear Network Coding (RLNC [10]). And the computation is implemented in a Galois field (e.g., $GF(2^8)$).

$$C[i] = \sum_{j=1}^{n} \alpha_{ij} p_j \; ; \quad i = 1, 2, 3, ..., m \tag{1}$$

2.3 TCP Functionality

As mentioned, TCP/NC must not interference the TCP operation; thus, it works transparently to other layers. Moreover, TCP functionalities have been studied and worked stably in a long history. TCP/NC should take all these advantages such as retransmission and congestion control mechanisms. The source must retransmit the packets when the number of packet losses is larger than the recovery capacity of NC sub-layer. In that situation, both the TCP layer and NC sub-layer at the source receive many duplicated ACK equaling the oldest "unseen" packet. The NC sub-layer only needs to wait for the retransmission from the TCP layer. Increasing or decreasing the CWND is also controlled by TCP layers, not NC sub-layer.

3 Proposed Scheme

3.1 Packet Reordering Estimation

Packet reordering issue causes goodput degradation as shown in the study [9]. When the sink receives packets out of order, all returning ACK packet will have the same ACK number with the ACK packet for the last in order packet. When

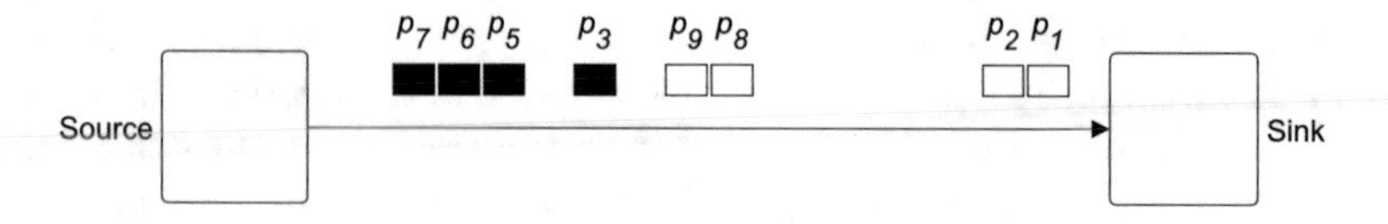

Fig. 3. Illustration of packet reordering length

the source receives too many duplicated ACKs, the TCP will enter the retransmission process and reduce the CWND by the half (e.g., in TCP NewReno). TCP/NC relies on the same mechanism for congestion control; hence, like TCP, unwanted duplicated ACKs affect the goodput of TCP/NC. However, unlike TCP, TCP/NC returns the ACK number based on the seen/unseen packets. Thus, we suggest increasing the size of the NC window sliding, which equals the number of packets in one combination, to overcome this problem. If the largest distance of two out of order packets less than NC window sliding size, the packet reordering is not sensed by the sink-side TCP that can send an ACK packet with an incremental ACK number.

To estimate the largest distance of two out of order packets (referred to as reordering length), we use the existing field Packet-ID (*Pid*) in the NC header which is the sequential number assigning to the combination for loss estimation purposes. The sink will perform that responsibility. If the *Pid* of the received combination packet ($Pid_{current}$) is less than the *Pid* of the previous combination packet ($Pid_{previous}$), the packet comes out of order. And, if the $Pid_{current}$ is higher than the $Pid_{previous}$ at least two packets, the reordering length equal $Pid_{current} - Pid_{previous} - 1$. However, it may contain the packet loss; thus, the sink must record which packet in this length will be received. The final value of reordering length equals the number of the packets will be received. Another scheme will handle these packet loss such as TCP/NCwLRLBE [7] that does not scope in this paper. Figure 3 shows a simple example to explain the reordering length. When the sink receives the p_8, the temporary length is set to 5. After that, when the p_3, p_5, p_6, and p_7 arrives to the sink, the reordering length is set to 4 because the p_4 is lost. The reordering length in average (L_u) is calculated periodically in every pre-configuration period (e.g., 5 s in this paper). Besides, we also get the average value (l_u) using the Simple Moving Average (SMA) method with the window length of 5 for preventing the suddenly substantial change. After determining l_u, the sink will send the ceiling of l_u to the source via ACK packet by using the *"reordering length"* field in the NC-ACK header shown in Fig. 4 and Table 1. The flag U (update indication) in the NC-ACK header is set to 1 also to let the source know the change to update the NC parameters. Noted that the NC header and NC-ACK header retained from our previous propose which is discussed in [3,7,8].

When the source receives the *"reordering length"* with the D flag, it will find the new NC parameters (n and m) which having the redundancy factor R approximate $\frac{n_{old}+k_{old}+l_u}{n_{old}}$. Noted that the maximum of k is 10.

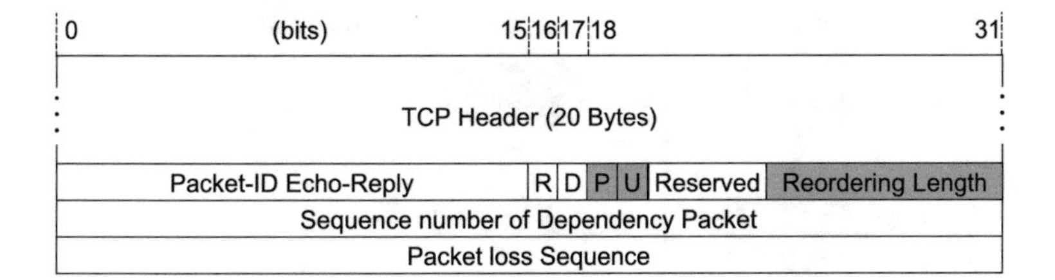

Fig. 4. NC-ACK header

Table 1. NC-ACK header fields description

Field name	Description
Packet-ID Echo-reply	The packet identity echo reply
R	The redundancy flag
D	The dependence flag
P	The Pause-ACK flag
U	The update indication flag
Reserved	Reserved for the future use
SN of the dependence pkt	The sequence number of the dependence packet at the sink. Using to notify the source to retransmit this packet
Packet loss Sequence	Store the status of the 32 previous packets start from the newest received packet having the *Pid* equal the *Pid Echo-Reply*
Reordering Length	Average value of the reordering length

3.2 Pause-ACK

Since a large k results in a lower goodput due to too many redundant packets, the average value l_u is used to limit k. However, the instant packet reordering length sometimes becomes larger than the average value, which still causes many duplicated ACKs. It will reduce the goodput, leading to the sending rate degradation. Keeping the stable sending rate plays an important role in this scheme because the reordering length depends on the sending rate proportionally. When the sending rate is low, the estimated l_u will be not accurate. We propose the mechanism called Pause-ACK to let the source delay some duplicated ACK to wait for the proper ACK. This act can help the sink limit the number of mistakenly reducing the CWND.

The source has responsibility for estimating the loss condition; thus, it needs to receive as much as the number of ACK packets possible. Therefore, delaying

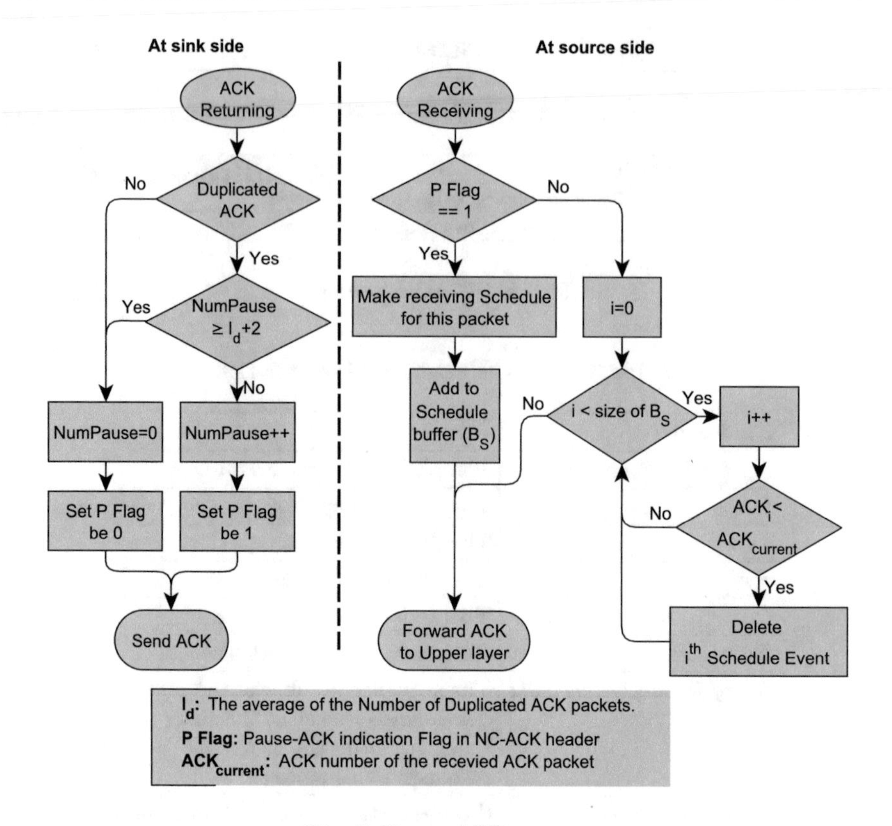

Fig. 5. Pause-ACK process

the ACK packet to send to the upper layer will perform at the source. But, determining which the delayed ACK packet is complete at the sink as shown on the left side of Fig. 5. The sink will indicate the delayed ACK packet by using P flag in the NC-ACK header. The number of the delayed packets calculated from the length of the duplicated ACK (referred to as the pause length). The pause length in average (L_d) is calculated periodically in every pre-configuration period. We also get the average value (l_d) using the SMA method with the window length of 5 for preventing the suddenly substantial change. Based on our simulation, we see that the number of the delayed packets (pause length) equal to $l_d + 2$ having the good goodput performance.

At the source, when it receives the ACK packet with a P flag, the source will store this packet to the Pause-ACK buffer and set the sending schedule to after 200 ms for this ACK packet. (at this point we following the setting of Delay-ACK of the regular TCP). Pause-ACK mechanism only works on duplicated ACK packets. Thus, if the source receives the new ACK packet having ACK number higher than that of ACK packets in the buffer, the source will erase these packet from the buffer. These process at the source is shown on the right flowchart in Fig. 5.

4 Simulation Result

4.1 Simulation Setup

The simulation is accomplished by Network Simulator 3 (ns-3) [11] which is a discrete-event network simulator for Internet systems. The topology of the simulation consists of a backbone with two arranged routers. One source and one sink are on either side of the backbone shown in Fig. 6. The simulation parameters is shown in Table 2. The configuration propagation delay is changed dynamically based on the proposed topology in [9]. The path delay changes in every "Inter-switch time" (δ). The propagation delay changes randomly around $200\tau+50$ with the standard deviation of $\frac{200\tau}{3}$ where $\tau\in[0,2]$. We investigate the variation of the goodput performance on four values of δ that are 50 ms, 250 ms, 500 ms, and 1000 ms.

TCP NewReno, TCP/NCwER, TCP/NCwLRLBE, TCP/NC with the Pause-ACK mechanism (TCP/NC with Pause-ACK), and the proposed scheme are compared in the time-averaged goodput. The detailed description is shown in Table 3. Note that TCP/NCwLRLBE protocol in this paper is an improved version developed in [8] for the ACK loss problem.

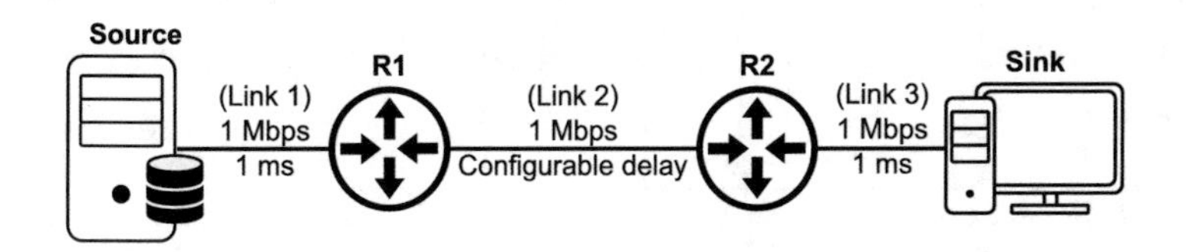

Fig. 6. Simulation topology

Table 2. Simulation parameters

Parameter	Value
Bandwidth of all links	1 Mbps
Propagation delay of Link 1 and 3	5 ms
Propagation delay of Link 2	Configurable
Buffer size in a router	100 packets
TCP protocol	TCP NewReno
Maximum CWND	65535 bytes
Payload size	536 bytes
Minimum of TCP Timeout	1 s
TCP Delayed-ACK	2 packets
Loss model	Random loss model of ns-3 simulator
Simulation iteration	20 times

Table 3. Protocols description

Protocol	Description
TCP	TCP NewReno
TCP/NCwER	TCP/NC with Enhanced Retransmission [3]
TCP/NCwLRLBE	TCP/NC with Loss Rate and Loss Burstiness Estimation [8]
TCP/NCwLRLBE with Pause-ACK	TCP/NCwLRLBE combined with Pause-ACK mechanisms
Proposed scheme	TCP/NCwLRLBE combined with Reordering Estimation/Adaptation and Pause-ACK mechanisms

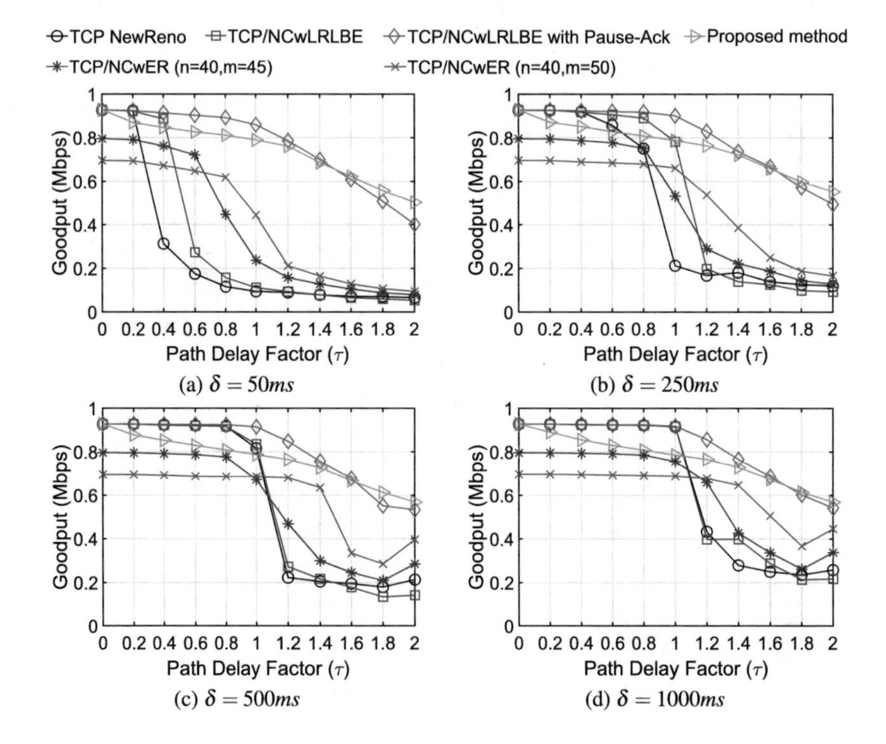

Fig. 7. Goodput comparison in different reordering cases (link loss rate is zero)

4.2 Goodput Evaluation in Reordering Case

4.2.1 No Link Loss Case

In this simulation (Fig. 7), we can see that only increasing k is not enough to overcome the packet reordering. TCP and TCP/NCwLRLBE have a worse goodput performance compared to other protocols. Because of no link loss,

TCP/NCwLRLBE keeps k is zero or a small value. In the TCP/NCwER cases, k is constant of 5 and 10; the goodput is stable in the low degree of packet reordering. But when τ increases, the goodput is decreased. For example of $k = 10$, the goodput start decreasing at τ of 0.8, 1, 1.4, and 1.4 corresponding to δ of 50, 250, 500, and 1000 ms.

The goodput performance of the proposed scheme is stable but smaller than that of TCP/NCwLRLBE with Pause-ACK. It is because the proposed scheme increases k to overcome the out-of-order packet arrivals. But in this case, the only Pause-ACK mechanism is enough when τ less than about 1.4. In future work, this issue can be solved by using more information to decide the value of k. The additional information which is packet reordering rate may be solved this problem.

4.2.2 Link Loss Case

In this simulation, the link loss rate set to 0.05 and 0.1. The results show in Figs. 8 and 9, respectively. The loss happens at the interface on R2 connected to R1 in the data sending direction.

Based on the results, we can see the advantage of the proposed scheme compared to the TCP/NCwLRLBE using only the Pause-ACK mechanisms. In Figs. 8 and 9, we can see that using only the Pause-ACK mechanism is not enough in the lossy case. The Pause-ack mechanism can slow down the retrans-

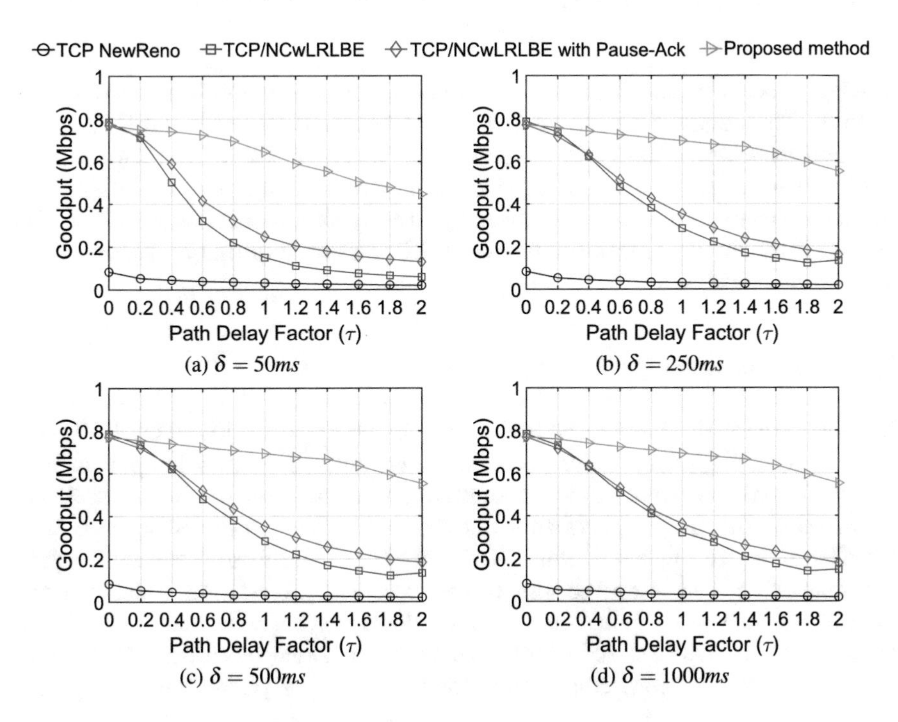

Fig. 8. Goodput comparison in different reordering cases (link loss rate is 0.05)

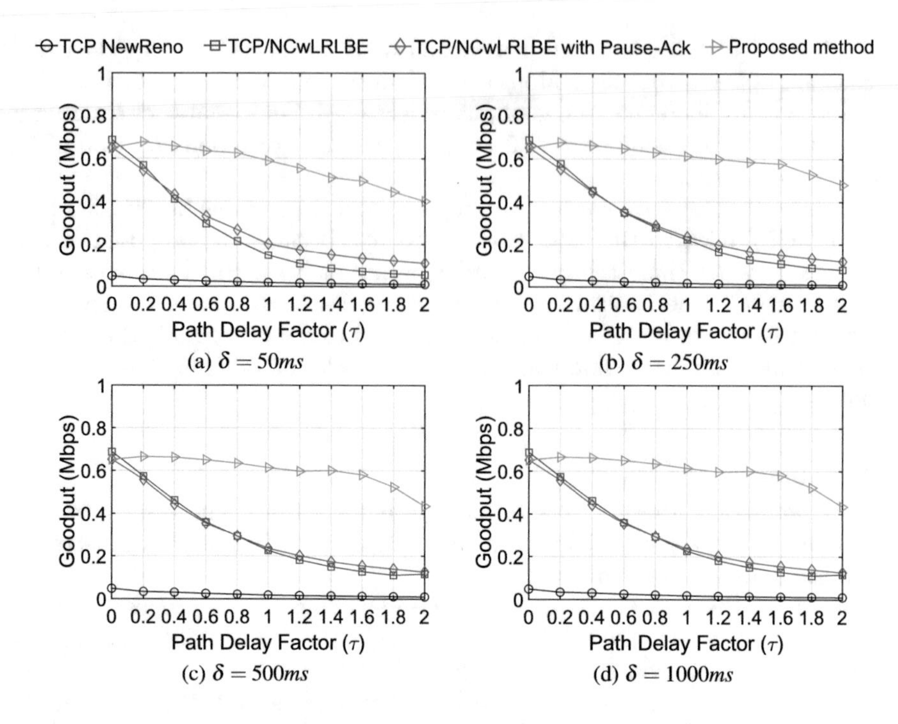

Fig. 9. Goodput comparison in different reordering cases (link loss rate is 0.1)

mission process of the TCP when it delays the proper duplicated ACK mistakenly. The goodput performance of TCP/NCwLRLBE with the Pause-ACK mechanism is better than TCP/NCwLRLBE but be not significant. Meanwhile, the proposed scheme gets a goodput performance to compare to the others. The proposed scheme increases k to overcome the packet reordering problem. This act can limit the number of duplicated ACK packets, resulting in decreasing the amount of Pause-ACK.

5 Conclusion

In this paper, we have proposed the scheme to help the TCP/NC work well on receiving out-of-order packets, which may sometimes happen in most practical complex network environments. The simulation results on ns-3 have shown that the proposed scheme outperforms other protocols such as TCP NewReno and the recent variant of TCP/NC. In the future, we will improve the scheme to well adapt to any packet reordering condition which depends on diverse network environments and affects seriously to the performance of the system. One possible approach is to consider more additional information on packet reordering estimation such as the packet reordering rate.

Acknowledgements. The research results have been achieved by the "Resilient Edge Cloud Designed Network (19304)," the Commissioned Research of National Institute of Information and Communications Technology (NICT), and by JSPS Grant-in-Aid for Scientific Research (KAKENHI) Grant number JP18H06467 and JP16K00130, Japan.

References

1. Mascolo, S., Casetti, C., Gerla, M., Sanadidi, M.Y., WangR.: TCP westwood: bandwidth estimation for enhanced transport over wireless links. In: Proceeding of the 7th Annual International Conference on Mobile Computing and Networking, pp. 287–297 (2001)
2. Sundararajan, J.K., Shah, D., Medard, M., Mitzenmacher, M., Barros, J.: Network coding meets TCP. In: Proceeding of the IEEE International Conference on Computer Comunication, pp. 280–288 (2009)
3. Ha, N.V., Kumazoe, K., Tsuru, M.: TCP network coding with enhanced retransmission for heavy and bursty loss. IEICE Trans. Commun. **E100–B**(2), 293–303 (2017)
4. Song, S., Li, H., Pan, K., Liu, J., Li, S.Y.R.: Self-adaptive TCP protocol combined with network coding scheme. In: Proceeding of the 6th Conference on Systems and Networks Communications, pp. 20–25 (2011)
5. Cheng, C.Y., Yi, H.Y.: Adaptive network coding scheme for TCP over wireless sensor networks. J. Comput. Commun. Control **8**(6), 800–811 (2013)
6. Vu, T.V., Boukhatem, N., Nguyen, T.M.T.: Dynamic coding for TCP transmission reliability in multi-hop wireless networks. In: Proceeding of the IEEE International Symposium on a World of Wireless, Mobile and Multimedia Networks, 6 p. (2014)
7. Ha, N.V., Kumazoe, K., Tsuru, M.: TCP network coding with adapting parameters for bursty and time-varying loss. IEICE Trans. Commun. **E101–B**(2), 476–488 (2018)
8. Ha, N.V., Tsuru, M.: TCP/NC performance in bi-directional loss environments. In: Proceeding of the International Conference on Electronics, Information, and Communication, 4 p. (to appear in ICEIC 2019, January 2019)
9. Leung, K., Li, V.O., Yang, D.: An overview of packet reordering in Transmission Control Protocol (TCP): problems, solutions, and challenges. IEEE Trans. Parallel Distrib. Syst. **18**(4), 522–535 (2007)
10. Ho, T., Koetter, R., Medard, M., Karger, D., Effros, M.: The benefits of coding over routing in a randomized setting. In: Proceeding of IEEE International Symposium on Information Theory, pp. 442–447 (2003)
11. Network simulator (ns-3). https://www.nsnam.org/. Accessed 20 Sept 2018

A Common Ontology Based Approach for Clinical Practice Guidelines Using OWL-Ontologies

Khalid Samara[1(✉)], Munir Naveed[2], Yasir Javed[2], and Mouza Alshemaili[1]

[1] Computer and Information Sciences, Higher College of Technology, Ras Al Khaimah, UAE
{ksamara,malshemaili}@hct.ac.ae

[2] Computer and Information Sciences, Higher College of Technology, Al'Ain, UAE
{mnaveed,yjaved}@hct.ac.ae

Abstract. The production and dissemination of clinical practice guidelines (CPG) is usually reliant upon the opinions and interventions of the physicians' knowledge that are presented in the form of text narratives. The knowledge utilized during the production of CPGs, is largely technical and procedural knowledge. However, the cognitive challenge encountered by the physician is to internalize this new guideline knowledge routinely into actions and clinical decisions. Ontologies have often been used to formalize and represent clinical guidelines. In this study, we propose an approach to the acquisition of CPG knowledge into computer-interpretable form to develop a semantically rich common ontology. To establish a comprehensive representation of CPGs we analyzed abstracts taken from the sub-domains of *HeartDiseases* related to its diagnosis, possible treatments, and interventions and structured them using the protégé-OWL formal modeling tool. The completeness, and expressiveness of the ontology are then validated using structured and unstructured queries.

1 Introduction

In the health domain, clinical decision-making is a cognitive process that relies mainly on clinical judgments about patient care. As research into the study of expert performance has shown, that the problematic nature of this form of knowledge is largely due to the technical and procedural knowledge that underlies scientific reasoning which is usually complex to infer into evidence-based care [1–3]. In order to exploit new knowledge for a potential concoction of best available clinical evidence, the codified elements are essential in conceptualizing the true meaning of clinical knowledge which can make it easier to translate into evidence-based guidelines [4, 5]. Clinical practice guidelines (CPGs) are a means of representing the best and up to date evidence-based care and for standardizing the way in which a particular pathology should be managed. The objective of these tools is to exhibit the latest knowledge and should be made available in a clear, concise and user-friendly format with the intent of influencing the codification of medical knowledge [6–8]. However, practitioners may not always

L. Barolli et al. (Eds.): EIDWT 2019, LNDECT 29, pp. 564–575, 2019.
https://doi.org/10.1007/978-3-030-12839-5_52

follow all the guidelines provided in a narrative form and thus their adherence to knowledge-intensive activities can be minimal, making it difficult to synthesize reliable clinical evidence [9–11]. These concerns are important for our understanding of expert knowledge and the development of technology for maintaining and instilling up to date evidence-based care. Health informatics researchers have demonstrated the effectiveness of computerized interpretable clinical guidelines to assist in the development of evidence-based care [12]. Ontologies can provide such a common semantic base in order to support computer-interpretable clinical guidelines [13–15]. The development of ontologies can also address the dissension between the physicians' expert knowledge that underlies routinized procedures for patient care and explicit evidence-based knowledge in a given medical environment [16]. However, as soon as ontologies are made available to support knowledge exchange between people, a fundamental problem occurs, different people will tend to use different ontologies [17]. This is because different domains have different data structures, syntaxes, and semantics which lead to complications in interoperability and communication. Common ontologies avoid the inevitable heterogeneity of knowledge that occurs during the intense acquisition and sharing of knowledge in a distributed environment [7, 16]. Given that clinical knowledge is represented in a wide variety of lexical forms ontologies can be used to model the exact elements in that domain and the relationships between those elements [13, 15]. Thus, the need of a common ontology for computer interpretable CPGs is not only to provide a foundation for knowledge acquisition, transfer, and creation but to reach a common understanding before ontology development could be made and to provide a language for a community to exchange knowledge about their domain [8–10].

The design of a common ontology that is described in this paper aims to identify the knowledge requirements that the ontology should fulfill in order to adequately represent CPGs into computer-interpretable form. The paper also suggest areas for improved coordination and consistency in their development and application. As a result, the paper is organized as follows. Section 2 develops and integrates the different perspectives from the literature on clinical CPG development and the constraints that occur to the implementation of common ontologies. Section 3, presents and formalizes a common ontology and the knowledge representation of CPGs into computer-interpretable form. The final section a conclusion and suggestion for further work is presented.

2 Barriers to Clinical Guideline Representation

In the health domain, the common techniques for managing clinical codified knowledge consists of recommendations or evidence in the form of clinical guidelines, clinical protocols and care pathways for the transfer of new clinical knowledge [1, 2, 4, 5]. The accepted practice in CPG development is the need for providing explicit recommendations and influence clinical practice through a formal process of disseminating advice on effective expert decisions. In most cases, physicians need to translate new explicit knowledge into well-rehearsed procedures that can be routinely implemented [17]. As a result, the proliferation of CPG development requires the possessions of human and domain-specific knowledge to create the transparency and coverage needed for decision-making [19–22]. These perspectives are known issues in the health domain, the increasing diversity, and complexity that concern enabling the knowledge representation needed for clinical guideline development. The effects of these challenges, however, have led to a general understanding that physicians do not always apply new knowledge of how to enhance patients' outcomes into practice to help make inferences quickly enough [17]. Furthermore, as medical knowledge continues to evolve over time, the prospect that new expert knowledge is integrated into narrative clinical guidelines to enhance patients' outcomes into practice is uncertain [18]. As a result, physicians need to acquire methods to integrate the evidence into a change in their practices and are expected to accomplish this within a congested and demanding clinic environment [24–27]. In the next section, the definitions of common ontologies are presented, followed by our development of a common ontology for representation of CPGs.

3 Common Ontologies

The definition of an ontology as prescribed by Gruber is an "explicit specification of a conceptualization" [22]. It consists of a set of formal terms with which one represents knowledge. According to Gruber, a common ontology is a vocabulary for representing the knowledge needed for some purposes in some domain [23]. As a result, a common ontology may be used by different ontologies which may support a user to extract and aggregate information to respond to user queries or as input data to other ontologies. The concept of a common ontology has been used in earlier studies [15, 17]. The concept known to a common ontology is generally confined to two separate perspectives. One perspective interprets common ontology, as an interaction between two or more knowledge-based systems without an agent committing to its internal encoding of knowledge [11, 18]. Another interpretation views common ontology "as a mapping between a language and something, which can be called ontology" [15]. In this context, the ability to create common, standardized CPGs is fundamental and ontologies are deemed to be key enabling technologies for that. Structured ontologies serve not only to make possible the reuse and sharing of existing knowledge but also to function as a mechanism for representing common knowledge amongst various users [7]. The real strength behind ontologies lies in the ability to create connections among classes and instances and to assign properties to those relationships to allow us to make inferences about them [25]. For example, in Fig. 1 the *Cardiovascular system disease* class has three subclasses: *Cardiac tuberculosis, Cardiomyopathy, Congestive heart failure. Congestive heart failure* class is also structured in two subclasses, *Cardiac arrest* and *Cor pulmonale.* This is a simple construction on which the ontology is personalized to a patient [26].

Gruber's [22, 23] work and that of other researchers led to the development of several formal ontology languages. The Resource Description Framework (RDF) and XML are the current standards for establishing semantic interoperability on the Web [15]. Ontology Web Language (OWL) designed by W3C Web Ontology can be used to develop Semantic Web applications [18]. Several studies have contributed to the

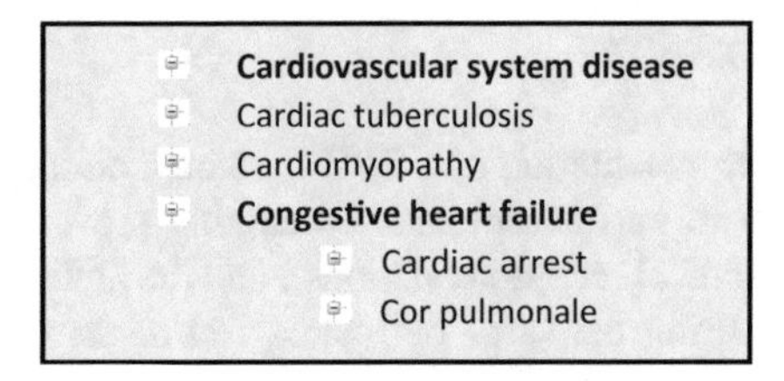

Fig. 1. Ontologies of cardiovascular system disease represent common properties across subclasses

development of clinical guideline ontologies. For example, Abu-Hanna et al. [27] used Protégé to construct a medical terminological system. El-Sappagh et al. [28] developed an ontology of concepts to support the modeling of CPGs for chronic disease management and as a standard medical terminology. Gorín et al. [29] explored an approach to model CPG of psychiatric care and for clinical decision support using OWL ontology. Similarly, Iqtidar [30] developed a structured and machine-readable genetic disease ontology and a clinical decision support system extracted from data sources. Moreover, recent studies have also realized the enormous potential of the Internet of Things (IoT) based on ontology development to resolve the problem of interoperability and the heterogeneous nature of clinical systems [31]. For example, studies have proposed that an IoT infrastructure can allow connectivity to be used to monitor or transmit in real-time through sensor network-enabled devices as a platform for monitoring patient activities [32]. All of these studies highlight the importance of ontology-driven decisions and the power and effectiveness of integrating and implementing a common ontology that gathers healthcare data from various sources.

4 Structural Representation for Common Ontologies

In this work, we use a structural approach to formulate knowledge representation. The structural entities are linked via mapping functions. The following structural entities and mapping functions are used to formulate the following representation:

Definition 1: The finite set of entities in a domain is represented by S.

Definition 2: A finite set of concepts in the domain is represented by C.

Definition 3: $d: S \times C \rightarrow k$ is a mapping function between domain entities and concepts such that if an entity s1 and concept c1 are of belong to same semantics, the k is always greater than 0 otherwise it remains 0.

Definition 4: P is a finite set of all possible properties for an entity.

Definition 5: $f: S \times S \rightarrow P$ is a mapping function between entities and properties in a domain.

Definition 6: $v: S \times P \times C \rightarrow k$ is a validation function which identifies same concepts for different entities and their properties.

5 Design Considerations

In achieving a detailed representation of a CPG ontology, an important consideration is that the ontology needs to be interoperable containing previous clinical data sources (e.g. clinical trials and clinical recommendations), and the ability to utilize and combine these sources while ensuring the same consistency in decision-making [33]. Accordingly, many items which are clinically beneficial for the generation of a CPG ontology (Heart Disease, Symptoms, Cardiologists) are held in an existing legacy system. Figure 2 illustrates the utilisation of the existing legacy clinical information for our CPG ontology.

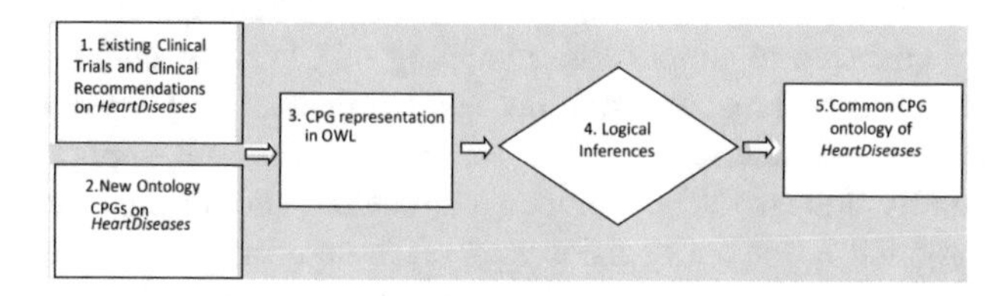

Fig. 2. A representation of the CPG ontology composed of five elements.

To support this, the method we propose is a reverse-engineering mapping approach to capture clinical history already stored in the legacy clinical system model to the target CPG ontology [33, 34]. Moreover, the process of reverse-engineering can support the recovery of conceptual structures. To achieve this, the CPG ontology requires logical inference for context reasoning, reuse and knowledge preservation from a legacy system that conceptualizes related parts of the target CPG ontology. This approach provides a well-defined process for integration with a preexisting clinical system and allows reuse of clinical information.

6 Ontology Development for Clinical Practice Guideline

The ontologies for CPGs are built using the protégé-OWL (Web Ontology Language) which is a formal modeling tool to represent computer interpretable CPGs [18, 22]. The PubMed and MEDLINE databases [26, 35–37] have been used to build ontologies for CPGs. The structural ontologies are built on conceptual entities like diseases. For example, as shown in Fig. 1, the main concept (the highest level in the structural hierarchy) is cardiovascular system disease. The purpose of such structural ontology is to provide a model based on possible diseases and a computer interpretable CPG environment to relate certain symptoms to possible set of diseases under a class of diseases. There are several ways to construct a common medical ontology [29], in our model, to build a hierarchical structure for ontological representation. The main benefit of such kind of ontologies is setting up causal link between the identification of specific cases and respond to treatments. Protégé-OWL provides a visual support in building ontology concepts (classes), class properties/relationships, taxonomies, constraints and object instances. In this study, we applied a top-down approach to construct the class hierarchy by extracting the core concepts (main classes) first and then extracting their corresponding subclasses. The classes are represented with respect to user-based preferences. There are five main structural elements in our representation approach: *Disease class*, *Symptom class*, *Diagnosis class*, *Treatment class* and *Patient class* as shown in Fig. 3 for class.

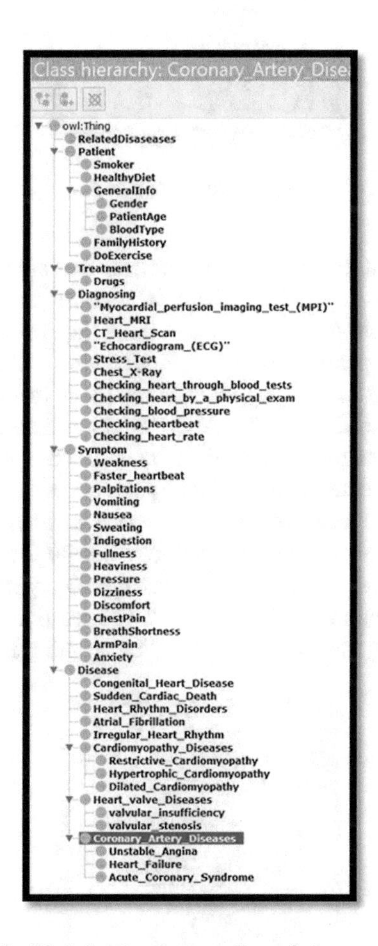

Fig. 3. High level representation of class structure

Object properties are the interfaces between different concepts and generalized relationships between various concepts that are related to each other e.g. *Cardiovascular disease* ***hasSymptoms***.

The object properties are shown in Fig. 4 where all possible properties or relationship in a domain are structured. Object properties establish relationships between any two classes and the datatype properties define an instance (individual) of a class and assign semantics according to a certain value e.g. a text, a number, binary, etc.). The data type properties and relationships are expressed in Fig. 5.

Fig. 4. Object properties

OWL classes are extracted as structural elements that are represented as individuals and are also called instances. Figure 5 displays some of the instances extracted for a CPG domain. The top concepts are decomposed using the individual instances. For example, the top-level concept *Disease* has individuals "*Cardiac Arrest*". Once the ontology is built, the reasoner is the main component in knowledge extraction and develops ontology expressed in OWL. After building instances, the tool applies the reasoner (for ontology construction e.g. Hermit) on the domain observation set to validate the consistency of the relationship among the classes and their relationship. The reasoner also explores the concept structural space to identify new properties from existing ones using inferences. The extracted relations by the reasoner could belong to different types of ontological properties like transitivity, symmetric, inverse and functional properties and use them to add new facts. The reasoner is applied for generating the outcomes of executing the queries such as "Description Logic (DL)" Queries.

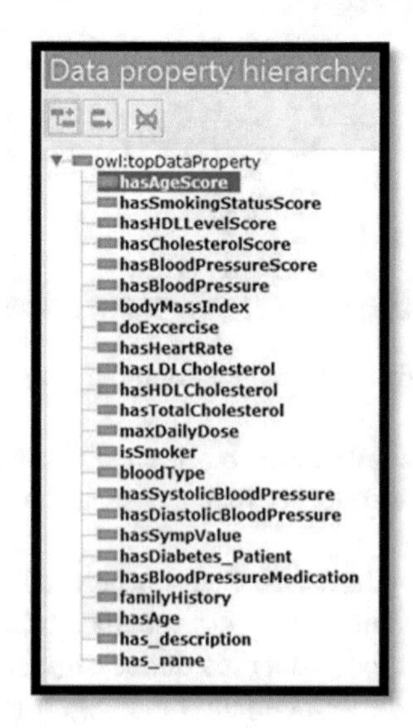

Fig. 5. Datatype properties

7 Results

To demonstrate the capabilities of structural representations, different queries are run on the newly built CPG. A set of queries are run using the object properties as shown in Fig. 6(a) and (b). Furthermore, the conceptual level is set at symptoms and decomposed from top level concept ("beta blockers") and queries are executed to search for symptoms of a disease. As a result, the outcomes are: "Coronary Artery Disease" and "Heart Valve Disease". Similarly, when we queried any disease related to "Heart Value Disease", the query results are "Shortness of breath, weakness and dizziness".

The physician also requires to query appropriate treatments e.g. to find a relevant procedure or medicine [21]. Therefore, the queries are also explored for the given data parameters as shown in Fig. 6(a) and (b) respectively. Keeping all the possible combinations that can be modeled using the asserted and inferred ontology, such systems are helpful in generating a representation of clinical guidelines.

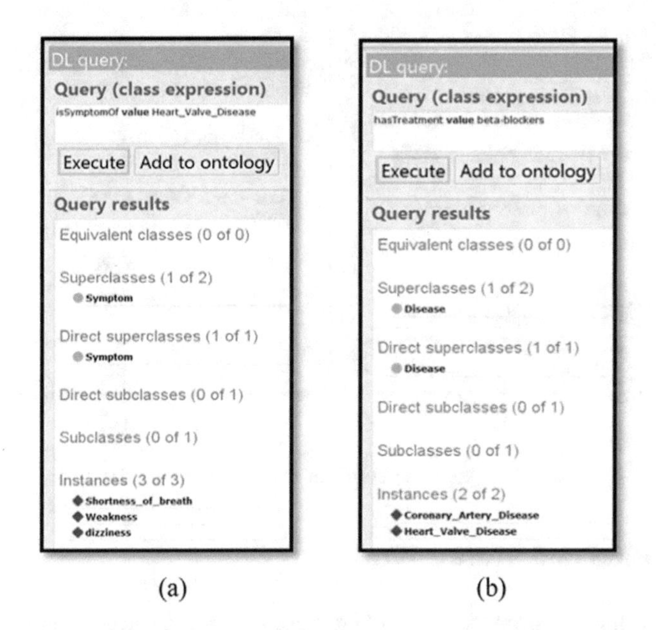

Fig. 6. Query structure for object properties

The query results also demonstrate the accuracy and completeness of the common ontology-based representation using OWL. As we have previously defined, the query function is an implementation of a mapping function f: $S \times S \rightarrow P$, traverses through the concept space and finds the answers in properties space for the queries as shown in Figs. 7(a) and 5(b). The query results also reveal that the mapping function (as a translator of each input) generates the complete and correct list of instances and avoids ambiguity. The results did not show any sign of ambiguity as the mapping function always give correct and complete answers for all queries. The query structure does not depend on concept hierarchy in the representation as shown in Fig. 7(a). The mapping functions for the ontology structure are the reasoning based processes and searches the hierarchy for a given query always generate an outcome (i.e. completeness). The query translator-based mapping function first navigates through the concept hierarchy to find the subclass and superclass for the given concept. The results also reveal that the domain representation have correct and unambiguous semantics for the common ontologies and it navigates structure via accurate decomposition to reach the relevant subclass.

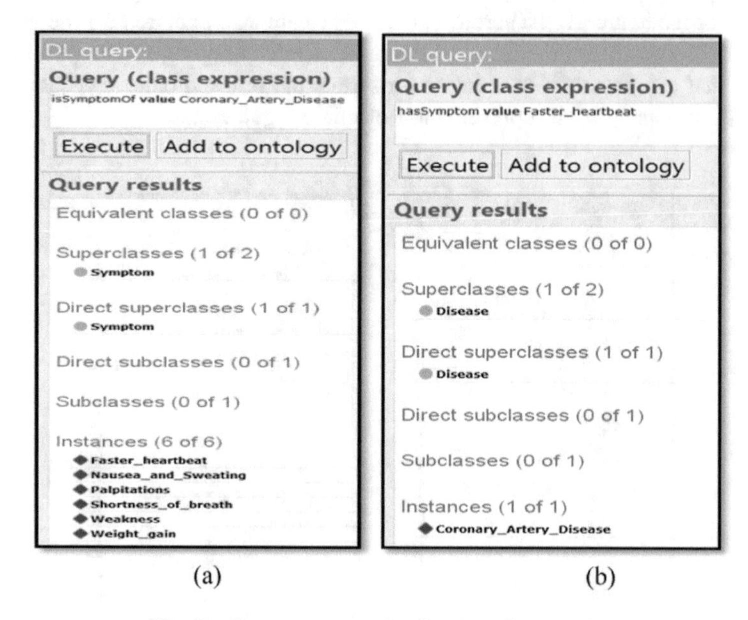

Fig. 7. Query structure for data based properties

8 Conclusion and Future Work

The common ontology proposed for CPG representation is to provide a model based on possible diseases and a shareable environment that relates to certain symptoms taking into account the possible diseases in a class of diseases. The work proposed in this study is focused on representing CPG ontology to define a common vocabulary and to share up to date evidence-based knowledge of patient conditions and interventions. We explored a formal representation of CPGs which is built using the information reported in medical abstracts in the PubMed/MEDLINE database. The preliminary investigation reveals that the translation of narrative documents into computer-interpretable CPGs can serve as a scalable and complete representation for domain-specific knowledge and the common ontology. The results show that such formal representation can be applied to query the expert domain knowledge where query structure is kept simple and user-friendly. Thus, the initial results show an evidence of successful use of OWL ontologies to represent expert knowledge in the domain of clinical guidelines. We also proposed an approach to capture important clinical data from the existing legacy clinical information to the CPG ontology. It is necessary that the utilisation of knowledge acquisition both from the domain expert and legacy clinical information are bridged so that a common ontology can be achieved more efficiently. We are continuing to develop the CPG ontology to extend and integrate the ontologies with clinical applications to run a combination of structured and unstructured queries. In the future, we will also build a formal model by using a set of predicates which will be integrated with an inference engine for decision-making. Furthermore, work will also be conducted to evaluate whether computer interpretable CPGs positively impact physicians' behaviour and adoption or the reasons for non-adoption to computer interpretable CPGs.

References

1. Klein, G.A., Calderwood, R., Macgregor, D.: Critical decision method for eliciting knowledge. IEEE Trans. Syst. Man Cybern. **19**(3), 462–472 (1989)
2. Nonaka, I.: Dynamic theory of organisational knowledge creation. Organ. Sci. **5**(1), 14–37 (1994)
3. Zhou, L., Nunes, M.B.: Knowledge sharing in healthcare sectors. In: Knowledge Sharing in Chinese Hospitals, pp. 19–38. Springer, Heidelberg (2015)
4. Irby, D.M.: Excellence in clinical teaching: knowledge transformation and development required. Med. Educ. **48**(8), 776–784 (2014)
5. Pourzolfaghar, Z., et al.: A technique to capture multidisciplinary tacit knowledge during the conceptual design phase of a building project. J. Inf. Knowl. Manage. **13**(2), 1450013 (2014)
6. Rizzo, J.: Patients' mental models and adherence to outpatient physical therapy home exercise programs. Physiother. Theory Pract. **31**(4), 253–259 (2015)
7. Tuan, L.T.: The role of CSR in clinical governance and its influence on knowledge sharing. Clin. Gov. Int. J. **18**(2), 90–113 (2013)
8. Straus, S.E., Tetroe, J., Graham, I.: Defining knowledge translation. Can. Med. Assoc. J. **181** (3–4), 165–168 (2009)
9. Kamsu-Foguem, B., Tchuenté-Foguem, G., Foguem, C.: Using conceptual graphs for clinical guidelines representation and knowledge visualization. Inf. Syst. Front. **16**(4), 571–589 (2014)
10. Parry, D.: A fuzzy ontology for medical document retrieval. In: Proceedings of the Second Workshop on Australasian Information Security, Data Mining and Web Intelligence, and Software Internationalisation, vol. 32. Australian Computer Society (2004)
11. Riaño, D., et al.: An ontology-based personalization of healthcare knowledge to support clinical decisions for chronically ill patients. J. Biomed. Inform. **45**(3), 429–446 (2012)
12. Wang, H.Q., et al.: Creating personalised clinical pathways by semantic interoperability with electronic health records. Artif. Intell. Med. **58**(2), 81–89 (2013)
13. Topaz, M., et al.: Developing nursing computer interpretable guidelines: a feasibility study of heart failure guidelines in homecare. In: AMIA Annual Symposium Proceedings. American Medical Informatics Association (2013)
14. Resnik, P.: Semantic similarity in taxonomy: an information-based measure and its application to problems of ambiguity in natural language. J. AI Res. **11**, 95–130 (1998)

15. McGuinness, D.L., Harmelen, F.V.: OWL web ontology language overview. W3C recommendation, vol. 10(10) (2004)
16. Hameed, A., Preece, A., Sleeman, D.: Ontology reconciliation. In: Handbook on Ontologies, pp. 231–250. Springer, Heidelberg (2004)
17. Green, L.A., Seifert, C.M.: Translation of research into practice: why we can't "just do it". J. Am. Board Fam. Pract. **18**(6), 541–545 (2005)
18. Santos, J.M., Santos, B.S., Teixeira, L.: Using ontologies and semantic web technology on a clinical pedigree information system. In: Digital Human Modeling. Applications in Health, Safety, Ergonomics and Risk Management, pp. 448–459. Springer, Cham (2014)
19. Jović, A., Gamberger, D., Krstačić, G.: Heart failure ontology. Bio Algorithms Med. Syst. **7**(2), 101–110 (2011)
20. Maarouf, H., et al.: An ontology-aware integration of clinical models, terminologies and guidelines: an exploratory study of the Scale for the Assessment and Rating of Ataxia (SARA). BMC Med. Inf. Decis. Mak. **17**(1), 159 (2017)
21. Lovering, R.C., et al.: Improving interpretation of cardiac phenotypes and enhancing discovery with expanded knowledge in the gene ontology. Circ. Genom. Precis. Med. **11**(2), e001813 (2018)
22. Gruber, T.R., Tenenbaum, J.M., Weber, J.C.: Toward a knowledge medium for collaborative product development. In: Artificial Intelligence in Design 1992, pp. 413–432. Springer, Dordrecht (1992)
23. Gruber, T.R.: A translation approach to portable ontology specifications. Knowl. Acquis. **5**(2), 199–220 (1993)
24. Dueñas, M., et al.: Relationship between using clinical practice guidelines for pain treatment and physicians' training and attitudes toward patients and the effects on patient care. Pain Pract. **18**(1), 38–47 (2018)
25. Jepsen, T.C.: Just what is an ontology, anyway? IT Prof. Mag. **11**(5), 22 (2009)
26. McMurray, J.J.V., et al.: ESC guidelines for the diagnosis and treatment of acute and chronic heart failure 2012. Eur. J. Heart Fail. **14**(8), 803–869 (2012)
27. Abu-Hanna, A., et al.: Protégé as a vehicle for developing medical terminological systems. Int. J. Hum. Comput. Stud. **62**(5), 639–663 (2005)
28. El-Sappagh, S., et al.: DMTO: a realistic ontology for standard diabetes mellitus treatment. J. Biomed. Semant. **9**(1), 8–16 (2018)
29. Gorín, D., Meyn, M., Naumann, A., Polzer, M., Rabenstein, U., Schröder, L.: Ontological modelling of a psychiatric clinical practice guideline. In: Joint German/Austrian Conference on Artificial Intelligence. Künstliche Intelligenz, pp. 300–308. Springer, Cham (2017)
30. Iqtidar, A., et al.: A biomedical ontology on genetic disease. In: Proceedings of the Second International Conference on Internet of Things and Cloud Computing. ACM (2017)
31. Gomez, J., Oviedo, B., Zhuma, E.: Patient monitoring system based on Internet of Things. Procedia Comput. Sci. **83**, 90–97 (2016)
32. Ganzha, M., Paprzycki, M., Pawłowski, W., Szmeja, P., Wasielewska, K.: Semantic interoperability in the Internet of Things: an overview from the INTER-IoT perspective. J. Netw. Comput. Appl. **81**, 111–124 (2017)
33. Bouamrane, M.M., Rector, A., Hurrell, M.: Using OWL ontologies for adaptive patient information modelling and preoperative clinical decision support. Knowl. Inf. Syst. **29**(2), 405–418 (2011)
34. Bouamrane, M.M., Rector, A., Hurrell, M.: Semi-automatic generation of a patient preoperative knowledge-based from a legacy clinical database. In: Proceedings of 8th International Conference on Ontologies, DataBases, and Applications of Semantics, ODBASE 2009, on the Move to Meaningful Internet Systems Conferences, Vilamoura, Algarve, Portugal. LNCS, vol. 5871, pp. 1224–1237. Springer, Heidelberg (2009)

35. Ramani, G.V., et al.: Chronic heart failure: contemporary diagnosis and management. Mayo Clin. Proc. **85**(2), 180–195 (2010)
36. Figueroa, M.S., Peters, J.I.: Congestive heart failure: diagnosis, pathophysiology, therapy, and implications for respiratory care. Respir. Care **51**(4), 403–412 (2006)
37. Guyatt, G.H., Devereaux, P.J.: A review of heart failure treatment. Mt. Sinai J. Med. **71**(1), 47–54 (2004)

Author Index

L. Barolli et al. (Eds.): EIDWT 2019, LNDECT 29, pp. 577–579, 2019.
https://doi.org/10.1007/978-3-030-12839-5